Biotechnology and Bioinformatics

Biotechnology and Bioinformatics

Editor: Lydell Norris

RCALLISTO
REFERENCE

www.callistoreference.com

Callisto Reference,
118-35 Queens Blvd., Suite 400,
Forest Hills, NY 11375, USA

Visit us on the World Wide Web at:
www.callistoreference.com

ISBN: 978-1-64116-155-8 (Hardback)

Trademark Notice: Registered trademark of products or corporate names are used only for explanation and identification without intent to infringe.

Cataloging-in-Publication Data

Biotechnology and bioinformatics / edited by Lydell Norris.
    p. cm.
Includes bibliographical references and index.
ISBN 978-1-64116-155-8
1. Biotechnology. 2. Bioinformatics. 3. Genetic engineering. I. Norris, Lydell.
TP248.2 .B563 2019
660.6--dc23

Preface

Over the recent decade, advancements and applications have progressed exponentially. This has led to the increased interest in this field and projects are being conducted to enhance knowledge. The main objective of this book is to present some of the critical challenges and provide insights into possible solutions. This book will answer the varied questions that arise in the field and also provide an increased scope for furthering studies.

The development of products or processes by using or modifying biological systems and living organisms for specific use falls under the scientific domain of biotechnology. It is a multidisciplinary field that integrates the studies of genomics, applied immunology, recombinant gene techniques, pharmaceutical therapies and diagnostic tools. Its applications are prevalent in a number of significant domains such as agriculture, health care, industries and environment. Bioinformatics is a branch of biotechnology. It is concerned with the application of statistics and computing to develop methods and software for analyzing biological data. The study of the genetic basis of disease, genetic diversity in populations and unique adaptations is possible because of the application of bioinformatics. It aids in furthering the studies in experimental molecular biology, genetics, genomics, etc. This book unravels the recent studies in the fields of biotechnology and bioinformatics. It also elucidates new techniques and their applications in a multidisciplinary manner. Students, researchers, experts and all associated with these fields will benefit alike from this book.

I hope that this book, with its visionary approach, will be a valuable addition and will promote interest among readers. Each of the authors has provided their extraordinary competence in their specific fields by providing different perspectives as they come from diverse nations and regions. I thank them for their contributions.

Editor

What can we learn from molecular dynamics simulations for GPCR drug design?

Christofer S. Tautermann, Daniel Seeliger, Jan M. Kriegl *

Boehringer Ingelheim Pharma GmbH & Co. KG, Lead Identification and Optimization Support, Birkendorfer Str. 65, D-88397 Biberach a.d. Riss, Germany

ARTICLE INFO

ABSTRACT

Recent years have seen a tremendous progress in the elucidation of experimental structural information for G-protein coupled receptors (GPCRs). Although for the vast majority of pharmaceutically relevant GPCRs structural information is still accessible only by homology models the steadily increasing amount of structural information fosters the application of structure-based drug design tools for this important class of drug targets. In this article we focus on the application of molecular dynamics (MD) simulations in GPCR drug discovery programs. Typical application scenarios of MD simulations and their scope and limitations will be described on the basis of two selected case studies, namely the binding of small molecule antagonists to the human CC chemokine receptor 3 (CCR3) and a detailed investigation of the interplay between receptor dynamics and solvation for the binding of small molecules to the human muscarinic acetylcholine receptor 3 (hM3R).

Keywords:
Molecular dynamics simulations
GPCR
Homology modeling
Water network
CC chemokine receptor 3
Muscarinic acetylcholine receptor 3

1. Introduction

G-protein coupled receptors (GPCRs) are key elements of eukaryotic signaling cascades. They transduce stimuli from the extracellular compartment into the interior of the cell, where further intracellular signaling events are triggered. Under physiological conditions GPCR signal transduction is initiated by their endogenous ligands, which range from lipids, fatty acids, neurotransmitters, cytokines, hormones, to metal ions or – in a figurative sense – even light. The opportunity to modify cellular signaling cascades by modulating the function of GPCRs makes them an attractive class of targets for pharmaceutical drug discovery and development efforts [1,2]. To date approximately 30% or even more of all marketed drugs target GPCRs [3,4], and still a substantial fraction of drugs that were recently approved by US regulatory authorities are GPCR drugs [5]. According to recent estimates ~350 GPCRs are of potential interest to treat human diseases [6]. There are still ~100 orphan receptors for which neither the natural ligand nor the physiological role is yet known [7].

A breakdown of the number of marketed GPCR drugs reveals that the number of unique GPCRs which are targeted is much less than what would be expected from the mere share of all marketed drugs: the GPCR drugs cover less than 10% of the entire target space which is addressed to date [5]. In other words, there are still ample opportunities to exploit hitherto unexplored GPCR targets with known ligands and function, but also orphan receptors. Besides novel molecular and cell biology techniques such as specifically engineered receptors [2] or the systematic generation of chemical probes [8–10] recent breakthroughs in the elucidation of GPCR structural information [11–13] expand the toolbox to discover and optimize novel ligands with therapeutic potential for this target class.

The increasing coverage of the GPCR phylogenetic tree with structural information offers the opportunity to apply structure-based drug design methodologies for this target class [14–17]. One approach which gained increasing attention in pharmaceutical industry over recent years is fragment-based drug discovery [18–20], which is nowadays reported to be applied also in GPCR drug discovery programs [21–25]. Despite the continuous progress in structure elucidation, the experimental determination of GPCR structures is still a cumbersome and slow process, which does not match the typical cycle times in lead optimization. Thus, only very few case studies are described in which ligand optimization is actually accompanied by experimentally solved GPCR-ligand complexes [26]. In most cases cost-efficient alternatives like homology modeling which allow rapid project support are reported [15]. Moreover, it is important to note that despite the continuous increase of solved GPCR structures for the majority of the GPCRs structural information is only accessible via homology modeling [15,27]. In many cases homology models are merely accurate enough to guide the overall direction of optimization efforts rather than to predict compound affinity [14,28,29], especially if the target to template similarity is low. Latest developments, however, show that homology models can be significantly improved by employing MD-methods [30,31], which is not surprising due to the flexible nature of GPRCs [32].

The availability of the β2 adrenergic receptor as active [33,34] and inactive [35] structure has prompted several research groups to model the

---

* Corresponding author
E-mail address: jan.kriegl@boehringer-ingelheim.com (J.M. Kriegl).

activation process of GPCRs on an atomistic level by using molecular dynamics simulations [36,37]. In other studies, ligand recognition or GPCR oligomerization has been investigated by employing MD simulations (Ref [38] and references therein). All these studies have been facilitated by the availability of experimental structural information, but also by a steady increase in compute power which is provided either by continuous advancements of hardware performance including GPUs, tailored computer architectures [39], or cloud computing approaches [36]. Nowadays, several microseconds of simulation data can be collected within few days. Since the microsecond timescale marks the lower border at which important biological function such as ligand binding occurs, atomistic simulations open novel opportunities for structure-based drug discovery.

In fact, molecular dynamics (MD) simulations are being more and more used in drug design [40–42]. The notion of the importance of receptor flexibility has fostered the usage of computational tools such as MD simulations to generate ensembles of energetically accessible conformations [43,44]. Talking specifically of GPCRs, recent developments include target specific scoring functions to identify MD snapshots which still retain the typical GPCR specific conserved geometric features [30] in order to avoid unphysical decoys. Hot spots in binding pockets and, more recently, on protein–protein interfaces are being postulated by solvent MDs. For this technique small organic molecules such as propane or benzene are added to the water box, and regions of high solute density in the simulation are used as indicators of protein site druggability [45]. Recently, thermodynamic integration and/or free energy perturbation methods have gained an increasing attention in predicting relative free energies of binding [46,47].

In the following we will discuss the application of MD simulations in GPCR drug design with the help of two case studies which were performed at Boehringer Ingelheim Pharma. We selected these examples because they illustrate the different levels of information which can be utilized for drug design efforts. In the first case study binding of small molecule antagonists to the human CC chemokine receptor 3 (CCR3) was investigated. Since to date no crystal structure of CCR3 is reported we generated homology models to provide structural information for a medicinal chemistry program. Complexes with manually modeled ligand binding poses were subjected to MD simulations to check the integrity of the binding pose. The final model was utilized to rationalize rodent selectivity data. The second case study addresses the binding of small molecule ligands to the human muscarinic acetylcholine receptor 3 (hM3R). In this case, detailed experimental structural information has been available which enabled us to investigate the interplay between receptor dynamics, water networks, and ligand–receptor interactions by MD simulations on a very detailed level.

2. Computational methods

2.1. Homology modeling and ligand placement for CCR3

CCR3 homology models were generated based on several templates. In the following we will describe the homology modeling procedure employing the X-ray structure of human CC chemokine receptor 5 (CCR5) complexed with maraviroc as a template (PDB code 4MBS). This structure was solved in 2013 [48]. In order to enhance crystallization, a rubredoxin entity was fused to intracellular loop 3. To generate a model of the human CCR3 receptor, a sequence alignment between the template X-ray structure and the target hCCR3 sequence was done in MOE [49] employing the BLOSUM50 matrix. By this the exact location of the rubredoxin insertion could be identified, and rubredoxin was removed manually from the template structure. The employed sequence alignment is shown in the Supplementary material (Fig. S1), revealing a target/template sequence identity of more than 50%. The comparative modeling step was done with Modeller [50,51], a standard homology modeling tool. Finally a stepwise optimization of the structure was performed. In a first step the two loose ends caused by the cut of rubredoxin were manually connected and the whole receptor was

protonated at pH 7.4 by the protonate3D procedure as implemented in MOE. In a next step an energy optimization of the receptor with high tethers on the heavy atoms was carried out (tether 1000). The tethers were then reduced to 100 and 10. In the final optimization step the side chains were energy minimized with fixed protein backbone atoms. All minimizations were done with the MMFF94x force field as implemented in MOE. The ligand (structure shown in Fig. 2A) was placed manually into the receptor. First, an ensemble of low energy conformations was generated. These conformations were then manually docked into the receptor such that the ionic interaction between the positively charged center of the ligand and the $E287^{7.39}$ on transmembrane helix (TM) 7 was enforced. Conformations which caused major clashes were discarded, and checks like the shape overlay with maraviroc in hCCR5 or the formation of an additional ionic interaction with $H97^{2.67}$ (for details we refer to ref. [52]) were employed to identify the most plausible pose. This pose underwent stepwise geometry optimization (as described before) and was ultimately subjected to a full equilibration and a 60 ns production run (MD setup as described below) to yield the binding mode shown in Fig. 3.

2.2. Molecular dynamics simulations

Molecular dynamics simulations were carried out using the gromacs-4.5 package [53]. The simulation system consists of the aforementioned CCR3 homology model or the hM3R model [54] with bound tiotropium (derived from PDB 4DAJ) embedded in DMPC lipid bilayer [55] and solvated in water. NaCl was added to achieve a 150 mM salt concentration. The amber99sb-ildn* force field [56,57], the SPC/E water model [58] and ion parameters from Joung et al. were used [59]. Force field parameters for the small molecule ligands were obtained according to the generalized amber force field (GAFF) procedure [60] with partial charges derived from quantum chemical calculations with Gaussian09 [61] at a HF/6-31G* level of theory. Amber topologies for the ligands were converted to gromacs format using acpype [62]. The membrane simulation system was built with g_membed [63]. After energy minimization, 50 ps equilibration with positional restraints on heavy atoms, and 1 ns equilibration with position restraints in z-direction on the phosphor atoms of DMPC followed. Trajectories were subsequently collected at 310 K with standard NPT ensemble settings (thermostat: velocity-rescaling [64,65]; barostat: Parrinello-Rahman [66], semi-isotropic coupling). Electrostatic interactions were calculated at every step with the particle-mesh Ewald method [67], short-range repulsive and attractive dispersion interactions were simultaneously described by a Lennard-Jones potential, which was cut off at 1.0 nm. The SETTLE [68] algorithm was used to constrain

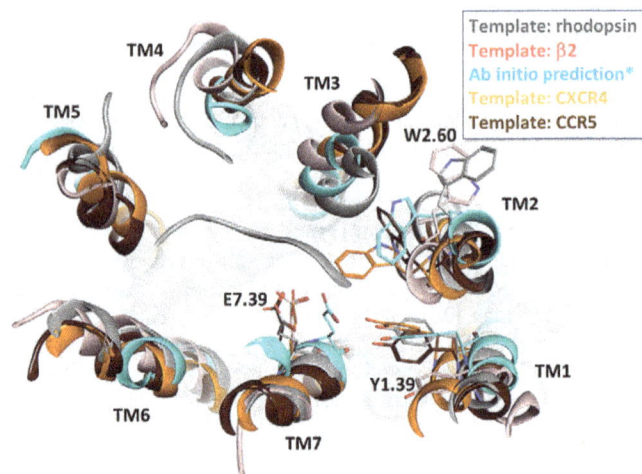

Fig. 1. Multiple versions of the CCR3 receptor homology model, color-coded according to the template that was utilized. Top view from the extracellular side.

bonds and angles of water molecules; LINCS [69] was used for all other bonds. Virtual sites [70] were introduced to remove other fast vibrating degrees of freedom, allowing a time step of 4 fs. In the hM3R case five trajectories (400 ns each) were recorded for every system. Individual analyses were always done based on the averages of the 5 independent trajectories per system.

3. Case Study 1: binding of small molecules to the CC chemokine receptor-3

CCR3 is a member of the chemokine receptor family. One of its endogenous ligands, the chemokine eotaxin-1/CCL11, is a chemoattractive protein which has been identified to recruit eosinophils under allergic conditions and binds exclusively to CCR3. It is therefore postulated that antagonizing CCR3 with small molecules is a viable approach to treat allergic diseases such as asthma, for which numerous in vivo studies suggest that they are closely linked to eosinophilia [71,72].

When optimizing antagonists for their potency on CCR3 in binding and functional cellular assays we observed a pronounced selectivity versus rodent receptor orthologues [52]. To explain the selectivity profiles and eventually identify opportunities to optimize potency especially on mouse and/or rat CCR3 we constructed a homology model of CCR3. Over the time the homology model underwent several refinements. They were triggered either by the release of a novel GPCR structure with a higher sequence similarity to CCR3 or – in one instance – by the availability of an ab-initio algorithm to predict GPCR 3D structures [73]. Fig. 1 shows an overlay of the different models, color-coded according to the template or modeling technique that was employed. In all models, E287$^{7.39}$ (the superscript corresponds to the Weinstein–Ballesteros

numbering scheme) on TM7, which is postulated as one key anchoring residue for small molecule CCR3 ligands [74] and which is highly conserved among other chemokine receptors and across species [75], points towards the interior of the transmembrane cavity in a very similar manner. The same observation was made e.g. for Y41$^{1.39}$ on TM1. The largest differences between the individual models were observed for TM2, which underwent in total a 90° rotation throughout all model versions. When rhodopsin or β2 was used as template structure, W90$^{2.60}$ protrudes into the lipid bilayer. We were not able to generate a plausible ligand binding mode with these two models since the cavity between TM2 and TM3 is not well defined. In the models that were either generated ab-initio or from closer related templates (CXCR4 and CCR5), W90$^{2.60}$ lines up the TM cavity such that a hydrophobic patch is offered for ligands to interact with. A closer inspection of the CXCR4 and CCR5 X-ray structures shows that TM2 displays a helical bulge in the extracellular part, thus causing this 90° rotation of the upper part, which is different to β2/rhodopsin based models.

3.1. Results of Case Study 1

We manually docked a compound (Fig. 2A) carrying the essential molecular topology and key functional groups of one of our CCR3 antagonists (ref [76]) into the CCR5-derived homology model as described above. In the initial pose ionic interactions with E287$^{7.39}$ and H97$^{2.67}$ and a hydrophobic interaction with W90$^{2.60}$ were established. In addition, the hydrophobic phenylsulfanylpropyl tail pointed to a region between helices 4 and 5 (Fig. 2B, conformations colored in red). The complex was subsequently submitted to a 60 ns MD simulation to check the stability of the proposed binding mode. The entire complex remained stable during the

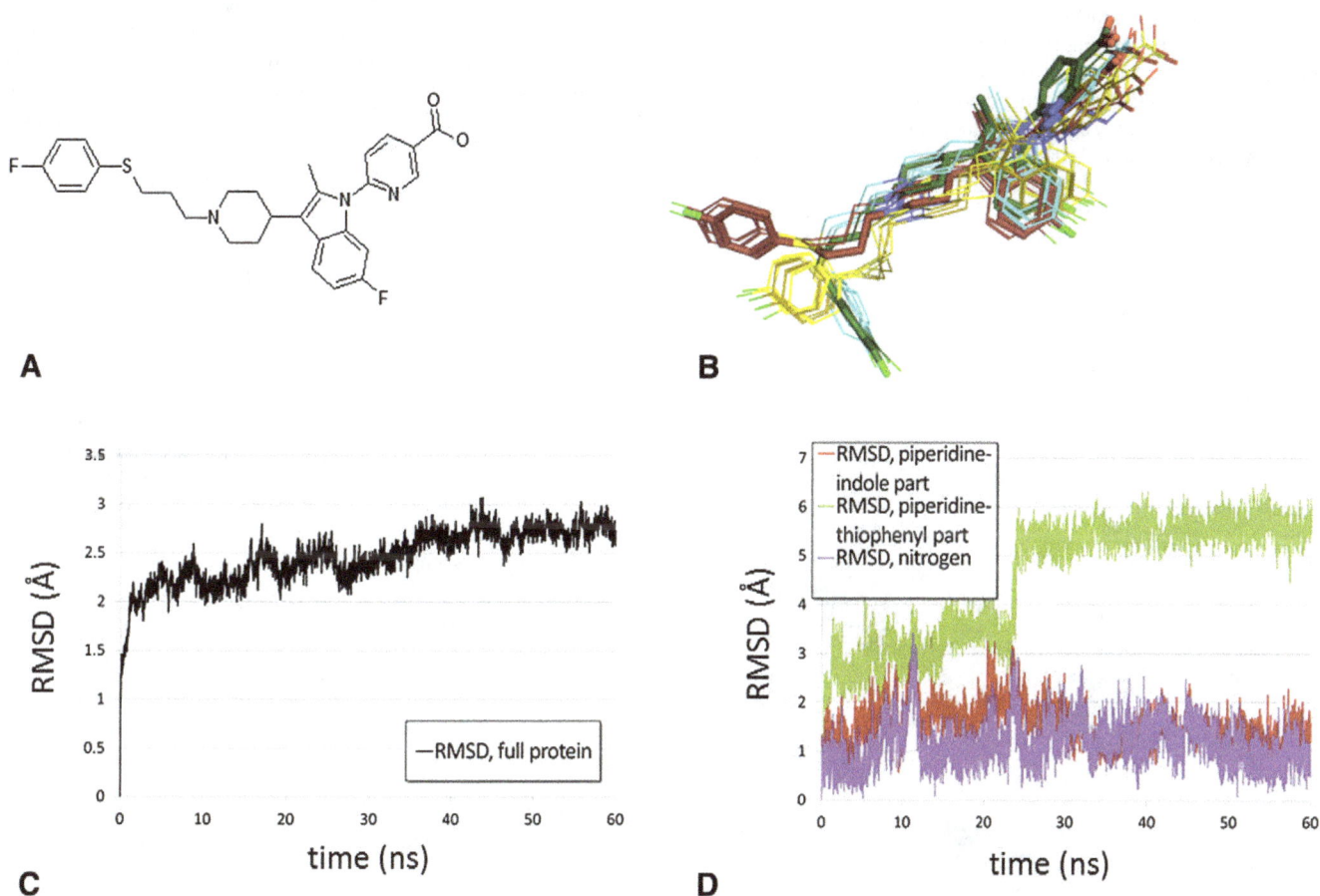

Fig. 2. A) 2D representation of the molecule that was embedded in CCR3. B) Conformations that were visited during the simulation. The starting conformation is shown in dark red, the final conformation in dark green. C) RMSD plot for the full protein. D) RMSD plots for certain parts of the ligand, illustrating the flexibility of these parts throughout the simulation. In panels C) and D) all RMSD values are calculated with respect to the starting geometry.

simulation, as indicated by the RMSD plot (Fig. 2C). The ligand did not show major rearrangements such as complete dissociation from the transmembrane cavity or flips around the center of the molecule. In particular, the positioning of the piperidine-indole part, which is engaged in directed interactions with receptor residues, did not vary substantially (c.f. Fig. 2D). On the other hand, the more flexible phenylsulfanylpropyl moiety fluctuated between various orientations. This is reflected in the corresponding RMSD plots (Fig. 2C). After approximately 25 ns the tail adopted a conformation in which it protruded deeply into the interior of the TM cavity, in close analogy to the X-ray binding mode of maraviroc in CCR5 (PDB entry 4MBS [48]). This conformation remained stable over the remaining 35 ns of the simulations. Representative snapshots are shown in cyan and green (Fig. 2B). After minimization of the last snapshot of the trajectory we ultimately came up with the pose shown in Fig. 3. This pose was utilized as template to place close analogs into the receptor and ultimately to rationalize species selectivity that we observed in this particular structural class. For more details we refer to ref. [52].

3.2. Discussion of Case Study 1

A retrospective analysis of the modeling procedure and the outcome of the MD simulations highlight two important aspects: first, the choice of the template for homology modeling efforts has a crucial impact on the quality of the results and hypotheses which are derived therefrom. Although the GPCR structures solved to date show a high degree of analogy in the TM region [13], subtle structural features such as unusual helix turns or side chain orientations can be expected to be accurately modeled only when the template for homology modeling is sufficiently close. Even with a model that was derived from the sequence-wise close CCR5 template we observed that the modeled binding mode underwent some changes during an MD simulation. Without further experimental information it is challenging (if not impossible) to pick the relevant binding mode from an MD trajectory alone. In that respect the MD simulation helped us to build up confidence in the general orientation of the ligand inside the TM cavity and the directed (ionic) interactions that are formed. A true validation of the binding mode can only be established by experimental data. In our case, we found agreement with species selectivity data. A more rigorous validation would be provided by site-directed mutagenesis data.

4. Case Study 2: binding of small molecules to hM3R

Bronchodilators are central to symptom management in airway diseases like asthma and COPD. While short-acting medications are

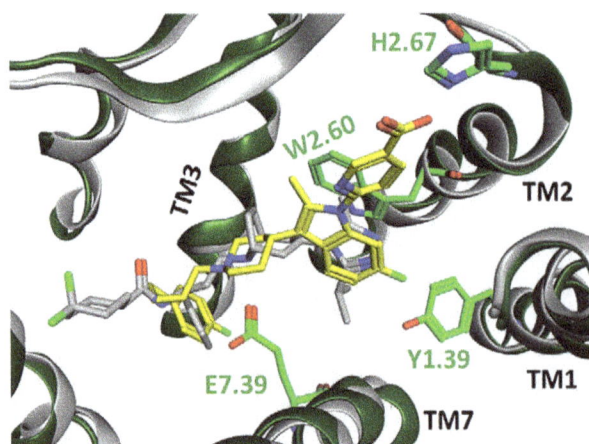

Fig. 3. Final modeled binding mode for molecule shown in Fig. 2A. CCR3 and key residues are shown in green. For comparison, CCR5 in complex with maraviroc (PDB entry 4MBS) is shown too (gray).

used for immediate symptom relief, long-acting drugs are essential for disease control and maintenance therapy [77,78]. Currently two major classes of long-acting bronchodilators are available: muscarinic antagonists and β2-adrenergic agonists. The use of long-acting muscarinic antagonists (LAMAs), like tiotropium, is the mainstay of current COPD treatment [77,78]. The structure of the complex of tiotropium and rat muscarinic M3 receptor (M3R) has been solved in 2012 [79], and this makes hM3R a well suited target for structure based drug design investigations. The binding mode of tiotropium in a homology model of hM3R is shown in Fig. 4. (The homology model of the human receptor is very closely based on the X-ray structure of rat M3R and has already been described elsewhere [54].) Tiotropium is binding in a deeply buried site in the receptor, which is accessed through a narrow channel. In the binding site the positively charged epoxytropane head group of the ligand is coordinated into an aromatic/ionic cage consisting of D148$^{3.32}$, Y149$^{3.33}$, W504$^{6.48}$, Y507$^{6.51}$, Y530$^{7.39}$, and Y534$^{7.43}$ by ionic and aromatic interactions. Remarkably Y149$^{3.33}$, Y507$^{6.51}$, and Y530$^{7.39}$ form a "lid" above tiotropium which blocks its way through the exit channel. In addition the ligand forms a strong double hydrogen bond to N508$^{6.52}$, and the thiophene groups fit into a hydrophobic crevice formed by W200$^{4.57}$, L226$^{5.33/EL2}$, T232$^{5.39}$, A236$^{5.43}$, and A239$^{5.46}$. Altogether, the ligand is very tightly bound, making all kinds of interactions – hydrophilic, hydrophobic and ionic – thus leading to a very high potency of about 10 pM [54].

In a recent study we have published residence times and binding affinities of tiotropium and related compounds to hM3R and mutated variants thereof [54]. The goal of the investigation was the elucidation of the key structural elements of tiotropium which leads to the very slow off-rate from hM3R. In the course of the study about 30 single point mutants have been generated, and the dissociation rates of tiotropium from the mutated receptors have been recorded. In most cases K_i values and the corresponding off-rates are directly related to each other (i.e., $k_{off} = const * K_i$), but mutation of residues along the exit channel accelerates the dissociation tremendously. Single amino acid mutation of Y149$^{3.33}$, Y507$^{6.51}$, N508$^{6.52}$, or Y530$^{7.39}$ results in off-rates which deviate from linearity by more than one order of magnitude. Modification of the ligand only leads to unexpected behavior if the hydroxy group is modified. As shown in Fig. 4 the hydroxy-group directly interacts with N508$^{6.52}$, thus prompting the assumption that the double hydrogen bond is a decisive structural element for long acting antimuscarinic drugs. To rationalize the effect of the N508$^{6.52}$A mutation several MD simulations have been performed. From them it has been proposed that the interaction of the ligand with N508$^{6.52}$ works like a snap-lock. Once the interaction is formed it keeps tiotropium in place and prevents its translation towards the exit-channel. These investigations have been a first step to understand the experimental findings on a molecular basis. Additional, more extensive simulations for closely related ligands are the subject of the present study. The systems that were investigated are described in Table 1. For each system a total of 2 µs simulation data was collected from five independent trajectories of 400 ns length.

4.1. Results of Case Study 2

MD trajectories were analyzed by focusing on various different receptor regions. Therefore the description of our observations will be split into four sections: (i) changes of the tyrosine cage, (ii) changes of the ligand–receptor interactions, (iii) exit channel flexibility, and (iv) water densities and networks. In the discussion the observations will be put into context to obtain a clear picture of the changes of receptor dynamics upon mutation of receptor or ligand.

4.1.1. Changes in the tyrosine cage

As described in the Introduction section and displayed in Fig. 4, the positively charged head group is surrounded by several tyrosine residues. The three most important tyrosines in dissociation experiments

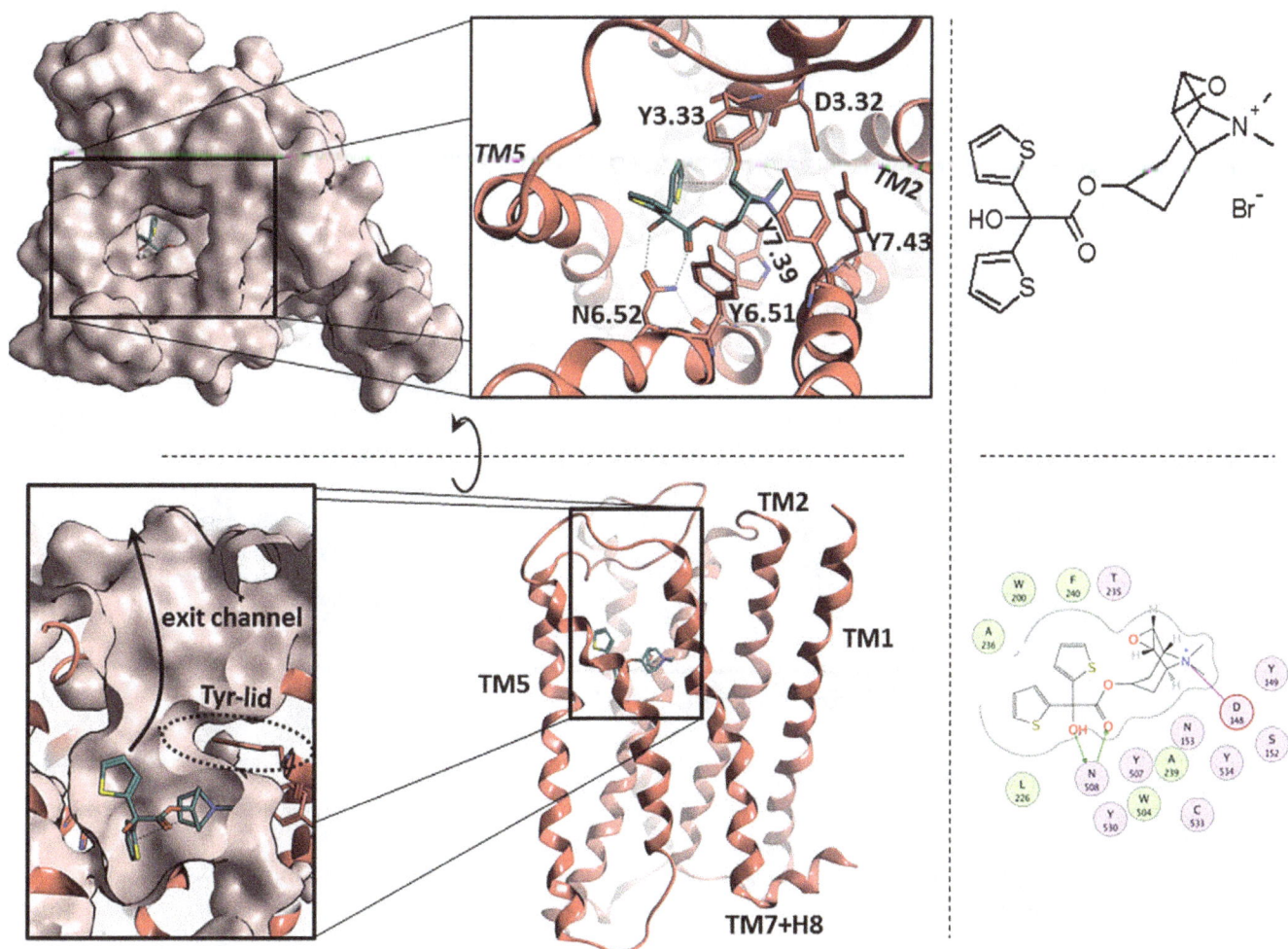

Fig. 4. Binding mode of tiotropium (cyan) in hM3R. Top left: top view from the extracellular side. Bottom left: side view with TM6 and TM7 in front. Right: structure and 2D-interaction plot of tiotropium.

(Y149$^{3.33}$, Y507$^{6.51}$, and Y530$^{7.39}$) form a lid above the ligand. This lid has to break up for ligand dissociation. In the case of tiotropium binding to wild type hM3R this lid is tightly closed [54]. In the apo-WT simulation it has been shown that the lid is open in a significant number of snapshots and thus enables quick ligand association. The correlations of movements of the lid-tyrosines in the systems under consideration are shown in Fig. 5. Ligand binding stabilizes the lid in a closed position for all ligands investigated, and in both apo simulations open as well as closed conformations are found. Surprisingly the apo-WT simulation still has the larger fraction of closed lid structures, whereas the apo simulation of the mutated receptor (apo-N508$^{6.52}$A) shows a predominant population of snapshots with an open lid.

In tiotropium bound simulations the tyrosine lid is very stable showing hardly any opening events. However, the distance between Y149$^{3.33}$ and Y530$^{7.39}$ is slightly elongated in the tio-N508$^{6.52}$A simulation compared to tio-WT, as displayed in Fig. 6. Upon binding of methyl-tio or hyd-tio the tyrosines are much more flexible. Especially for hyd-tio the tyrosine lid is open in many snapshots. An example for a snapshot with an open lid configuration is shown in the inset in Fig. 6. We find both tyrosines (Y149$^{3.33}$ and Y530$^{7.39}$) in a different rotameric state compared to the M3R X-ray structure, and the ligand is directly accessible by the solvent.

To summarize the observations for the tyrosine lid flexibility: in apo simulations a dynamic equilibrium between open and closed lids is observed. Ligand binding stabilizes the tyrosine lid, however, ligands

Table 1
Systems under investigation in the present study.

	Ligands:	Receptor	Abbreviation of the system
1	Tiotropium (R = OH)	Wild type hM3R	Tio-WT
2	Des-OH tiotropium (R = H) ["hyd-tio"]	Wild type hM3R	Hyd-tio-WT
3	Methyl-des-OH-tiotropium (R = CH$_3$) ["methyl-tio"]	Wild type hM3R	Methyl-tio-WT
4	Tiotropium (R = OH)	N508$^{6.52}$A hM3R	Tio-N508$^{6.52}$A
5	None	Wild type hM3R	Apo-WT
6	None	N508$^{6.52}$A hM3R	Apo-N508$^{6.52}$A

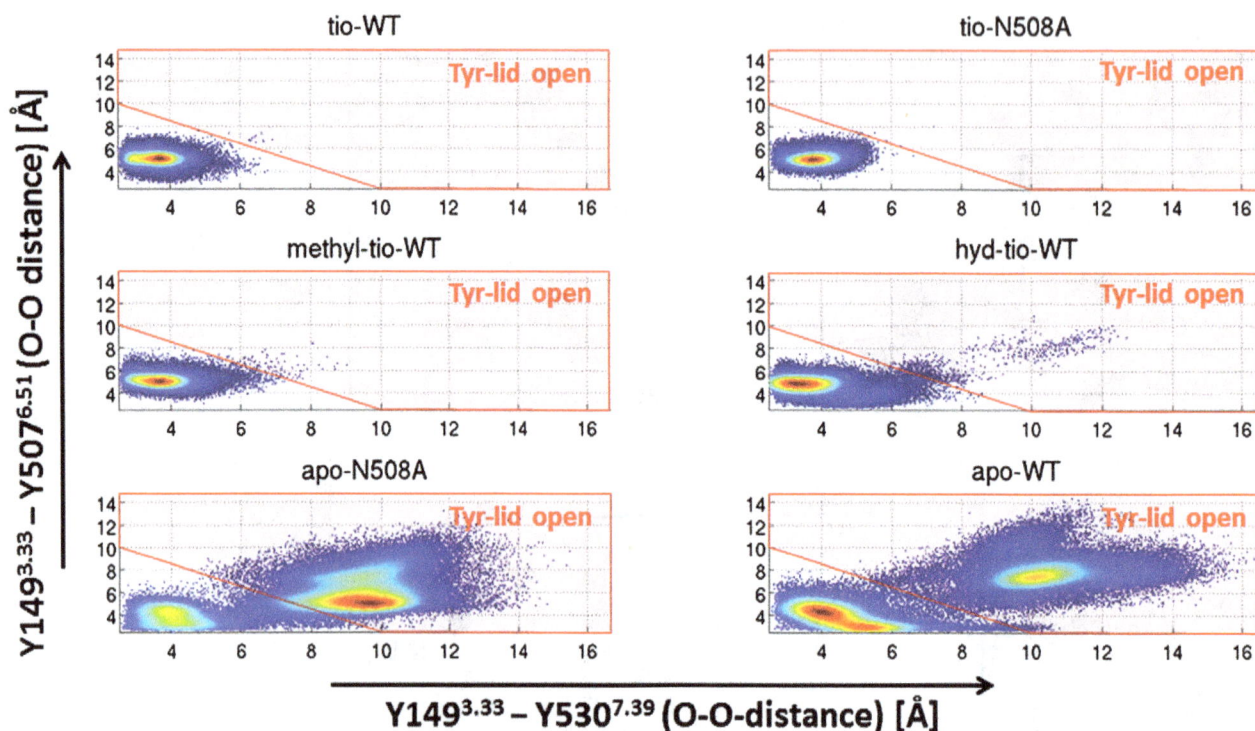

Fig. 5. Correlation plots for 100.000 snapshots of all trajectories. Red shaded areas correspond to snapshot conformations, where the tyrosine lid opened by breaking at least one of the tyrosine hydrogen–bond interactions.

without the OH-group are not able to stabilize the lid over the full simulation time to the same extent as tiotropium.

4.1.2. Changes in the ligand–receptor interactions

Tiotropium is strongly bound to hM3R by ionic interactions with the charged epoxytropane head group, by hydrogen bonds with the ester carbonyl group as well as the hydroxy group, and by lipophilic interactions with the two thiophenes. In this chapter we investigate the changes of tiotropium movement if the hydrogen bonds are weakened or even entirely removed. In Fig. 7 the distances of the charged nitrogen of the ligand to $D148^{3.32}$ are monitored as a measure for the flexibility of the head group. It can be observed that neither removal of the

hydroxy group of tiotropium nor the $N508^{6.52}A$ mutation of the receptor has any significant effect on the distance of the ionic interaction. Not surprisingly, the mobility of the quaternary carbon of the ligand is strongly affected by distorting the double hydrogen bond. We plot the distance to the $C\beta$ atom of $N508^{6.52}$ (we chose to use the $C\beta$ to be able also to include tio-$N508^{6.52}A$ in the analysis) as a measure for the flexibility of the bis-thiophene substructure in relation to the decisive $N508^{6.52}$ interaction partner. Although both modified ligands, i.e., hyd-tio and methyl-tio, still do have the ability to form hydrogen bonds to $N508^{6.52}$, these bonds are often strongly elongated or even broken. Especially methyl-tio shows very long distances, where the hydrogen bond is mediated by an additional water molecule, as shown in Fig. 7.

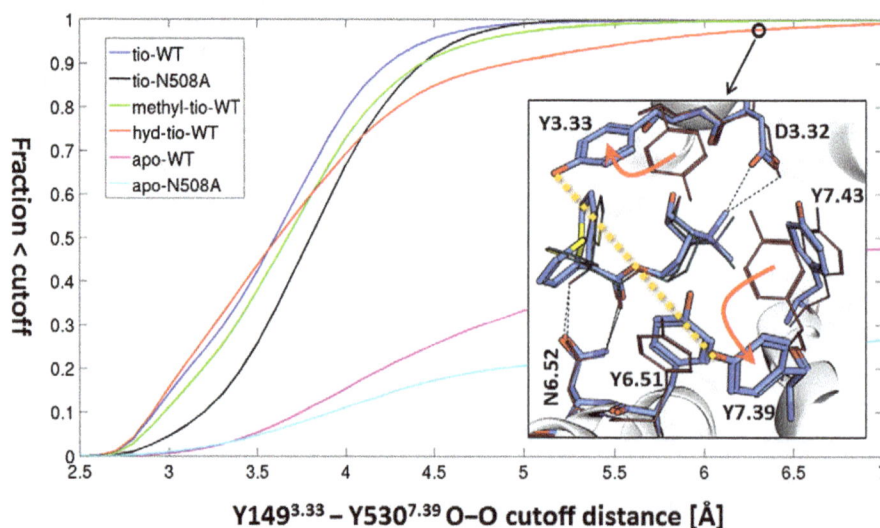

Fig. 6. Cumulative distribution function of the $Y149^{3.33}$–$Y530^{7.39}$ (oxygen–oxygen) distance (orange dashed line in the insert). Inset: snapshot from the hyd-WT simulation (blue sticks) overlaid with the M3R crystal structure (orange lines). The red arrows mark lid-opening movements of $Y149^{3.33}$ and $Y530^{7.39}$.

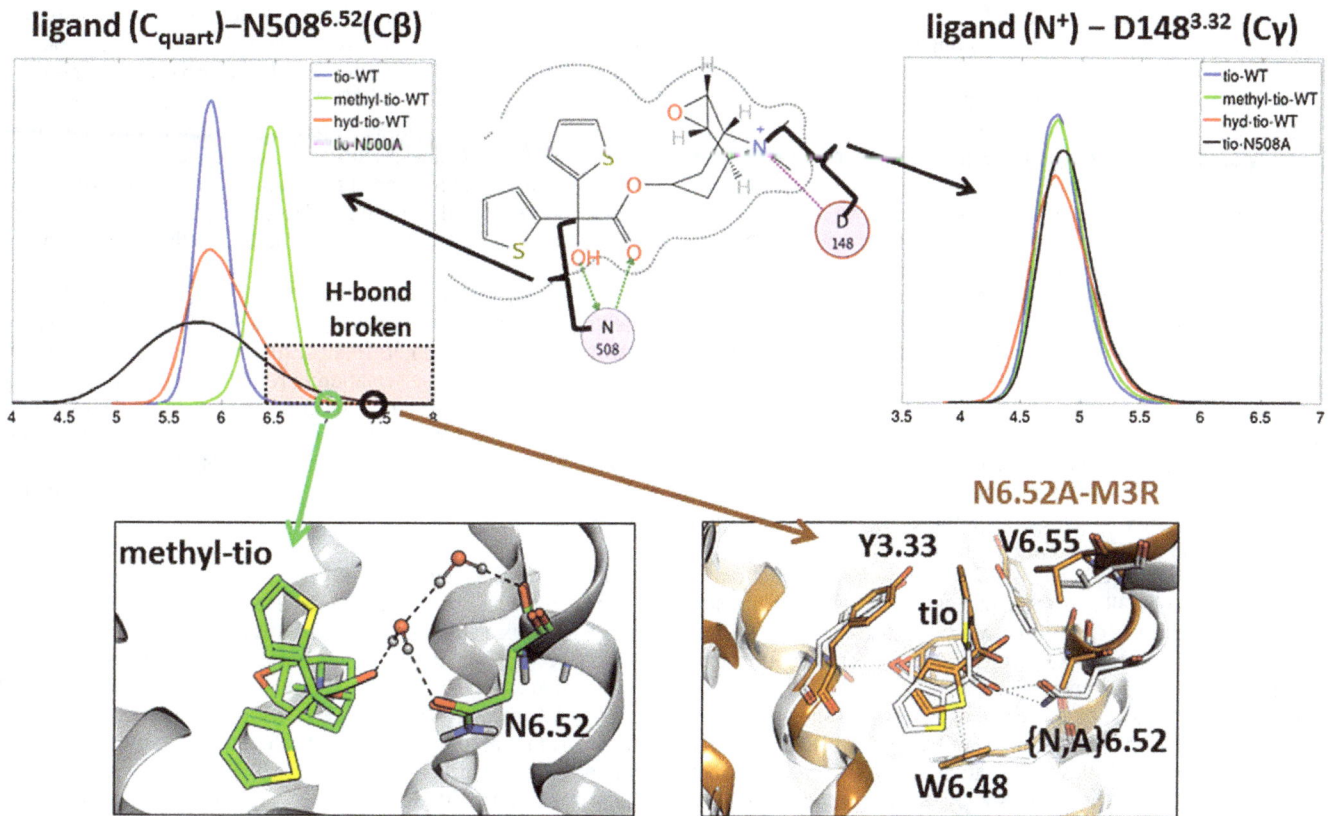

Fig. 7. Top: Distances of the ligands to the interaction partners on the receptor side. Bottom: snapshot from the methyl-WT simulation, where the ligand–N508$^{6.52}$ hydrogen bond is broken (left) and snapshot from the tio-N508$^{6.52}$A simulation to demonstrate the shift of tiotropium towards the exit channel (white structure: X-ray, brown: snapshot from the tio-N508$^{6.52}$A simulation).

The movements of tiotropium in N508$^{6.52}$A-hM3R reveal a high flexibility due to the lack of the hydrogen bond partner N508$^{6.52}$. Tiotropium's hydroxy group forms an intramolecular hydrogen bond to the ester carbonyl group (Fig. S2, Supplementary material). Calculation of the RMSF of different parts of the molecule confirms this observation, as shown in Fig. S5 in the Supplementary material. Additional analyses show that the quaternary carbon is located much more towards the extracellular part of the exit channel compared to the tio-WT simulation (Fig. S6,

Supplementary material). A snapshot to illustrate this ready-to-exit behavior is also shown in Fig. 7.

To summarize the observations for the ligand–receptor interactions: for the modified ligands (hyd-tio and methyl-tio) the hydrogen bond between the ligand and receptor is elongated and especially for methyl-tio often broken. Tiotropium in N508$^{6.52}$A-hM3R reveals pronounced dislocations of the bis-thiophene substructure of the molecule in the receptor and is pushing towards the exit channel.

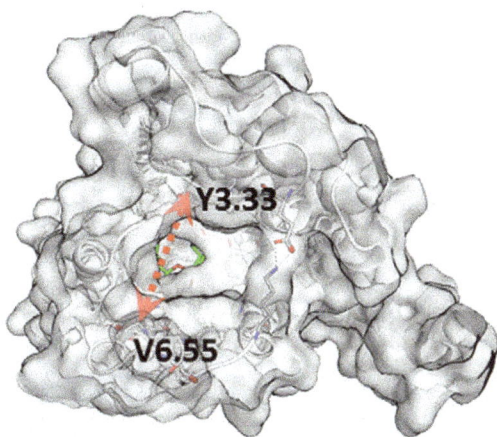

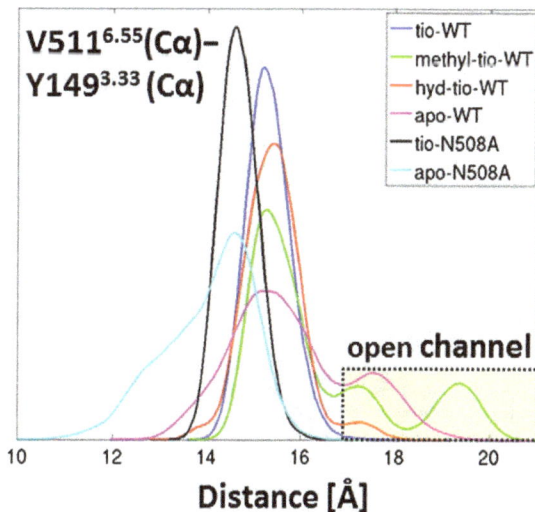

Fig. 8. Left: location of the Cα atoms of V511$^{6.55}$ and Y149$^{3.33}$ as indicator for the diameter of the exit channel. Right: distribution functions of the distances between the Cα atoms of V511$^{6.55}$ and Y149$^{3.33}$.

4.1.3. Flexibility of the entry/exit channel

So far only the direct binding site of tiotropium has been investigated. Considerations of dissociation dynamics also need to include the investigation of the dissociation pathway. As shown in Fig. 4, the binding site is accessed through a narrow tunnel. We investigate the accessibility of the entry channel by monitoring the distance of the Cα atoms of V511$^{6.55}$ and Y149$^{3.33}$. These two amino acids are located at opposing locations of the channel. Their location is shown in Fig. 8.

The distribution functions of the distances show some very surprising results. Tiotropium does not allow too much flexibility of the exit channel. When simulating methyl-tio or hyd-tio in WT hM3R, the picture changes entirely, because these ligands induce open-channel states. Especially in the methyl-tio-WT simulation very large values for the diameter of the channel are observed. Similar open-channel structures are observed in the apo-WT simulation. In sharp contrast to that, the apo-N508$^{6.52}$A simulation shows a preference for a tightly closed channel. The difference between the open state in methyl-tio-WT and apo-N508$^{6.52}$A can be as large as 10 Å. In line with this, the channel is also slightly tighter closed in the tio-N508$^{6.52}$A simulation than in the tio-WT simulation. Fig. 9 shows the correlation of the channel diameter and the solvation of N508$^{6.52}$, which sits at the bottom of the channel. In WT hM3R tiotropium prevents N508$^{6.52}$ from being solvated. A closer water site can be observed at a distance of about 7 Å. Removal of tiotropium's hydroxy group allows water to access the exit channel and to directly interact with N508$^{6.52}$. Therefore, the close interaction is only feasible if the channel opens up — as shown in Fig. 9 (middle panel). For methyl-tio there are two distinct open states of the channel, where both allow a solvation of N508$^{6.52}$ and with that also water insertion into the ligand-N508$^{6.52}$ hydrogen bond. When observing the trajectories over time (Fig. S3, Supplementary material), the opening and closing motions resemble a "breathing" like behavior of the receptor. Also for hyd-tio the solvation of N508$^{6.52}$ comes with an opening of the channel, but no distinct second (widely) open state is observed. In the WT-apo simulation N508$^{6.52}$ is, as expected, always solvated, independently from the opening state of the channel. The

simulations of the N508$^{6.52}$A mutants have to be interpreted with caution, because the hydrophilic N508$^{6.52}$ is substituted by an apolar amino acid, which does not require solvation to achieve a low free energy. Tiotropium bound N508$^{6.52}$A-hM3R shows a similar pattern to tiotropium bound WT, but apo-N508$^{6.52}$A completely deviates from apo-WT. First, the channel is always closed and second, A508$^{6.52}$ is solvated only in a small fraction of the snapshots. All these observations strongly suggest that the opening and closing of the channel are directly connected to the water network at the extracellular surface of the receptor.

To summarize the observations for the channel flexibility: the N508$^{6.52}$A variants of the receptor induce a more tightly closed conformation of the exit channel. This is caused by the lack of the water network required for the solvation of N508$^{6.52}$. Tiotropium stabilizes the channel in a closed state, thus preventing water to interfere with the double hydrogen bond between N508$^{6.52}$ and the ligand. Methyl-tio and hyd-tio enable openings of the channel, always accompanied by the entry of water, which competes with the ligand–N508$^{6.52}$ hydrogen bond.

4.1.4. Water densities and networks

In the previous section we have reported that the channel opening is always accompanied by the approach of water towards N508$^{6.52}$. In this section we are comparing water networks in the simulations by means of calculated water densities. Fig. 10 shows the differences in the tio-WT and hyd-WT/methyl-WT water grids. The most important difference is the occurrence of some water density close to N508$^{6.52}$ in the methyl-WT and also to a smaller extent in the hyd-WT grids. These water locations correspond to those events in which the channel is open and water gets close to the ligand–N508$^{6.52}$ hydrogen bond. Within the ligand binding site no water mediated contacts are observed. Towards the extracellular region, the water densities are becoming more and more bulk-like, however, close to the receptor very distinct water shapes are observed. This means that close to the protein surface water networks are well conserved. In Fig. 8 very strong differences in the channel

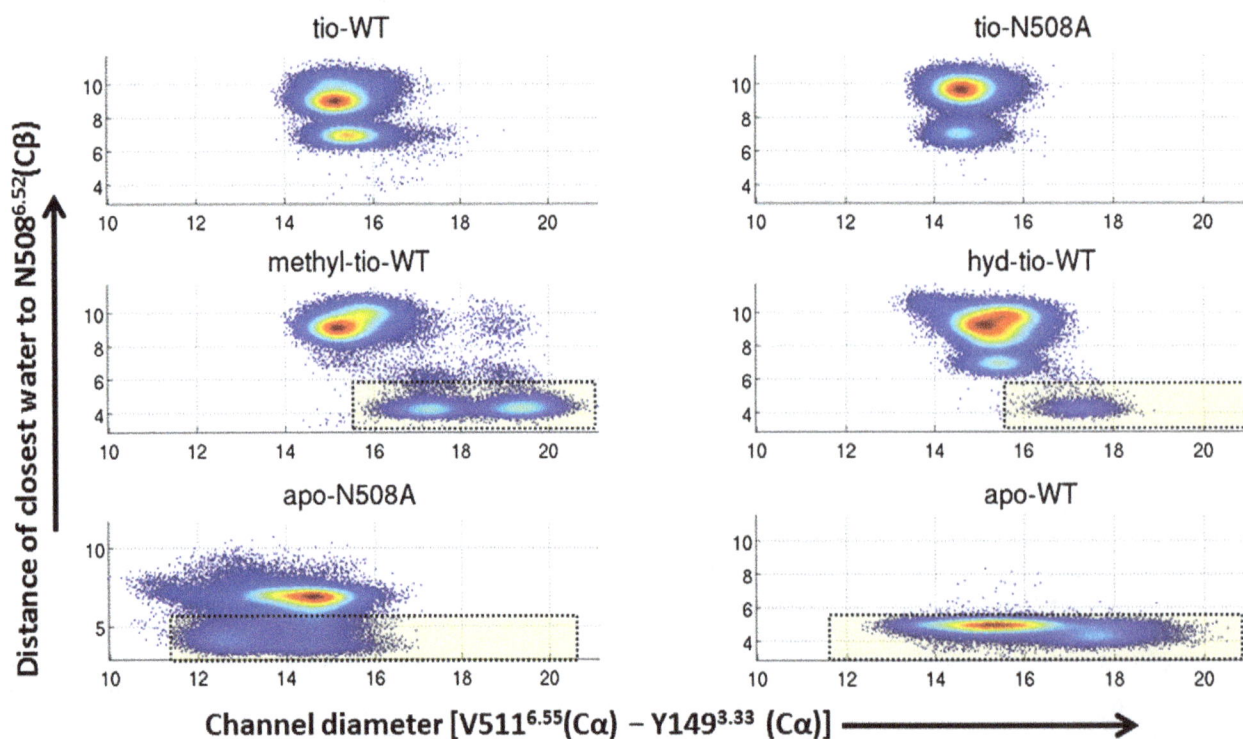

Fig. 9. Density plots of the correlation between the channel diameter (V511$^{6.55}$–Y149$^{3.33}$) and the distance of the closest water to the Cβ of N508$^{6.52}$. The yellow boxes mark the regions of direct interaction of N508$^{6.52}$ with water.

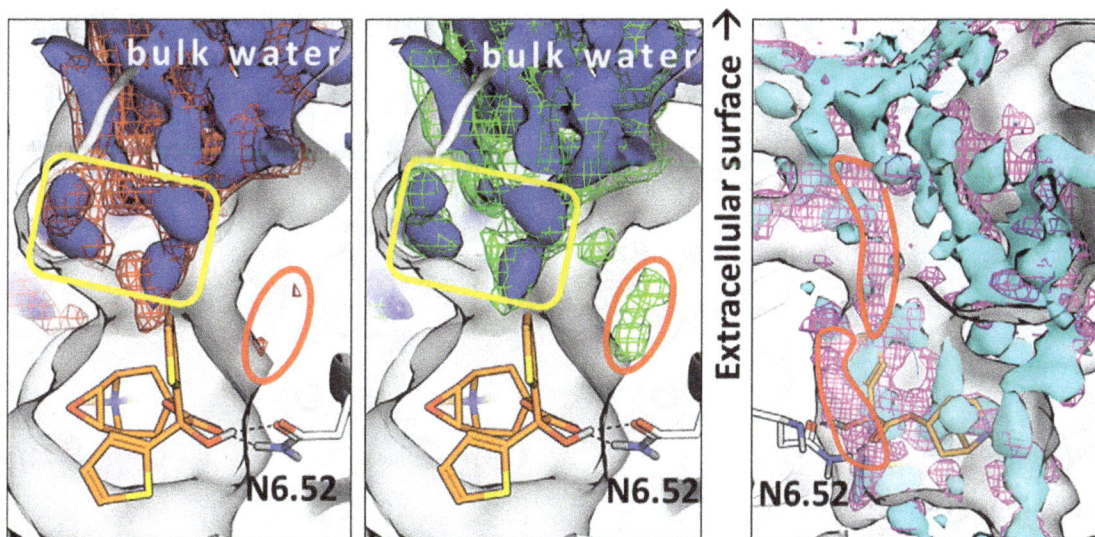

Fig. 10. Comparison of the water (oxygen) densities in the tio-WT and the hyd-WT (left panel) and the methyl-WT (middle panel). The tio-WT water densities are displayed as blue solid surfaces, hyd-tio-WT and methyl-tio-WT water densities as red and green mesh, respectively. The most important difference in the water densities is encircled in red. Regions of conserved water networks are marked by a yellow box. Right panel: overlay of apo-WT (magenta mesh) with N508$^{6.52}$A-apo (cyan solid surface) water grids. Regions which are nearly exclusively hydrated in apo-WT are marked by red contours. All grids are overlaid onto the tiotropium bound WT hM3R.

geometry between apo-WT and apo-N508$^{6.52}$A are shown. The corresponding water grids are also displayed in Fig. 10 (right). In the apo-WT simulation water density is found throughout the channel and also close to N508$^{6.52}$. In contrast the apo-N508$^{6.52}$A simulation shows essentially a "dry" channel and no water density close to A508$^{6.52}$. In other regions of the binding site, i.e., in the tyrosine cage the water grids are similar for both simulations. This means that the lack of N508$^{6.52}$ causes the interruption of the water network connecting the ligand binding site with bulk water. The tiotropium bound simulations (tio-WT and tio-N508$^{6.52}$) do not show any significant difference in the water network densities (Fig. S4, Supplementary material).

To summarize the observations for the water networks: tiotropium binding to either WT or N508$^{6.52}$A-hM3R keeps the channel closed, and no water density is observed in the exit channel. For the modified ligands (methyl-tio and hyd-tio) a distinct water density is observed in the WT-simulation, solvating the ligand-N508$^{6.52}$ hydrogen bond. The N508$^{6.52}$A mutation causes a completely different water density pattern as hardly any water density is found in the exit channel.

4.2. Discussion of Case Study 2

Putting together all these findings from the MD simulations we can now understand the differences in the binding dynamics of tiotropium analogs and the effect of the N508$^{6.52}$A mutant. Experimentally we have shown that the removal of one ligand–receptor hydrogen bond leads to overproportionally quick dissociation, no matter if the hydrogen bond partner was removed on the ligand side (hyd-tio, methyl-tio) or on the receptor side (N508$^{6.52}$A). Thus we observed experimentally that the N508$^{6.52}$A mutation of the receptor has exactly the same effect as the removal of the hydroxy group from the ligand. Experimental data however, did not give any hint towards differences in the mechanism which causes accelerated dissociation, whereas insights obtained from MD simulations suggest that there is a major discrepancy between the underlying mechanisms. Upon removal of the hydroxy-group of tiotropium (hyd-tio, methyl-tio) the ligand–receptor hydrogen bond becomes water accessible. This is supposed to have a significant effect on the off-rates. Schmidtke et al. [80] have investigated buried polar atoms in protein binding sites that form ligand–protein hydrogen bonds which are shielded from water. Forming and breaking of this kind of hydrogen bonds lead to an energetically unfavorable transition state because it occurs asynchronously with hydration. In consequence,

water-shielded hydrogen bonds are exchanged at slower rates. When tiotropium binds to the receptor the hydrogen bond to N508$^{6.52}$ is water shielded, thus breaking more slowly. For the derivatives hyd-tio and methyl-tio the hydrogen bond is not shielded any longer as water enters into the channel, and the exchange occurs faster. This is now a plausible explanation for the observed rate enhancement upon hydroxy removal from the ligand.

The explanation for the accelerated rate upon N508$^{6.52}$A mutation of hM3R is not that straightforward, because no hydrogen bond needs to be broken and the line of argumentation from above is not applicable. Tiotropium forms an intramolecular hydrogen bond (hydroxy to ester) and no solvation of the hydroxy group of the bound ligand is observed. Indeed in the tio-N508$^{6.52}$A simulation the exit channel is as tightly closed as in the tio-WT simulation. The most striking difference is found in the location of the ligand, where tiotropium frequently progresses into the exit channel in the tio-N508$^{6.52}$A simulation. We thus propose that the first steps of dissociation of tiotropium from N508$^{6.52}$A hM3R do occur through a different mechanism than the dissociation from WT hM3R. In WT a solvation of the hydrogen bond is the first step for dissociation, but in N508$^{6.52}$A hM3R the ligand's motion into the exit channel is not supported by extracellular water. It may be seen as a mechanical event of impinging against the exit channel, which occurs much more often if the hydrogen bond to the receptor is missing.

Obviously the cavity around D148$^{3.32}$ has to be re-solvated upon ligand dissociation (for both cases, WT and N508$^{6.52}$A), but this site is directly connected to the receptor spanning water network which is conserved across class A GPCRs [81]. Therefore there is no need for the synchronous movement of extracellular solvent into the binding site upon ligand dissociation.

To sum up, the MD-simulations provide deeper insight in the mechanisms which cause accelerated dissociation from the M3 receptor. The lack of a hydrogen bond between the ligand and N508$^{6.52}$ of hM3R causes an enhanced off-rate, but the underlying mechanism is different. If the hydroxy group of tiotropium is removed, the hydrogen bond between N508$^{6.52}$ and the ester group of the ligand gets exposed to water and is not shielded any more. However, if N508$^{6.52}$ is mutated to alanine, tiotropium pushes towards the exit channel. In this case solvation of the exit channel does not seem to be important. Without MD simulations we would not be able to elucidate the differences in the dissociation enhancements. The remaining question is if we would have been able to predict behavior like this by MD and thus identify crucial ligand–protein interactions. For systems like the present one there is currently no method available which allows a quantification of the effects. We expect that in the future the application of recently developed simulation and analysis techniques based on Markov State Models together with increased computer power will lead to reliable predictions of off-rates also for highly potent ligands which show slower association rates [36,42,82,83]. For potent ligands, at present, we can use MD simulations only to rationalize the experiments or qualitatively investigate if some modifications of the ligands cause major rearrangements of the receptor or of the water network.

5. Summary and outlook

We described two case studies in which MD simulations were employed to investigate the interactions between a GPCR and small molecule ligands. However, both studies were carried out within totally different initial situations: in the case of CCR3, structural information about the ligand–receptor complex had to be generated via homology modeling. We observed that the template which is chosen to derive the homology model had a significant impact on key characteristics of the putative transmembrane binding pocket. For some template/alignment combinations it was even impossible to postulate a reasonable binding mode. With the availability of an X-ray structure of a sufficiently closely related receptor we were able to generate binding hypotheses which were in line with observed species differences and SAR. MD simulations helped to exclude receptor and ligand conformations which might not reflect the most stable state under the conditions that were modeled. At the same time the MD simulations delivered an ensemble of conformations of the receptor–ligand complex which are all accessible. Without additional experimental or computed information it is not possible to predict which configuration is biologically relevant. Experimental information such as structure activity relationship data, species selectivity data as in our case, or – ideally – site-directed mutagenesis data is key to establish a validated binding mode which is suitable for prospective ligand design efforts. Although we were finally able to rationalize species selectivity data and postulate regions for chemical modifications on a very coarse-grained level, it is important to emphasize that the final ligand–receptor model was not accurate enough to assess the impact of chemical modifications on the free energy of binding in a quantitative manner. Also, the accuracy of the model does not allow us to investigate the role of water molecules for ligand binding in detail.

In contrast to the CCR3 case study the computational study of ligand binding to hM3R is based on very detailed experimental (X-ray) structural information of a receptor–ligand complex. This allowed us to investigate the influence of solvation and desolvation and receptor motions on the binding of a congeneric series of small molecules to this receptor on a molecular level. Complemented by site-directed mutagenesis studies, a microscopic model of binding and dissociation could be derived which is in agreement with experimental data collected so far. MD simulations were key to elucidate differences between the behavior of closely related ligands in wild-type and mutated receptors.

In both case studies MD simulations turned out to be instrumental to come up with models which are able to rationalize experimental data in a retrospective fashion. However, the true impact of MD simulations on GPCR drug discovery programs can only be evaluated by prospective applications and subsequent experimental challenging of the predictions. Either synthesis and testing of compounds with proposed structural modifications or measuring ligand binding in mutated receptors will reveal if in silico predictions have been correct. Ideally, a combination of both approaches will be employed. The speed in elucidation of novel GPCR structural information together with further growth in compute power will enable us to carry out computational experiments on a large scale. From a careful analysis of these studies and

a rigorous assessment of the agreement or disagreement between computational predictions and experimental outcomes we can hope to derive general rules which can guide the prospective application of MD simulations in GPCR ligand design. This might also be expanded beyond pure binding to e.g. computational modeling of GPCR modulation.

Acknowledgments

We thank Matthias Röhm, Johannes Koppe and Peter Stauffert for excellent technical assistance.

References

[1] Lundstrom K. Latest development in drug discovery on G protein-coupled receptors. Curr Protein Pept Sci 2006;7:465–70.
[2] Giguere PM, Kroeze WK, Roth BL. Tuning up the right signal: chemical and genetic approaches to study GPCR functions. Curr Opin Cell Biol 2014;27:51–5.
[3] Overington JP, Al-Lazikani B, Hopkins AL. How many drug targets are there? Nat Rev Drug Discov 2006;5:993–6.
[4] Wise A, Gearing K, Rees S. Target validation of G-protein coupled receptors. Drug Discov Today 2002;7:235–46.
[5] Garland SL. Are GPCRs still a source of new targets? J Biomol Screen 2013;18:947–66.
[6] Fredriksson R, Lagerstrom MC, Lundin LG, Schioth HB. The G-protein-coupled receptors in the human genome form five main families. Phylogenetic analysis, paralogon groups, and fingerprints. Mol Pharmacol 2003;63:1256–72.
[7] Lin SHS, Civelli O. Orphan G protein-coupled receptors: targets for new therapeutic interventions. Ann Med 2004;36:204–14.
[8] Alvarez-Curto E, Ward RJ, Milligan G. Novel assay technologies for the discovery of G protein-coupled receptor drugs. Neuromethods 2011;60:231–53.
[9] Beets I, Lindemans M, Janssen T, Verleyen P. Deorphanizing G protein-coupled receptors by a calcium mobilization assay. Methods Mol Biol 2011;789:377–91.
[10] Boehm M, Hepworth D, Loria PM, Norquay LD, Filipski KJ, Chin JE, et al. Chemical probe identification platform for orphan GPCRs using focused compound screening: GPR39 as a case example. ACS Med Chem Lett 2013;4:1079–84.
[11] Lundstrom K. An overview on GPCRs and drug discovery: structure-based drug design and structural biology on GPCRs. Methods Mol Biol 2009;552:51–66.
[12] Stevens RC, Cherezov V, Katritch V, Abagyan R, Kuhn P, Rosen H, et al. The GPCR Network: a large-scale collaboration to determine human GPCR structure and function. Nat Rev Drug Discov 2013;12:25–34.
[13] Tautermann CS. GPCR structures in drug design, emerging opportunities with new structures. Bioorg Med Chem Lett 2014;24:4073–9.
[14] Congreve M, Langmead C, Marshall FH. The use of GPCR structures in drug design. Adv Pharmacol 2011;62:1–36.
[15] Tautermann CS. The use of G-protein coupled receptor models in lead optimization. Futur Med Chem 2011;3:709–21.
[16] Topiol S. X-ray structural information of GPCRs in drug design: what are the limitations and where do we go? Expert Opin Drug Discov 2013;8:607–20.
[17] Weyand S, Tate CG. Structure-based drug design applied to GPCRs: the holy grail realized. Biochemist 2014;36:23–7.
[18] Congreve M, Chessari G, Tisi D, Woodhead AJ. Recent developments in fragment-based drug discovery. J Med Chem 2008;51:3661–80.
[19] Joseph-McCarthy D, Campbell AJ, Kern G, Moustakas D. Fragment-based lead discovery and design. J Chem Inf Model 2014;54:693–704.
[20] Murray CW, Verdonk ML, Rees DC. Experiences in fragment-based drug discovery. Trends Pharmacol Sci 2012;33:224–32.
[21] Andrews SP, Brown GA, Christopher JA. Structure-based and fragment-based GPCR drug discovery. ChemMedChem 2014;9:256–75.
[22] Visegrády A, Keser GM. Fragment-based lead discovery on G-protein-coupled receptors. Expert Opin Drug Discov 2013;8:811–20.
[23] Christopher JA, Brown J, Doré AS, Errey JC, Koglin M, Marshall FH, et al. Biophysical fragment screening of the beta1-adrenergic receptor: identification of high affinity arylpiperazine leads using structure-based drug design. J Med Chem 2013;56:3446–55.
[24] Congreve M, Rich RL, Myszka DG, Figaroa F, Siegal G, Marshall FH. Fragment screening of stabilized G-protein-coupled receptors using biophysical methods. Methods Enzymol 2011;493:115–36.
[25] Chen D, Errey JC, Heitman LH, Marshall FH, Ijzerman AP, Siegal G. Fragment screening of GPCRs using biophysical methods: identification of ligands of the adenosine A(2A) receptor with novel biological activity. ACS Chem Biol 2012;7:2064–73.
[26] Congreve M, Andrews SP, Doré AS, Hollenstein K, Hurrell E, Langmead CJ, et al. Discovery of 1,2,4-triazine derivatives as adenosine A(2A) antagonists using structure based drug design. J Med Chem 2012;55:1898–903.
[27] Schlyer S, Horuk R. I want a new drug: G-protein-coupled receptors in drug development. Drug Discov Today 2006;11:481–93.
[28] Bertheleme N, Chae PS, Singh S, Mossakowska D, Hann MM, Smith KJ, et al. Unlocking the secrets of the gatekeeper: methods for stabilizing and crystallizing GPCRs. Biochim Biophys Acta 2013;1828:2583–91.
[29] Costanzi S. Modeling G protein-coupled receptors and their interactions with ligands. Curr Opin Struct Biol 2013;23:185–90.

[30] Heifetz A, Barker O, Morris GB, Law RJ, Slack M, Biggin PC. Toward an understanding of agonist binding to human Orexin-1 and Orexin-2 receptors with G-protein-coupled receptor modeling and site-directed mutagenesis. Biochemistry 2013;52:8246–60.

[31] Latek D, Pasznik P, Carlomagno T, Filipek S. Towards improved quality of GPCR models by usage of multiple templates and profile-profile comparison. PLoS One 2013;8:e56742.

[32] Preininger AM, Meiler J, Hamm HE. Conformational flexibility and structural dynamics in GPCR-mediated G protein activation: a perspective. J Mol Biol 2013;425:2288–98.

[33] Rasmussen SG, DeVree BT, Zou Y, Kruse AC, Chung KY, Kobilka TS, et al. Crystal structure of the beta2 adrenergic receptor-Gs protein complex. Nature 2011;477:549–55.

[34] Rasmussen SG, Choi HJ, Fung JJ, Pardon E, Casarosa P, Chae PS, et al. Structure of a nanobody-stabilized active state of the beta(2) adrenoceptor. Nature 2011;469:175–80.

[35] Cherezov V, Rosenbaum DM, Hanson MA, Rasmussen SG, Thian FS, Kobilka TS, et al. High-resolution crystal structure of an engineered human beta2-adrenergic G protein-coupled receptor. Science 2007;318:1258–65.

[36] Kohlhoff KJ, Shukla D, Lawrenz M, Bowman GR, Konerding DE, Belov D, et al. Cloud-based simulations on Google Exacycle reveal ligand modulation of GPCR activation pathways. Nat Chem 2014;6:15–21.

[37] Dror RO, Arlow DH, Maragakis P, Mildorf TJ, Pan AC, Xu H, et al. Activation mechanism of the beta2-adrenergic receptor. Proc Natl Acad Sci 2011;108:18684–9.

[38] Bruno A, Costantino G. Molecular dynamics simulations of G protein-coupled receptors. Mol Inform 2012;31:222–30.

[39] Dror RO, Dirks RM, Grossman JP, Xu H, Shaw DE. Biomolecular simulation: a computational microscope for molecular biology. Annu Rev Biophys 2012;41:429–52.

[40] Zhao H, Caflisch A. Molecular dynamics in drug design. Eur J Med Chem 2014. http://dx.doi.org/10.1016/j.ejmech.2014.08.004.

[41] Durrant JD, McCammon JA. Molecular dynamics simulations and drug discovery. BMC Biol 2011;9:71.

[42] Harvey MJ, De Fabritiis G. High-throughput molecular dynamics: the powerful new tool for drug discovery. Drug Discov Today 2012;17:1059–62.

[43] Cozzini P, Kellogg GE, Spyrakis F, Abraham DJ, Costantino G, Emerson A, et al. Target flexibility: an emerging consideration in drug discovery and design. J Med Chem 2008;51:6237–16.

[44] Feixas F, Lindert S, Sinko W, McCammon JA. Exploring the role of receptor flexibility in structure-based drug discovery. Biophys Chem 2014;186:31–45.

[45] Guvench O, Mackerell Jr AD. Computational fragment-based binding site identification by ligand competitive saturation. PLoS Comput Biol 2009;5:e1000435.

[46] Christ CD, Fox T. Accuracy assessment and automation of free energy calculations for drug design. J Chem Inf Model 2014;54:108–20.

[47] Hansen N, Van Gunsteren WF. Practical aspects of free-energy calculations: a review. J Chem Theory Comput 2014;10:2632–47.

[48] Tan Q, Zhu Y, Li J, Chen Z, Han GW, Kufareva I, et al. Structure of the CCR5 chemokine receptor–HIV entry inhibitor maraviroc complex. Science 2013;341:1387–90.

[49] Chemical Computing Group. MOE — Molecular operating environment. Montreal: Chemical Computing Group; 2010.

[50] Marti-Renom MA, Stuart AC, Fiser A, Sanchez R, Melo F, Sali A. Comparative protein structure modeling of genes and genomes. Annu Rev Biophys Biomol Struct 2000;29:291–325.

[51] Sali A, Blundell TL. Comparative protein modelling by satisfaction of spatial restraints. J Mol Biol 1993;234:779–815.

[52] Kriegl JM, Martyres D, Grundl MA, Anderskewitz R, Dollinger H, Rast G, et al. Rodent selectivity of piperidine-4-yl-1H-indoles, a series of CC chemokine receptor-3 (CCR3) antagonists: insights from a receptor model. Bioorg Med Chem Lett 2014. http://dx.doi.org/10.1016/j.bmcl.2014.11.063.

[53] Hess B, Kutzner C, van der Spoel D, Lindahl E. GROMACS 4: algorithms for highly efficient, load-balanced, and scalable molecular simulation. J Chem Theory Comput 2008;4:435–47.

[54] Tautermann CS, Kiechle T, Seeliger D, Diehl S, Wex E, Banholzer R, et al. Molecular basis for the long duration of action and kinetic selectivity of tiotropium for the muscarinic M3 receptor. J Med Chem 2013;56:8746–56.

[55] Berger O, Edholm O, Jähnig F. Molecular dynamics simulations of a fluid bilayer of dipalmitoylphosphatidylcholine at full hydration, constant pressure, and constant temperature. Biophys J 1997;72:2002–13.

[56] Best RB, Hummer G. Optimized molecular dynamics force fields applied to the helix–coil transition of polypeptides. J Phys Chem B 2009;113:9004–15.

[57] Lindorff-Larsen K, Piana S, Palmo K, Maragakis P, Klepeis JL, Dror RO, et al. Improved side-chain torsion potentials for the Amber ff99SB protein force field. Proteins 2010;78:1950–8.

[58] Berendsen HJC, Grigera JR, Straatsma TP. The missing term in effective pair potentials. J Phys Chem 1987;91:6269–71.

[59] Joung IS, Cheatham III TE. Determination of alkali and halide monovalent ion parameters for use in explicitly solvated biomolecular simulations. J Phys Chem B 2008;112:9020–41.

[60] Wang J, Wolf RM, Caldwell JW, Kollman PA, Case DA. Development and testing of a general amber force field. J Comput Chem 2004;25:1157–74.

[61] Frisch MJ, Trucks GW, Schlegel HB, Scuseria GE, Robb MA, Cheeseman JR, et al. Gaussian 09, Revision B.01. Gaussian 09, Revision B.01. Wallingford CT: Gaussian, Inc.; 2009.

[62] Sousa da Silva A, Vranken W. ACPYPE — AnteChamber PYthon Parser interfacE. BMC Res Notes 2012;5:367.

[63] Wolf MG, Hoefling M, Aponte-Santamaría C, Grubmüller H, Groenhof G. g_membed: efficient insertion of a membrane protein into an equilibrated lipid bilayer with minimal perturbation. J Comput Chem 2010;31:2169–74.

[64] Berendsen HJC, Postma JPM, van Gunsteren WF, DiNola A, Haak JR. Molecular dynamics with coupling to an external bath. J Chem Phys 1984;81:3684–90.

[65] Bussi G, Donadio D, Parrinello M. Canonical sampling through velocity rescaling. J Chem Phys 2007;126:014101–7.

[66] Parrinello M, Rahman A. Polymorphic transitions in single crystals: a new molecular dynamics method. J Appl Phys 1981;52:7182–90.

[67] Essmann U, Perera L, Berkowitz ML, Darden T, Lee H, Pedersen LG. A smooth particle mesh Ewald method. J Chem Phys 1995;103:8577–93.

[68] Miyamoto S, Kollman PA. Settle: an analytical version of the SHAKE and RATTLE algorithm for rigid water models. J Comput Chem 1992;13:952–62.

[69] Hess B, Bekker H, Berendsen HJC, Fraaije JGEM. LINCS: a linear constraint solver for molecular simulations. J Comput Chem 1997;18:1463–72.

[70] Feenstra KA, Hess B, Berendsen HJC. Improving efficiency of large time-scale molecular dynamics simulations of hydrogen-rich systems. J Comput Chem 1999;20:786–98.

[71] Viola A, Luster AD. Chemokines and their receptors: drug targets in immunity and inflammation. Annu Rev Pharmacol Toxicol 2008;48:171–97.

[72] Bahl A, Springthorpe B, Riley R. Chemokine CCR3 antagonists. In: Hansel TT, Barnes PJ, editors. New drugs and targets for asthma and COPD. 39th ed. Basel: Karger AG; 2010. p. 153–9.

[73] Trabanino RJ, Hall SE, Vaidehi N, Floriano WB, Kam VW, Goddard 3rd WA. First principles predictions of the structure and function of g-protein-coupled receptors: validation for bovine rhodopsin. Biophys J 2004;86:1904–21.

[74] Wise EL, Duchesnes C, da Fonseca PC, Allen RA, Williams TJ, Pease JE. Small molecule receptor agonists and antagonists of CCR3 provide insight into mechanisms of chemokine receptor activation. J Biol Chem 2007;282:27935–43.

[75] Surgand JS, Rodrigo J, Kellenberger E, Rognan D. A chemogenomic analysis of the transmembrane binding cavity of human G-protein-coupled receptors. Proteins 2006;62:509–38.

[76] Martyres D, Hoffmann M, Seither P, Bouyssou T. Novel substituted piperidyl-propane-thiolsWO2008/049875 A1[WO2008/049875] ; 2008.

[77] Global Strategy for the Diagnosis, Management and Prevention of COPD. Global Initiative for Chronic Obstructive Lung Disease (GOLD); 2014 Available from: http://www.goldcopd.org/.

[78] Cazzola M, Page CP, Calzetta L, Matera MG. Pharmacology and therapeutics of bronchodilators. Pharmacol Rev 2012;64:450–504.

[79] Kruse AC, Hu J, Pan AC, Arlow DH, Rosenbaum DM, Rosemond E, et al. Structure and dynamics of the M3 muscarinic acetylcholine receptor. Nature 2012;482:552–6.

[80] Schmidtke P, Luque FJ, Murray JB, Barril X. Shielded hydrogen bonds as structural determinants of binding kinetics: application in drug design. J Am Chem Soc 2011;133:18903–10.

[81] Katritch V, Fenalti G, Abola EE, Roth BL, Cherezov V, Stevens RC. Allosteric sodium in class A GPCR signaling. Trends Biochem Sci 2014;39:233–44.

[82] Pande VS, Beauchamp K, Bowman GR. Everything you wanted to know about Markov State Models but were afraid to ask. Methods 2010;52:99–105.

[83] Prinz JH, Wu H, Sarich M, Keller B, Senne M, Held M, et al. Markov models of molecular kinetics: generation and validation. J Chem Phys 2011;134:174105.

2

Vibrational analysis on the revised potential energy curve of the low-barrier hydrogen bond in photoactive yellow protein

Yusuke Kanematsu [a,b], Hironari Kamikubo [c], Mikio Kataoka [c], Masanori Tachikawa [b,*]

[a] Graduate School of Information Science, Hiroshima City University, 3-4-1 Ozuka-Higashi, Asa-Minami-Ku, Hiroshima 731-3194, Japan
[b] Quantum Chemistry Division, Yokohama City University, Seto 22-2, Kanazawa-ku, Yokohama 236-0027, Japan
[c] Nara Institute of Science and Technology, 8916-5 Takayama, Ikoma, Nara 630-0192, Japan

ARTICLE INFO

ABSTRACT

Keywords:
Low-barrier hydrogen bond
Photoactive yellow protein
Vibrational analysis
ONIOM
PCM

Photoactive yellow protein (PYP) has a characteristic hydrogen bond (H bond) between p-coumaric acid chromophore and Glu46, whose OH bond length has been observed to be 1.21 Å by the neutron diffraction technique [Proc. Natl. Acad. Sci. 106, 440–4]. Although it has been expected that such a drastic elongation of the OH bond could be caused by the quantum effect of the hydrogen nucleus, previous theoretical computations including the nuclear quantum effect have so far underestimated the bond length by more than 0.07 Å. To elucidate the origin of the difference, we performed a vibrational analysis of the H bond on potential energy curve with O...O distance of 2.47 Å on the equilibrium structure, and that with O...O distance of 2.56 Å on the experimental crystal structure. While the vibrationally averaged OH bond length for equilibrium structure was underestimated, the corresponding value for crystal structure was in reasonable agreement with the corresponding experimental values. The elongation of the O...O distance by the quantum mechanical or thermal fluctuation would be indispensable for the formation of a low-barrier hydrogen bond in PYP.

1. Introduction

Photoactive yellow protein (PYP) is a water-soluble photosensor protein found in halophilic photosynthetic bacteria. This protein is known to play an important role in the photocycle that regulates the negative phototaxis behavior of the bacteria [1]. PYP from *H. halophila* has especially drawn substantial attention due to the characteristic hydrogen bond (H bond) in the site. Yamaguchi et al. reported the detailed crystal structure of dark-state PYP with the coordinates of hydrogen atoms by the neutron diffraction technique [2]. The structure of PYP is shown in Fig. 1. They have assigned a hydrogen bond between Glu46 and p-coumaric acid (pCA) chromophore of PYP with significantly long O-H bond of 1.21 Å as low-barrier hydrogen bond (LBHB), which had never been directly observed in proteins until then. They have suggested the roles of LBHB in PYP to stabilize the negative charge around the chromophore in the protein interior, and to mediate the fast proton transfer during the photocycle. It has also been found that Arg52 located near the chromophore was deprotonated (neutral), whereas it had been believed to be protonated to act as counterion for the negative chromophore according to X-ray crystallography[3] and electronic structure calculation [4].

It was, however, claimed by Saito and Ishikita that Arg52 should be protonated and the H-bond between pCA and Glu46 was not LBHB but a normal H-bond according to the potential energy profile analysis with the conventional QM/MM calculation, and the comparison between the experimental chemical shifts in solution [5] and their computational values under the previously mentioned condition [6,7]. Hirano and Sato compared the potential energy profiles for systems with and without the protonation of Arg52 using the ONIOM method [8,9], which is an efficient method to calculate large systems such as proteins by dividing the system into several layers, and found the low barrier height for deprotonated Arg52 model with respect to hydrogen transfer coordinates in H-bond [10].

For the computation of the molecular geometry for H-bonded systems, we should pay careful attention to the protonic or deuteronic fluctuation in H-bonds due to nuclear quantum effect [11] and not only to the electrostatic contribution of the surrounding environment. Kita et al. calculated the isolated cluster model of H-bonding center in PYP by multicomponent quantum mechanics (MC_QM) method [12–14] and found that O-H bond length becomes longer due to the nuclear quantum effect [15]. Nadal-Ferret et al. showed that long O-H bonds could be experimentally observed by taking account of the nuclear quantum fluctuation of the hydrogen nuclei, and by local vibration analysis using QM/MM for PYP with deprotonated Arg52 [16]. By comparison of

---
* Corresponding author

a

b   Tyr42 ──⬡── O . H   ⬡ pCA
                     ⋮
                     O
                     ⋮
                     H
                     ⋮
        Glu46 ── O
                   ‖
                   O

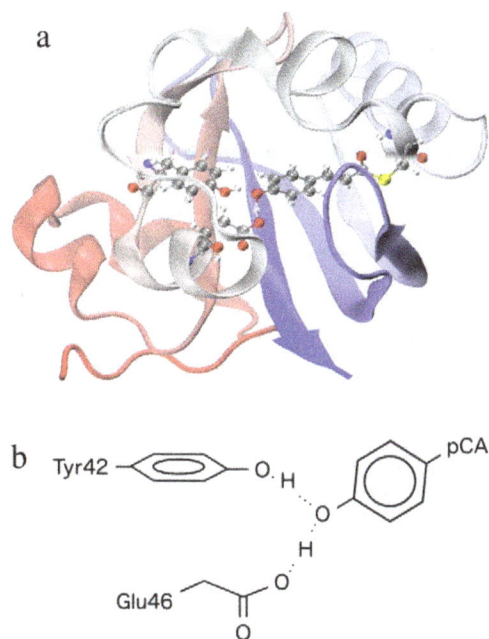

Fig. 1. (a) The entire structure and (b) the active center of photoactive yellow protein. OH bond focused on in the present work is also indicated.

2. Computational detail

We performed a numerical one-dimensional vibrational analysis by solving the nuclear Schrödinger equation on the basis of the Born–Oppenheimer approximation. The analyses dealt with two initial structures of PYP, one of which was the crystal structure (PDB ID: 2ZOI; temperature, 295 K; resolution, 1.50 Å) obtained by the neutron diffraction technique [2], and the latter was the equilibrium structure that has been optimized by conventional ONIOM calculation in our previous work [17]. The missing atoms in the crystal structure have been compensated for by AmberTools [18,19], resulting 1929 atoms with a total charge of 6– for the entire PYP. We utilized the ONIOM Electronic Embedding (ONIOM-EE) method [8,9] with the computational condition of "System 3-dp" of ref. [17], which includes Ile31, Tyr42, Glu46, Thr50, Cys69, pCA, and Arg52 in deprotonated form inside the QM region with CAM-B3LYP/6-31 + G(d,p) [20–22] level of calculation, and the other residues are in molecular mechanical level of calculation with AMBER ff99 and GAFF parameters [23]. As in ref. [16] and the Appendix in ref. [17], unrelaxed one-dimensional potential energy curves for the migration of the hydrogen nucleus along the direction vector q from Glu46 to pCA have been constructed for the vibrational analysis.

3. Results and discussion

3.1. The vibrational analysis on the equilibrium structure

At first, we would like to focus on the results of the vibration analysis on the equilibrium structure. Fig. 2 shows the potential energy curve and the corresponding vibrational distributions of the ground and the first excited states of a proton and a deuteron. The corresponding energy levels and the vibrationally averaged OH bond lengths are shown in

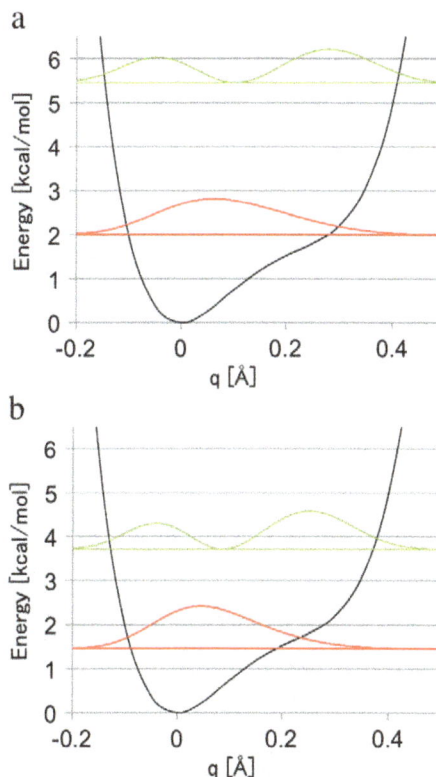

the experimental and the computational geometrical parameters with and without the protonation of Arg52, they suggested that the formation of LBHB between pCA and Glu46 would be possible if Arg52 was deprotonated. Recently, we have also analyzed this geometry by the combination of ONIOM and MC_QM [ONIOM (MC_QM:MM)], and validated the above suggestion by comparison with the corresponding experimental values [17].

Table 1 lists the experimental and the computational OH bond lengths in various previous papers. Each computational work in Table 1 shows that the nuclear quantum effect of hydrogen nuclei elongates the OH bond length from the equilibrium structure. There, however, remains a difference between experimental and theoretical OH bond lengths of more than 0.07 Å, and the experimental observation was not yet fully supported by the computations.

In our previous work, we have examined several computational conditions for ONIOM calculations [17]. We found the expansion of the QM region can involve the degradation of the barrier height, which reasonably agreed with all electron ONIOM (QM:QM) calculations. It could be expected that the vibrational analysis on this potential energy curve would provide longer OH bonds than that of previous works. Therefore, the present paper will be devoted to the vibrational analysis for the model system of PYP with expanded QM region in order to elucidate the origin of the difference between the experimental measurement and the theoretical computation.

Table 1
The experimental and the computational OH bond lengths of Glu46 in the deuterated PYP from the previous works.

Experimental	(ref. [2])	R	1.21
Computational	(ref. [15])	R_{eq}	1.02
		$\langle R \rangle$	1.05
	(ref. [16])	R_{eq}	1.05
		$\langle R \rangle$	1.10
	(ref. [17])	R_{eq}	1.08
		$\langle R \rangle$	1.14

Units are in Å. For computational lengths, both the equilibrium (R_{eq}) and the vibrationally averaged $\langle R \rangle$ values are shown.

a

b

Fig. 2. The potential energy curve (black) and the corresponding vibrational distribution of the ground (red) and the first excited (green) states of (a) a proton and (b) a deuteron in the hydrogen bond between Glu46 and pCA of PYP at the equilibrium structure. The origin of the coordinate q was set on the equilibrium bond length (1.08 Å).

Table 2

Vibrational energy levels (ε in kcal/mol) for the first three states and the averaged OH bond lengths ($\langle R \rangle$ in Å) for a proton (H) and a deuteron (D) in hydrogen bond between Glu46 and pCA of PYP at the equilibrium structure.

	State	ε	$\langle R \rangle$
H	2	10.14	1.21
	1	5.46	1.23
	0	2.02	1.17
	R_T		1.17
D	2	6.52	1.22
	1	3.73	1.24
	0	1.47	1.15
	R_T		1.15

The bond length at the energy minimum is 1.08 Å [17]. Thermally averaged bond lengths (R_T in Å) at room temperature are also shown.

Table 2. We can see the anharmonic single-well potential energy curve with no barrier, and the corresponding widespread vibrational distribution in Fig. 2. The shape of the potential energy curve was drastically different with those of the double-well curve in the previous reports [10,16,17]. Such difference could be mainly attributed to the short O... O distance of 2.47 Å between the H-bond donor and the acceptor, which was about 0.1 Å shorter than that of the experimental value. Table 2 indicates that the bond lengths were extended from the equilibrium lengths, as in the previous reports [10,16,17]. The averaged bond lengths of ground states reasonably agreed with those of our previous work with the corresponding computational condition; 1.14 Å for DD and 1.15 Å for HH isotopologues [17].

In order to include the thermal effect on the OH bond, we can utilize the Boltzmann averaging for the temperature T according to the equation:

$$R_T = \frac{\sum_i \langle R \rangle_i \exp[-\epsilon_i / k_B T]}{\sum_i \exp[-\epsilon_i / k_B T]} , \qquad (1)$$

where ϵ_i's are the energies of the respective vibrational states and k_B is the Boltzmann constant. We calculated the thermally averaged OH bond lengths for the first three vibrational states at room temperature ($T = 300$ K). The resulting values are 1.15 Å for a deuteron and 1.17 Å for a proton, and corresponded with those of ground states in the shown digits. This is because the excitation energies are higher than 2.0 kcal/mol, and the contributions of the excited states against the thermal averaging are less than 3% of the ground states at the room temperature. As a result, there still remained a difference of 0.06 Å from the experimental value even though the thermal effect was included.

3.2. The vibrational analysis on the crystal structure

Here, we would like to focus on the experimental crystal structure. Fig. 3 shows the potential energy curve and the corresponding vibrational distributions of the ground and the first excited states of a proton and a deuteron. For the deuteron, the vibrational distribution of the second excited state is also shown. The potential energy curve represents a double-well shape at the crystal structure with the O...O distance of 2.56 Å, as already mentioned. The barrier height is 2.36 kcal/mol at $q = 0.10$ Å, where the OH bond length is 1.31 Å. The corresponding energy levels and the vibrationally averaged OH bond lengths are shown in Table 3.

We can see that the zero-point vibrational energy for a proton is almost comparable with the barrier height, and the proton overcame the barrier to widely spread. On the other hand, there is an energy gap of about 1 kcal/mol between the barrier height and the zero-point energy of a deuteron, and the ground state deuteron tends to localize around the minimum. This isotopic difference of the vibration results in a

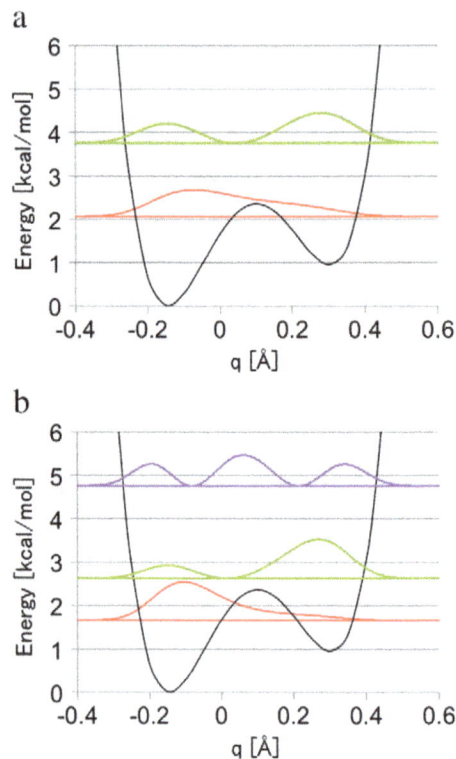

Fig. 3. The potential energy curve (black) and the corresponding vibrational distribution of the ground (red) and the first (green) and second (purple) excited states of (a) a proton and (b) a deuteron in the hydrogen bond between Glu46 and pCA of PYP at the crystal structure. The origin of the coordinate q was set on the crystal bond length (1.21 Å).

difference of the average bond length of the ground state by 0.06 Å, which is three times larger than that of the equilibrium structure in Table 2.

Another point worth mentioning is that the energy gap between ground and the first excited states for the crystal structure is lower than those of the equilibrium structure. Particularly, the deuteron only has a gap of 0.96 kcal/mol, which allows vibrational excitation at room temperature. The thermally averaged OH bond length was extended from the ground state by 0.03 Å for the deuteron, while it did not change in the shown digits for the proton. In other words, the OH bond length for the proton elongated by 0.17 Å from the energy minimum due almost only to the nuclear quantum effect, while that for the deuteron elongated by 0.11 Å due to the nuclear quantum effect and by 0.03 Å due to the thermal effect. The resulting length for the deuteron agreed with the experimental value within the error of 0.01 Å.

Whereas reasonable agreement was obtained, it should be noted here that the present analysis against one-dimensional partial vibration neglected the fluctuations of the other atoms and the motion of the

Table 3

Vibrational energy levels (ε in kcal/mol) for the first three states and the averaged OH bond lengths ($\langle R \rangle$ in Å) for a proton (H) and a deuteron (D) in hydrogen bond between Glu46 and pCA of PYP at the crystal structure.

	State	ε	$\langle R \rangle$
H	2	7.36	1.29
	1	3.76	1.34
	0	2.08	1.23
	R_T		1.23
D	2	4.77	1.28
	1	2.63	1.39
	0	1.67	1.17
	R_T		1.20

The bond length at the energy minimum is 1.06 Å. Thermally averaged bond lengths (R_T in Å) at room temperature are also shown.

hydrogen nucleus coupled with them. It cannot be said that the present analysis is more accurate or realistic than the previous analyses, although the reasonability of the one-dimensional potential energy curve used in the present analysis has been confirmed [17]. Higher-dimensional vibrational analysis would be desirable to acquire more quantitative reliability.

Finally, let us consider the difference of the equilibrium and the crystal structures. As already mentioned, they significantly differ in the H-bonding O…O distance of Glu46 and pCA by about 0.1 Å, involving the drastic difference of the potential energy curve. The agreement between the present results for the equilibrium structure and our previous result by ONIOM (MC_QM:MM) implies that the underestimation of the OH bond length by ONIOM (MC_QM:MM) could be possibly associated with the underestimation of the H-bonding O…O distance, taking into account that the vibrational analysis on the crystal structure with large O…O distance reasonably reproduced the experimental OH bond length. The underestimation would be owing to the lack of the O…O stretching by the nuclear quantum or thermal effect. Nadal-Ferret et al. demonstrated that a significant fluctuation of the O…O distance can arise from the thermal effect [16]. The crystal structure can be regarded as a thermally averaged structure, which may provide better energy profile for the vibrational analysis in the present work. It should also be noted that the error in the equilibrium structure from the CAM-B3LYP functional for the evaluation of H-bonding distance is not negligible [24], although the functional was reasonably selected for the evaluation of the energy profile of proton migration by Nadal-Ferret et al. [16]

4. Conclusion

We analyzed the vibrational states of the hydrogen nucleus in a hydrogen bond of photoactive yellow protein (PYP). We found that the potential energy curve was a single-well shape for the equilibrium structure, while it was a double-well shape with low energy barrier for the crystal structure. The vibrationally averaged OH bond length reasonably agreed with that of the deuterated PYP. The isotopic difference of the temperature dependency was suggested by the Boltzmann thermal averaging. It was implied that the extension of O…O distance by the quantum or thermal effect from the equilibrium structure would be indispensable for the formation of the law-barrier hydrogen bond with drastically extended OH bond length as seen in the neutron diffraction measurement.

Acknowledgement

This work is partly supported by Grants-in-Aid for Scientific Research (KAKENHI) of the Ministry of Education, Culture, Sports, Science and Technology (MEXT), Grant Numbers 25102003 for HK and Grant Numbers 26620013, 26102539, and 15KT0067 for MT. Theoretical calculations were partly performed at the Research Center for Computational Science, Institute for Molecular Science, Japan and Center of Computational Materials Science, Institute for Solid State Physics, The University of Tokyo, Japan.

References

[1] Sprenger WW, Hoff WD, Armitage JP, Hellingwerf KJ. The eubacterium *Ectothiorhodospira halophila* is negatively phototactic, with a wavelength dependence that fits the absorption spectrum of the photoactive yellow protein. J Bacteriol 1993;175:3096–104.
[2] Yamaguchi S, Kamikubo H, Kurihara K, Kuroki R, Niimura N, Shimizu N, et al. Low-barrier hydrogen bond in photoactive yellow protein. Proc Natl Acad Sci U S A 2009;106:440–4. http://dx.doi.org/10.1073/pnas.0811882106.
[3] Borgstahl GE, Williams DR, Getzoff ED. 1.4 Å structure of photoactive yellow protein, a cytosolic photoreceptor: unusual fold, active site, and chromophore. Biochemistry 1995;34:6278–87.
[4] Gromov EV, Burghardt I, Ko H, Cederbaum LS. Electronic Structure of the PYP Chromophore in Its Native Protein Environment; 2007[6798–800].
[5] Sigala PA, Tsuchida MA, Herschlag D. Hydrogen bond dynamics in the active site of photoactive yellow protein. Proc Natl Acad Sci U S A 2009;106:9232–7. http://dx.doi.org/10.1073/pnas.0900168106.
[6] Saito K, Ishikita H. Energetics of short hydrogen bonds in photoactive yellow protein. Proc Natl Acad Sci U S A 2012;109:167–72. http://dx.doi.org/10.1073/pnas.1113599108.
[7] Saito K, Ishikita H. H atom positions and nuclear magnetic resonance chemical shifts of short H bonds in photoactive yellow protein. Biochemistry 2012;51:1171–7. http://dx.doi.org/10.1021/bi201877e.
[8] Dapprich S, Komáromi I, Byun KS, Morokuma K, Frisch MJ. A new ONIOM implementation in Gaussian98. Part I. The calculation of energies, gradients, vibrational frequencies and electric field derivatives. J Mol Struct THEOCHEM 1999;461–462:1–21.
[9] Vreven T, Byun KS, Komáromi I, Dapprich S, Montgomery JA, Morokuma K, et al. Combining quantum mechanics methods with molecular mechanics methods in ONIOM. J Chem Theory Comput 2006;2:815–26. http://dx.doi.org/10.1021/ct050289g.
[10] Hirano K, Sato H. A theoretical study on the electronic structure of PYP chromophore in low barrier hydrogen bonding model. Chem Phys 2013;419:163–6. http://dx.doi.org/10.1016/j.chemphys.2013.01.040.
[11] Isotope Effects In Chemistry and Biology. In: Kohen A, Limbach H-H, editors. Boca Raton: Taylor & Francis Group, LLC; 2006. http://dx.doi.org/10.1201/9781420028027.
[12] Tachikawa M. Multi-component molecular orbital theory for electrons and nuclei including many-body effect with full configuration interaction treatment: isotope effects on hydrogen molecules. Chem Phys Lett 2002;360:494–500. http://dx.doi.org/10.1016/S0009-2614(02)00881-3.
[13] Kreibich T, Gross E. Multicomponent density-functional theory for electrons and nuclei. Phys Rev Lett 2001;86:2984–7. http://dx.doi.org/10.1103/PhysRevLett.86.2984.
[14] Udagawa T, Tachikawa M. H/D isotope effect on porphine and porphycene molecules with multicomponent hybrid density functional theory. J Chem Phys 2006;125:244105. http://dx.doi.org/10.1063/1.2403857.
[15] Kita Y, Kamikubo H, Kataoka M, Tachikawa M. Theoretical analysis of the geometrical isotope effect on the hydrogen bonds in photoactive yellow protein with multicomponent density functional theory. Chem Phys 2013;419:50–3. http://dx.doi.org/10.1016/j.chemphys.2012.11.022.
[16] Nadal-Ferret M, Gelabert R, Moreno M, Lluch JM. Are there really low-barrier hydrogen bonds in proteins? The case of photoactive yellow protein. J Am Chem Soc 2014;136:3542–52. http://dx.doi.org/10.1021/ja4116617.
[17] Kanematsu Y, Tachikawa M. Theoretical analysis of geometry and NMR isotope shift in hydrogen-bonding center of photoactive yellow protein by combination of multicomponent quantum mechanics and ONIOM scheme. J Chem Phys 2014;141:185101. http://dx.doi.org/10.1063/1.4900987.
[18] Case DA, Cheatham TE, Darden T, Gohlke H, Luo R, Merz KM, et al. The Amber biomolecular simulation programs. J Comput Chem 2005;26:1668–88. http://dx.doi.org/10.1002/jcc.20290.
[19] Wang J, Wang W, Kollman PA, Case DA. Automatic atom type and bond type perception in molecular mechanical calculations. J Mol Graph Model 2006;25:247–60.
[20] Yanai T, Tew DP, Handy NC. A new hybrid exchange–correlation functional using the Coulomb-attenuating method (CAM-B3LYP). Chem Phys Lett 2004;393:51–7. http://dx.doi.org/10.1016/j.cplett.2004.06.011.
[21] Ditchfield R. Self-consistent molecular-orbital methods. IX. An extended Gaussian-type basis for molecular-orbital studies of organic molecules. J Chem Phys 1971;54(724). http://dx.doi.org/10.1063/1.1674902.
[22] Frisch MJ, Pople J. a., Binkley JS. Self-consistent molecular orbital methods 25. Supplementary functions for Gaussian basis sets. J Chem Phys 1984;80:3265. http://dx.doi.org/10.1063/1.447079.
[23] Wang J, Wolf RM, Caldwell JW, Kollman PA, Case DA. Development and testing of a general amber force field. J Comput Chem 2004;25:1157–74. http://dx.doi.org/10.1002/jcc.20035.
[24] Peach MJG, Helgaker T, Sałek P, Keal TW, Lutnaes OB, Tozer DJ, et al. Assessment of a Coulomb-attenuated exchange-correlation energy functional. Phys Chem Chem Phys 2006;8:558–62. http://dx.doi.org/10.1039/b511865d.

3

A case study for cloud based high throughput analysis of NGS data using the globus genomics system

Krithika Bhuvaneshwar [a], Dinanath Sulakhe [b,c], Robinder Gauba [a], Alex Rodriguez [b], Ravi Madduri [b,c], Utpal Dave [b,c], Lukasz Lacinski [b,c], Ian Foster [b,c], Yuriy Gusev [a], Subha Madhavan [a,*]

[a] Innovation Center for Biomedical Informatics (ICBI), Georgetown University, Washington, DC 20007, USA
[b] Computation Institute, University of Chicago, Argonne National Laboratory, 60637, USA
[c] Globus Genomics, USA

ARTICLE INFO

ABSTRACT

Next generation sequencing (NGS) technologies produce massive amounts of data requiring a powerful computational infrastructure, high quality bioinformatics software, and skilled personnel to operate the tools. We present a case study of a practical solution to this data management and analysis challenge that simplifies terabyte scale data handling and provides advanced tools for NGS data analysis. These capabilities are implemented using the "Globus Genomics" system, which is an enhanced Galaxy workflow system made available as a service that offers users the capability to process and transfer data easily, reliably and quickly to address end-to-endNGS analysis requirements. The Globus Genomics system is built on Amazon's cloud computing infrastructure. The system takes advantage of elastic scaling of compute resources to run multiple workflows in parallel and it also helps meet the scale-out analysis needs of modern translational genomics research.

Keywords:
Next generation sequencing
Galaxy
Cloud computing
Translational research

1. Introduction

1.1. Background

The popularity of next generation sequencing (NGS) grew exponentially since 2007 due to faster, more accurate and affordable sequencing [1]. Initial studies were focused on comparing data and analysis results from NGS technologies with those from traditional polymerase chain reaction (PCR) and Sanger sequencing methods. Since then, we have come a long way in understanding how different it is from traditional methods and genome wide association studies (GWAS). The potential of NGS is now being tapped in a wide variety of applications including re-sequencing, functional genomics, translational research, and clinical genomics [2,3].

Focusing on NGS applications for translational research, the most basic use cases involve comparison of two cohorts — a case and control group with added complexity for longitudinal studies and meta-analyses. Such use cases require medium to large sample sizes, ranging from hundreds to thousands of samples, to be able to derive statistically significant results [4]. As these large-scale genomic studies become a

reality, high throughput data storage, management and computation for large sample sizes are becoming increasingly challenging.

Current high performance computing (HPC) solutions in the genomics area involve clusters and grids, which are distributed systems targeted towards users who prefer a command line interface. These HPC solutions are not cheap because they require support and maintenance. University based clusters are shared resources with many competing users. To support maximum usage of these expensive clusters, the jobs are queued, and it becomes a buffer for managing IT capacity. For NGS applications that use medium to large sized samples, researchers would have to wait until enough resources become available; the time needed to complete processing becomes unpredictable. Users could potentially avoid queues by using grids, which are a collection of resources from different locations; but the cost of constructing a grid is high and its architecture and management is complex. Cloud computing leverages virtual technology to provide computational resources to users and this virtualization helps better utilize resources [5]. Its shared computing environment and pay-as-you-go storage can greatly benefit geographically dispersed teams working on the same dataset. There are a number of providers that offer cloud based solutions, some of them include Amazon [6], Google [7], and Microsoft [8]. The need for cloud computing for genomic analysis has been well-described by leaders in bioinformatics and computational biology [4,9,10] due to its flexibility, scalability and lower costs. This has been proven by the fact that many medical institutes and centers in the US and around the world have

* Corresponding author at: Innovation Center for Biomedical Informatics (ICBI), Georgetown University Medical Center, 2115 Wisconsin Ave NW, Suite 110, Washington, DC 20007, USA.
E-mail address: sm696@georgetown.edu (S. Madhavan).

already embraced it [11–16]. NGS analyses are well-suited for the cloud since data upload (of input files) to an Amazon cloud instance does not incur any extra charge and data download (of output files) becomes relatively inexpensive as only a small percentage of output is needed for downstream analysis [17,18]. There are several cloud service models: (a) Infrastructure as a service (IaaS) offers compute, storage and network resources as a service, (b) Platform as a service (PaaS) that runs applications on the cloud and hides infrastructure implementation details from the user, and (c) Software as a service (SaaS) that provides software and databases as a service. SaaS eliminates the need to install and maintain the software. It also allows users to run HPCprograms on the cloud through graphical interfaces, and may be a promising solution for NGS analysis for biologists and researchers [5,19].

While a few large genomics sequencing centers such as the National Institutes of Health (NIH) and major academic centers have developed custom solutions relying on significant investment in local computation infrastructure, an increasing number of universities and academic institutions across the US are facing challenges due to increasing interest and demand from researchers to utilize NGS technology. These small to medium size biomedical research entities neither have the capabilities to implement local computing infrastructures, nor are they able to rapidly expand their capabilities depending on sequencing data management needs. Additionally, there is an increasingly urgent need for adequate software support and management systems capable of providing reliable and scalable support for the ever-increasing influx of NGS data. Some academic centers have been developing customized software solutions, which are often coupled with commercial computing infrastructures such as Mercury [20] utilizing Amazon Web Services cloud via the DNAnexus [21] platform. However there is clearly a lack of standardized and affordable NGS management solutions on the cloud to support the growing needs of translational genomics research.

1.2. Existing commercial and non-commercial solutions

Before choosing the Globus Genomics system [22] for our case study, we briefly explored various commercial systems that offer solutions including Partek [23], DNAnexus [21], CLC Bio [24], DNASTAR[25], Maverix Biomics [26], Seven Bridges [27] and Golden Helix [28]. At the time we explored these commercial tools, only a few of these systems had cloud based solutions for large scale batch processing and such solutions were too expensive for an academic center to adopt. Galaxy, however is an open source web based platform for bioinformatics analysis [29,30]. It provides users with an easy-to-use web interface that allows users to create complex biological workflows by simply dragging-and dropping tools into its "workflow canvas". The settings and parameters for each tool can be customized by the user. After upload of data, the workflow gets submitted to their backend analysis server. The completed analysis results can be viewed, published (made public), or shared with other users. Galaxy has an expanding repository of tools in its "Tool Shed" [31]. It provides an extensible framework and allows many software tools to be integrated into the platform. An active community of developers ensures that the latest tools are available through the Galaxy Tool Shed. The biggest advantage of the Galaxy framework is that it automatically and transparently tracks analysis details, and allows results to be documented, downloaded, shared, and published with complete provenance, guaranteeing transparency and reproducibility.

A public Galaxy instance operated by Penn State University [32]-allows thousands of users to perform hundreds of thousands of analyses each month. This is a great solution for biologists analyzing small genomes, but the free public resource has data transfer and compute usage limits and hence is not suitable for large datasets. A CloudMan framework helps researchers run their own Galaxy server on a cloud infrastructure [33]. However, CloudMan still requires users to

understand the operating complexities of cloud computing, an expertise that most researchers lack. Although Galaxy is easy to use, it has data upload, storage and data manipulation bottlenecks, especially for large datasets. It can analyze only sample at a time, and does not take complete advantage of the elastic cloud compute capabilities (Supplementary File 1a and 1b). This limitation of Galaxy is due to its dependence on a single shared file system. When processing large datasets across distributed compute resources, this limitation represents a significant bottleneck [22].

1.3. Motivation

This paper presents a case study for using a cloud based computational environment for the processing and analysis of terabyte scale NGS data. The paper is designed to provide guidance to the users of NGS analysis software on how to address the scalability and reproducibility issues with the existing NGS pipelines when dealing with very large volumes of translational research data.

Analyzing whole genome, exome, or transcriptome sequencing data for a large number of human subjects samples requires the ability to transfer data from multiple samples into the analysis system (batch processing) and run them simultaneously (parallel processing) so as to complete the analysis in a few hours as opposed to days or weeks on a compute-intensive resource that could scale elastically (i.e., increasing and decreasing compute capacity in response to changing demand). The Globus Genomics system has these necessary features designed for AWS, and is the focus of this case study.

This case study covers an Amazon cloud based data management software solution for next generation sequencing using the Globus Genomics architecture, which extends the existing Galaxy workflow system to overcome the barrier of scalability. We present three NGS workflows to illustrate the data management and sharing capabilities of the Globus Genomics system, and the novel cloud scheduling architecture that can scale analyses elastically across a dynamic pool of cloud nodes. The NGS workflows involve medium to large scale genomics data presented through the Globus Genomics architecture; providing a fast and scalable solution for pre-processing, analysis, and sharing of large NGS data sets typical for translational genomics projects.

The Globus Genomics system was developed at the Computation Institute, University of Chicago.The Innovation Center for Biomedical informatics (ICBI) at Georgetown University has collaborated with the Globus Genomics team on a pilot project to develop and test several NGS workflows and has summarized our experiences in this paper.

2. Methods

2.1. The globus genomics system overview

The Globus Genomics system is a data management and analysis platform built on top of the Galaxy platform to take advantage of Galaxy's best features, and overcome Galaxy's data transfer, storage and data manipulation bottlenecks and limitations. It also provides additional features such as faster computation times, advanced data security, and support and maintenance of the system. It is offered as a Software as a service (SaaS) that eliminates the need to install and maintain the software, and allows users to run HPC workflows on the cloud through graphical interfaces; so users don't have to worry about any operating complexities [22,34]. By leveraging Galaxy, which is an existing, functional platform with multiple users in the translational research community, the Globus Genomics system maximizes the use of existing capabilities while adding multiple new features that will enable a wider community use, not just for NGS analysis but all other types of datasets as well. Fig. 1 shows a summary of architecture diagram of the system.

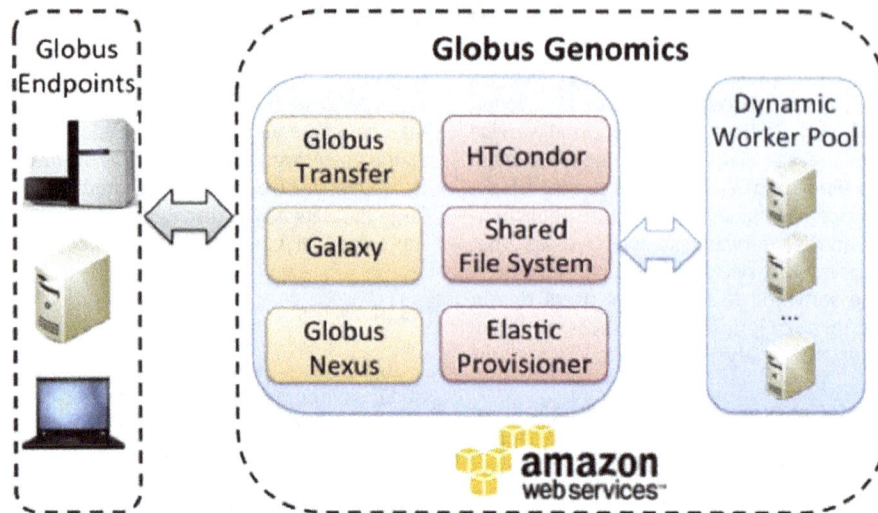

Fig. 1. Architecture of the Globus Genomics system. The orange colored components indicate the three distinct components of the system (at a higher level), and the pink colored components are additional features added by the Globus Genomics team.

2.1.1. How the globus genomics system provides faster computation times

The Globus Genomics system is implemented using Amazon's cloud computing infrastructure. One of the important features of the system is the optimization for selecting the right instance types for the analytical tools. An Amazon web services (AWS) instance type comprises varying combinations of multi-core processors, memory, storage, and networking capacity [35,36].

As part of the managed service, the Globus Genomics team creates computational profiles for various analytical tools used within the platform to ensure optimal and efficient execution on the AWS. When any new tool is added to the platform, all the critical details required for best performance of the tool, such as the number of compute cores and memory requirements, are collected and documented as a computational profile for that tool. For example, for BWA alignment tool [37], a compute instance with 16 cores and 32 GB RAM was found to provide best performance for the tool. These computational profiles are used to dynamically launch appropriate compute nodes (AWS Spot instances [38]) for a given analytical tool thus making sure the node can run the tool efficiently and within the best possible execution time.

The system takes advantage of elastic scaling of compute clusters using Amazon (Elastic Compute Cloud) EC2 [25]. Elastic scaling refers to the automatic scaling up or down of compute resources based on demand and pre-defined conditions to maximize performance, and minimize costs [39]. The Globus Genomics system provides parallelism at the workflow level, such that multiple workflows can be submitted in parallel, and new compute resources are added to the pool on demand. It also allows tools to use multi-threaded parallelism by launching the appropriate multi-core nodes as per the profile for that tool. The system uses HTCondor [26], a queue based scheduler for efficient scheduling of these pipelines over many processors and can run multiple tasks simultaneously for faster computation [34,40].

2.1.2. How the globus genomics system provides improved data transfer capabilities

Efficient and reliable data transfer is a critical feature in handling large volumes of sequence data. In addition to data transfer, we need robust authentication and authorization mechanisms in place to ensure data security. In order to address these requirements, the Globus Genomics system is integrated with Globus Transfer [41] and Globus Nexus [42] services for transfer and identity and group management capabilities.

Globus Transfer is a service that provides high-performance and secure data transfer between endpoints. An "endpoint" refers to the point where data transfer occurs to and from the Globus Genomics system, and can be a local desktop, data center, external hard drive, or Amazon storage buckets (Amazon S3). Globus Transfer provides managed transfer capabilities (users don't have to wait and manage the transfers and the service provides automated fault recovery), tuning parameters to maximize bandwidth, managing security configurations, and notifications service for error and success [23]. In addition to the transfers, it also provides sharing capability to share data in place without the overhead of moving data to the cloud. Within the Globus Genomics system, the Globus Transfer service has been integrated with Galaxy using the REpresentational State Transfer Application Programming Interface (REST API). This enables users to perform large-scale data transfers between remote source endpoints and the Amazon cloud where Galaxy is hosted.

The Globus Genomics system leverages the Globus Nexus' identity and group management services. Globus Nexus integration handles the authentication operations ensuring secure access to data. It provides Single Sign On (SSO) across the entire infrastructure and when transferring data to/from other endpoints thus allowing Globus Genomics users to sign in using their preferred identity. Globus Genomics also uses the groups within Globus Nexus to control access to a particular project's instance or to limit access to data, applications and workflows.

User authentication in the Globus Genomics system follows the typical OAuth2 workflow where by a user is redirected to authenticate using Globus Nexus (where they can use their preferred identity provider), and then the user is redirected back to the Globus Genomics instance with a limited time access token which is mapped to the Galaxy session and the Globus username. Thus users don't have to create new account with the Galaxy component and their Globus username is used across various components of the system (Transfer and Galaxy). This mapped information is used by Globus transfer service to perform data transfer on the user's behalf.

Globus Transfer leverages Globus GridFTP [43] an open source, standards-based [44] technology for reliable, high performance, secure data transfer; and its superiority over other technologies has been well-established [45–47]. Supplementary File 2 shows a performance comparison of a number of data transfer technologies done by the Globus Genomics team.

These Globus platform services are used by many large computing facilities including XSEDE[48], KBase [49], and other national centers including Semel Institute at UCLA,NYU Langone Medical Center, STAR Experiment at Brookhaven National Lab, University of Colorado, and NERSC (National Energy Research Scientific Computing Center) [50]. The 1000 Genomes project [51,52] and EBI's European Nucleotide Archive [53] now offer data download options using the Globus Transfer system. As of September 2014, there are about 25,000 Globus Platform users that have transferred about 1 billion files which is about 60PBs of data.

2.1.3. Additional features – batch submission

For NGS applications for translational research, it becomes a necessity to be able to process batches of samples together. If the computational infrastructure, storage and data transfer capabilities are not powerful and fast enough, it may take many weeks or months to process NGS data. The Globus Genomics team has implemented a batch submission capability that allows users to submit large batches of samples for analysis in parallel.

Called the "batch submit" workflow, it has been implemented as a Galaxy tool within the Globus Genomics system and leverages Galaxy APIs to submit batches of input sequences. Users are required to complete a tab-delimited file template file for each analytical pipeline, where rows represent the different samples to be submitted and columns represent the parameters to be set at run-time. When "batch submit" is submitted, the desired workflow is executed on each sample in parallel. Using the computational profile, each tool in the workflow is optimized to run in the best available compute node (i.e. compute intensive jobs can be submitted to a multiple core node and memory intensive jobs can be executed on high RAM nodes). Thus, multiple samples can use multiple core nodes in parallel to efficiently execute the analysis. The tool also takes advantage of Galaxy's workflow tracking system, and once the batch is submitted successfully, users can track the analysis of each sample separately in its own history within Galaxy.

Another important feature of batch submission is that the data transfers can also be included as part of the workflows. Thus, there is no need to pre-stage the data and each run in the batch can transfer its own input and output data to and from a remote endpoint using Globus Transfer.

This combination of on-demand cloud computing resources and batch submission capabilities makes the Globus Genomics system a powerful platform for NGS data analysis at scale.

2.1.4. Maintenance and support

The Globus Genomics team has adopted a Software-As-A-Service SaaS [54] delivery model so that researchers can access sophisticated analysis functionality without requiring any software to be installed locally. All interaction with the software occurs through web browsers and APIs. This centrally deployed software is updated, operated and supported, a service provided by the Globus Genomics team.

2.1.5. Taking advantage of the galaxy platform for NGS analysis

The Globus Genomics system not only uses Galaxy's workflow and tracking system, but also its pipeline design tool where new pipelines can be designed by end users and deployed on the infrastructure. The Galaxy tool shed has a comprehensive collection of tools to be able to create a wide variety of workflows.

Upon request by a user, the Globus Genomics team can add tools that are not present in Galaxy's tool shed, so the user can take advantage of the latest tools without waiting for a new release of Galaxy. So where necessary, custom pipelines can be developed and deployed for scientists. Even though there is flexibility in creating one's own workflows, there is convenience and time saving in reusing already established public workflows.

ICBI has created and provided three ready-to-use common NGS workflows for a convenient and hassle free experience for the user without having to spend time creating workflows. These computational

pipelines are widely used best practices for whole genome, whole exome and whole transcriptome data. Some well-known tools used in the best practices include Tophat [55], Cufflinks [56], RSEM[57], GATK[58], Samtools [59], and others; many of which have been reviewed [60,61]. These standard workflows include data transfer of raw sequencing files into the system, alignment to genome, variant calling and other steps. The processed output files are variant calls or gene/isoform expression data that can be easily exported from the system and used for biological interpretation and drive hypothesis generation for personalized medicine research.

These workflows have been made public, and can be imported and shared within the Globus Genomics system. To demonstrate usability and efficiency, we ran these workflows on publicly available datasets, evaluated their performance and have made the results public.

2.2. NGS analysis using the globus genomics system – a case study

For a typical translational genomics project, DNA or mRNA extracted from multiple samples of blood/tissue is subjected to library preparation. The libraries will then undergo, for example, Illumina HiSeq sequencing, which outputs raw data in the form of fastq files. After an investigator obtains the raw sequencing files from the vendor or core lab, a number of processing steps are needed to get meaningful results for biological interpretation.

First, the user would have to manage the large amount of data that would arrive from the sequencing center via hard drives, FTP, or other means, which is a nontrivial task. Secondly, the user would have to determine the processing steps, tools, and the appropriate analysis workflow for a given data type. Even knowledgeable users who are familiar with Unix or Python would have to find a local cluster or a high performance computing environment that could handle such large data, install the tools required, and run the analysis. Depending on the sample sizes and computational power of a local machine, this process would take anywhere from a few days to weeks. And this does not include the time required to identify the appropriate set of tools, install the tools, write the necessary scripts to execute the target workflow and secure the level of resources needed for the eventual analysis. Both a novice or knowledgeable user may not want to bother with these implementation details for translational genomics research; a solution such as the Globus Genomics system can save significant time and cost.

In this case study, we ran the three readymade ICBI workflows for Whole exome sequencing (WES) data (b) Whole genome sequencing (WGS) data and (c) mRNA sequencing (RNA-seq) data on the Globus Genomics system on publicly available datasets, and evaluated their performance (cost, time and CPU). Fig. 2 shows what is required of the user to run one of the ready-madeNGS workflows on the Globus Genomics system. Detailed steps are shown in Supplementary File 3a.

The three analytical pipelines are: (a) Whole exome sequencing (WES) workflow (b) Whole genome sequencing (WGS) workflow and (c) mRNA sequencing (RNA-seq) workflow. These workflows are currently designed for Illumina HiSeq platforms. We are currently in the process of creating workflows for other platforms and other NGS data types.

2.2.1. Whole Exome Wequencing (WES) and Whole Genome Sequencing (WGS) workflow

The workflow for pre-processing of WES and WGS is the same, the difference being that WES only sequences the exome region, while in WGS; the entire genome is sequenced as seen in the difference in size and content of the fastq files. (Fig. 3a shows a schematic block diagram of the workflow and Fig. 3b shows the same workflow created in the Globus Genomics system).

The fastq files are filtered based on quality using Sickle [62]. Sickle accepts gzipped fastq files as input and works effectively on paired end data for both WES and WGS data. The filtered output is aligned to

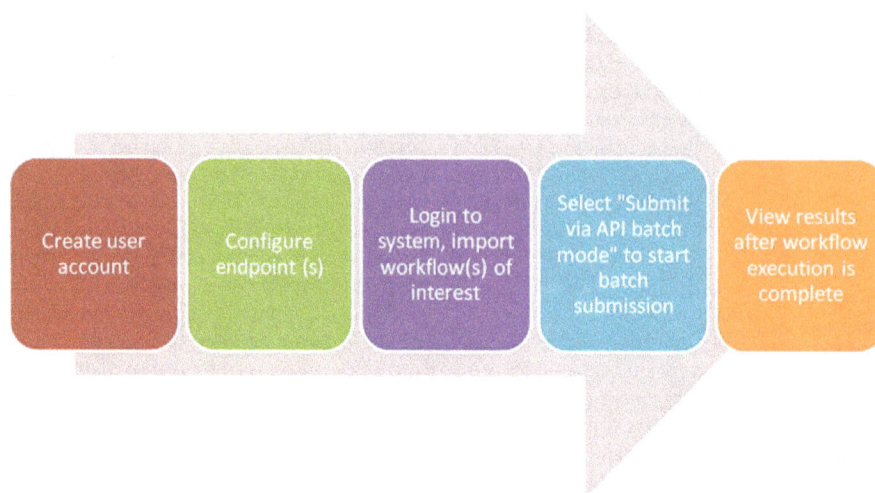

Fig. 2. How to run a ready-made NGS workflow in the Globus Genomics system.

a reference human genome using Bowtie2 [63], an ultrafast, memory efficient short read aligner to create alignment files in BAM format. The BAM files are re-ordered and read groups are added using Picard [64]. PCR duplicates removed using Samtools [59]. Variants are called using Genome Analysis Toolkit (GATK) [58]. VCF-tools[65] are used to separate the SNPs from the indels and produce two variant call format (VCF) files for each sample. These VCF files are small in size (MB range) and can be easily exported from the Globus system. Once exported, the VCF files can be used for further case–control association tests that provide statistically significant variants, which can then be filtered to obtain a short list of non-synonymous, potentially deleterious markers. These variants can then be mapped to genomic regions and further aggregated at the levels of gene, pathways, and biological processes relevant to disease outcome.

2.2.2. Whole transcriptome sequencing (RNA-seq) workflow

For this workflow in the Globus Genomics system, RNAseq fastq files are pre-processed for quality checks using Sickle, and input to RSEM[57] a software package that uses Bowtie for alignment and estimates gene and isoform expression levels. Fig. 4a shows a schematic block diagram of the workflow and Fig. 4b shows the workflow in Globus Genomics. Variants are extracted from this data using Picard, GATK and VCF-tools as mentioned above in the form of VCF files. The advantage of variants extracted from RNA-seq data is that these have already undergone transcription and is a validation of variants from WGS data. The output of the workflow are the gene and isoform expression data and the VCF files which can be exported from the Globus system and further analyzed at the level of gene, pathways and biological processes relevant to disease outcome.

For the WES, WGS and RNA-seq workflows created for this case study, the downstream analyses steps have not been included; as the filtering and settings for downstream analysis may vary depending on the biological question in mind. Most of the downstream analysis steps can be added and executed by the user through the Galaxy interface of the Globus Genomics system.

3. Results

3.1. Performance evaluation

3.1.1. WES workflow performance

We ran the WES pipeline on a batch of 78 samples from a lung cancer study obtained from the European Bioinformatics Institute's Sequencing Read Archive (SRA) [66], from which we downloaded the fastq files.

First, we executed the workflow on a single sample of average input size (6.5 GB compressed paired-end fastq files) to set the baseline, which completed in 4 h. Next, we executed the workflow on all samples, which ran in parallel and completed analysis in 40 h generating between 20–120 GB of data per sample depending on the size of the fastq files. The actual execution time for the batch was about 10 times higher than running a single sample of average input size due to the I/O (disk usage for input/output files) bottlenecks. This bottleneck is introduced by the Galaxy component that requires a shared file system wherein all the jobs from multiple workflows that are run simultaneously need to read the input data from and write the intermediate outputs to the same shared file system [22]. Due to this high I/O nature of the analysis, the Globus Genomics team was able to determine that the servers being used were not optimal for this type of analysis. They switched to a more I/O intensive node (e.g. h1.4x large) and were able to reduce the total execution time for all 78 samples to about 12 h. The I/O intensive node uses provisioned I/O on the Elastic Block Storage (EBS) [67] when building the shared file system, which significantly improved the read/write performance. Each sample was analyzed in an average time of 10 h, which was closer to baseline. The input data totaled to about 400 GB, and the amount of data generated from running the pipeline was 2.7 TB. The total data handled by the system for this dataset was about 3.1 TB.

Fig. 5 shows summary of cost, time and total data generated for the analysis of 78 lung cancer samples through the exome-seq workflow executed on a single multi-core Amazon instance (non-optimal run). Fig. 6 shows summary of cost, time and total data generated for the analysis of 78 lung cancer samples through the exome-seq workflow (optimal run). It shows improvement in CPU and execution time as compared to the non-optimal run. For both figures, we can see that larger input files (fastq files) generate larger intermediate and output sizes, which is typical for NGS analysis.

Supplementary Files 4 and 5 show run times for each sample in the batch job run (non I/O optimized and I/O optimized). It shows a large amount of data generated by intermediate files.

3.1.2. WGS workflow performance

To demonstrate this workflow, we ran the WGS workflow on a human breast cancer cell line dataset. We were unable to obtain fastq files for medium-large sized public WGS dataset on Illumina platform and hence chose this small dataset. This fastq file was of 80 GB size, it took 12 h to produce variants (VCF) files in a compute intensive cluster instance (cr1.8x large). Details of run time for this sample is shown in Supplementary File 6.

a

b

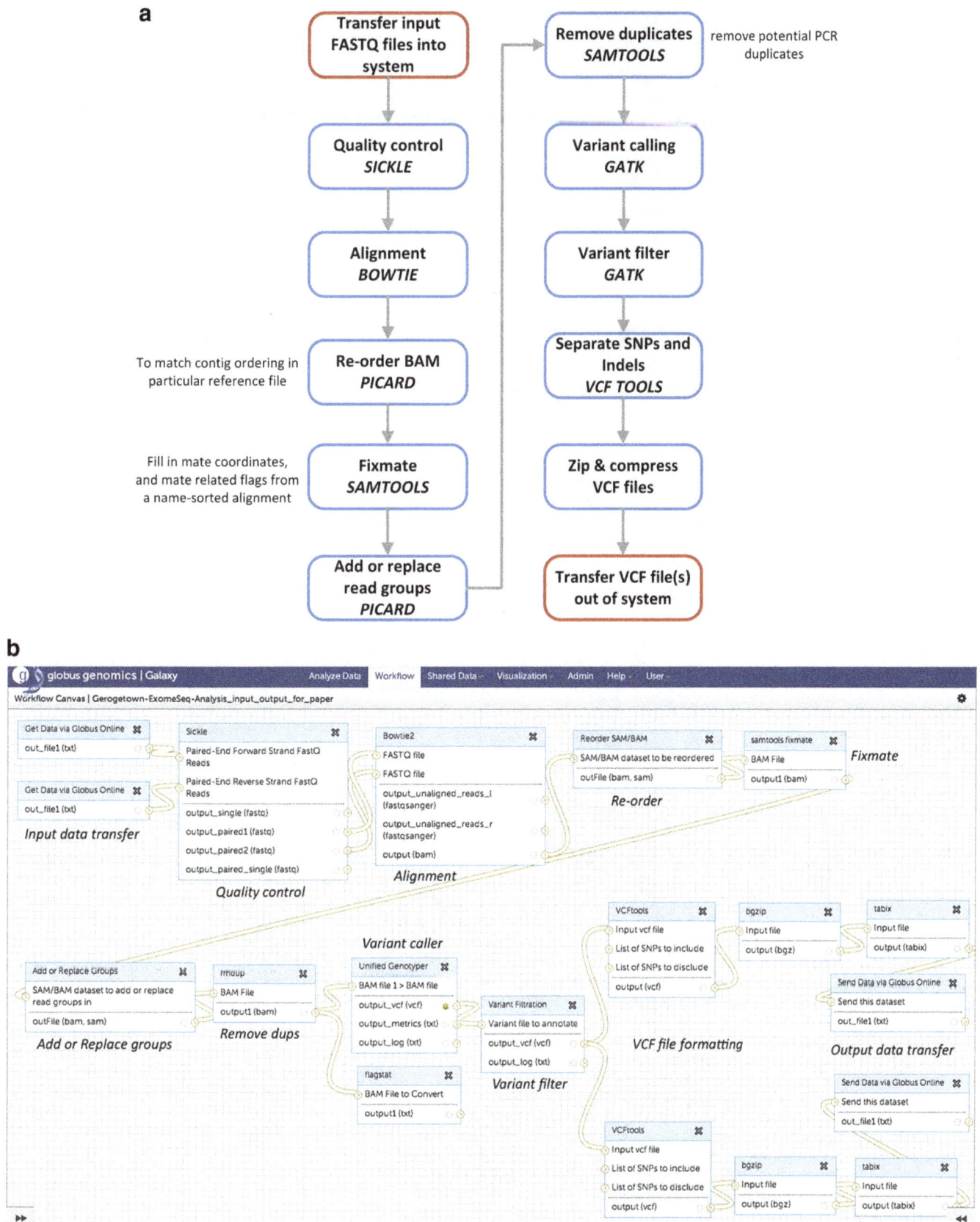

Fig. 3. a. Schematic diagram of the whole Genome and whole exome analysis workflow. b. Whole genome and exome analysis workflow inside the Globus Genomics system.

3.1.3. RNA-seq workflow performance

We ran this workflow on The Cancer Genome Atlas' (TCGA's) ovarian cancer samples. We downloaded raw files from the Cancer Genomic Hub (CGhub) archive [68] and extracted fastq files from the raw files. This study has 25 samples in all, and we applied the workflow to 21 samples as 4 samples did not pass quality check. Each sample ran

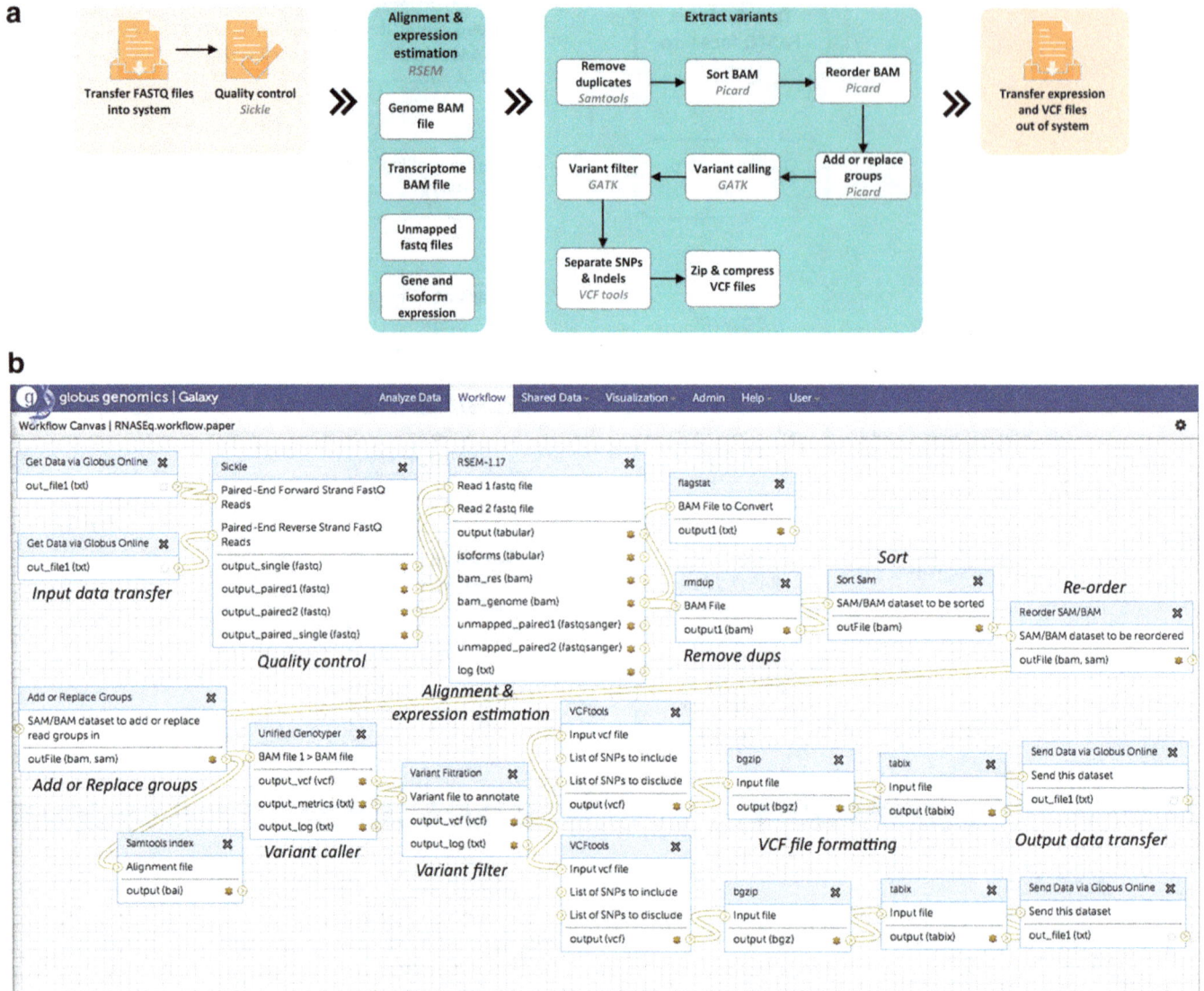

Fig. 4. a. Schematic diagram of the whole transcriptome (RNA-seq) analysis workflow. b. Whole transcriptome (RNA-seq) analysis workflow inside the Globus Genomics system.

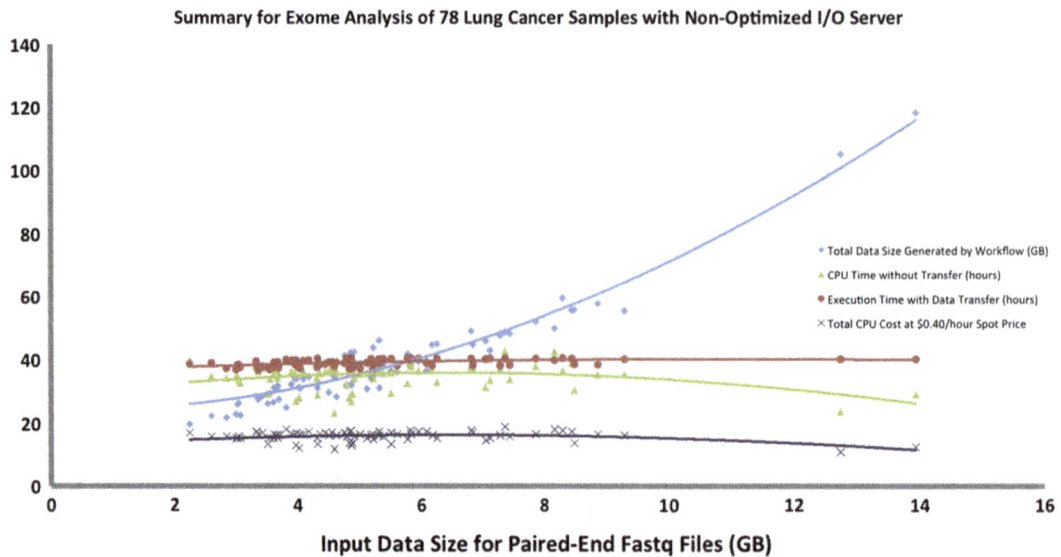

Fig. 5. Summary for analysis of 78 lung cancer samples through the exome-seq workflow. Execution time was not optimal due to the high nature of I/O in the workflow. "Spot Price" as mentioned in the figure key refers to the price of the AWS spot instance [38].

Summary for Exome Analysis of 78 Lung Cancer Samples with Optimized I/O Server

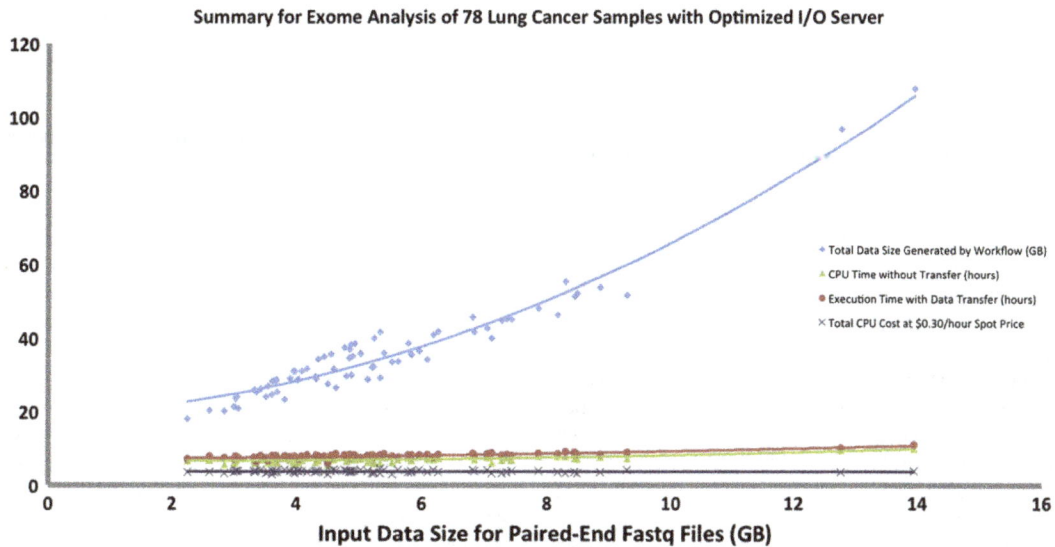

Fig. 6. Summary of the 78 lung cancer samples in an I/O optimized server.
"Spot price" refers to the price of the AWS spot instance [38].

in parallel based on the settings in the computational profiles taking about 20–22 h for each sample to generate expression files and variants, generating about 150 GB of data depending on size of fastq files. The intermediate files contribute the most to the overall size of data. The 21 samples were completed within 24 h from the time the first sample was submitted to the time the last sample completed. Overall, the input data totaled to about 480 GB, and the amount of data generated from running the pipeline is 2.9 TB. The total data the system handled for this dataset was about 3.2 TB.

Fig. 7 shows a summary of the RNA-seq analysis for the 21 samples. The Amazon spot instance [38] used for this run (cr1.8x large instance) cost $0.34 per hour. Supplementary file 7 shows run time details for each sample in the batch run.

The graphs in Figs. 5, 6, and 7 show a linear relationship between the input size and data generated by the workflow, while for CPU time, workflow execution time with data transfer, and cost the relationship is non-linear. This is mostly due to heavy I/O utilization especially

when multiple samples are written to the same disk space. As smaller samples get completed, the larger samples have less I/O issues and thus can be executed faster. This issue can be resolved by using a more I/O intensive node as previously explained.

4. Discussion

In a typical translational research setting a core genomics or a bioinformatics laboratory is facing the challenge of processing and analyzing a massive volume of next generation sequencing data in studies amounting hundreds of DNA or RNA samples. ICBI in collaboration with the Globus Genomics team has conducted a case study aimed at testing a data management solution by running fast, standard, scalable and reproducible bioinformatics pipelines on an enhanced Galaxy platform called the Globus Genomics system built on the Amazon cloud.

Summary for RNA-Seq Analysis of 21 Samples from TCGA Database

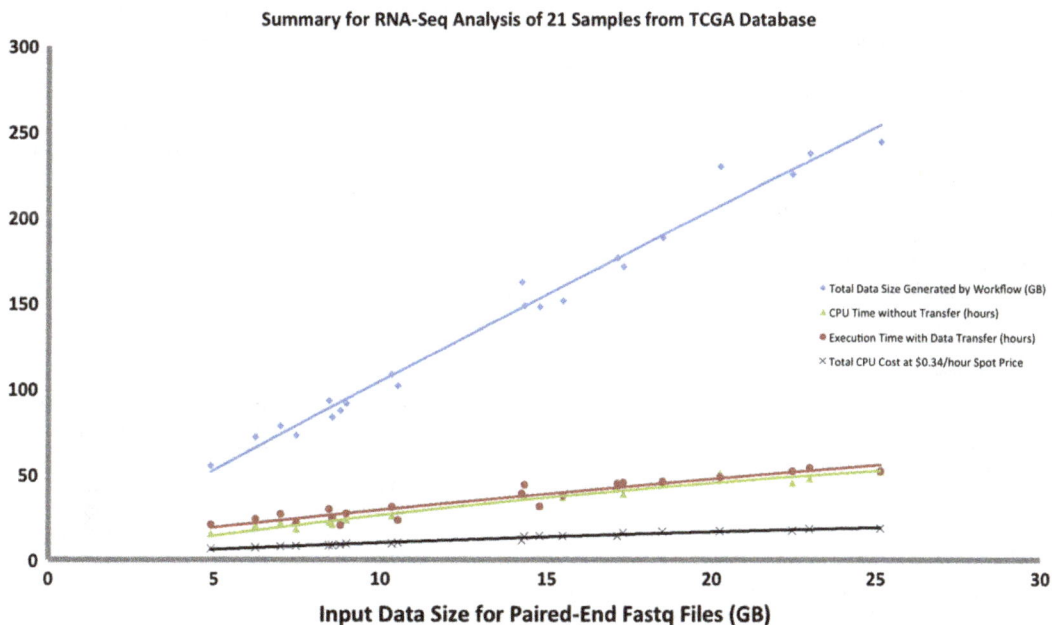

Fig. 7. Summary for RNA-Seq Analysis of 21 TCGA samples of varying input sizes.
"Spot price" refers to the price of the AWS spot instance [38].

4.1. User experience from case study

After running the case study at Georgetown ICBI, we found pros and cons with the Globus Genomics system. The main advantage was that the system was user friendly — its user-interface is suitable for scientists who don't have programming experience. The system is especially suited for genomics cores that need to process medium to large volumes of NGS data in a short amount of time, and have to share the processed results with their respective clients. Other advantages of the system include (a) it was convenient to use since it's available on the web, we did not have to worry about updates and maintenance of system, (b) the upload of the file template into the system and batch execution for the analysis of 21 whole transcriptome files and 78 whole exome samples was not difficult, (c) we were able to track the progress in processing of each sample. The workflows could be run overnight without any supervision. Most samples completed processing overnight, which was very convenient as compared to non-cloud based systems.

We found the system to have bottlenecks as well. We had first tested the RNAseq workflow and then the Exome seq workflow. So when we scaled the analysis from 21 samples to 78 samples, we encountered I/O related issues mentioned previously. We learned that the Globus Genomics I/O becomes a bottleneck when multiple concurrent applications start accessing the same file system thus deteriorating the performance. As demonstrated in the results, using provisioned I/O on the EBS[67] when building the shared file system significantly improves the performance. While provisioned I/O can help scale the number of parallel jobs to a couple of hundred, there is a natural upper limit in the number of concurrent jobs that can be handled by the shared file system. The Globus Genomics team is currently working on better load balancing techniques and is working closely with engineers from AWS for larger scalability.

Researchers that have cited the Globus Genomics system include: the Cox lab [69] and Olopade lab [70] at University of Chicago, and the Dobyns lab at Seattle Children's Research Institute [71]. Other users of the system include Kansas University Medical Center [72], Inova Translational Medicine Institute, and the Genome Sciences Institute at Boston University [73]. As of September 2014, there are about 20 institutions/research groups actively using the Globus Genomics platform.

4.2. Economics of running the analysis pipelines on the cloud

The Globus Genomics team has adopted a Software-As-A-Service SaaS [54] delivery model so that researchers can access sophisticated analysis functionality without requiring any software to be installed locally. Although this model offers cost savings over traditional approaches with multiple local software installations, some costs remain including running the service on Amazon Web Services (AWS), as well as providing any ongoing technical support.

To recover these types of costs, the Globus Genomics team has adopted a subscription model, whereby users are charged for components of usage such as cloud compute and cloud storage as well

as operational and technical support. Fortunately, with the continuous reduction in costs of cloud resources driven by economies of scale and gains in efficiency, public cloud infrastructure becomes increasingly cost effective and most importantly, provides the flexibility of on-demand resource scaling. Advantages for users include lower cost of development as only a single platform is supported, accelerated feature delivery, transparent and frequent software updates, subscription based licensing, pay-as-you-go usage, collaborative and social integration (the option to publish and rate the workflows, so that other experts or users in the field can also rate these published workflows thus leading to best practices), and intuitive and easy to use interfaces for users.

Table 1 shows actual costs for executing five workflows commonly used in NGS analysis using the Globus Genomics system. To minimize compute costs, the Globus Genomics team created computational profiles of the tools (as described earlier in the System Overview section) used in the analysis workflows and matched them with appropriate Amazon resources to achieve the best price/performance balance during workflow execution. The team also used spot instances [38] to scale-up to the required compute levels with the lowest cost resources.

The Globus Genomics team accounts for AWS storage costs mentioned in Table 1. This allows storage of the computation results for a month, and also accounts for outbound I/O costs from moving the intermediate and final results from Amazon to local resources for downstream analysis or local archiving. While AWS charges for outbound I/O, users can transfer these intermediate and final results of analysis to their own S3 buckets or other AWS storage with no I/O costs, though they may have to pay for the actual storage itself.

At the end, 21 RNA seq samples ran in parallel (average input size 13.5 GB each paired-end set compressed) based on the settings in the computational profiles in about 20–22 h. The total data handled by the system for this dataset was about 3.2 TB. 78 WES samples (average input size 5.5 GB each paired-end set compressed) completed execution on about 12 h. The total data handled by the system for this dataset was about 3.1 TB. One WGS cell line sample of 80 GB size completed execution in about 12 h. This will hopefully allow users to roughly predict the time required to complete the analysis given the workflow and size of data.

In summary, the Globus Genomics system achieves a high degree of end-to-end automation that encompasses every stage of the data analysis lifecycle from initial data retrieval (from remote sequencing center or database by the Globus file transfer system) to on-demand resource acquisition (on Amazon EC2); specification, configuration, and reuse of multi-step processing pipelines (via Galaxy); and efficient scheduling of these pipelines over many processors (via the HTCondor scheduler [74]). The system allows researchers to perform rapid analysis of terabyte scale NGS datasets using just a web browser in a fully automated manner, with no software installation.

4.3. Conclusion and future work

The Globus Genomics architecture extends the existing Galaxy workflow system adding not only superior data management

Table 1
Sample workflow run costs including compute, temporal storage and outbound I/O[a].

Workflow	Input data size	Storage size reqs (GBs)	Amazon storage costs	Compute requirement (node hours)	Amazon compute costs	Data download (GBs)	Amazon outbound I/O costs	Total amazon costs
DNA copy number	.070 GB	0.03	<$0.01	0.15	$0.05	0.003	<$0.01	$0.05
microRNA Seq	0.3 GB	1	<$0.01	0.5	$0.17	0.1	$0.01	$0.18
RNA Seq	10 GB (~5 Gbp)	70	$0.12	20	$6.80	7	$0.70	$7.62
WES	6 GB (~5 Gbp)	50	$0.08	6	$2.04	5	$0.50	$2.62
WGS	72 GB (~35 Gbp)	320	$0.53	30	$10.20	32	$3.20	$13.93

[a] The analysis presented in Table 1 was carried out under the following assumptions: (a) Input data are compressed in GZ format, paired-end Illumina reads (b) RNA-seq analysis includes variant analysis as well: Sickle QC, RSEM (singleton and paired), sort, rmdup, fixmate, picard reorder, picard add or replace groups, GATK Unified Genotyper, GATK recalibration, and GATK variant filtering (c) WES analysis includes: BWA, sort, rmdup, fixmate, picard reorder, picard add or replace groups, GATK Unified Genotyper, GATK recalibration, and GATK variant filtering (d) WGS analysis includes: Bowtie2, sort, rmdup, fixmate, picard reorder, picard add or replace groups, and GATK Unified Genotyper (e) Reference genome used for all analyses is hg19.

capabilities but also a novel cloud scheduling architecture that can scale analyses elastically across a dynamic pool of cloud nodes [22].

We present three NGS workflows for medium to large scale genomic data in a Galaxy based system built on the cloud that executes these workflows across high performance compute systems. We believe that Globus Genomics is a valuable system that provides a hassle free and fast solution for pre-processing and analysis of large NGS data sets typical for translational genomics projects.

We hope to expand this system to support other NGS workflows and platforms in the future. The Globus Genomics team is also developing new features to enable cataloging of dynamic collections of data and metadata including provenance metadata. Another future direction is to provide sophisticated search capabilities to discover and analyze datasets based on user-defined and automatically extracted metadata.

Funding

This work was supported by in part by the NHLBI grant for Globus Genomics: The Cardiovascular Research Grid [R24HL085343] and by the U.S. Department of Energy under contract [DE-AC02-06CH11357]. We are grateful to the generous support from Amazon, Inc., for Amazon Web Services credits that facilitated early experiments.

Acknowledgments

We thank Globus Genomics users for their invaluable feedback and contributions. We thank Dr. Laura Sheahan for editing the manuscript.

References

[1] Schuster SC. Next-generation sequencing transforms today's biology. Nat Methods 2008;5:16–8.
[2] Koboldt DC, Steinberg KM, Larson DE, Wilson RK, Mardis ER. The next-generationsequencing revolution and its impact on genomics. Cell 2013;155:27–38.
[3] Park JY, Kricka LJ, Fortina P. Next-generation sequencing in the clinic. Nat Biotechnol 2013;31:990–2.
[4] Baker M. Next-generation sequencing: adjusting to data overload. Nat Methods 2010;7:495–9.
[5] Church P, Goscinski A. A survey of cloud-based service computing solutions for mammalian genomics. IEEE Trans Serv Comput 2014;1–1.
[6] Amazon web services. http://aws.amazon.com/; June 30 2014.
[7] Google cloud platform. https://cloud.google.com/; Oct. 10 2014.
[8] Microsoft Azure. https://azure.microsoft.com/en-us/; Oct. 10 2014.
[9] Stein LD. The case for cloud computing in genome informatics. Genome Biol 2010; 11:207.
[10] Answers to genome analysis may be in the clouds. http://www.genome.gov/27538886; Oct. 21 2014.
[11] AWS case study — Harvard Medical School. http://aws.amazon.com/solutions/case-studies/harvard/; Oct. 8 2014.
[12] AWS use case — Genomic Medicine Institute, Seoul National University College of Medicine, Korea. http://aws.amazon.com/solutions/case-studies/gmi/; Oct. 8 2014.
[13] AWS case study — Icahn School of Medicine at Mount Sinai. http://aws.amazon.com/solutions/case-studies/mt-sinai/; Oct. 8 2014.
[14] AWS case study: New York University Langone Medical Center. http://aws.amazon.com/solutions/case-studies/new-york-university/; Oct. 8 2014.
[15] AWS case study: Penn State Biological Engineering Department. http://aws.amazon.com/solutions/case-studies/penn-state/; Oct. 8 2014.
[16] AWS case study: University of California Berkeley AMP Lab's Genomics Research Project. http://aws.amazon.com/solutions/case-studies/university-of-california-berkeley-amp-lab-genomics-research/; Oct. 8 2014.
[17] Dudley JT, Pouliot Y, Chen R, Morgan AA, Butte AJ. Translational bioinformatics in the cloud: an affordable alternative. Genome Med 2010;2:51.
[18] Sulakhe D, Rodriguez A, Prozorovsky N, Kavthekar N, Madduri R, Parikh A, et al. Distributed tools deployment and management for multiple galaxy instances in globus, genomics; 2013 106–11.
[19] Cloud computing service models. http://en.wikipedia.org/wiki/Cloud_computing#Service_models; Oct. 10 2014.
[20] Reid JG, Carroll A, Veeraraghavan N, Dahdouli M, Sundquist A, English A, et al. Launching genomics into the cloud: deployment of Mercury, a next generation sequence analysis pipeline. BMC Bioinforma 2014;15:30.
[21] DNAnexus. https://www.dnanexus.com/; June 30 2014.
[22] Madduri RK, Sulakhe D, Lacinski L, Liu B, Rodriguez A, Chard K, et al. Experiences building Globus Genomics: a next-generation sequencing analysis service using Galaxy, Globus, and Amazon Web Services. Concurrency and Computation: Practice and Experience: n/a-n/a; 2014.
[23] Partek Flow for NGS Analysis. http://www.partek.com/; June 30 2014.
[24] CLC Genomics Workbench. http://www.clcbio.com/products/clc-genomics-workbench/; June 30 2014.
[25] DNASTAR Lasergene Genomics Suite. http://www.dnastar.com/t-nextgenhome.aspx; June 30 2014.
[26] Maverix Biomics. http://www.maverixbio.com/; June 30 2014.
[27] Seven Bridges Genomics. https://www.sbgenomics.com/; June 30 2014.
[28] Golden Helix SNP & Variation Suite 8. http://www.goldenhelix.com/SNP_Variation/index.html; June 30 2014.
[29] Blankenberg D, Gordon A, Von Kuster G, Coraor N, Taylor J, Nekrutenko A, et al. Manipulation of FASTQ data with Galaxy. Bioinformatics 2010;26:1783–5.
[30] Blankenberg D, Hillman-Jackson J. Analysis of next-generation sequencing data using Galaxy. Methods Mol Biol 2014;1150:21–43.
[31] Galaxy tool shed. https://wiki.galaxyproject.org/Toolshed; June 30 2014.
[32] Galaxy. https://usegalaxy.org/; June 30 2014.
[33] Afgan E, Baker D, Coraor N, Chapman B, Nekrutenko A, et al. Galaxy CloudMan: delivering cloud compute clusters. BMC Bioinforma 2010;11(Suppl. 12):S4.
[34] Madduri Ravi K, Dave Paul, Sulakhe Dinnanath, Lacinski Lukasz, Liu Bo, Foster Ian T. Experiences in building a next-generation sequencing analysis service using galaxy, globus online and Amazon web service. New York: ACM; 2013.
[35] Amazon instance types. http://aws.amazon.com/ec2/instance-types/; June 30 2014.
[36] Marx V. Genomics in the clouds. Nat Methods 2013;10:941–5.
[37] Li H, Durbin R. Fast and accurate short read alignment with Burrows–Wheeler transform. Bioinformatics 2009;25:1754–60.
[38] Amazon spot instances http://aws.amazon.com/ec2/purchasing-options/spot-instances/
[39] Autoscaling.
[40] Liu B, Sotomayor B, Madduri R, Chard K, Foster I. Deploying bioinformatics workflows on clouds with Galaxy and Globus Provision. High Performance Computing, Networking, Storage and Analysis (SCC), 2012 SC Companion; 2012 1087–95.
[41] Allen Bryce, Bresnahan John, Childers Lisa, Foster Ian, Kandaswamy Gopi, Kettimuthu Raj, et al. Software as a service for data scientists. Communications of the ACM: ACM; 2012.
[42] Ananthakrishnan RB J, Chard K, Foster I, Howe T, Lidman M, Tuecke S. Globus Nexus: an identity, profile, and group management platform for science gateways and other collaborative science applications. IEEE; Sep. 23–27 2013.
[43] Allcock W, Bresnahan J, Kettimuthu R, Link M. The Globus striped GridFTP framework and server; 2005 [54-54].
[44] Allcock W. GridFTP: Protocol Extensions to FTP for the Grid. Global Grid Forum GFD-R-P.020; 2003.
[45] Brightwell P. High performance file transfer over IP networks. EBU Tech Rev; 2010 [(http://tech.ebu.ch/techreview): BBC http://downloads.bbc.co.uk/rd/pubs/whp/whp-pdf-files/WHP183.pdf].
[46] Mattmann CA, Kelly S, Crichton DJ, Hughes JS, Hardman S, Ramirez P, et al. A classification and evaluation of data movement technologies for the delivery of highly voluminous scientific data products. NASA/IEE Conference on Mass Storage Systems and Technologies (MST 2006). Pasadena, CA: Jet Propulsion Laboratory, National Aeronautics and Space Administration; 2006.
[47] Esposito R, PM, Tortone G, Taurino FM. Standard FTP and GridFTP protocols for international data transfer in Pamela Satellite Space Experiment; Computing in High Energy and Nuclear Physics 2003 Conference Proceedings. San Diego, California: La Jolla; 2003 24-28 March.
[48] The Extreme Science and Engineering Discovery Environment (XSEDE). https://www.xsede.org/overview; June 30 2014.
[49] KBase. http://kbase.us/about/about/; June 30 2014.
[50] Users of Globus for data movement. https://www.globus.org/case-studies; Oct. 10 2014.
[51] The 1000 Genomes project now offers their FTP site as data transfer point through the Globus Transfer service. http://www.1000genomes.org/announcements/1000-genomes-ftp-site-now-available-through-globus-online-2014-06-17; Oct. 17 2014.
[52] FAQ: Can I access 1000 genomes data with Globus Online? http://www.1000genomes.org/faq/can-i-access-1000-genomes-data-globus-online; Oct. 17 2014.
[53] Read data through Globus GridFTP; 2014. http://www.ebi.ac.uk/about/news/service-news/read-data-through-globus-gridftp.
[54] Dubey A, Wagle D. Delivering software as a service. Web Exclusive: The McKinsey Quarterly; 1-12 May 2007.
[55] Trapnell C, Pachter L, Salzberg SL. TopHat: discovering splice junctions with RNA-Seq. Bioinformatics 2009;25:1105–11.
[56] Trapnell C, Williams BA, Pertea G, Mortazavi A, Kwan G, van Baren MJ, et al. Transcript assembly and quantification by RNA-Seq reveals unannotated transcripts and isoform switching during cell differentiation. Nat Biotechnol 2010;28:511–5.
[57] Li B, Dewey CN. RSEM: accurate transcript quantification from RNA-Seq data with or without a reference genome. BMC Bioinforma 2011;12:323.
[58] McKenna A, Hanna M, Banks E, Sivachenko A, Cibulskis K, Kernytsky A, et al. The Genome Analysis Toolkit: a MapReduce framework for analyzing next-generationDNA sequencing data. Genome Res 2010;20:1297–303.
[59] Li H, Handsaker B, Wysoker A, Fennell T, Ruan J, Homer N, et al. The Sequence Alignment/Map format and SAMtools. Bioinformatics 2009;25:2078–9.
[60] Pabinger S, Dander A, Fischer M, Snajder R, Sperk M, Efremova M, et al. A survey of tools for variant analysis of next-generation genome sequencing data. Brief Bioinform 2014;15:256–78.
[61] Del Fabbro C, Scalabrin S, Morgante M, Giorgi FM. An extensive evaluation of read trimming effects on Illumina NGS data analysis. PLoS One 2013;8:e85024.
[62] Sickle — Windowed Adaptive Trimming for fastq files using quality. https://github.com/najoshi/sickle; June 30 2014.

[63] Langmead B, Salzberg SL. Fast gapped-read alignment with Bowtie 2. Nat Methods 2012;9:357–9.

[64] Picard. http://picard.sourceforge.net.

[65] Danecek P, Auton A, Abecasis G, Albers CA, Banks E, DePristo MA, et al. The variant call format and VCFtools. Bioinformatics 2011;27:2156–8.

[66] Lung Cancer Sequencing Project Exome sequencing of lung adenocarcinomas and their normal counterparts. ERP001575 http://www.ebi.ac.uk/ena/data/view/ERP001575; July 1 2014.

[67] Amazon EBS. http://aws.amazon.com/ebs/; July 1 2014.

[68] Cancer Genomics Hub. https://cghub.ucsc.edu/; June 30 2014.

[69] Trubetskoy V, Rodriguez A, Dave U, Campbell N, Crawford EL, Cook EH, et al. Consensus Genotyper for Exome Sequencing (CGES): improving the quality of exome variant genotypes. Bioinformatics 2014.

[70] Zheng TW Y, Yoshimatsu F, Lee M, Gulsuner S, Casadei S, Rodriguez A, et al. A profile of inherited predisposition to breast cancer among Nigerian women. 64th Annual Meeting of the American Society of Human Genetics. California: San Diego Convention Center (SDCC) in San Diego; 2014.

[71] Pediatric brain research laboratory uses Globus Genomics to overcome IT hurdles. https://www.globus.org/genomics/resources/case-study-dobyns-pediatric-brain-research-lab.pdf; Oct. 10 2014.

[72] Rama Raghavan DV, Fridley Brooke L. Globus Genomics: A medical center's bioinformatics core perspective. Chicago: Globus World; 2014.

[73] Globus Genomics NGS Analysis. http://www.bumc.bu.edu/gsi/next-generation-sequencing/globus-genomics/; Oct. 10 2014.

[74] Litzkow MJ, Livny M, Mutka MW. Condor — a hunter of idle workstations. 8th International Conference on Distributed Computing Systems; 1988. p. 104–11.

4

Application of the Protein Maker as a platform purification system for therapeutic antibody research and development

Geneviève Hélie [a], Marie Parat [a], Frédéric Massé [b], Cory J. Gerdts [c], Thomas P. Loisel [a], Allan Matte [a,*]

[a] Protein Purification, Human Health Therapeutics, National Research Council Canada, 6100 Royalmount Ave. Montreal, QC H4P 2R2, Canada
[b] Primary Assays, Human Health Therapeutics, National Research Council Canada, 6100 Royalmount Ave. Montreal, QC H4P 2R2, Canada
[c] Protein BioSolutions Inc., Suite 280, 401 Professional Drive, Gaithersburg, MD, 20879, USA

ARTICLE INFO

ABSTRACT

Keywords:
Parallelized protein purification
Antibody
Protein A
Protein G
Hybridoma
Process development

Within the research and development environment, higher throughput, parallelized protein purification is required for numerous activities, from small scale purification of monoclonal antibodies (mAbs) and antibody fragments for *in vitro* and *in vivo* assays to process development and optimization for manufacturing. Here, we describe specific applications and associated workflows of the Protein Maker liquid handling system utilized in both of these contexts. To meet the requirements for various *in vitro* assays, for the identification and validation of new therapeutic targets, small quantities of large numbers of purified antibodies or antibody fragments are often required. Reducing host cell proteins (HCP) levels following capture with Protein A by evaluating various wash buffers is an example of how parallelized protein purification can be leveraged to improve a process development outcome. Stability testing under various conditions of in-process intermediates, as an example, the mAb product from a clarified harvest, requires parallelized protein purification to generate concurrent samples for downstream assays. We have found that the Protein Maker can be successfully utilized for small-to-mid scale platform purification or for process development applications to generate the necessary purified protein samples. The ability to purify and buffer exchange up to 24 samples in parallel offers a significant reduction in time and cost per sample compared to serial purification using a traditional FPLC system. By combining the Protein Maker purification system with a TECAN Freedom EVO liquid handler for automated buffer exchange we have created a new, integrated platform for a variety of protein purification and process development applications.

1. Introduction

Purification of antibodies and antibody fragments are key activities in the generation of critical reagents for various *in vivo* and *in vitro* assays as part of biotherapeutic lead identification and process development. Often, it is necessary to purify large numbers of antibodies with milligram yield, relatively quickly and at minimal cost. Various strategies are available to achieve such purification outcomes, and can involve to various extents both automated and manual methods [1,2]. While parallelized purification methods yielding sub-milligram quantities of pure proteins based on packed columns, 96-well plates containing small quantities of chromatographic resins or ligands immobilized to the surfaces of membranes have been developed, there are relatively

fewer options available for generating purified quantitates of protein in the intermediate (5–100) milligram scale. A few examples of customized solutions to this problem exist, involving integration of existing purification platforms such as the ÄKTA Purifier with a CETAC autosampler [3], ÄKTA Pure [4] or liquid handling robotics [5] have been reported. Other solutions include the design and fabrication of customize robotics platform, including the Protein Expression and Purification Platform [6]. While some commercial instruments for purification of small quantities of protein have been developed, such as the QIAcube for purification of His-tagged proteins [7], there are few examples of commercial instruments that can be utilized for platform purification at milligram scale.

In the context of process development applications, various commercially available scale-down protein purification products have been developed, including Predictor plates (GE) and Robo-columns (GE and Atoll Bio). While very useful for early-stage screening of various chromatographic conditions, the maximum size of the columns possible in these platforms (600 μL bed volume) results in a considerable gap in the scale between screening and further optimization of process

Abbreviations: CHO, Chinese Hamster Ovary; DPBS, Dulbecco's phosphate buffered saline; HC, IgG heavy chain; HCP, host cell protein; IMAC, immobilized metal ion affinity chromatography; LC, IgG light chain; mAb, monoclonal antibody; OAA, one-armed antibody.
* Corresponding author.
E-mail address: allan.matte@nrc-cnrc.gc.ca (A. Matte).

conditions. Some examples of higher throughput, automated solutions to purification process development have been reported [8,9]. While automated, sequential purification of samples is possible using a chromatography system connected to an auto-sampler, this cannot be parallelized using a single instrument, thereby reducing the possible number of samples processed.

A specific instrument which has been designed around accomplishing the task of parallelized, medium scale purification is the Protein Maker system, originated by Emerald BioStructures [10] and subsequently developed and marketed by Protein BioSolutions. The Protein Maker is an automated protein purification platform designed for purification of feed volumes of various sizes, from ~10 mL to 1 L (~1 mg to 100 mg) or more utilizing up to 24 chromatography columns, each with an independent flow path. The main components of the system are (i) the syringe pumps with the associated 9-port valve, mixing syringe and sample lines, which together form the initial portion of the flow path, (ii) the column gantry, columns and associated tubing from the syringe pumps, which form the subsequent portion of the flow path and (iii) the deck, which contains up to 19 positions for SBS format plates and a dedicated waste position.

While purification of a variety of proteins from any number of sources is in principle possible with the instrument, the focus herein are examples of purification of antibodies and their fragments generated from mammalian expression systems. We have utilized the Protein Maker as a key component of a platform purification system that integrates automated buffer exchange implemented on a TECAN Freedom EVO liquid handler. This protein purification platform can be used for both parallelized, small-medium scale purification of antibodies and their fragments, as well in various process development applications.

2. Materials and methods

2.1. Antibody production

Murine IgG samples were produced in hybridoma culture in IMDM supplemented with 10% heat-inactivated FBS and mouse IL-6 by a procedure previously described [11]. For some antibodies, cultures were performed transiently in Chinese Hamster Ovary (CHO) cells as previously described [12]. Productions were harvested by centrifugation or filtration (0.22 μm or 0.45 μm) and IgG containing supernatants stored at 4 °C until purified.

2.2. Purification of mAbs and Fabs

For development of Protein Maker purification methods, protein samples were purified using 1 mL HiTrap columns (GE Life Sciences), including Protein G HP, MabSelect SuRe™ (Protein A) and Ni Sepharose Excel™ mounted on an ÄKTA Purifier 10/100 system. Chromatographic profiles were monitored at 280 nm. Columns were equilibrated in Dulbecco's Phosphate Buffered Saline Solution (DPBS, HyClone Laboratories), the sample applied at the appropriate residence time (1 min for Ni Sepharose Excel or 3 min for MabSelect SuRe or Protein G HP) and a portion of the flow-through fraction collected for subsequent non-reducing SDS-PAGE analysis. Columns were washed with DPBS and bound proteins eluted with two column volumes (CV) of sodium-citrate buffer pH 3.6 (MabSelect SuRe), two CV of 100 mM glycine-HCl pH 2.6 (Protein G HP) or one CV of DPBS with 500 mM imidazole pH 8 (IMAC purification). For proteins eluted from Protein A or Protein G columns, samples were pH adjusted using 1 M solutions of sodium HEPES or Tris–HCl buffer to a final pH of 6–7.

Platform, parallelized purification experiments were performed using the Protein Maker running the Protein Maker v2.0 software (Protein BioSolutions). Purification runs were performed using the 1 mL HiTrap columns and the chromatography conditions (residence time, column washing and sample elution) established using the ÄKTA purification system. Sample and buffer lines were cleaned in place with 0.5 M NaOH and equilibrated in DPBS or appropriate buffer solutions. During purification, a portion of the flow-through fraction was collected for subsequent non-reducing SDS-PAGE analysis. Protein samples were eluted in three steps, consisting of a pre-elution volume, elution volume and post-elution volume. Protein concentration measurements (A_{280} nm) on these fractions were used to establish the final pooled sample.

2.3. Process development for mouse IgG2a purification

The Protein Maker system was used to purify in parallel five murine IgG2a samples from mouse Hybridomas using MabSelect SuRe and Protein G HP 1 mL HiTrap columns. For each mouse IgG2a, 20–22 mL of supernatant (~0.5 to 2 mg of mouse IgG2a, depending on titer) was purified on either column using the purification method described above. For protein G purifications, elution was performed in two steps, first with 100 mM citrate buffer pH 3.6 and then with 100 mM glycine-HCl buffer pH 2.6. Elution fractions were neutralized using 1 M Tris. The quantity of IgG2a contained in elution fractions were determined based on A_{280} nm. For protein G purification, the quantity of IgG2a obtained from the two elution steps was summed for calculating the yield.

2.4. Development of a post-load wash step to improve HCP removal during protein A purification

Data were obtained with three different antibodies expressed in CHO cells. The Protein Maker system was used to perform parallel purifications using MabSelect SuRe 1 mL HiTrap columns. For each antibody, eight wash conditions were tested. For each tested condition, 10 mL of supernatant (16.3 to 17.5 mg of antibodies) were loaded at a residence time of 3 min. HiTrap columns were then washed and antibodies were eluted using 2.5 CV of 100 mM citrate buffer pH 3.0. Elution fractions were neutralized using 1 M HEPES. The quantity of antibody in elution fractions was determined by A_{280} nm. The quantity of HCP in elution fractions was measured using a CHO HCP ELISA kit (Cygnus Technologies).

2.5. Buffer exchange and aseptic filtration

Buffer exchange into DPBS following affinity purification was performed manually either using Zeba-spin columns (Thermo-Fisher Scientific) by centrifugation or using PD-10 desalting columns (GE Healthcare) by gravity according to the manufacturer's instructions. Alternatively, sample buffer exchange using PD Miditrap G-25 columns was automated on a TECAN Freedom EVO150® liquid handler according to gravity protocols from GE Healthcare. The Freedom EVO150® was equipped with a liquid displacement Liquid Handler (LiHa) configured with 8 channels (4 disposable tips and 4 washable tips), a Robotic Manipulator (RoMa), a Tecan Vacuum (TeVac), carriers for 11 microplates, shelf for 4 microplates, one reservoir position for elution buffer, and tip carriers for hanging tips. All tubing and components of the liquid displacement system were cleaned in-place with 0.5 M NaOH for at least 15 min and rinsed with sterile water prior to operations. A script was developed with the flexibility to process from 24 to 96 samples at once. Before starting, the storage solution from PD MidiTrap G25 columns was removed manually, the columns placed in a 24 position custom holder and the rack positioned on the Freedom EVO150® worktable. Purified protein samples from the Protein Maker were stored in 24 deep well microplate (Seahorse Bioscience). The system liquid was replaced with DPBS (HyClone Laboratories) and the script started. First, the RoMa arm brought the column racks onto the TeVac and columns were equilibrated with three bed volumes using the Freedom EVO150® system liquid (DPBS). The equilibration buffer was allowed to enter the packed bed completely and the flow-through discarded in the TeVac waste. Samples were pipetted onto the columns using disposable filter tips (Tecan) and time was allowed for them to enter the packed bed by gravity. Column racks were tapped 4× times on the Te-Vac by the RoMa arm to remove any droplets and transported on top of a 24 deep well block for elution. The LiHa pipetted 1.5 mL of elution buffer (DPBS) to each MidiTrap, respectively, using disposable tips and the eluate containing the protein of interest was collected by gravity. Subsequent aseptic filtration was performed by centrifugation using either sterile Multiscreen 0.22 μm 96-well plates (Millipore) or deep well 0.22 μm 96-well plates (Corning) stacked with a sterile receiver plate.

2.6. Analytical methods

Proteins were quantitated based on A_{280} nm values obtained using a NanoDrop 2000 spectrophotometer (Thermo Scientific) and concentration values were corrected based on calculated extinction coefficients derived from the protein sequence. Non-reducing SDS-PAGE analysis was performed using 4–12% Bis-Tris NuPage gels (Novex, Thermo-Fisher scientific) and stained with Sypro Ruby protein gel stain (Thermo-Fisher Scientific) as recommended by the manufacturer. SDS-PAGE gels were imaged using a ChemiDoc MP imaging system (Bio-Rad Laboratories).

3. Results

3.1. Development and implementation of protein maker platform purification methods

Our strategy for the development and implementation of affinity purification methods is summarized in Fig. 1. Key factors that influence the purification process, including residence time for product capture and elution volume are determined using the ÄKTA purification runs. The essential value of these runs is to provide the absorbance trace during the purification in order to quickly converge to appropriate starting conditions for purification. The essential features of the different affinity purification methods are similar for Protein G, Protein A and IMAC based purification method development. Overall, we observed comparable results in terms of product yield and sample purity when using the same purification method on an ÄKTA purification platform and the Protein Maker (results not shown).

A key factor in establishing the capture step is to determine the residence time to achieve optimal capture of the product on the column of interest. In the case of Protein G affinity chromatography, using a residence time of 3 min, no IgG was found in the flow-through of hybridoma-generated mAb samples as determined by non-reducing SDS-PAGE with Sypro Ruby staining. For larger volumes of feeds containing small quantities of product, it is necessary to split samples over two or more columns in order to reduce the total time required for the binding step. Using such an approach, ~200–250 mL of product (~30 mg of protein in the case of Protein A) can be passed over 1 mL columns in an overnight run.

We have found that using the approach of collecting elution fractions in three steps (pre-elution (0.5 CV), main elution (0.5–1.5 CV) and post-elution (1.5 + CV) volumes) allows for optimization of the quantity of purified product in a minimum volume, allowing for maximum product concentration. At the elution step, the practical outcome is to obtain ≥80% of the purified protein in a minimum volume suitable for manual or automated buffer exchange as the second step in the workflow shown in Fig. 2.

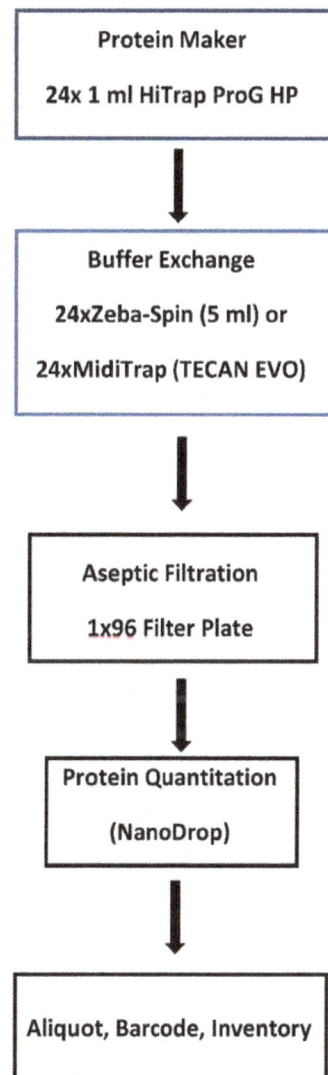

Define critical purification parameters

(residence time, load, elution conditions)

↓

Test and refine conditions using ÄKTA platform

(generate chromatograms, measure recovery, determine purity)

↓

Transfer test method and evaluate on the Protein Maker

(correct for difference in system dead volume, test 2-4 columns in parallel, evaluate method reproducibility)

↓

Implement method on Protein Maker in platform mode

(up to 24 columns run in parallel)

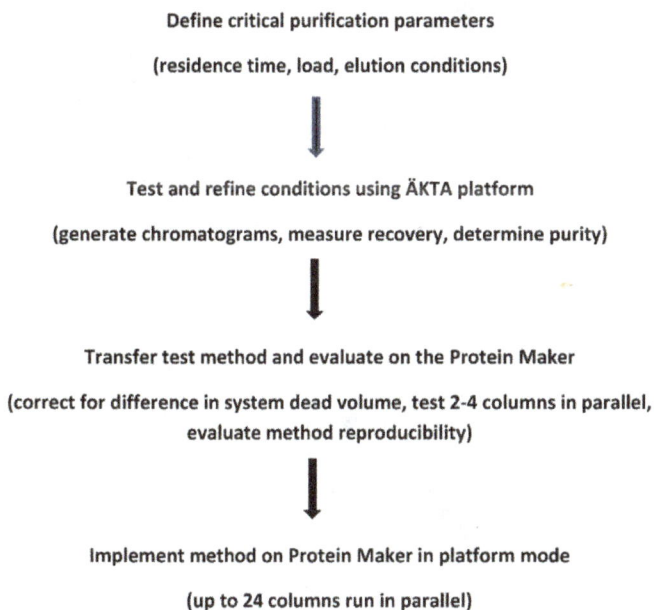

Fig. 1. Protein Maker purification development strategy.

| Protein Maker |
| 24x 1 ml HiTrap ProG HP |

↓

| Buffer Exchange |
| 24xZeba-Spin (5 ml) or |
| 24xMidiTrap (TECAN EVO) |

↓

| Aseptic Filtration |
| 1x96 Filter Plate |

↓

| Protein Quantitation |
| (NanoDrop) |

↓

| Aliquot, Barcode, Inventory |

Fig. 2. Protein G purification workflow for purification of mAbs from hybridoma supernatants.

3.2. Platform purification applications

3.2.1. Small scale purification of mAbs from hybridoma supernatants

In order to generate purified mAbs for cell-based assays, we have developed a Protein G based purification workflow using Protein G HP as the capture resin (Fig. 2). Each purification cycle permits 23 mAb samples along with one murine IgG control sample, to be purified using Protein G, buffer exchanged into DPBS and aseptically filtered prior to aliquoting in bar coded tubes for storage. The purpose of the control sample is to act as a sentinel for the purification run *via* determination of the recovery of purified protein. In principle, this sample will help to troubleshoot problems in executing the purification method, as for example, a change in column binding capacity resulting from excessive numbers of clean in place (CIP) cycles.

In most cases, approximately 0.1–1 mg of purified mAb were obtained per hybridoma supernatant, mainly with a concentration range of approximately 0.75 to 0.95 mg/mL. Non-reducing SDS-PAGE followed by staining with Sypro Ruby, a high-sensitivity protein stain, revealed no IgG in the flow-through fraction (Fig. 3B). The purity of the resulting purified mAbs was also verified by non-reducing SDS-PAGE (Fig. 3A).

In early experiments, samples were buffer exchanged manually using Zeba-spin centrifugal columns. While this approach minimizes sample dilution, it is manually intensive, and sub-optimal within a high throughput protein purification workflow (Fig. 2). Using PD MidiTrap G-25 columns in gravity mode, we implemented an automated approach to buffer exchange samples following affinity purification, reducing the manual labor, time and cost associated with this step. A consequence of this change was increased dilution of the purified sample by ~1.5-fold compared to the Zeba-spin buffer exchange.

3.2.2. Antibody fragments — purification of Fabs

We applied the Protein Maker platform to the purification of a series of his-tagged Fab constructs generated for *in vitro* studies. Other than the buffer composition for column washing and sample elution, the overall strategy is similar to that applied to Protein G or Protein A purification methods. Using this approach, up to 20 mg of purified Fab from ~200 mL CHO culture has been obtained. Recoveries post-buffer exchange are similar to that obtained for mAbs purified using Protein

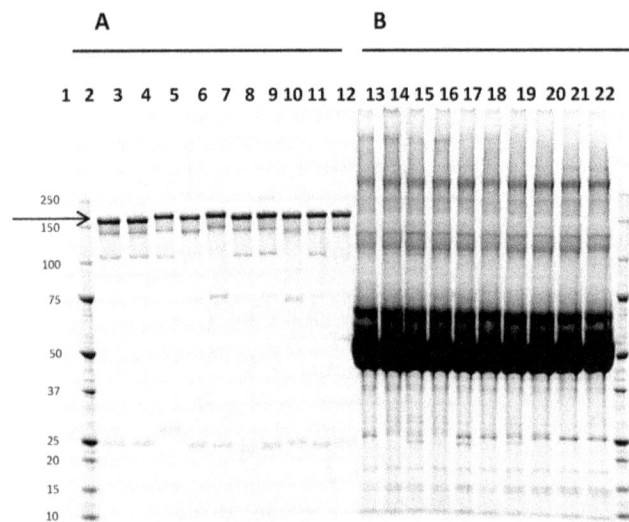

Fig. 4. Non-reducing SDS-PAGE and Sypro Ruby staining of CHO-produced and IMAC purified Fab samples using the Protein Maker. Lanes 1, 9, 17, molecular weight markers, in kDa; lanes 2, 10, CHO supernatants of expressed Fabs; Lanes 3,4,11,12, flow-through fractions; lanes 5,6,13,14, wash fractions; lanes 7,8,15,16, elution fractions. The ~50 kDa species in the F/T fraction represents light chain (LC) dimers.

G or Protein A. The purity of representative Fab samples as determined by non-reducing SDS-PAGE is shown in Fig. 4.

3.3. Process development applications

3.3.1. Evaluation of protein A vs protein G resin for purification of murine IgG2a

Protein G resin is typically considered as the first choice to purify antibodies produced from mouse Hybridomas, especially when the IgG subclass is unknown. Indeed, Protein G has strong affinity for all mouse IgG subclasses, whereas Protein A has strong affinity for only certain subclasses, specifically IgG2a and IgG2b. For purification of mouse IgG2a, the choice of one resin *vs* the other is not straightforward. According to some manufacturers, mouse IgG2a has strong binding affinity for both Protein A and Protein G, however, protein A columns have a higher binding capacity and could be a better choice especially for larger scale purification. To determine which of Protein A or Protein G is the better choice to purify mouse IgG2a, we used the Protein Maker

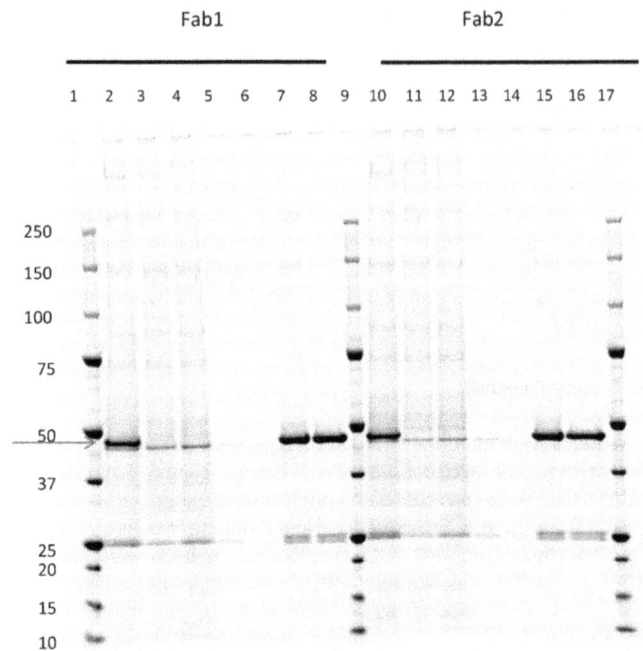

Fig. 3. Non-reducing SDS-PAGE and Sypro Ruby staining of ten different hybridoma-produced antibodies purified using Protein G HP with the Protein Maker (A) purified mAbs, (lanes 2–11)(B) flow-through fractions (lanes 12–21), revealing no unbound mAb (arrow). The mAb species (2HC + 2LC) is indicated with an arrow, and minor species include 2HC (~100 kDa), half-antibody (~75 kDa) and free LC (~25 kDa) are visible for some samples. Molecular weight markers (lanes 1, 22) are shown in kDa.

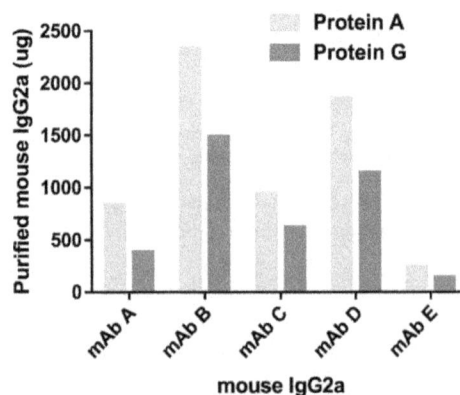

Fig. 5. Quantities of mouse IgG2a (μg) purified from mouse hybridoma supernatants using either Protein A or Protein G.

to purify five mouse IgG2a samples in parallel on these columns. The total quantity of mouse IgG2a obtained with either resin following purification is summarized in Fig. 5. For all tested mouse IgG2a, the quantity of purified protein obtained from Protein A was at least 50% higher than those obtained from Protein G purification.

3.3.2. Development of a post-load wash step to improve HCP removal during protein A purification

Removing impurities post-Protein A chromatography still represents a significant challenge to purification process development in order to achieve the required drug substance specifications suitable for patient administration (HCP < 100 ppm). Using a post-load wash step is a key means to achieve HCP clearance, since it has been demonstrated that HCP associates with antibodies and co-elute during the elution step. Basic pH and wash additives such as arginine seem to improve HCP removal during the Protein A chromatography wash step by disrupting interactions between the antibody and the HCPs [13]. Here we evaluated both Tris and phosphate-based arginine wash buffers at different pH values (7 to 9) in comparison to conventional wash buffer (DPBS or citrate pH 5.0) using 1 mL HiTrap MabSelect SuRe columns. These various wash conditions (Table 1) were tested in parallel with 3 different antibodies (CHO supernatants) using the Protein Maker. Yield and HCP levels obtained for each wash condition are presented in Fig. 6. The use of basic wash buffers containing arginine showed improved HCP removal (1.7 to 2.4-fold) compared to the conventional DPBS wash for the three antibodies tested, with detrimental effects on purification yields only for one antibody (mAb3).

3.3.3. Stability hold of clarified harvest for a mAb

In a series of experiments, a clarified mAb harvest was held at either 2–8 °C or 19–23 °C and sampled at various time points corresponding to t = 0, 1, 2, 3, 5 and 7 days before purification using MabSelect SuRe followed by buffer exchange to DPBS in order to evaluate changes in the product using an analytical assay panel. The purification performance at time points, t = 0 and t = 7 days, is shown in Table 2. Purification recoveries based on the measured product titer were at least 85%. Analysis of charge variants and glycosylation profiles at each time point at either temperature revealed a significant decrease in acidic charge variants upon storage at 19–23 °C, corresponding to a drop in the sialic acid content of the mAb (results not shown). This assessment made it possible to define the manufacturing hold time duration necessary for process control during GMP manufacturing.

4. Discussion

During early-stage therapeutic antibody R&D projects, it is often necessary to purify large numbers of samples that will be used as reagents for *in vivo* or *in vitro* screening assays. The nature of the purification strategy employed is determined by the quantity and final concentration of purified protein required; the number of samples that need to be purified per unit time (throughput), the supernatant volume and initial product titer, as well as the availability, if any, of automated liquid handling instrumentation. While clearly there is more than one possible solution to achieve the purification objectives,

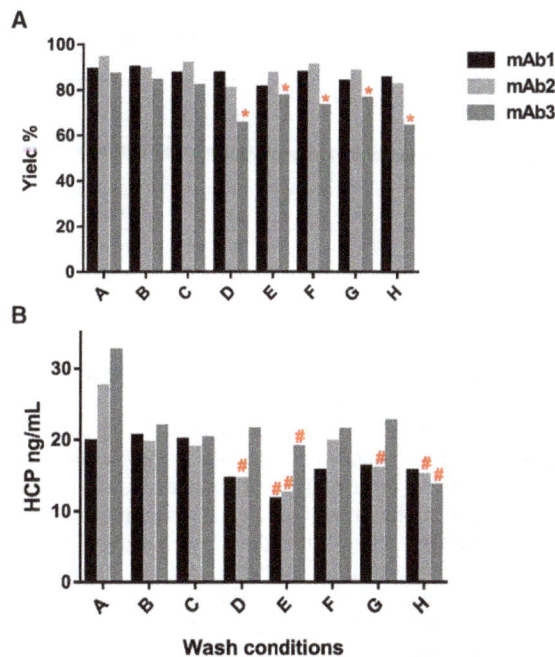

Fig. 6. Effect of wash buffers on (A) purification yields and (B) HCP levels. Wash conditions — A: PBS; B: 100 mM Citrate, 50 mM NaCl pH 5.0; C: 50 mM Tris, 100 mM Arginine, 50 mM NaCl pH 7.0; D: 50 mM Tris, 100 mM Arginine, 50 mM NaCl pH 8.0; E: 50 mM Tris, 100 mM Arginine, 50 mM NaCl pH 9.0; F: 10 mM Phosphate, 100 mM Arginine, 50 mM NaCl pH 7.0; G: 10 mM Phosphate, 100 mM Arginine, 50 mM NaCl pH 8.0; H: 10 mM Phosphate, 100 mM Arginine, 50 mM NaCl pH 9.0. *: Yields below 80% — #: HCP reduction *vs* PBS greater than 1.7-fold.

we selected the Protein Maker as a platform using HiTrap columns, including Protein G (Protein G HP), Protein A (MabSelect SuRe) and IMAC (Ni Sepharose Excel), and have applied these to feeds from either hybridoma or CHO supernatants ranging from 10 mL to 200 mL volumes. A key outcome has been for purification of small quantities (<1 mg) of purified mAbs using the Protein G method described here, with greater than 1000 mAbs purified and utilized as reagents for *in vitro* screening assays.

For the development of platform-based purification methods on the Protein Maker, several parameters that influence the overall purification outcome required evaluation. One of the most critical parameters is the residence time used in the binding step, as most of the overall time required to execute the method involves binding of the product to the packed bed. Through several purification campaigns utilizing Protein G HP with murine IgG's, we have established that a three-minute residence time ensures capture of the product, as revealed by non-reducing SDS-PAGE analysis. Another critical parameter is to establish the elution volume range for the product. The elution volume will depend on the column volume and the quantity of product bound to the packed bed, with larger elution volumes being required as one approaches the dynamic binding capacity limit of the column. The optimal elution volume is a balance between the desired product concentration, yield and recovery of the purified product. One rule of thumb

Table 1
Wash conditions tested for Protein A post-load HCP removal.

	Wash 1 (5 CV)	Wash 2 (5 CV)	Wash 3 (5 CV)
1	PBS	PBS	PBS
2	PBS	100 mM Citrate, 50 mM NaCl pH 5.0	PBS
3	PBS	50 mM Tris, 100 mM Arginine, 50 mM NaCl pH 7.0	PBS
4	PBS	50 mM Tris, 100 mM Arginine, 50 mM NaCl pH 8.0	PBS
5	PBS	50 mM Tris, 100 mM Arginine, 50 mM NaCl pH 9.0	PBS
6	PBS	10 mM Phosphate, 100 mM Arginine, 50 mM NaCl pH 7.0	PBS
7	PBS	10 mM Phosphate, 100 mM Arginine, 50 mM NaCl pH 8.0	PBS
8	PBS	10 mM Phosphate, 100 mM Arginine, 50 mM NaCl pH 9.0	PBS

Table 2
Chromatographic Performance Table — MabSelect SuRe purification, clarified harvest stability hold study for t = 0 and t = day 7. The global process purification yield is calculated as the ratio of the final yield over the initial load on the column.

| Storage temperature (°C) | Time point (Day) | HiTrap Mab Select SuRe | | | | PD-10 buffer exchange and aeptic filtration | | | Global process yield (%) |
| | | Load | | Elution | | Load | Final | | |
		Volume (mL)	Quantity (mg)	Quantity (mg)	Step yield (%)	quantity (mg)	quantity (mg)	Step yield (%)	
N/A	0	50	16.6	14.8	89	13.6	13.6	99.9	82.0
2–8 °C	7	50	16.5	14.0	85	12.8	12.6	98.7	76.6
19–23 °C	7	50	16.6	15.3	92	14.0	13.7	97.5	82.6

would be to establish the elution volume such that ≥80% of the purified protein is captured in one fraction.

In many antibody affinity purification work-flows, purification is followed by buffer exchange into a formulation buffer, including PBS, in order to minimize protein aggregation as a result of unfavorable conditions of pH or ionic strength. While it is possible to perform the buffer exchange step in high-throughput mode using 96-well plates [14], this only applies to samples having a small volume, typically less than 130 µL. Initially, we performed this step in a more manually intensive manner using Zeba-spin buffer exchange columns. In order to increase the throughput and operational efficiency of the buffer exchange step, a more automated approach was developed using PD MidiTrap G-25 columns on a TECAN Freedom EVO 150 liquid handling system. The integration of an automated buffer exchange step in a 24-sample format, the same format as the Protein Maker, increases the overall sample throughput, decreases manual manipulations and enhances the consistency of results.

In addition to platform-based purification of full-size IgG's, there is often a requirement to purify IgG fragments, including Fabs and scFv's, as well as to evaluate the effectiveness of different affinity purification resins for purification of the same product. For purification of murine IgG2a, the choice of Protein A vs Protein G purification requires experimental verification prior to establishing the final process to be used. By performing all of the purification experiments in parallel using the Protein Maker, a few hours were sufficient to determine that protein A is the better choice for purification of murine IgG2a samples. Indeed, the quantity of purified protein obtained from Protein A was at least 50% higher than those obtained from Protein G purification, although in the absence of additional data we cannot offer a definitive explanation for this result. Executing the same experiment using one ÄKTA purification system would require approximately five-fold more time than performing the purification experiments in parallel with the Protein Maker. It is noteworthy that these 5 antibodies were subsequently purified at larger (2 L) scale using protein A resin with greater than 80% recovery for four out of the five antibodies processed. In the

purification scheme presented here, Fabs are purified via his-tags using IMAC, although Fab purification can be achieved in some instances using protein A and more commonly using the CH1 domain of the Heavy Chain (HC) or the conserved domain of the Light Chain (LC) as capture modes. Other therapeutic antibody formats, including one-armed antibodies, hybrids, bi-specifics and various Fc-fusion molecules can also be purified in platform mode with the appropriate affinity resin. Examples of readily available affinity chromatography resins and how they can be applied to various antibody purification requirements is summarized in Table 3.

Both the post-load wash step for optimization of HCP removal from Protein A as well as the clarified harvest hold stability study offer examples of how the Protein Maker can be effectively utilized in the context of protein purification process development. Minimizing host cell protein levels at the Protein A capture step through modification of wash buffer composition can improve the purification process [13, 15,16]. One cycle of parallelized purification using the Protein Maker, requiring approximately 3 h, was sufficient to purify 24 samples (eight different wash conditions for three different antibodies). Executing the same experiment using one ÄKTA purification system (i.e. 24 sequential purifications on an ÄKTA Purifier) would have required approximately 37.5 h. In this experiment, the Protein Maker not only increased purification throughput by 12.5-fold but also permitted purification of samples in an unbiased manner by having the capacity to purify all of the samples at the same time. Indeed, under conditions where a mAb is unstable in the clarified harvest, the ability to purify multiple samples in parallel offers a distinct advantage over serial purification using conventional purification equipment, minimizing misinterpretation of data due to sample degradation. As the effectiveness of optimized wash buffers and their effect on purification yields seem to vary depending on the antibody, the same wash conditions cannot be used for all samples. Rather, post-load wash conditions should be optimized for each antibody as part of purification process development. Due to its ability to process multiple samples in parallel, the Protein Maker represents a preferred instrument for performing wash buffer screening. Moreover, as multiple wash conditions can be evaluated in parallel, it is possible to minimize variability in sample handling that could bias data interpretation.

Therapeutic antibody products often pose various challenges during the development of upstream and downstream processes, including degradation, modification of Fc or Fd glycans, reduction of intramolecular disulfides, deamidation of Asn and Gln residues, as well as aggregation [17]. In order to de-risk the overall process, an important component of the transition between upstream and downstream processing is to understand the stability of the product post-clarification but prior to the initial capture purification step. The removal of sialic acid from mAbs by the action of extracellular CHO sialidase is an example of the kinds of post-production, pre-purification modifications that are possible [18]. The utility of the Protein Maker in this application is its ability to purify several samples in parallel, for example, if the clarified harvest is to be held at various temperatures or at different pH values, or if multiple harvests are to be tested in parallel and purified under the same controlled conditions.

Table 3
Possible platform affinity purification modes for antibodies and antibody fragments using the Protein Maker.

Format	Isotype	Species	Capture mode	Resin
mAb, bispecific	IgG1, IgG2, IgG4	Human	Fc	Protein A
mAb, bispecific	IgG2	Murine	Fc	Protein A
mAb, bispecific	IgG1, IgG2, IgG3, IgG4	Human	Fc	Protein G
mAb, bispecific	IgG1, IgG2, IgG3	Murine	Fc	Protein G
OAA	N/A	Human	Fc	Protein A, G
scFv		Human	LC (V_L)	Protein L
Hybrid		Human	Fc	Protein A, G
Hybrid		Human	HC (CH1)	CH1
Hybrid		Human	LC (κ, C_L)	Kappa-Select
Fab		Human	HC (CH1)	CH1
Fab		Human	LC (κ, C_L)	Kappa-Select
Fab		Human	LC (λ, C_L)	Lambda-Fab Select

5. Conclusions

We have found the Protein Maker to be a versatile tool for a number of purification problems, ranging from small scale mAb and Fab purification to various applications in process development. We have developed platform purifications methods using Protein G, Protein A and IMAC and have applied these to several projects. Use of the Protein Maker and TECAN Freedom EVO150 together has resulted in reducing both the required personnel time and creation of an integrated workflow that includes both protein purification and automated buffer exchange in a 24-sample format. Future implementation of the UV monitoring capability will add an additional dimension to the capabilities of this instrument, permitting 24-channel monitoring of the protein absorbance signal, thereby expediting utilization of the instrument for various process development applications.

Acknowledgements

We wish to thank our colleagues within the Protein Purification Team who have provided valuable feedback on the various methods that have been developed. We also wish to thank our colleagues in various projects who have provided us with materials with which we have developed the various methods described. In particular we thank Maria Jaramillo and Anne Marcil for their encouragement. This research was supported by the National Research Council Canada.

References

[1] Cummins E, Luxenberg DP, McAleese F, Widom A, Fennell BJ, Damanin-Sheehan A, et al. A simple high-throughput purification method for hit identification in protein screening. J Immunol Methods 2008;339:38–46.
[2] Su B, Hrin R, Harvey BR, Wang YJ, Ernst RE, Hampton RA, et al. Automated high-throughput purification of antibody fragments to facilitate evaluation in functional and kinetic based assays. J Immunol Methods 2007;322:94–103.
[3] Yoo D, Provchy J, Park C, Schulz C, Walker K. Automated high-throughput protein purification using an AKTApurifier and a CETAC autosampler. J Chromatogr A 2014;1344:23–30.
[4] Holenstein F, Eriksson C, Erlandsson I, Norrman N, Simon J, Danielsson A, et al. Automated harvesting and 2-step purification of unclarified mammalian cell-culture broths containing antibodies. J Chromatogr A 2015;1418:103–9.
[5] Alm T, Steen J, Ottosson J, Hober S. High-throughput protein purification under denaturing conditions by the use of cation exchange chromatography. Biotechnol J 2007;2:709–16.
[6] Gonzalez R, Jennings LL, Knuth M, Orth AP, Klock HE, Ou W, et al. Screening the mammalian extracellular proteome for regulators of embryonic human stem cell pluripotency. Proc Natl Acad Sci U S A 2010;107:3552–7.
[7] McGraw J, Tatipelli VK, Feyijinmi O, Traore MC, Eangoor P, Lane S, et al. A semi-automated method for purification of milligram quantities of proteins on the QIAcube. Protein Expr Purif 2014;96:48–53.
[8] Kelley BD, Switzer M, Bastek P, Kraimarczyk JF, Molnar K, Yu T, et al. High-throughput screening of chromatographic separations: IV. Ion-exchange. Biotechnol Bioeng 2008;100:950–63.
[9] Teeters M, Bezila D, Alread P, Velayudhan A. Development and application of an automated, low-volume chromatography system for resin and condition screening. Biotechnol J 2008;3:1212–23.
[10] Smith ER, Begley DW, Anderson V, Raymond AC, Haffner TE, Robinson JI, et al. The Protein Maker: an automated system for high-throughput parallel purification. Acta Crystallogr F 2011;67:1015–21.
[11] Yokoyama WM, Li Z. Monoclonal antibody supernatant and ascites fluid production. Curr Protoc Immunol 2001 [40:II:2.6:2.6.1–2.6.9].
[12] Raymond C, Robotham A, Spearman M, Butler M, Kelly J, Durocher Y. Production of α2,6-sialylated IgG1 in CHO cells. MAbs 2015;7:571–83.
[13] Chollangi S, Parker R, Singh N, Li Y, Borys M, Li Z. Development of robust antibody purification by optimizing protein-A chromatography in combination with precipitation methodologies. Biotechnol Bioeng 2015;112:2292–304.
[14] Ying W, Levons JK, Carney A, Gandhi R, Vydra V, Rubin AE. Semiautomated sample preparation for protein stability and formulation via buffer exchange. J Lab Autom May 12 2015 [pii: 2211068215585484].
[15] Aboulaich N, Chung WK, Thompson JH, Larkin C, Robbins D, Zhu M. A novel approach to monitor clearance of host cell proteins associated with monoclonal antibodies. Biotechnol Prog 2014;30:1114–24.
[16] Sisodiya V, Lequieu J, Rodriguez M, McDonald P, Lazzareschi KP. Studying host cell protein interactions with monoclonal antibodies using high throughput protein A chromatography. Biotechnol J 2012;7:1233–41.
[17] Vasquez-Ray M, Lang DA. Aggregates in monoclonal antibody manufacturing processes. Biotechnol Bioeng 2011;108:1494–508.
[18] Gramer MJ, Goochee CF, Chock VY, Brousseau DT, Sliwkowski MB. Removal of sialic acid from a glycoprotein in CHO cell culture supernatant by action of an extracellular CHO cell sialidase. Biotechnology (N Y) 1995;13:692–8.

Unbiased Mitoproteome Analyses Confirm Non-canonical RNA, Expanded Codon Translations

Hervé Seligmann

Unité de Recherche sur les Maladies Infectieuses et Tropicales Émergentes, Faculté de Médecine, URMITE CNRS-IRD 198 UMER 6236, Université de la Méditerranée, Marseille, France

ARTICLE INFO

Keywords:
Frameshift
Bijective transformation
Digestive enzymes
Unbiased analyses
RNA–DNA difference

ABSTRACT

Proteomic MS/MS mass spectrometry detections are usually biased towards peptides cleaved by experimentally added digestion enzyme(s). Hence peptides resulting from spontaneous degradation and natural proteolysis usually remain undetected. Previous analyses of tryptic human proteome data (cleavage after K, R) detected non-canonical tryptic peptides translated according to tetra- and pentacodons (codons expanded by silent mono- and dinucleotides), and from transcripts systematically (a) deleting mono-, dinucleotides after trinucleotides (delRNAs), (b) exchanging nucleotides according to 23 bijective transformations. Nine symmetric and fourteen asymmetric nucleotide exchanges (X ↔ Y, e.g. A ↔ C; and X → Y → Z → X, e.g. A → C → G → A) produce swinger RNAs. Here unbiased reanalyses of these proteomic data detect preferentially non-canonical tryptic peptides despite assuming random cleavage. Unbiased analyses couldn't reconstruct experimental tryptic digestion if most detected non-canonical peptides were false positives. Detected non-tryptic non-canonical peptides map preferentially on corresponding, previously described non-canonical transcripts, as for tryptic non-canonical peptides. Hence unbiased analyses independently confirm previous trypsin-biased analyses that showed translations of del- and swinger RNA and expanded codons. Accounting for natural proteolysis completes trypsin-biased mitopeptidome analyses, independently confirms non-canonical transcriptions and translations.

1. Introduction

Protein sequences are more complex than texts written in natural human languages [1]. This implies that genes include superimposed information, overprinted on the classical protein coding gene; for example, the three frames of the shortest self-replicating circular RNA virusoid code for proteins [2]. Cryptic coding revealed by frameshifts also implies that a punctuation code regulates ribosomal frame translation. This role seems fulfilled by the natural circular code X, a set of 20 regular codons overrepresented in the coding frame of genes versus other frames. X possesses peculiar mathematical properties that enable retrieval of translational frame [3–10]. This punctuation code can also be considered as cryptic superimposed information.

The natural circular code seems to prevent unwanted ribosomal frameshifts. In addition, the structure of the genetic code implies a further mechanism against frameshifted translation. This mechanism, rather than preventing ribosomal slippage before it occurs, as assumed for X, minimizes translation after ribosomal frameshifts. This is because the genetic code's codon-amino acid assignments are such that they maximize off frame stop codons [11], avoiding metabolic waste after ribosomal slippages [12–16].

Superimposed coding is also indicated by other peculiar genetic code properties. The genetic code includes symmetries, such as Rumer's symmetry [17–20], where transformations of nucleotides into other nucleotides along specific rules reveal symmetries between codon-amino acid families.

Rumer's symmetry implies that after applying that specific transformation, all codons coding for a given amino acid are transformed into codons coding for another amino acid [21]. These theoretical observations seem related to the following empirical observations. Recent transcriptomic and proteomic findings show that genetic information is revealed by each systematic frameshifting and nucleotide transformations. Here I develop these issues and present analyses that strengthen the proteomic evidence for translation of proteins coded by overprinting associated with systematic frameshifts and systematic nucleotide transformations.

Previously detected peptides match predictions of regular translations of non-canonical mitochondrial RNAs, and non-canonical translations of codons expanded by silent mono- and dinucleotides (detailed in Fig. 1). These results seem overall robust as detected non-canonical peptides mapped on the human mitogenome with corresponding non-canonical RNAs [22,23]. Hence existences of non-canonical RNAs and peptides are validated by independent detections of non-canonical RNAs and peptides, and by convergences (associations) between detected non-canonical RNAs and peptides.

E-mail address: hselig1@gmail.com.

Systematic transformations during transcription (alternating codons underlined)
A) Original sequence AAA<u>CCC</u>TTT<u>GGG</u>
   Translated K   P   F   G
B) A<->C-swinger transformed sequence CCC<u>AAA</u>TTT<u>GGG</u>
   Translated P   K   F   G
C) Expanded to tetracodons CCC<u>AA</u>AATTTGGG
   Translated P   K   W
D) Systematic mononucleotide deletions CCC_AAT_TGG_
   Translated P   K   W
E) Expanded to pentacodons CCCAA<u>ATTT</u>GGG
   Translated P   I   G
F) Systematic dinucleotide deletions CCC___ATT__GG
   Translated P   I   G

Fig. 1. Sequence (A) and its systematic transformations and corresponding translations (B–F). B) A ↔ C systematic nucleotide exchange of sequence in A; C) assuming systematic codon expansion by silent mononucleotides; D) assuming systematic mononucleotide deletion after each trinucleotide (translation identical to that in C); E) assuming systematic codon expansion by silent dinucleotides; F) assuming systematic dinucleotide deletion after each trinucleotide (translation identical to that in E). RNAs and peptides corresponding to these alternative transcriptions and translations have been previously described for human mitochondria [22,23]. For swinger transformations, A ↔ C is only one among 23 possibilities, nine symmetric of type X ↔ Y, and 14 asymmetric, of type X → Y → Z → X. Systematic deletions of mono- and dinucleotides after each trinucleotide are annotated as delRNA$_{3-1}$ and delRNA$_{3-2}$. Systematic deletions can start at the 5′ extremity of a sequence, which is indicated by delRNA$_{3-1.0}$ and delRNA$_{3-2.0}$, deletion frames can be shifted by 0–2 and 0–3 nucleotides for delRNA$_{3-1}$ and delRNA$_{3-2}$, respectively, which can be indicated by corresponding indices.

However, MS/MS matching between observed and predicted mass spectra is biased. Hence the unconventional natures of non-canonical transcriptions and translations presumably producing these peptides require careful evaluation of proteomic analyses. Below I review the different types of non-canonical transcriptions and translations, and relevant previous results. Previous conclusions about non-canonical peptides are then re-evaluated according to analyses that account for overfitting that could have affected the previously used proteomic search algorithm [22,23]. These new analyses strengthen previous conclusions on non-canonical mitochondrial transcriptions and translations.

1.1. Non-Canonical Transcriptions: RNA–DNA Differences

Transcription is not always perfectly accurate, but in some cases, RNA–DNA differences (RDDs) are not random and are systematically detected at some specific positions, either in the form of nucleotide substitutions [24], also observed on mitochondrion-encoded RNAs [25–27]) or deletions [28]. These punctual differences between transcript and DNA occur shortly after transcripts exit the RNA polymerase, suggesting posttranscriptional RNA editing [29]. These single nucleotide modifications produce non-canonical transcripts.

In other types of non-canonical RNAs, modifications occur systematically for all nucleotides for the complete (or almost complete) RNA. Two types of systematic transformations occur: (a) systematic deletions of mono- and dinucleotides after each trinucleotide, producing delRNAs in human mitochondria [22]; and 23 types of systematic exchanges between nucleotides (nine symmetric exchanges, type X ↔ Y, e.g. A ↔ C; and fourteen asymmetric exchanges, type X → Y → Z → X, e.g. A → C → G → A, [30–33]), producing swinger RNAs.

Swinger- and delRNAs are probably not due to RNA edition, unlike the punctual RDDs. This is because transformations frequently occur systematically on sequences longer than 100 nucleotides. They seem produced by the same RNA polymerase as regular RNA, presumably after the RNA polymerase stabilizes in a hypothetical mode similar to that causing punctual nucleotide misinsertions [32–34]. This is also indicated by contiguity between regular and swinger sequences in the few detected chimeric RNAs, DNAs and peptides that consist of regular and swinger sequences [35,36].

1.2. Non-canonical Transcriptions: Systematic Deletions

Three independent lines of evidence suggest del-transcription, transcription systematically deleting/jumping mono- and dinucleotides after each trinucleotide. (a) Contiguous short RNA reads in the human transcriptome match the mitogenome transformed by systematic (mono- and dinucleotide) deletions after each trinucleotide [22]. (b) Peptides corresponding to these delRNAs were detected in proteomic data [22]. (c) The human mitogenome, after del-transformations, has more inverted palindromes potentially forming stem-loop hairpins than comparable randomly shuffled sequences [23]. Excess palindromes after del-transformations suggest biological roles for del-transformed sequences. Palindromes in del-transformed sequences apparently down-regulate del-transcriptions [37]. Convergences between these different evidences suggest actual biological roles for del-transformed DNA/RNA.

1.3. Non-canonical Transcriptions: Systematic Nucleotide Exchanges

Homology between some RNAs and the 'parent' DNA can only be detected when assuming systematic exchanges between nucleotides along the complete RNA length. This process is called 'swinger' transcription. Several independent evidences show that swinger polymerizations occasionally occur. (a) Swinger DNA has been detected [39–40], in addition to swinger RNA [30–33]. (b) Peptides corresponding to detected swinger RNAs also occur [23]. (c) The human mitogenome includes swinger repeats, meaning repeats of other parts of the mitogenome, at the condition one assumes a given swinger transformation. These swinger repeats are more numerous than for comparable randomized sequences [38,41]. (d) The swinger-transformed human mitogenome has more inverted repeats potentially forming stem-loop hairpins than randomly shuffled sequences [42]. (e) Chimeric DNA/RNA [35] and peptides [36] exist. These nucleotide and peptide sequences consist of at least two contiguous parts, where one part is 'regular' (= untransformed), the other is swinger-transformed according to one of the 23 potential bijective transformations of nucleotide sequences [19,20].

These various material evidences converge with one another: detected swinger peptides map on detected swinger RNAs [23]; mitochondrial swinger RNA abundances increase with abundances of swinger repeats in the mitogenome; and palindromes formed by swinger-transformed mitosequences associate with swinger RNA detection [42]. This association between transcripts and hairpins for swinger RNA is expected because regular mitochondrial post-transcriptional RNA processing depends on secondary structure formation for regular RNAs, a process called tRNA punctuation of mitochondrial posttranscriptional RNA processing [43]. Hence swinger RNA processing resembles regular RNA processing.

In addition, mitochondrial swinger RNA has been detected within datasets produced by classical Sanger sequencing [30–33], and by massive next generation Illumina sequencing [23]. Swinger RNA properties converge between RNAs sequenced by these different methods [23].

1.3.1. Systematic Nucleotide Exchanges and the Natural Circular Code

A specific property of the genetic code is that it includes a 'punctuation' code which enables retrieval of the protein coding frame, called the natural circular code X [44,45], putatively by interactions between mRNAs and the ribosomal decoding center [8–10]. X consists of 20 codons that are over-represented in the coding frame of genes, as compared to non-coding frames, and as a group, have several strong mathematical properties that enable detecting the coding frame.

The 23 bijective transformations (or swinger transformations), when applied to X, produce also circular codes [34,44]. The reading frame retrieval capacity (RFR) of circular codes can be quantified [5]. The RFRs of these transformations of X correlate with properties of corresponding detected swinger RNAs [34]. This means that strictly theoretical considerations predict swinger transcription properties. Swinger RNA abundances are proportional to the invariance of circular code properties of sequences after corresponding bijective transformations.

Associations between empirical observations of swinger transformations and theoretical properties derived from X are strong evidence that swinger transformations increased the coding potential of short protogenomes. This is because X, shared by almost all organisms [45], is very ancient. Hence swinger transformations were embedded within the polymerization machinery since its earliest inception.

1.3.2. Swinger Transformations and tRNA-Replication Origins

A peculiar observation on palindromes formed by some human mitogenome sequences after specific swinger-transformations also suggests, among others, that swinger transformed sequences are integrated in the genome, and participate in creation of new functional sequences.

Mitochondrial light strand replication typically originates at the OL, the light strand replication origin, a stem-loop hairpin located within the largest tRNA gene group in vertebrate mitogenomes [46]. The OL loop contains the recognition and initial binding site of the mitochondrial DNA polymerase [47,48]. In several taxa, such as most birds, the OL is totally missing [49], suggesting that its function is performed by adjacent tDNAs, which form OL-like structures [50–58].

No clear homology between mitochondrial tRNAs and the OL has been observed, despite functional indications suggesting some interchangeabilities between tRNA and OL functions. These include aminoacylation of RNA corresponding to the OL [59] and similarities between tRNA (and tRNA-related enzymes) and elements of the replicational machineries of ancient viruses [60,61]. Only recent analyses searching for inverted palindromes in the swinger-transformed human mitogenome detected ten nucleotide long complementarity between the human mitochondrial OL loop and the D-arm of mitochondrial tRNA Ala [42].

Eight swinger transformations which form the group 2 bijective transformations [19], create this OL-tRNA palindrome. Hence swinger transformations reveal the previously presumed OL-tRNA homology. This suggests unsuspected evolutionary implications for swinger transformations in the context of de novo creation of functional structural RNAs [62]. It also confirms the above considerations that swinger polymerizations occurred since the onset of the molecular machinery of life.

1.4. Peptides Matching Translation of Codons Expanded by Silent Mono- and Dinucleotides

Several observations indicate that sequences code for many more proteins than usually assumed. For example, activity of stop-suppressor (or antitermination) tRNAs [63–66] presumably templated by the antisense sequence of regular mitochondrial tRNAs [55,65,66] might enable translation of supposed non-coding frames that include stop codons [67–71]. This is also suggested by coevolution between predicted mitochondrial suppressor tRNAs and predicted mitochondrial off-frame coding regions in several taxonomic groups (primates [67,68]; Drosophila [68,69]; turtles [70]; and chaetognatha [71]). These analyses assume a change in genetic code where stop codons are reassigned to code for unknown amino acid(s) [72]. This stop-codon reassignment is also suggested by comparisons between mitochondrial and other genetic codes [72–81].

Translation by another type of tRNAs, tRNAs with expanded anticodons, unleashes further coding potential. This is indicated by coevolution between predicted mitochondrial tRNAs with expanded anticodons and predicted coding sequences translated from stretches of tetracodons, codons expanded by a silent fourth nucleotide [56,57,82,83].

Presumably, regular tRNA translation of delRNAs produces the same peptides as regular RNAs translated by unusual tRNAs with expanded anticodons [84–90]. Expanded codons are compatible with symmetry and error-correcting properties of the tessera, a subset of 64 among the 264 tetracodons. Tessera are the presumed ancestors of the vertebrate mitochondrial genetic code [91]. The tessera hypothesis is compatible with the fact that regular codon–anticodon interactions are too weak for peptide elongation without ribosome, which presumably evolved after primordial translation mechanisms [92–94].

Some mitochondrial peptides match translations according to tetra- and pentacodons, including for translations of swinger-transformed versions of the human mitogenome [22,23]. This type of peptide translation is particularly peculiar. For now it is deduced from (a) coevolution between predicted tRNAs with expanded anticodons with predicted tetracoding sequences, (b) empirical matches between predicted and observed MS/MS spectrometry data, and (c) associations between detected peptides and corresponding detected non-canonical swinger RNAs. Fig. 1 shows examples of translation according to tetra- and pentacodons. Hence re-analyses are designed to avoid some biases present in previous analyses. These reanalyses confirm the validity of previously described non-canonical peptides, particularly those coded by expanded codons [22,23].

1.5. Supervised versus Unsupervised Analyses

Proteomic analyses characterize protein expression patterns from mass spectrometry data of cell proteome extracts. These typically match numerous MS/MS spectra predicted from the annotated genes in genomes with observed spectra (e.g. for the bacterium Tropheryma whipplei, agent of Whipple's disease [95]). This approach is biased: it optimizes the fit between observed and expected MS/MS datasets. Such supervised/biased analyses always imply some false positive detections due to overfitting between observation and prediction, particularly for large datasets [96], including microarray analyses [97–99]. Deliberate biases in analyses also improve estimations [100], but overfitting remains a problem, especially for detection of unknown phenomena.

A known bias that affects classical proteomic search algorithms is that predicted protein sequences are matched to observed data, assuming specific cleavage according to cleavage by the digestion enzyme used during protein/proteome extraction and preparation. Hence if trypsin was used during sample preparation, the amino acid at the carboxyl extremity of peptides is a priori supposed tryptic, specific cleavage after K or R.

Hence searches are usually biased towards peptides matching the cleavage rules of digestion enzyme(s) used during sample preparation, because this limits greatly cleavage options, saves computational machine time. Detections of non-canonical peptides would be validated if unsupervised analyses that do not predefine specific cleavage rule(s) detect mainly tryptic peptides. Unbiased analyses can only recover experimental tryptic conditions if most non-canonical peptides detections are accurate.

Proteomic search algorithms usually enable analyses assuming random cleavage when fitting observed and expected mass spectra, by options indicating 'no enzyme' or 'no specific cleavage'. This option is rarely used, because it increases search times enormously.

Here analyses assume random cleavage of actually tryptic proteomes. These should preferentially detect peptides ending by K or R, as compared to other amino acids. This biased result for unbiased analyses would validate conclusions from previous trypsin-biased analyses. The latter detected numerous peptides matching non-canonical RNAs and translations [22,23]. Here I aim at confirming these previous results using unbiased analyses assuming random cleavage.

1.6. Unsupervised Analyses and Natural Proteolysis

Numerous natural proteases are active in cells, including in mitochondria, forming the mitodegradome [101]. Natural proteolysis interferes with proteomic analyses based on artificial additions of digestion enzymes [102–104]. Analyses accounting for natural proteolysis can complete proteome descriptions [105–111]. Hence unsupervised analyses assuming random cleavage might detect some actual non-tryptic peptides produced by natural proteolysis or spontaneous protein degradation (especially during sample preparation), potentially completing descriptions of non-canonical mitochondrial peptidomes.

Analyses examining associations between non-tryptic non-canonical peptides and previously detected corresponding non-canonical RNAs [22,23] could test whether non-tryptic peptides are false positives. Positive results would validate the existence of non-canonical transcriptions and translations, independently of the expected bias for tryptic peptides among peptide populations detected by unbiased analyses.

1.7. Hypotheses and Predictions

Here unbiased proteomic analyses search tryptic human proteome data [112] for peptides matching translations of del- and swinger-transformed versions of the human mitogenome, as done by previous biased analyses that assumed tryptic digestion [22,23]. Unbiased analyses assume random protein cleavage. They are applied to the same proteomic data as previous tryptic-biased analyses that detected non-canonical peptides that match del- and swinger-transformed versions of the human mitogenome (the latter according to three codon sizes, tri-, tetra- and pentacodons). Properties of detected non-canonical peptides are compared to those detected by classical, trypsin-biased analyses.

The working hypothesis predicts that unbiased analyses detect peptide populations biased towards trypsin-digestion. This result would mean that non-canonical peptides are not false positives. Unbiased analyses could not reconstruct tryptic experimental conditions unless a majority of detected peptides were true detection. The second aspect of the working hypothesis is that detected non-tryptic non-canonical peptides result from natural proteolysis, and hence are not false positives. In that case, these should map preferentially on detected non-canonical RNAs, as previously observed for tryptic non-canonical peptides.

The primary aim is to test whether conclusions from previous results obtained by trypsin-biased analyses can be qualitatively reproduced by unbiased (unsupervised) analyses, considering potential natural proteolysis/spontaneous protein degradation, rather than experimentally added trypsin. Confirming natural proteolysis and expanding the coverage of the non-canonical mitoproteome are secondary aims. Analyses are restricted to predictions of peptides encoded by the mitogenome and its various systematic del- and swinger transformations.

2. Materials and Methods

Materials and methods are essentially identical to the corresponding sections for peptides translated from del-transformed versions of the human mitogenome [22], and those translated from the swinger-transformed versions of the human mitogenome [23]. The only difference is in the fact that the proteomic search software Proteome Discoverer 1.3 (Thermo Fisher Scientific, Illkirch) is set to analyze proteomic data digested by 'no enzyme'. The same data as previously are analyzed [112].

As for previous analyses [22,23], associations between detected non-canonical peptides and corresponding detected non-canonical RNAs are based on human transcriptomic data [113], as previously presented (del-RNAs [22], therein Tables 1 and 2; swinger-RNAs [23], therein Table 1 and supplement). For swinger-transformed versions of the human mitogenome, predicted peptides are translated according to each tri-, tetra- and pentacodons, as previously described [22,23].

All frames of transformed sequences were translated according to the vertebrate genetic code, three, four and five frames for each positive and negative strands, for codon sizes three, four and five, respectively. For codon sizes above three, codons are translated according to the genetic code, expanding the codon by silent mono- and dinucleotides, respectively. The next codon in these cases does not include the silent nucleotide(s) (see Fig. 1). The hypothetical peptides translated from non-canonical mitogenome transformations and along non-canonical

codon sizes were trypsinized *in silico*, to create the fasta file containing predicted peptides.

Stop codons are translated by the letter 'X', which the software Proteome Discoverer recognizes as Leu or Ile (not distinguishable by mass spectrometry because of equal masses). Each predicted peptide including at least one stop is represented 19 times in the input database of hypothetical predicted peptides, replacing all stops by one among the 18 remaining amino acid species, excluding Leu and Ile.

Consensus searches were handled with the Sequest (Thermo Fisher Scientific, Illkirch) algorithm with molecular mass tolerances: Parent = 1 Da and Fragment = 0.5 Da (monoisotopic masses). I activated fixed carbamidomethyl (C) and variable Oxydation (M) modifications, as well as the lysine → pyrrolysine modification.

2.1. Why Include Lysine to Pyrrolysine Modifications?

An anonymous reviewer notes that pyrrolysine is not a lysine modification, but is usually encoded by UAG stop codons [114–116]. It is presumably not encoded in eukaryotes. There are several reasons for allowing lysine → pyrrolysine modifications, despite that this probably increases search times. The first reason is methodological: results of the present analyses have to be comparable to previous searches, which allowed this modification. The second reason is that non-canonical peptides might result from mechanisms that are relicts from the mitochondrion's bacterial ancestors, which probably did translate UAG by pyrrolysine.

Hence allowing this modification might enable detecting peptides that otherwise would not be detected, when stops are assumed translated by lysine. The software does not differentiate between 'regular' lysine and lysine translated by stops. Analyses presented here do not explore issues implied by modifications, but presented data include that information for future analyses.

2.2. Unbiased Analyses Can't Include Nucleus-Encoded Proteins

This anonymous reviewer also indicates that analyses should ideally include the predicted canonical human nuclear-encoded proteome, including the mitochondrial nuclear-encoded proteins imported from the cytosol. These canonical proteins would provide valuable controls for analyses designed to detect non-canonical peptides.

First, they would prevent spurious matches between observed mass spectra and predicted non-canonical peptides resembling canonical peptides. Secondly, one expects much fewer detections of non-canonical than canonical peptides, an additional prediction that can be tested. Third, such analyses would enable to test the hypothesis that higher proportions of non-canonical than canonical peptides are non-tryptic (versus tryptic ones). This hypothesis assumes directed natural proteolysis of non-canonical, hence probably dysfunctional, peptides.

The first point is handled by a different, less time-consuming analysis described in the Results section, which shows that such spurious results are unlikely. Unfortunately, the nucleus-enocoded canonical proteome can't be included in unbiased analyses. This is not only because results from unbiased analyses have to be compared to previous biased analyses that did not include the canonical nucleus-encoded proteome (reanalyses including them are planned).

Unbiased analyses including the much larger canonical nucleus-encoded proteome are technically impossible with available computing capacities. Including these canonical proteins would manifold increase numbers of predicted peptides to be matched with observed mass spectra. This would render analyses impractical, to unknown extents, as searches excluding canonical nucleus-encoded proteins last 10 days. Hence inclusion of these controls will have to wait for commercial availability of computers and software with parallel processing capacities greater than those used now (I use a machine that has 32 parallel processors, regular PCs have 2 processors). It is adequate to remind here that analyses reported here are already control analyses for previous

Table 1

Abundances of residues at carboxyl extremities of non-canonical peptides detected by unbiased analyses. Analyses assume random cleavage of tryptic human mitoproteome. Peptides are translated from the del-, swinger-transformed human mitogenome, for codons expanded by 0–2 silent nucleotides. Column 1 indicates the residue. Columns 2, 6, 10 and 14 are numbers of detected peptides with residue indicated in 1, for each analysis assuming different transcription/translation (del-, swinger-, tetra- and pentacodon); 3, 7, 11 and 15 indicate total number of that residue in corresponding translations of the mitogenome; 4, 9, 12 and 16 indicate the bias of detecting peptides with that residue in carboxyl terminus position considering the total frequency of the residue in the corresponding translation of the mitogenome; 5, 9, 13 and 17 indicate numbers of peptides mapping on corresponding detected non-canonical RNAs. The two last lines compare results when merging tryptic vs other peptides, numbers of non-canonical peptides mapping on non-canonical RNAs are followed by expected numbers assuming random mapping.

1	2	3	4	5	6	7	8	9	10	11	12	13	14	15	16	17
AA	Del				Swinger				Swinger				Swinger			
	Tri	Genome	Bias	RNA	Tri	Genome	Bias	RNA	Tetra	Genome	Bias	RNA	Penta	Genome	Bias	RNA
A	10	8328	1.49	0	2	46,838	0.30	0	6	47,048	0.75	2	9	46,178	1.33	0
C	4	5680	0.87	0	0	22,697	0.00		5	22,836	1.29	0	0	22,108	0.00	0
D	1	4662	0.27	0	3	21,545	0.97	0	0	21,752	0	1	3	20,795	0.98	0
E	6	6281	1.18	0	4	24,954	1.12	0	7	25,290	1.63	0	4	24,200	1.13	0
F	5	7868	0.79	0	4	30,878	0.91	0	4	30,982	0.76	0	4	29,626	0.92	1
G	12	12,857	1.16	0	9	57,120	1.10	1	12	57,648	1.22	0	9	55,452	1.11	0
H	0	6578	0.00		4	22,775	1.23	0	7	22,836	1.80	1	5	21,803	1.56	0
IL	26	27,890	1.15	3	11	97,382	0.79	0	15	97,914	0.90	1	6	93,464	0.44	0
K	16	8142	2.43	2	10	30,603	2.29	0	14	30,982	2.65	0	14	29,608	3.22	2
M	6	7581	0.98	0	2	22,553	0.62	0	4	22,836	1.03	0	4	21,660	1.26	0
N	5	7723	0.80	0	7	26,449	1.85	0	6	26,660	1.32	0	5	25,427	1.34	1
P	9	12,857	0.87	3	6	57,489	0.73	0	7	57,648	0.71	0	7	55,452	0.86	0
Q	7	5647	1.53	2	6	24,126	1.74	0	13	24,206	3.15	0	4	23,422	1.16	0
R	9	5876	1.90	0	11	46,786	1.64	0	18	47,048	2.25	0	10	45,626	1.49	0
S	4	16,856	0.29	1	5	68,457	0.51	0	2	68,800	0.17	3	5	65,983	0.52	0
T	3	11,954	0.31	0	9	46,822	1.34	0	1	47,047	0.13	0	4	44,813	0.61	0
V	11	11,954	1.14	2	10	46,591	1.50	1	4	47,047	0.50	4	7	44,813	1.06	1
W	9	6838	1.63	1	3	23,960	0.88	0	4	24,206	0.97	0	1	23,118	0.30	0
Y	5	7630	0.81	0	0	23,238	0.00	0	2	22,836	0.51	0	4	21,364	1.28	0
Tot	148	183,202		14/6.31	106	741,263		2/3.02	127	745,622		12/3.38	105	714,912		5/2.94
Tryps	25	14,018	2.21	2/1.07	21	77,389	1.90	0/0.68	32	78,030	2.41	0/0.19	24	75,234	2.24	2/0.70
Others	123	169,184	0.90	12/5.24	85	663,874	0.90	2/2.34	95	667,592	0.84	12/3.19	81	639,678	0.85	3/2.24

results. Hence inclusion of further controls, though valuable, has always an arbitrary component, besides the above noted technical problems.

2.3. Peptide Detection Criteria

False discovery rates FDR [117–119] were estimated against a reverse decoy database using the Percolator algorithm. No protein grouping was allowed since the database only contained non-redundant entries. Peptides are considered detected with FDR q < 0.05 and Xcorr > 1.99. FDR is calculated by comparing Xcorr obtained from expected and observed MS/MS mass spectra with those obtained for a decoy database of false negative predicted peptides.

Xcorr is a cross-correlation statistic that compares observed and predicted MS/MS data. It sums the products between observed (y) and

expected (x) values for series of data. In this case, these are the observed and expected mass spectrometry data [120]:

$$Xcorr = \sum_{i=0}^{n-1} x(i) * y(i + \tau),$$

where τ is a displacement (lag) between observed and expected data for position i, with n positions in the data.

Peptide posterior error probabilities (PEP) are also indicated. PEP estimates confidence in detections of specific individual peptides. This approach differs from q, designed to estimate confidence for groups of detected peptides. The latter optimizes between false positive and false negative rates. PEP should be used with caution because it inflates false negatives [117]. Analyses focus on peptide populations and hence do not integrate PEP, but PEP is indicated because it could be useful in the context of future analyses focusing on specific peptides.

3. Results

3.1. Unsupervised Analyses Assuming Random Cleavage

Proteomic analyses of the 96 human proteome extracts [112], when no specific cleavage enzyme is specified, lasted about 10 days for each unsupervised analysis. Four such unbiased 10-day long searches were performed, for peptides matching translations of the nine del-transformations of the human mitogenome (four del-transformations deleting a mononucleotide after each trinucleotide, and five del-transformations, deleting a dinucleotide after each trinucleotide), and of the 23 swinger-transformed versions of the human mitogenome. The latter are translated according to three codon sizes: regular tricodons, tetra- and pentacodons, which are regular codons expanded by silent mono- and dinucleotides.

Comparable analyses of these data and predicted peptides, but trypsin-biased, last 8–9 h for each analysis. The longer times required

Table 2

Pearson correlation coefficient r between abundances of non-canonical peptides detected by unsupervised proteomic analyses of trypsin-digested human mitochondrial proteomic MS/MS data and abundances of corresponding, previously detected non-canonical RNAs [22,23]. Correlations are calculated separately for tryptic peptides (carboxyl extremity K or R) and other peptides. Non-canonical transcripts are del-and swinger-transformations of the human mitogenome, the latter translated along codons expanded by 0, 1 and 2 silent nucleotides. P values are one tailed, expecting positive correlations. Fisher's method for combining P values sums the $-2 \times \log Pi$, where i runs from 1 to k. This sum follows a chi-square statistic distribution with $2 \times k$ degrees of freedoms, where k is the number of Ps combined (here k = 4). Bold indicates statistical significance at P < 0.05.

Pearson r	Unbias Tryps		Other		
Transformation	r	P	r	P	All
Del	0.358	0.172	0.270	0.241	0.401
Swinger	**0.446**	0.016	0.171	0.217	0.253
Swinger tetra	0.109	0.310	0.099	0.327	0.143
Swinger penta	0.192	0.190	0.306	0.078	0.186
Combined chi	**17.41**	0.026	13.24	0.104	

for analyses reflect much greater potential cleavage combinations when comparing observed and expected peptide mass spectra for unbiased analyses.

3.1.1. Unsupervised Analyses: Bias for Tryptic Peptides

All four unsupervised searches matching observed MS/MS mass spectra with predicted ones detected preferentially tryptic peptides (carboxyl-extremity K or R, Table 1). Hence without *a priori* biasing searches, populations of detected non-canonical peptides are for these analyses biased towards tryptic peptides.

Biases for residue identity at the carboxyl extremity of detected peptides are calculated as the frequency of observing a given amino acid at that position in detected peptides, divided by that amino acid's frequency in hypothetical peptides translated from the corresponding complete transformed human mitogenome. The highest bias favors lysine (K) for translation of delRNAs, swinger RNAs, and swinger RNAs according to pentacodons, and second highest (after bias for Q) for swinger RNAs according to tetracodons. Lysine is one among 19 possibilities, so the probability to obtain the strongest bias for K is $1/19 = 0.053$, for tetracodons the result has $P = 0.11$.

According to Fisher's method for combining independent P values [121], the overall result for bias in favor of K across all four analyses is $P = 0.0046$. Hence results show a strong bias favoring K at the end of detected peptides, despite assuming random cleavage.

Bias for arginine (R) at the carboxyl extremity of detected peptides is the second highest for peptides matching del-transformed versions of the human mitogenome (excluding K, this has $P = 1/18 = 0.056$), the fourth highest for regular translations of the 23 swinger-transformed versions of the human mitogenome ($P = 0.17$), and the third highest for their translation according to tetra- and pentacodons ($P = 0.11$, each). According to Fisher's method for combining independent P values [121], the overall result for bias in favor of K across all four analyses is $P = 0.02$.

Bias for combined K and R ending peptides is highly significant according to chi-square tests for each of the four independent analyses: delRNAs, $P = 0.00023$; swinger RNAs according to tricodons, $P = 0.0016$; tetracodons, $P = 0.00000006$; and pentacodons, $P = 0.000038$.

This means that unsupervised searches for non-canonical peptides detect preferentially non-canonical peptides that match the known tryptic sample preparation. This result could not be obtained if the majority of detections were false positives. Hence these biased results obtained from unbiased analyses confirm that the various populations of non-canonical peptides (peptides translated from delRNAs, and from swinger RNAs, these translated according to regular tri-, tetra and pentacodons) exist. This result is not trivial, and independently confirms conclusions from previous trypsin-biased analyses [22,23].

3.2. Search Bias and Absolute Versus Relative Majority of Tryptic Peptides

An anonymous reviewer notes that tryptic peptides are only a relative majority among detected peptides, rather than an absolute majority. This might to some extent contradict the above conclusion of bias towards tryptic peptides. This issue can be understood by comparing the tryptic bias obtained from unbiased analyses, to that obtained for chymotrypsin-biased analyses.

These chymotrypsin-biased analyses were for the same proteomic data and the same predicted non-canonical peptides ([122], therein supplementary data). They differ from previous analyses assuming tryptic digestion, and the current unbiased ones, because analyses assumed chymotryptic digestion. Analyses assuming chymotryptic digestion detected 479 non-canonical peptides, among which 131 (27.35%) had carboxyl terminal residues matching chymotryptic digestion (W, Y and F). All other peptides (72.65%) were tryptic. Hence the absolute majority of peptides detected by these analyses biased towards three possible non-tryptic carboxyl terminal residues are tryptic peptides, as expected.

These chymotrypsin-biased analyses were completed by three separate analyses biased to detect only peptides with a specific, chymotryptic ending, hence separately W, Y, or F. In these analyses set to detect peptides with only one possible chymotryptic residue at its carboxyl terminus, tryptic peptides represent on average across all analyses 87.3% of all detected peptides.

In order to compare these results with those from unbiased analyses presented here, I calculated the bias between tryptic peptides and peptides matching each other possible (non-tryptic) carboxyl terminus for data in Table 1. Tryptic peptides are on average 82.76% of the peptides in Table 1 when considering only one alternative residue at the carboxyl terminus. This average value is very comparable to the above mentioned 87.3% tryptic peptide majority obtained for chymotrypsin-biased analyses searching for only one of the three chymotryptic carboxyl termini.

The meaning of this is mathematically trivial. Tryptic bias decreases the more other options are allowed. It is highest when analyses are biased towards tryptic peptides: all detected peptides were tryptic. This bias decreases when analyses are biased towards a single different possibility (separating W, Y and F). Tryptic bias further decreases when all three chymotryptic carboxyl termini are considered (W, Y and F). Tryptic bias is lowest, yet still statistically significant, when analyses are unbiased, as when considering all results in Table 1. When averaging abundances of non-tryptic carboxyl termini in Table 1, the tryptic bias is comparable to that obtained for analyses biased towards only one non-tryptic carboxyl terminus, as obtained for analyses biased separately towards W, Y or F.

The issue of relative versus absolute tryptic majority is only a matter of doing adequate comparisons. I agree that adding canonical nucleus-encoded proteins in the analyses would probably yield valuable further insights in this context, regarding potential biases for tryptic canonical peptides, versus more non-tryptic natural digestions for non-canonical peptides. Currently such analyses are technically impossible in the context of unbiased analyses.

3.3. Associations between Non-canonical RNA and Peptide Abundances

Peptides with non-tryptic carboxyl extremity could represent false positive detections. They might alternatively result from natural proteolysis/spontaneous degradation of proteins. The bias for tryptic peptides (previous section) corresponds to experimental trypsin-preparation of extracts. Remaining non-tryptic peptides are not necessarily false positives.

This can be tested by exploring associations between non-canonical peptides and corresponding non-canonical RNAs, as previously described for trypsin-biased analyses [22,23]. Two independent methods are used in this respect: (a) Pearson correlation analyses between abundances of detected non-canonical peptides and corresponding RNAs, expecting positive correlations between abundances; (b) precise mapping of individual peptides and RNAs, which expects that more detected peptides map on detected RNAs than expected by chance. These positive associations between non-canonical peptides and RNAs would show the regular causal link between RNA and peptides, for the various non-canonical transcriptions and translations.

Numbers of non-canonical peptides are counted from lists of peptides in the supplementary data, mitogenome coverages by non-canonical RNA for delRNAs and for swinger RNAs are from previous publications (delRNAs, [22], therein Tables 1 and 2, for systematic mono- and dinucleotide deletions, respectively; swinger RNAs [23], therein Table 1). These non-canonical RNAs had been detected from human transcriptome data [113], using blastn with Megablast default alignment parameters [123].

For delRNAs, correlation analyses of associations between abundances depend on nine observations, based on coverages of the del-transformed human mitogenome. There are four and five observations, according to the four frames of systematic mononucleotide deletions, and the five frames of systematic dinucleotide deletions, respectively (data in [22], therein Tables 1 and 2). Correlation analyses for swinger transformations are based on 23 observations, for each swinger transformation of the human mitogenome (RNA coverage data from [23], therein Table 1).

For tryptic peptides detected by unbiased analyses, abundances of detected swinger peptides coded by regular codons are proportional to corresponding swinger RNA coverage of the human mitogenome ($r = 0.45$, $P = 0.016$, one tailed test). Correlations are positive but not statistically significant at $P < 0.05$ for swinger analyses of other codon sizes, and del-transformations. Combining the four P values using Fisher's method for combining P values [121] yields an overall significant positive association ($P = 0.026$). These results from unbiased analyses confirm previous trypsin-biased analyses [22,23].

Correlation analyses for non-tryptic peptides detected by unbiased analyses are also positive, though never statistically significant at $P < 0.05$, also not after combining P values ($P = 0.104$). These weaker associations between peptide abundances and RNA coverage for non-tryptic peptides suggest that a greater proportion of these peptides could be false positive detections, though the overall positive trends are rather compatible with them resulting from natural proteolysis. Local mappings on RNAs below test this point.

3.4. Local Mapping of Non-canonical RNA and Peptides

Some peptides detected by unsupervised proteomic analyses map on previously detected, corresponding non-canonical RNAs (Table 1). Previous similar analyses for non-canonical peptides detected by trypsin-biased searches showed that detected peptides map more frequently than expected by chance on corresponding detected non-canonical RNAs, for del- and swinger-transformations [22,23].

Analyses across unsupervised analyses for del- and swinger-transformations find that 4 among 102 detected non-canonical tryptic peptides (3.9%) map on previously detected non-canonical RNAs. Only 2.64 among these 102 detected tryptic peptides should map on detected RNAs if mapping is random. Small sample size does not enable statistical testing, but suggests a non-significant difference corresponding to the expected association between non-canonical RNAs and peptides. Though this result is not statistically significant, it should be considered as confirmative as it is in line with previous, statistically significant results for tryptic peptides detected by analyses biased towards tryptic peptides.

For non-tryptic peptides detected by unsupervised analyses, 29 among 388 (7.5%) map on previously detected RNAs. This is statistically significantly more than the 13.01 expected according to random mapping (chi-square test, $P = 0.000009$). Rates of mapping on RNAs do not differ between tryptic and non-tryptic peptides according to a chi-square test. Hence overall, there are not more false positive detections of non-tryptic than tryptic peptides.

In total, 33 among the 490 non-canonical peptides detected by unbiased analyses map on non-canonical RNAs (6.7%), which is statistically significantly more than the 15.65 expected by chance (chi-square test, $P = 0.000012$).

These results for non-tryptic peptides detected by unbiased analyses confirm conclusions about non-canonical transcriptions and translations, independently of previous results for tryptic peptides detected by trypsin-biased proteomic analyses. Notably, results from independent analyses strengthen conclusions that swinger RNAs are also translated according to tetra- and pentacodons.

3.5. Unique Versus Multiple Detections of Tryptic Peptides

Unbiased analyses confirm previous trypsin-biased analyses in two ways. First, they detect preferentially tryptic peptides. This corresponds to the tryptic experimental design. Second, detected peptides associate with previously detected RNAs, for tryptic and other peptides, as found in previous publications for tryptic peptides detected by trypsin-biased analyses [22,23].

These results independently confirm trypsin-biased analyses because most tryptic peptides detected by trypsin-biased analyses differ from those detected by the unbiased analyses presented here. Only some tryptic peptides detected by trypsin-biased analyses are also detected by unbiased analyses (one or two peptides). Hence analyses confirm independently of previous analyses, the previous results for trypsin-biased analyses.

Note that analyses of the same data, testing the same hypotheses, and assuming chymotryptic digestion, yield similar conclusions. These analyses detect majorities of tryptic peptides. Both tryptic and chymotryptic peptides associate with detected RNAs [122]. Hence unbiased analyses yield a third independent confirmation of results obtained by tryptic- and chymotryptic-biased analyses.

3.6. Negative Control: Residues after the Carboxyl-Terminal of Detected Peptides

A further analysis shows a peculiar unknown fact about natural mitochondrial proteolysis. Unbiased analyses yield peptide populations with diverse residues at their carboxyl extremity, which might mainly reflect proteolysis by naturally occurring digestion enzymes in human mitochondria. The alternative hypothesis (or rather the null hypothesis) is that peptides were actually randomly cleaved, which would also be compatible with random peptide detections, possibly due to majorities of false detections.

Biases for tryptic peptides overall falsify the random cleavage hypothesis, but remaining peptide populations, after excluding tryptic peptides, might nevertheless fit random cleavage. Here control analyses test for this by recording the first amino acid expected according to non-canonical translations, after the carboxyl extremity of the detected peptides. According to the compilation by ExPaSy PeptideCutter (http://web.expasy.org/peptide_cutter/peptidecutter_enzymes.html, accessed 6VI2016), the majority of listed specific cleavage rules relate to the carboxyl extremity of peptides, rather than the N-terminal of the next peptide. Nevertheless, the possibility that populations of detected peptides include biases for N-terminal cleavage in relation to the 'downstream'-encoded amino acid is plausible.

Table 3 presents biases, calculated as for Table 1 by using total abundances of amino acids, for amino acids at the N-terminal of the peptide located after the detected peptides, for the various unsupervised analyses (peptides translated from del- and swinger-transformed human mitogenome, and translated according to tetra- and pentacodons for the swinger-transformed versions).

These biases do not resemble biases detected for the carboxyl extremity of the detected peptides, when considering the same amino acid species. Bias distributions are systematically less extreme for N-terminals of the next undetected peptide than for the carboxyl extremity of detected peptides. For the N-terminal of the next peptide, the lowest bias is 0.44, 0.40, 0.33 and 0.42 for detected peptides translated from del-, swinger-transformed versions of the human mitogenome, and for swinger-transformed mitogenome translations according to tetra- and pentacodons. Such biases below "1" indicate cleavage avoidance. For the carboxyl terminal of detected peptides, corresponding minimal biases are 0 for each non-canonical translation, the strongest possible negative bias.

Maximal biases for N-terminals of undetected peptides next to detected peptides are 1.59, 1.98, 1.64 and 2.31. For carboxyl extremities of detected peptides, corresponding maximal biases are 2.43, 2.29, 3.15 and 3.22. Overall, distributions for biases of amino acid identities at the N-terminal of next peptides are much closer to the value '1', indicating no bias, and seem random around this value. This suggests that there is no evidence for N-terminal specific cleavage in these data for the human mitochondrial proteome.

Table 3
Bias in amino acid identity at the N-terminal (column 1) of the peptide after detected peptides, for unbiased analyses assuming random cleavage. Analysis search for peptides matching translations of the del- (columns 2–4) and swinger-transformed human mitogenome (columns 5–7), and translations of the swinger mitogenomes according to tetra- (columns 8–10) and pentacodons (columns 11–13). Columns 2, 5 , 8 and 11 inicate numbers of detections. 'Genome' (columns 3, 6, 9, 12) indicates abundances of that residue in the corresponding hypothetical translations of the complete mitogenome after transformations and non-canonical translations. Biases (columns 4, 7, 10, 13) do not resemble those for carboxyl-extremities of detected peptides (Table 1) and are less extreme. Overall they match random distributions around '1', indicating lack of bias. This suggests that there is no or very little natural proteolysis with cleavage specificity related to the N-terminal of peptides after detected peptides.

1	2	3	4	5	6	7	8	9	10	11	12	13
AA	Del			Swinger			Swinger			Swinger		
	Tri	Genome	Bias	Tri	Genome	Bias	Tetra	Genome	Bias	Penta	Genome	Bias
A	8	8328	1.19	6	46,838	0.79	10	47,048	1.09	3	46,178	0.42
C	4	5680	0.87	3	22,697	0.82	6	22,836	1.34	8	22,108	2.31
D	6	4662	1.59	3	21,545	0.86	7	21,752	1.64	5	20,795	1.54
E	6	6281	1.18	8	24,954	1.98	7	25,290	1.41	2	24,200	0.53
F	5	7868	0.79	2	30,878	0.40	5	30,982	0.82	2	29,626	0.43
G	13	12,857	1.25	8	57,120	0.87	8	57,648	0.71	4	55,452	0.46
H	6	6578	1.13	4	22,775	1.09	4	22,836	0.90	2	21,803	0.59
IL	20	27,890	0.89	20	97,382	1.27	25	97,914	1.30	18	93,464	1.23
K	5	8142	0.76	4	30,603	0.81	5	30,982	0.82	4	29,608	0.86
M	5	7581	0.81	6	22,553	1.64	5	22,836	1.12	4	21,660	1.18
N	9	7723	1.44	2	26,449	0.47	6	26,660	1.15	3	25,427	0.75
P	11	12,857	1.06	5	57,489	0.54	12	57,648	1.06	11	55,452	1.27
Q	2	5647	0.44	4	24,126	1.24	4	24,206	0.84	4	23,422	1.09
R	3	5876	0.61	8	46,786	1.06	3	47,048	0.33	13	45,626	1.82
S	12	16,856	0.88	8	68,457	0.72	19	68,800	1.41	11	65,983	1.06
T	11	11,954	1.14	12	46,822	1.58	9	47,047	0.98	8	44,813	1.14
V	9	11,954	0.93	13	46,591	1.72	5	47,047	0.54	6	44,813	0.96
W	7	6838	1.27	2	23,960	0.52	2	24,206	0.42	2	23,118	0.55
Y	6	7630	0.97	2	23,238	0.53	4	22,836	0.90	2	21,364	0.60
Tot	148	183,202		120	741,263		146	745,622		112	714,912	

These analyses show non-random patterns in cleavages for detected non-canonical peptide populations, for carboxyl termini. In this respect, results for N-termini function as negative controls and strengthen confidence in results.

3.7. Few Nuclear Contaminations: Peptides Follow the Mitochondrial Vertebrate Code

Eukaryotic nuclear genomes include numerous inserts of the mitogenome. Hence detected non-canonical peptides could originate from non-canonical transcriptions and translations of such nuclear mitogenome inserts, or from translations of nuclear sequences that by chance resemble the transformed mitogenome. This possibility is tested by translating the transformed mitogenome using the nuclear genetic code, and by checking whether detected non-canonical peptides are compatible with translation according to the nuclear genetic code.

Considering that coding assignments of 60 among 64 codons (93.75%) are identical for the nuclear and the vertebrate mitochondrial genetic codes, I calculated numbers of peptides, considering lengths of detected peptides, expected to match also nuclear genetic code translation. I used equation $N \times 0.9375^{-k}$, where N is the number of detected peptides with k residues. This equation expresses the fact that all codons coding for the peptide must be among those invariant between the two genetic codes, when one or more codons belong to the four codons differing between these two genetic codes, the detected peptide is incompatible with the nuclear genetic code.

There are 177.86 peptides expected compatible with both codes across all analyses. This is far more than the 91 detected non-canonical peptides with translations identical according to both genetic codes. Comparisons between expected and observed peptides compatible with translation according to the nuclear genetic code, separately according for the four different non-canonical transcriptions and translations, follow the same principle: observed numbers of peptides compatible also with the nuclear genetic code are far fewer than expected (Table 4).

This bias means that observed non-canonical peptides match specifically more than expected by chance translation according to the mitochondrial vertebrate genetic code. This result also excludes that

detections of non-canonical peptides are incorrect, that these mass spectra actually correspond to similar, nucleus-encoded canonical peptides. This is because the analysis reported in this section accounts for the extreme and plausible situation where sequences identical to the mitogenome were translated. The fact that analyses differentiate between nuclear versus mitochondrial translations of the mitogenome is incompatible with nuclear contaminations.

4. General Discussion

Analyses presented here are mainly designed to test conclusions from previous analyses of the human mitochondrial peptidome (data from [112]), where non-canonical peptides matching translations of del- and swinger-transformed versions of the human mitogenome were detected, including translations of expanded codons. Del-transformations assume transcription that systematically deletes mono- and dinucleotides after every third transcribed nucleotide [22]. Swinger-transformed RNAs result presumably from systematic nucleotide exchanges, during transcription along 23 exchange rules, also called bijective transformations [34]. The human proteome includes peptides matching detected swinger RNA, translated according to tri-, tetra- and pentacodons (expanded by silent mono- and dinucleotides) [23].

Table 4
Observed (column 4) and expected (column 5) numbers of detected non-canonical peptides compatible with translations according to each nuclear and mitochondrial vertebrate genetic codes. Predictions account for peptide length (mean length and standard deviation in columns 2 and 3), considering that translation of 60/64 (0.9375) codons is identical between these genetic codes. Results indicate strong biases against detection of peptides compatible with both genetic codes, showing that detected populations of peptides are specifically translated according to the mitochondrial vertebrate genetic code. This systematic bias excludes that detected non-canonical peptides have cytosolic origins.

1	2	3	4	5
Transformation	AAs	Sd	Obs	Exp
Del	18.28	5.96	23	48.75
Swinger tri	21.23	9.75	27	36.26
Swinger tetra	17.37	7.38	19	52.45
Swinger penta	17.37	7.79	23	40.40

These previous analyses assumed tryptic proteome preparation [112]. Hence the first set of analyses was biased by information corresponding to sample preparation. Here analyses of the same data were repeated without using that information on tryptic-digestion, but assuming random cleavage. Results indicate a positive bias towards detection of tryptic non-canonical peptides by unsupervised analyses. This result is a strong confirmation that overall, populations of detected non-canonical peptides are not false positives: otherwise, unbiased analyses would not detect positive bias for tryptic peptides. This implies that these non-canonical transcriptions and translations are a biological reality.

Results of unbiased analyses also suggest the possibility that the proteome underwent other specific cleavages, presumably resulting from natural proteolytic activity in the biological sample, such as described for chymotrypsin [116]. Overall, tryptic and non-tryptic non-canonical peptides associate with previously detected corresponding non-canonical RNAs [22,23,116]. Convergences between peptide and RNA detections are further evidence that overall, tryptic and other detected peptides are not false positives.

In addition, detected non-canonical peptides preferentially match translation according to the vertebrate mitochondrial genetic code: fewer than expected by chance are compatible with translation according to the nuclear genetic code, considering that 93.75% of codons follow the same translation rules according to both genetic codes. This result is incompatible with detection of peptides originating from the cytosol, even for nuclear DNA sequences identical to the mitogenome.

4.1. Statistical Considerations and Peptide Detection

One can argue that non-canonical peptides were detected by chance and are false positives, due to a very large number of comparisons between predicted peptides and observed mass spectra. If it was so, (a) peptide detections would not be biased towards independently detected RNAs, (b) towards translation specific to the vertebrate mitochondrial genetic code, and (c) peptide populations detected by unsupervised analyses would not be biased towards experimental tryptic cleavage. In addition, peptide detections are confirmed by false detection rates q, based on decoy peptides that function as negative controls. FDR takes into account sample sizes (as do usual P values), but also the number of statistical tests done.

The last point in this argument is because analyses account that at stops, every possible amino acid could be inserted. Hence matching observed and expected peptides based on their molecular weight is not sufficient to ascertain the sequence of the peptide: the program can adjust any MS/MS spectrum with a close weight to one of the 19 peptides produced by sequences including at least one stop.

This point does not consider that MS/MS spectrometry accounts not only for total mass, but also for masses of secondary fragments. The simple example of peptide EFG can be helpful here. EFG has the same molecular weight as peptides EGF, FEG, FGE, GEF and GFE and can't be differentiated from these five other peptides by its total mass. However, the estimate of that mass is typically combined with estimates of secondary fragments. Only two peptides, EFG and GEF are compatible with detection of the mass of EF. The same point is valid for observing a mass corresponding to FG, which is also compatible with two peptides, EFG and FGE. The combined observation of masses corresponding to these two fragments characterizes the entire peptide sequence.

In addition, mass spectrometry analyses consider separately b and y ions. Hence the same sequence characterization may occur independently according to both ion types. The score Xcorr integrates these pieces of information, and is the statistic on base of which FDR is calculated to minimize false positives. In fact, numerous tryptic peptides were not detected by original analyses assuming that tryptic digestion was detected by analyses assuming non-tryptic digestion. In addition, previous analyses showed that tryptic peptides detected twice, by

analyses assuming tryptic and chymotryptic digestions do not differ in detection accuracy from those detected only by one of these analyses [116]. This suggests that the methodology used for peptide detections is rather prone to false negatives, rather than false positives. False positives are probably a small minority reduced to few individual cases that would not qualitatively alter conclusions.

4.2. Potential Confounding Factors: Nuclear Contaminations

The first detected swinger-transformed sequences are RNA and DNA sequences in Genbank's databases (EST data for RNA) longer than 100 nucleotides. These were detected by blast using default megablast alignment search parameters for input sequences consisting of in silico swinger transformed mitogenome versions [30–33]. The detected GenBank sequences aligning with high identity levels with in silico produced swinger mitogenome versions (>90% identity) were sequenced by the classical Sanger technology. Similar searches in Genbank's human transcriptome SRA (sequence read archives) data produced by RNA seq (Illumina) next generation sequencing technology using blastn (also with default search parameters) confirmed the relative abundances of swinger RNAs [23].

Further megablast analyses could not detect alignments between any human nuclear chromosome sequence and del-, swinger-transformed mitogenome versions. However, blastn analyses detected such alignments that could potentially confound several alignments between the transformed mitogenome and RNA seq data. Nevertheless, the majority of RNA seq alignments are due to RNAs originating from the mitochondrion, and are not nuclear, because identities between the transformed mitogenome versions and RNA seq sequences are greater than with corresponding nuclear chromosome sequences, and this for each del- and swinger-transformed mitogenomes [22,23].

Note that nuclear chromosome sequence alignments with the del- and swinger-transformed mitogenome imply that besides regular mitochondrial mitogenome inserts in nuclear chromosomes (numts, [124–141]), transformed versions of the mitogenome (or part of) occur in the nuclear genome. Alternatively, regular numts are transcribed according to del and swinger non-canonical systematic transformations. At this point, the main issue is the existence of polymerizations producing systematic transformations, independently of cell compartment where these occur, or whether produced by replication, reverse transcription or transcription. Hence answering with certitude to these questions beyond explained above, though important, is secondary at this point.

In addition, nuclear contaminations are at most minor for peptides presented here, because detected non-canonical peptides are less frequently compatible with both nuclear and vertebrate mitochondrial genetic codes than expected by chance. This bias suggests high specificity for mitochondrial origin of detected peptides.

4.3. Potential Confounding Factors: Heteroplasmy

Heteroplasmy [142–144] is a further known phenomenon that could explain results. However, single nucleotide substitutions can't explain observations of long, non-canonical peptides. Hence only length heteroplasmies, especially those resulting from insertions, could by chance explain non-canonical peptides predicted from systematic mitogenome transformations.

However, the most common length heteroplasmies are relatively few and mainly located in the mitochondrial control region [145], while the peptides detected for the various non-canonical transcriptions and translations are distributed all around the mitogenome. This excludes length heteroplasmies as a major confounding factor for detections of non-canonical peptides.

4.4. Potential Confounding Factors: Fused Transcripts

Some transcripts result from fusion of RNA transcribed from DNA regions that are not contiguous [146,147]. This can result from reverse-transcription artifacts during cDNA production [148]. Fused swinger RNAs also exist [35]. Fusions of regular RNAs are unlikely to produce RNAs that would mimic products of systematically transforming transcriptions. Hence only few single detected non-canonical peptides could by chance correspond to RNA fusions. Artificial transcript fusions during cDNA production could not have produced detected peptides.

4.5. Natural Proteolysis of Canonical Versus Non-canonical Peptides

An anonymous reviewer suggested that proteomic analyses should include classical, canonical proteins. This would enable comparing results between canonical and non-canonical peptides, expecting fewer non-canonical peptides than canonical ones. In addition, the reviewer expected that non-canonical peptides would more frequently match non-tryptic, hence natural proteolysis, than canonical peptides. The rationale behind this prediction is that one could expect that non-canonical products are preferentially digested as waste than products of canonical genes.

Practical reasons prevented me from performing these tests. These additional analyses require including among predicted peptides the complete human proteome (corresponding to more than 20,000 genes). This increases numbers of predicted peptides to extents that, for unbiased analyses, are incompatible with current computing capacities. For these reasons, previous and current analyses have been restricted to peptides encoded by the human mitogenome, excluding nucleus-encoded mitochondrial proteins, which are imported from the cytosol into the mitochondrion [149–152].

A possible solution to this technical problem is to sample the canonical proteome. Analyses searching for peptides matching the swinger-transformed versions of the human mitogenome, translated according to regular tricodons, included such a control. These analyses included peptides predicted according to the regular translation of the untransformed human mitogenome, with the canonical mitochondrion-encoded genes. Fifteen among the detected peptides correspond to translation of the untransformed human mitogenome, among which a single tryptic peptide (6.7% of detected peptides encoded by the untransformed mitogenome). However, 20.9% of the remaining non-canonical peptides are tryptic.

This difference is not compatible with the hypothesis that natural proteolysis digests preferentially non-canonical peptides. However, this qualitative result is not statistically significant, due to small sample size. In addition, the fifteen peptides translated from the regular mitogenome are not restricted to canonical translation of the 13 proteins encoded by the human mitogenome. They include translations of other frames of these genes, and of other sequences (e.g. rRNAs etc). This hypothesis requires analyses specifically designed to test its predictions, which are beyond the frame of present analyses.

4.6. Amino Acids Inserted at Stops

A further useful comment by a reviewer suggested to investigate which amino acids are detected inserted at stops. Table 5 shows the distribution of amino acids inserted at stops for the various types of investigated non-canonical peptides, those translated from delRNAs, swinger RNAs, and from the latter, translated according to tetra- and pentacodons. These distributions overall resemble each other, hence biases for insertion of specific amino acid species at stops are explored for the sum of amino acids across all types of non-canonical peptides.

This distribution is compared to the distribution of amino acids in the 13 canonical, mitogenome-encoded proteins. Chi-square tests detect statistically significant positive biases for five amino acids, in decreasing order of bias: K, Q, C, D and E. The two first amino acids are

Table 5

Distributions of amino acids inserted at stops in detected non-canonical peptides (columns Del, Swinger tri, Swinger tetra and Swinger penta), compared to the distribution of amino acids in canonical proteins encoded by the human mitogenome (Mito). Bias is the ratio between the frequency of the amino acid across all non-canonical peptides (column All) and its frequency in canonical proteins. P values are calculated using a chi-square test. Statistically significant results at P < 0.05 are underlined, and in bold when these are positive biases indicating greater than expected insertions at stop codons.

AA	Mito	Del	Swinger tri	Swinger tetra	Swinger penta	All peptides	Bias	P
A	225	1	10	7	1	19	0.65	0.062
C	22	3	3	3	0	9	3.15	**0.002**
D	66	2	5	10	0	17	1.98	**0.010**
E	88	5	9	4	2	20	1.75	**0.020**
F	216	3	2	4	1	10	0.36	0.0006
G	212	16	9	12	1	38	1.38	0.058
H	97	0	4	4	0	8	0.64	0.208
I,L	963	19	6	19	46	90	0.72	0.0006
K	95	18	22	22	11	73	5.92	**4×10^{-40}**
M	208	8	9	6	5	28	1.04	0.853
N	164	8	7	5	3	23	1.08	0.723
P	219	5	3	9	5	22	0.77	0.237
Q	90	18	7	8	10	43	3.68	**2×10^{-14}**
R	63	4	2	2	3	11	1.35	0.359
S	274	9	5	10	10	34	0.96	0.796
T	351	10	8	6	5	29	0.64	0.013
V	167	5	5	5	2	17	0.78	0.328
W	104	2	2	3	3	10	0.74	0.356
Y	135	2	1	7	4	14	0.80	0.414

identical to regular amino acids found most frequently inserted at stops by Aerni et al. [153]. This is a further indication that results presented here are not due to random false detections. In addition, this suggests that the mitochondrial system for translating stops resembles that found in bacteria, at least that from *Escherichia coli*.

4.7. Associations Between Independent Transcriptomic and Proteomic Data

A further important point raised by an anonymous reviewer relates to the origins of transcriptomic data, which are from patients with myeloid leukemia, versus the origins of proteomic data, which are from healthy patients. I previously discussed this issue for analyses of these data [26] along the following lines.

It is clear that if RNA and peptide data were obtained from the same cells, associations between RNA and peptide data would be strongest. The strength of the association would decrease if RNA and peptide were from the same tissues of the same individual(s), but not the same cells. Along that rationale, they would further decrease if RNA and peptide data were obtained from different individuals with similar backgrounds (e.g. all healthy).

Current analyses were done on data that were available to this author, in formats readily analyzable by available software, and for adequate quantities of data. The RNA and peptide data differ in cells, tissues, individuals and backgrounds. This means that statistically significant associations were repeatedly detected between RNA and peptide data despite a number of confounding factors that could mask RNA-peptide associations. The fact that associations between non-canonical RNAs and peptides were nevertheless repeatedly detected implies that the actual phenomenon is much stronger than evaluated in these suboptimal conditions.

A noisy background is more likely to mask than create statistically significant signals. In addition, noise would only occasionally create spurious associations, but associations were repeatedly detected. In fact, discrepancies between RNA and peptide origins explain why relatively few detected peptides map on detected RNAs. Nevertheless, these discrepancies could not prevent detecting associations between non-canonical RNAs and corresponding peptides.

5. Conclusions

- Unbiased analyses assuming random cleavage for tryptic data yield results biased towards tryptic peptides for peptides translated from non-canonical RNAs and along non-canonical translations. Results confirm previous trypsin-biased analyses that detected non-canonical peptides.
- Detected non-canonical RNAs associate with tryptic and non-tryptic peptides.
- Detected non-canonical peptides are overwhelmingly incompatible with translation according to the nuclear genetic code, and specifically match the mitochondrial vertebrate genetic code.
- Overall, results confirm translation of non-canonical RNAs (del- and swinger RNAs), and along expanded codons, in addition to detections of other types of non-canonical peptides, such as peptides translated from contiguous regular and swinger-transformed RNA [36].
- Proteomic analyses assuming random cleavage detect non-canonical peptides digested by natural proteolysis, expand proteomic coverage.

Acknowledgments

This work has been carried out thanks to the support of the A*MIDEX project (no ANR-11-IDEX-0001-02) funded by the "Investissements d'Avenir" French Government program, managed by the French National Research Agency (ANR).

References

[1] Popov O, Segal DM, Trifonov EN. Linguistic complexity of protein sequences as compared to texts of human languages. Biosystems 1996;38:65–74.
[2] AbouHaidar MG, Venkataraman S, Golshani A, Liu B, Ahmad T. Novel coding, translation, and gene expression of a replicating covalently closed circular RNA of 220 nt. Proc Natl Acad Sci U S A 2014;111:14542–7.
[3] Arquès DG, Michel CJ. A complementary circular code in the protein coding genes. J Theor Biol 1996;182:45–58.
[4] Ahmed A, Frey G, Michel CJ. Frameshift signals in genes associated with the circular code. In Silico Biol 2007;7:155–68.
[5] Ahmed A, Frey G, Michel CJ. Essential molecular functions associated with the circular code evolution. J Theor Biol 2010;264:613–22.
[6] Michel CJ. Circular code motifs in transfer and 16S ribosomal RNAs: a possible translation code in genes. Comput Biol Chem 2012;37:24–37.
[7] Michel CJ. Circular code motifs in transfer RNAs. Comput Biol Chem 2013;45:17–29.
[8] El Soufi K, Michel CJ. Circular code motifs in the ribosome decoding center. Comput Biol Chem 2014;52:9–17.
[9] El Soufi K, Michel CJ. Circular code motifs near the ribosome decoding center. Comput Biol Chem 2015;59a:158–76.
[10] El Soufi K, Michel CJ. Circular code motifs in genomes of eukaryotes. J Theor Biol 2016;408:198–212.
[11] Itzovitz S, Alon U. The genetic code is nearly optimal for allowing additional information within protein-coding sequences. Genome Res 2007;17:405–12.
[12] Seligmann H, Pollock DD. The ambush hypothesis: hidden stop codons prevent off-frame gene reading. Midsouth Computational Biology and Bioinformatics Society; 2003[Abstract 36].
[13] Seligmann H, Pollock DD. The ambush hypothesis: hidden stops prevent off-frame gene reading. DNA Cell Biol 2004;23:701–5.
[14] Seligmann H. Cost minimization of ribosomal frameshifts. J Theor Biol 2007;249:162–7.
[15] Seligmann H. The ambush hypothesis at the whole organism level: off frame, 'hidden' stops in vertebrate mitochondrial genes increase developmental stability. Comput Biol Chem 2010;34:80–5.
[16] Seligmann H. Coding constraints modulate chemically spontaneous mutational replication gradients in mitochondrial genomes. Curr Genomics 2012;13:18–36.
[17] Jestin J-L, Soulé C. Symmetries by base substitutions in the genetic code predict 2(′) or 3(′) aminoacylation of tRNAs. J Theor Biol 2007;247:391–4.
[18] Jestin J-L. A rationale for the symmetries by base substitutions of degeneracy in the genetic code. Biosystems 2010;99:1–5.
[19] Fimmel E, Giannerini S, Gonzalez DL, Strüngmann L. Dinucleotide circular codes and bijective transformations. J Theor Biol 2015;386:159–65.
[20] Gumbel M, Fimmel E, Danielli A, Strüngmann L. On models of the genetic code generated by binary dichotomic algorithms. Biosystems 2015;128:9–18.
[21] Kozyrev SV, Khrnnikov AY. 2-Adic numbers in genetics and Rumer's symmetry. Dokl Math 2010;81:128–30.
[22] Seligmann H. Codon expansion and systematic transcriptional deletions produce tetra-, pentacoded mitochondrial peptides. J Theor Biol 2015;387:154–65.
[23] Seligmann H. Translation of mitochondrial swinger RNAs according to tri-, tetra- and pentacodons. Biosystems 2016;140:38–48.
[24] Li M, Wang IX, Li Y, Bruzel A, Richards AL, Toung JM, Cheung VG. Widespread RNA and DNA sequence differences in the human transcriptome. Science 2011;333:53–8.

[25] Bar-Yaacov D, Levin AG, Richards AL, Hachen N, Rebolledo Jaramillo B, Nekrutenko A, Zarivach R, Mishmar D. RNA–DNA differences in human mitochondria restore ancestral form of 16S ribosomal RNA. Genome Res 2013;23:1789–96.
[26] Hodgkinson A, Idaghdour Y, Gbeha E, Grenier JC, Hip-Ki E, Bruat V, Goulet JP, de Malliard T, Awadalla P. High resolution genomic analysis of human mitochondrial RNA sequence variation. Science 2014;344:413–5.
[27] Moreira S, Valachi M, Aoulad-Aissa M, Otto C, Burger G. Novel modes of RNA editing in mitochondria. Nucleic Acids Res 2016;44:4907–19.
[28] Chen C, Bundschuh R. Systematic investigation of insertional and deletional RNA–DNA differences in the human transcriptome. BMC Genomics 2012;13:616.
[29] Wang IX, Core LJ, Kwak H, Brady L, Bruzel A, McDaniel L, Richards AL, Wu M, Grunseich C, Lis JT, Cheung VG. RNA–DNA differences are generated in human cells within seconds after RNA exits polymerase II. Cell Rep 2014;6:906–15.
[30] Seligmann H. Overlapping genes coded in the 3′-to-5′-direction in mitochondrial genes and 3′-to-5′ polymerization of non-complementary RNA by an 'invertase'. J Theor Biol 2012;318:38–52.
[31] Seligmann H. Polymerization of non-complementary RNA: systematic symmetric nucleotide exchanges mainly involving uracil produce mitochondrial RNA transcripts coding for cryptic overlapping genes. Biosystems 2013;111:156–74.
[32] Seligmann H. Systematic asymmetric nucleotide exchanges produce human mitochondrial RNAs cryptically encoding for overlapping protein coding genes. J Theor Biol 2013;324:1–20.
[33] Seligmann H. Triplex DNA:RNA, 3′-to-5′ inverted RNA and protein coding in mitochondrial genomes. J Comput Biol 2013;20:1–12.
[34] Michel CJ, Seligmann H. Bijective transformation circular codes and nucleotide exchanging RNA transcription. Biosystems 2014;118:39–50.
[35] Seligmann H. Swinger RNAs with sharp switches between regular transcription and transcription systematically exchanging ribonucleotides: case studies. Biosystems 2015;135:1–8.
[36] Seligmann H. Chimeric peptides from contiguous regular and swinger RNA. Comput Struct Biotechnol J 2016;14:283–97.
[37] Seligmann H. Systematically frameshifting by deletion of every 4th or 4th and 5th nucleotides during mitochondrial transcription: RNA self-hybridization regulates delRNA expression. Biosystems 2016;142:43–51.
[38] Seligmann H. Mitochondrial swinger replication: DNA replication systematically exchanging nucleotides and short 16S ribosomal DNA swinger inserts. Biosystems 2014;125:22–31.
[39] Seligmann H. Species radiation by DNA replication that systematically exchanges nucleotides? J Theor Biol 2014;363:216–22.
[40] Seligmann H. Sharp switches between regular and swinger mitochondrial replication: 16S rDNA systematically exchanging nucleotides A ↔ T + C ↔ G in the mitogenome of Kamimuria wangi. Mitochondrial DNA a DNA Mapp. Seq Anal 2015;27:2440–6.
[41] Seligmann H. Systematic exchanges between nucleotides: genomic swinger repeats and swinger transcription in human mitochondria. J Theor Biol 2015;384:70–7.
[42] Seligmann H. Swinger RNA self-hybridization and mitochondrial non-canonical swinger transcription, transcription systematically exchanging nucleotides. J Theor Biol 2016;399:84–91.
[43] Ojala D, Montoya J, Attardi G. tRNA punctuation model of RNA processing in human mitochondria. Nature 1981;290:470–4.
[44] Michel CJ, Pirillo G. Identification of all trinucleotide circular codes. Comput Biol Chem 2010;34:122–5.
[45] Michel CJ. The maximal C(3) self-complementary trinucleotide circular code X in genes of bacteria, eukaryotes, plasmids and viruses. J Theor Biol 2015;380:156–77.
[46] Clayton DA. Replication of animal mitochondrial DNA. Cell 1982;28:693–705.
[47] Hixson JE, Wong TW, Clayton DA. Both the conserved stem-loop and divergent 5′-flanking sequences are required for initiation at the human mitochondrial origin of light-strand DNA replication. J Biol Chem 1986;261:2384–90.
[48] Wanrooij S, Falkenberg M. The human mitochondrial replication fork in health and disease. Biochim Biophys Acta 1797/2010:1378–88.
[49] Desjardins P, Morais R. Nucleotide sequence and evolution of coding and noncoding regions of a quail mitochondrial genome. J Mol Evol 1991;32:153–61.
[50] Seligmann H, Krishnan NM. Mitochondrial replication origin stability and propensity of adjacent tRNA genes to form putative replication origins increase developmental stability in lizards. J Exp Zool B Mol Dev Evol 2006;306:433–49.
[51] Seligmann H, Krishnan NM, Rao BJ. Possible multiple origins of replication in primate mitochondria: alternative role of tRNA sequences. J Theor Biol 2006;241:321–32.
[52] Seligmann H, Krishnan NM, Rao BJ. Mitochondrial tRNA sequences as unusual replication origins: pathogenic implications for Homo sapiens. J Theor Biol 2006;243:375–85.
[53] Seligmann H. Hybridization between mitochondrial heavy strand tDNA and expressed light strand tRNA modulates the function of heavy strand tDNA as light strand replication origin. J Mol Biol 2008;379:188–99.
[54] Seligmann H. Mitochondrial tRNAs as light strand replication origins: similarity between anticodon loops and the loop of the light strand replication origin predicts initiation of DNA replication. Biosystems 2010;99:85–93.
[55] Seligmann H. Pathogenic mutations in antisense mitochondrial tRNAs. J Theor Biol 2011;269:287–96.
[56] Seligmann H. Pocketknife tRNA hypothesis: anticodons in mammal mitochondrial tRNA side-arm loops translate proteins? Biosystems 2013;113:165–76.
[57] Seligmann H. Putative anticodons in mitochondrial tRNA sidearm loops: pocketknife tRNAs? J Theor Biol 2014;340:155–63.
[58] Seligmann H, Labra A. The relation between hairpin formation by mitochondrial WANCY tRNAs and the occurrence of the light strand replication origin in Lepidosauria. Gene 2014;542:248–57.
[59] Yu CH, Liao JY, Zhou H, Qu LH. The rat mitochondrial Ori L encodes a novel small RNA resembling an ancestral tRNA. Biochem Biophys Res Commun 2008;372:634–8.

[60] Maizels N, Weiner AM. Phylogeny from function: evidence from the molecular fossil record that tRNA originated in replication, not translation. Proc Natl Acad Sci U S A 1994;91:6729–34.

[61] Maizels N, Weiner AM. Phylogeny from function: the origin of tRNA is in replication, not translation. Chapter 2. In: Fitch WM, Ayala FJ, editors. Tempo and mode in evolution: genetics and paleontology 50 Years after Simpson. Washington: National Academies Press; 1995. p. 25–40 [http://www.ncbi.nlm.nih.gov/books/NBK232211/].

[62] Seligmann H, Raoult D. Unifying view of stem-loop hairpin RNA as origin of current and ancient parasitic and non-parasitic RNAs, including in giant viruses. Curr Opin Microbiol 2016;31:1–8.

[63] Capone JP, Sharp PA, RajBhandary UL. Amber, ochre and opal suppressor tRNA genes derived from a human serine tRNA gene. EMBO J 1985;4:213–21.

[64] Beier H, Grimm M. Misreading of termination codons in eukaryotes by natural nonsense suppressor tRNAs. Nucleic Acids Res 2001;29:4767–82.

[65] Seligmann H. Avoidance of antisense antiterminator tRNA anticodons in vertebrate mitochondria. Biosystems 2010;101:42–50.

[66] Seligmann H. Undetected antisense tRNAs in mitochondria? Biol Direct 2010;5:39.

[67] Seligmann H. Two genetic codes, one genome: frameshifted primate mitochondrial genes code for additional proteins in presence of antisense antitermination tRNAs. Biosystems 2011;105:271–85.

[68] Seligmann H. Putative protein-encoding genes within mitochondrial rDNA and the D-loop region. In: Lin Z, Liu W, editors. Ribosomes: molecular structure, role in biological functions and implications for genetic diseases. Nova Science Publishers; 2013. p. 67–86.

[69] Seligmann H. An overlapping genetic code for frameshifted overlapping genes in *Drosophila* mitochondria: antisense antitermination tRNAs UAR insert serine. J Theor Biol 2012;298:51–76.

[70] Seligmann H. Overlapping genetic codes for overlapping frameshifted genes in Testudines, and *Lepidochelys olivacea* as a special case. Comput Biol Chem 2012; 41:18–34.

[71] Barthélémy R-M, Seligmann H. Cryptic tRNAs in chaetognath mitochondrial genomes. Cryptic tRNAs in chaetognath mitochondrial genomes. Comput Biol Chem 2016;62:119–32.

[72] Massey SE, Garey JR. A comparative genomics analysis of codon reassignments reveals a link with mitochondrial proteome size and a mechanism of genetic code change via suppressor tRNAs. J Mol Evol 2007;64:399–410.

[73] Seligmann H. Phylogeny of genetic codes and punctuation codes within genetic codes. Biosystems 2015;129:36–43.

[74] Knight RD, Landweber LF, Yarus M. How mitochondria redefine the code. J Mol Evol 2001;53:299–313.

[75] Sengupta S, Yang X, Higgs PG. The mechanisms of codon reassignments in mitochondrial genetic codes. J Mol Evol 2007;64:662–88.

[76] Ring KL, Cavalcanti AR. Consequences of stop codon reassignment on protein evolution in ciliates with alternative genetic codes. Mol Biol Evol 2008;25:179–86.

[77] Vallabhaneni H, Fan-Minogue H, Bedwell DM, Farabaugh PJ. Connection between stop codon reassignment and frequent use of shifty stop frameshifting. RNA 2009;15:889–97.

[78] Johnson LJ. Pseudogene rescue: an adaptive mechanism of codon reassignment. J Evol Biol 2010;23:1623–30.

[79] Johnson LJ, Cotton JA, Lichtenstein CP, Elgar GS, Nichols RA, Polly PD, Le Comber SC. Stops making sense: translational trade-offs and stop codon reassignment. BMC Evol Biol 2011;11:227.

[80] Campbell JH, O'Donoghue P, Campbell AG, Schwientek P, Sczyrba A, Woyke T, Söll D, Podar M. UGA is an additional glycine codon in uncultured SR1 bacteria from the human microbiota. Proc Natl Acad Sci U S A 2013;110:5540–5.

[81] Ivanova NN, Schwientek P, Tripp HJ, Rinke C, Pati A, Huntemann M, Visel A, Woyke T, Kyrpides NC, Rubin EM. Stop codon reassignment in the wild. Science 2014;344: 909–13.

[82] Seligmann H. Putative mitochondrial polypeptides coded by expanded quadruplet codons, decoded by antisense tRNAs with unusual anticodons. Biosystems 2012; 110:84–106.

[83] Seligmann H, Labra A. Tetracoding increases with body temperature in Lepidosauria. Biosystems 2013;114:155–63.

[84] Riddle DL, Roth JR. Frameshift suppressors. 3. Effects of suppressor mutations on transfer RNA. J Mol Biol 1972;66:495–506.

[85] O'Connor M, Gesteland RF, Atkins JF. tRNA hopping: enhancement by an expanded anticodon. EMBO J 1989;8:4315–23.

[86] Tuohy TM, Thompson S, Gesteland RF, Atkins JF. Seven, eight and nine-membered anticodon loop mutants of tRNA(2Arg) which cause +1 frameshifting. Tolerance of DHU arm and other secondary mutations. J Mol Biol 1992;228:1042–54.

[87] Walker SE, Frederick K. Recognition and positioning of mRNA in the ribosome by tRNAs with expanded anticodons. J Mol Biol 2006;360:599–609.

[88] Dunham CM, Selmer M, Phelps SS, Kelley AC, Suzuki T, Joseph S, Ramakrishnan V. Structures of tRNAs with an expanded anticodon loop in the decoding center of the 30S ribosomal subunit. RNA 2007;13:817–23.

[89] Maehigashi T, Dunkle JA, Miles SJ, Dunham CM. Strucural insights into +1 frameshifting promoted by expanded or modification-deficient anticodon stem loops. Proc Natl Acad Sci U S A 2014;111:12740–5.

[90] Beznosková P, Gunišová S, Valášek LS. Rules of UGA-N decoding by near-cognate tRNAs and analysis of readthrough on short uORFs in yeast. RNA 2016;22:456–66.

[91] Gonzalez DL, Giannerini S, Rosa R. On the origin of the mitochondrial genetic code: towards a unified mathematical framework for the management of genetic information. Nature Precedings; 2012. http://dx.doi.org/10.1038/npre.2012.7136.1.

[92] Baranov PV, Venin M, Provan G. Codon size reduction as the origin of the triplet genetic code. PLoS One 2009;4:e5708.

[93] Root-Bernstein M, Root-Bernstein R. The ribosome as a missing link in the evolution of life. J Theor Biol 2015;367:130–58.

[94] Root-Bernstein R, Root-Bernstein M. The ribosome as a missing link in prebiotic evolution II: ribosomes encode ribosomal proteins that bind to common regions of their own mRNAs and rRNAs. J Theor Biol 2016;397:115–27.

[95] Kowalczewska M, Villard C, Lafitte D, Fenollar F, Raoult D. Global proteomic pattern of *Tropheryma whipplei*: a Whipple's disease bacterium. Proteomics 2009;9:1593–616.

[96] Guerra L, McGarry LM, Robles V, Bielza C, Larrañaga P, Yuste R. Comparison between supervised and unsupervised classifications of neuronal cell types: a case study. Dev Neurobiol 2011;71:71–82.

[97] Brazma A, Vilo J. Gene expression data analysis. FEBS Lett 2000;480:17–24.

[98] Causton H, Quackenbusch J, Brazma A. Analysis of gene expression data matrices. Chapter 4. In: Causton H, Quackenbusch J, Brazma A, editors. Microarray Gene expression data analysis: a beginner's guide. ISBN 1-40510-682-4; 2003. p. 72–133.

[99] Paik S, Kim C. Evolving role of pathology in modern oncology. Chapter 2. In: Bonadonna G, Hortobagyi GN, Valagussa P, editors. Textbook of breast cancer: a clinical guide to therapy3rd ed. ; 2006. p. 17–31 [ISBN 13:978-1-4822-0287-8].

[100] Efron B. Biased versus unbiased estimation. Adv Math 1975;16:259–77.

[101] Quirós PM, Langer T, López-Otin C. New roles for mitochondrial proteases in health, ageing and disease. Nat Rev Mol Cell Biol 2015;16:345–59.

[102] Timerbaev AR, Buchberger W. Inorganic analysis and speciation. Chapter 22. In: Deyl Z, Miksik I, Tagliaro F, Tesatova E, editors. Advanced chromatographic and Electromigration methods in biosciences. Elseviers Science; 1998. p. 963–1012.

[103] Abonnenc M, Mayr M. Proteomics of atherosclerosis. Chapter 13. In: Wick G, Grundtman C, editors. Inflammation and atherosclerosis. Wien: Springer; 2012. p. 249–66.

[104] Piatkov KI, Oh J-H, Liu Y, Varshavsky A. Calpain-generated natural protein fragments as short-lived substrates of the N-end rule pathway. Proc Natl Acad Sci U S A 2014;111:E817–26.

[105] Schmidt W, Egbring R, Havemann K. Effect of elastase-like and chymotrypsin-like natural proteases from human granulocytes on isolated clotting factor XIII. Thromb Res 1975;6:315–29.

[106] Andrews AT, Alichanidis E. Proteolysis of caseins and the proteose-peptone fraction of bovine milk. J Dairy Res 1983;50:275–90.

[107] Rietschel B, Arrey TN, Meyer B, Bornemann B, Schuerken M, Karas M, Poetsch A. Elastase digests. New ammunition for shotgun membrane proteomics. Mol Cell Proteomics 2009;8:1029–43.

[108] Wildes D, Wells JA. Sampling the N-terminal proteome of human blood. Proc Natl Acad Sci U S A 2010;107:4561–6.

[109] Leonelli L, Pelton J, Schoeffler A, Dahlbeck D, Berger J, Wemmer DE, Staskawicz B. Structural elucidation and functional characterization of the Hyaloperonospora arabidopsidis effector protein ATR13. PLoS Pathog 2011;7:e1002428.

[110] Venter E, Smith RD, Payne SH. Proteogenomic analysis of bacteria and Archaea: a 46 organism case study. PLoS One 2011;6:e27587.

[111] Volkmann G, Volkmann V, Liu XQ. Site-specific protein cleavage in vivo by an intein-derived protease. FEBS Lett 2012;586:79–84.

[112] Gueugneau M, Coudy-Gandilhon C, Gourbeyre O, Chambon C, Combaret L, Polge C, Taillandier D, Attaix D, Friguet B, Maier AB, Butler-Browne G, Béchet D. Proteomics of muscle chronological ageing in post-menopausal women. BMC Genomics 2014; 15:1165.

[113] Garzon R, Volinia S, Papaioannou D, Nicolet D, Kohlschmidt J, Yan PS, Mrozek K, Bucci D, Carroll AJ, Baer MR, Wetzler M, Carter TH, Powell BL, Kolitz JE, Moore JO, Eisfeld AK, Blachly JS, Blum W, Caliguiri MA, Stone RM, Marcucci G, Croce CM, Byrd JC, Bloomfield CD. Expression and prognostic impact of lncRNAs in acute myeloid leukemia. Proc Natl Acad Sci U S A 2014;111:18679–84.

[114] Lobanov AV, Turanov AA, Hatfield DL, Gladyshev VN. Dual functions of codons in the genetic code. Crit Rev Biochem Mol Biol 2010;45:257–65.

[115] O'Donoghue P, Prat L, Heinemann IU, Ling J, Odoi K, Liu WR, Söll D. Near-cognate suppression of amber, opal and quadruplet codons competes with aminoacyl-tRNAPyl for genetic code expansion. FEBS Lett 2012;583:3931–7.

[116] Odoi KA, Huang Y, Rezenom YH, Liu WR. Nonsense and sense suppression abilities of original and derivative *Methanosarcina mazei* pyrrolysyl-tRNA synthetase-tRNA(Pyl) pairs in the *Escherichia coli* BL21(DE3) cell strain. PLoS One 2013;8: e57035.

[117] Käll L, Storey JD, MacCoss MJ, Noble WS. Posterior error probabilities and false discovery rates: two sides of the same coin. J Proteome Res 2008;7:40–4.

[118] Brosch M, Yu L, Hubbard T, Choudhary J. Accurate and sensitive peptide identification with MascotPercolator. J Proteome Res 2009;8:3176–81.

[119] Spivak M, Weston J, Bottou L, Käll L, Noble WS. Improvements to the percolator algorithm for peptide identification from shotgun proteomics data sets. J Proteome Res 2009;8:3737–45.

[120] Eng JK, McCormack AL, Yates III JR. An approach to correlate tandem mass spectral data of peptides with amino acid sequences in a protein database. J Am Soc Mass Spectrom 1994;5:976–89.

[121] Fisher RA. Combining independent tests of significance. Am Stat 1648;2:30–1.

[122] Seligmann H. Natural chymotrypsin-like-cleaved human mitochondrial peptides confirm tetra-, pentacodon, non-canonical RNA translations. Biosystems 2016; 147:78–93.

[123] Altschul SF, Madden TL, Schäffer AA, Zhang J, Zhang Z, Miller W, Lipman DJ. Gapped BLAST and PSI-BLAST: a new generation of protein database search programs. Nucleic Acids Res 1997;25:3389–402.

[124] Lopez JV, Yukhi N, Masuda R, Modi W, O'Brien SJ. Numt, a recent transfer and tandem amplification of mitochondrial DNA to the nuclear genome of the domestic cat. J Mol Evol 1994;39:174–90.

[125] Lopez JV, Cumver M, Stephens JC, Johnoson WE, O'Brien SJ. Rates of nuclear and cytoplasmic mitochondrial DNA sequence divergence in mammals. Mol Biol Evol 1997;14:277–86.

[126] Lopez JV, Stephens JC, O'Brien SJ. The long and short nuclear mitochondrial DNA (Numt) lineages. Trends Ecol Evol 1997;12:114.

[127] Zhang DX, Hewitt GM. The long and short of nuclear mitochondrial DNA (Numt) lineages reply from D-X. Trends Ecol Evol 1997;12:114.

[128] Bensasson D, Zhang D, Hartl DL, Hewitt GM. Mitochondrial pseudogenes: evolution's misplaces witnesses. Trends Ecol Evol 2001;16:314–21.

[129] Tourmen Y, Baris O, Dessen P, Jacques C, Malthièry Y, Reynier P. Structure and chromosomal distribution of human mitochondrial pseudogenes. Genomics 2002;80: 71–7.

[130] Bensasson D, Feldman MW, Petrov DA. Rates of DNA duplication and mitochondrial DNA insertion in the human genome. J Mol Evol 2003;57:343–54.

[131] Ricchetti M, Tekala F, Dujon B. Continued colonization of the human genome by mitochondrial DNA. PLoS Biol 2004;2:E273.

[132] Thalman O, Hebler J, Poinar HN, Päabo S, Vigilant L. Unreliable mtDNA data due to nuclear insertions: a cautionary tale from analysis of humans and other great apes. Mol Ecol 2004;13:321–35.

[133] Richly E, Leister D. NUMTs in sequenced eukaryotic genomes. Mol Biol Evol 2004; 21:1081–4.

[134] Schmitz J, Piskurek O, Zischler H. Forty million years of independent evolution: a mitochondrial gene and its corresponding nuclear pseudogene in primates. J Mol Evol 2005;61:1–11.

[135] Thalman O, Serre D, Hofreiter M, Lukas D, Eriksson J, Vigilant L. Nuclear insertions help and hinder inference of the evolutionary history of gorilla mtDNA. Mol Ecol 2005;14:179–88.

[136] Yao YG, Kong QP, Salas A, Bandelt HJ. Pseudomitochondrial genome haunts disease studies. J Med Genet 2008;45:769–72.

[137] Hazkani-Covo E, Zeller RM, Martin W. Molecular poltergeists: mitochondrial DNA copies (numts) in sequenced nuclear genomes. PLoS Genet 2010;6:e1000834.

[138] Ramos A, Barbena E, Mateiu L, del Mar González M, Mairal Q, Lima M, Montiel R, Aluja MP, Santos C. Nuclear insertions of mitochondrial origin: database updating and usefulness in cancer studies. Mitochondrion 2011;11:946–53.

[139] Tsuji J, Frith MC, Tomii K, Horton P. Mammalian NUMT insertion is non-random. Nucleic Acids Res 2012;40:9073–88.

[140] Soto-Calderón ID, Lee EJ, Jensen-Seaman MI, Anthony NM. Factors affecting the relative abundance of nuclear copies of mitochondrial DNA (numts) in hominoids. J Mol Evol 2012;75:102–11.

[141] Soto-Calderón ID, Clark NJ, Wildschutte JV, DiMattio K, Jensen-Seaman MI, Anthony NM. Identification of species-specific nuclear insertions of mitochondrial DNA (numts) in gorillas and their potential as population genetic markers. Mol Phylogenet Evol 2014;81:61–70.

[142] Smigrodzki RM, Khan SM. Mitochondrial microheteroplasmy and a theory of aging and age-related disease. Rejuvenation Res 2005;8:172–98.

[143] Rose G, Passarino G, Scornaienchi V, Romeo G, Dato S, Bellizzi D, Mari V, Feraco E, Maletta R, Bruni A, Franceschi C, De Benedictis G. The mitochondrial DNA control region shows genetically correlated levels of heteroplasmy in leukocytes of centenarians and their offspring. BMC Genomics 2007;8:293.

[144] Stefano GB, Kream RM. Mitochondrial DNA heteroplasmy in human health and disease. Biomed Rep 2016;4:259–62.

[145] Ramos A, Santos C, Mateiu L, del Mar Gonzalez M, Alvarez L, Azevedo L, Amorim A, Pilar Aluja M. Frequency and pattern of heteroplasmy in the complete human mitochondrial genome. PLoS One 2013;8:e74636.

[146] Frenkel-Morgenstern M, Gorohovski A, Lacroix V, Rogers M, Ibanez K, Boullosa C, Andres Leon E, Ben-Hur A, Valencia A. ChiTaRS: a database of human, mouse and fruit fly chimeric transcripts and RNA-sequencing data. Nucleic Acids Res 2013; 41:D142–51.

[147] Yang W, Wu J-M, Bi A-D, Ou-yang Y, Shen H-H, Chim G-W, Zhou J-H, Weiss E, Holman EP, Liao DJ. Possible formation of mitochondrial-RNA containing chimeric or trimeric RNA implies a post-transcriptional and post-splicing mechanism for RNA fusion. PLoS One 2013;8:e77016.

[148] Xie B, Yang W, Chen L, Jiang H, Liao Y, Liao DJ. Two RNAs or DNAs may artificially fuse together at a short homologous sequence (SHS) during reverse transcription or polymerase chain reactions, and thus reporting an SHS-containing chimeric RNA requires extra caution. PLoS One 2016;11:e0154855.

[149] Allen JF. Why chloroplasts and mitochondria retain their own genomes and genetic systems: colocation for redox regulation of gene expression. Proc Natl Acad Sci U S A 2015;112:10231–8.

[150] Bauer NC, Doetsch PW, Corbett AH. Mechanisms regulating protein localization. Traffic 2015;16:1039–61.

[151] Horvath SE, Rampelt H, Oeljeklaus S, Warscheid B, van der Laan M, Pfanner N. Role of membrane contact sites in protein import into mitochondria. Protein Sci 2015; 24:277–97.

[152] Kunze M, Berger J. The similarity between N-terminal targeting signals for protein import into different organelles and its evolutionary relevance. Front Physiol 2015; 6:259.

[153] Aerni HR, Shifman MA, Rogulina S, O'Donoghue P, Rinehart J. Revealing the amino acid composition of proteins within an expanded genetic code. Nucleic Acids Res 2015;43:e8.

6

Detection of Side Chain Rearrangements Mediating the Motions of Transmembrane Helices in Molecular Dynamics Simulations of G Protein-Coupled Receptors

Zied Gaieb, Dimitrios Morikis *

Department of Bioengineering, University of California, Riverside 92521, USA

ARTICLE INFO

ABSTRACT

Keywords:
Molecular dynamics
Change-point detection
Side chain reorganization
Helical domain motion
Intramolecular network
Membrane proteins
GPCR
GPCR computational modeling
GPCR allostery

Structure and dynamics are essential elements of protein function. Protein structure is constantly fluctuating and undergoing conformational changes, which are captured by molecular dynamics (MD) simulations. We introduce a computational framework that provides a compact representation of the dynamic conformational space of biomolecular simulations. This method presents a systematic approach designed to reduce the large MD simulation spatiotemporal datasets into a manageable set in order to guide our understanding of how protein mechanics emerge from side chain organization and dynamic reorganization. We focus on the detection of side chain interactions that undergo rearrangements mediating global domain motions and vice versa. Side chain rearrangements are extracted from side chain interactions that undergo well-defined abrupt and persistent changes in distance time series using Gaussian mixture models, whereas global domain motions are detected using dynamic cross-correlation. Both side chain rearrangements and global domain motions represent the dynamic components of the protein MD simulation, and are both mapped into a network where they are connected based on their degree of coupling. This method allows for the study of allosteric communication in proteins by mapping out the protein dynamics into an intramolecular network to reduce the large simulation data into a manageable set of communities composed of coupled side chain rearrangements and global domain motions. This computational framework is suitable for the study of tightly packed proteins, such as G protein-coupled receptors, and we present an application on a seven microseconds MD trajectory of CC chemokine receptor 7 (CCR7) bound to its ligand CCL21.

1. Introduction

Protein function is encoded into its dynamics as a large ensemble of conformations that can be grouped into distinct conformational states according to their function, free energy, and three-dimensional arrangement [1,2]. These conformational states are accessed at different equilibrium sampling probabilities in response to outside perturbation such as ligand-binding, amino acid mutation, post translational modification, or environmental changes (pH, ionic strength, temperature, etc.) [3]. In many cases, ligand-free proteins that favor their inactive state, may still briefly sample their intermediate or active states [1]. However, external perturbations, such as ligand-binding, result in an equilibrium shift where the protein favors its active state.

As a mechanism to regulate its transitions and sampling of conformational states upon external perturbation, allosteric function plays an important role in transmitting information between distant functional sites of the protein [1,2,4]. To comprehend such mechanism, we must understand how the mechanics of protein structures emerge from the rearrangement of their constituent parts, specifically, side chain interactions within structured regions of proteins. Molecular dynamics (MD) simulation is one of the major techniques that has played a key role in studying protein dynamics at atomic level [2]. Several recent advances in enhanced sampling methods, simulation speed, and accuracy have allowed us to reach biologically relevant timescales that are sampled in the hundreds of nanosecond to microseconds and capture the transitioning of a protein between different states; and consequently, allow the study of allostery [2,5–7]. Accordingly, several studies have explored the folding mechanism of a number of fast folding proteins [8] and captured protein state transitions [9,10]. To extract biologically-relevant protein motions, long MD simulations have been analyzed through manual and visual inspection of large biological datasets of inter-atomic distance and Cartesian coordinate time series [7,9–14]. These extracted protein motions have consisted of abrupt changes in intramolecular interaction distance time series that show a transition between two stable inter-residue distances and the collective motion of many residues in different domains of the protein (transmembrane helices in our case). Despite the major advances in our

* Corresponding author.
  E-mail address: dmorikis@ucr.edu (D. Morikis).

understanding of protein dynamics, the MD analysis scientific community has not yet reached a consensus method to extract biologically-relevant conformational changes in proteins.

Many MD analysis tools have been developed, but still come short in detecting all relevant side chain and backbone rearrangements. Widely used methods involve the detection of global conformational changes, and include principal component analysis (PCA) and dynamic cross-correlation (DCC) applied to the three-dimensional Cartesian coordinates of simulated protein structures [15–17]. PCA, which is used to extract the dominant collective protein motions, tend to neglect less-dominant collective motions that are critical to unravel the complex details orchestrating protein transitions between conformational states. A heat map generated through DCC of aligned atomic Cartesian coordinates results in critical protein motions with low correlation coefficients (less than 0.6) due to noise introduced by atomic fluctuations and superimposition of the atomic coordinates, making it difficult to distinguish between false positives and false negatives [9]. Other methods revolve around the detection of abrupt changes in spatiotemporal data comprising of inter-atomic distances or three-dimensional coordinate time series [18–20]. The most recent method, SIMPLE, is designed to favor the detection of collective change-points, depending on a sensitivity parameter [20]. Despite the advances in event detection made possible by SIMPLE, this method still comes short in detecting all relevant side chain and backbone rearrangements. Depending on the sensitivity parameter used, many critical protein motions can either be obscured by the large number of detected change-points (large number of false positives) when using a low sensitivity parameter, or omitted (large number of false negatives) when using a high sensitivity parameter.

Due to the aforementioned challenges in biological event detection, many studies rely on manual and visual analysis of MD data [20]. These measures are non-systematic, are labor intensive, and may not provide a complete analysis due to the overwhelming amount of data output by the MD simulations. Systematic detection of protein motions is a critical step in understanding the molecular mechanism of protein allostery and is a challenging problem for many reasons. First, MD simulations output an insurmountable amount of dynamics information that can be daunting to analyze due to the high fluctuating and complex nature of protein dynamics. Second, side chain and domain rearrangements have very different dynamics behaviors, where amino acid residue side chains involve more fluctuations and sporadic movements than the larger domain movements of the protein [21]. Third, functional side chain rearrangements are subtle and manifest themselves as a single inter-residue interaction rearrangement that can be obscured by the several fluctuating and unstable inter-residue interactions. These challenges have prompted a need to reduce the large simulation data into a compact representation of the dynamic conformational space of biomolecules to guide scientists in their analysis of the complex MD simulation data.

In this work, we reduce the protein dynamics to its constitutive dynamic components. To carry their dynamics, proteins involve two major types of motions: side chain and global domain conformational changes. These motions constitute the dynamic components that facilitate the transmission of signals between distant sites in a protein [1,2]. In the framework presented here, we start by screening for side chain rearrangements and global domain motions separately using Gaussian mixture models (GMMs) and DCC, respectively. All extracted components are then projected into a network based on their inter-component absolute average DCC coefficient and compartmentalized into different communities of correlated dynamics. The different network communities decompose the protein dynamics into its constitutive dynamic behaviors that are localized to different sectors of the protein, and comprise of side chain distance time series that are correlated (or anti-correlated) to the global domain motions of the protein. To illustrate the application of our computational framework, we apply our method to a previously published MD trajectory of a chemokine ligand, CCL21, bound to CC chemokine receptor 7 (CCR7) (Gaieb et al. REF).

Essentially, our method reduces the dynamic interaction space of G protein coupled receptors (GPCRs) to a manageable space composed of protein sectors with different dynamic behaviors. The communities of dynamic components present a unified picture of the complex behavior of the protein and will guide the user to further analyze the subgraphs and communities to provide an understanding of how side chain rearrangements mediate the global motions of the protein, which eventually facilitates transitioning between functional states.

2. Materials and Methods

Our computational framework is designed to systematically reduce the MD Cartesian coordinate time series of GPCRs to a few communities composed of coupled dynamic components (Fig. 1). This is done by first extracting side chain rearrangements and global domain motions from the protein's MD simulation trajectory.

Side chain rearrangements are often localized to a single inter-residue side chain interaction, which could be obscured by global domain motions when extracted from a large MD data set of inter-atomic distance time series. Therefore, both dynamic components, side chain (Fig. 1A) and backbone dynamics (Fig. 1B), are extracted separately using different methods: GMMs and DCC, respectively. Given the dynamic nature of proteins, only a fraction of the protein's extracted side chain dynamics is considered to contribute to regulating the global protein dynamics. Therefore, side chain rearrangements (Fig. 1A) are further reduced by extracting those that are correlated to the global domain motions (Fig. 1B). This is done by projecting all dynamic components into a network that is connected based on the absolute average inter-component correlation coefficient and then categorized into different communities, where domain motions and side chain dynamics within the same community show correlated time series (Fig. 1C).

2.1. Detection of Side Chain Contact Rearrangements From MD Simulations

Extracting all side chain rearrangements from MD simulations involves the identification of side chain interactions that experience abrupt and persistent changes in their distance time series, indicating a transition between substates. We extract such inter-residue interactions by fitting a GMM to the probability density of each interaction distance time series. GMMs are weighted sums of Gaussian densities and are used here as a parametric model of the probability density function of inter-residue time series (Gaussian densities are implemented in *scikit-learn*, a machine learning package in python) [22]. Stable non-varying interactions show a unimodal distribution (Fig. 2A), and multi-substate interactions show multi-modal distributions (Fig. 2B). The optimal number of Gaussians was efficiently determined using the Bayesian information criterion using *scikit-learn* [22], and GMM parameters were estimated using the iterative expectation-maximization algorithm, where the number of Gaussians is predetermined. This section of the computational framework is designed to systematically extract all interactions that show contact formation and breaking at any point during the simulations, as such contacts can be deemed critical in mediating global domain motions. GMMs are fitted to all distance time series representing van der Waals and polar interaction (listed below) distances between interacting side chain residues. Interacting residues used to calculate the distance time series are at least three residues apart in sequence and came into contact (a distance of at least 5 Å between all non-hydrogen side chain atoms) at any point during the simulation. To ensure complete formation and breaking of the side chain contacts, we calculate the inter-residue side chain distance time series using the minimum distance between all non-hydrogen side chain atoms of each of the amino acids. Similarly, polar interactions are also calculated using the minimum distance between all non-hydrogen polar head group atoms of interacting polar amino acids (atoms N_ε, C_ζ, $N_{\eta 1}$, or $N_{\eta 2}$ for R; atoms C_γ, $O_{\delta 1}$, or $N_{\delta 2}$ for N; atoms C_γ, $O_{\delta 1}$, or $O_{\delta 2}$ for D; atom S_γ for C; atoms C_δ, $O_{\varepsilon 1}$, or $N_{\varepsilon 2}$ for Q; atoms C_δ,

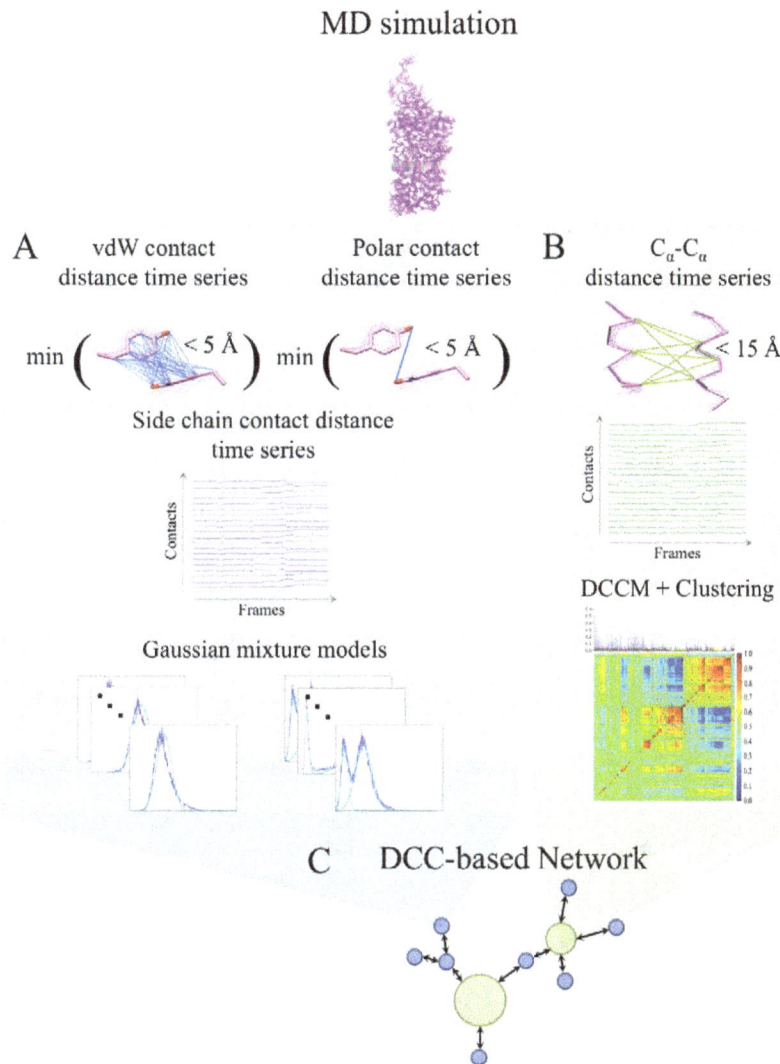

Fig. 1. Schematic of our computational framework to extract coupled side chain rearrangements and global domain motions in proteins. (A) Van der Waals and polar interactions that sample a maximum distance of 5 Å during the simulation are used to calculate distance time series from the MD simulation 3-dimentional coordinate data. The minimum distance between all side chain or polar atoms are used to extract inter-residue side chain distance time series. Probability density of each time series are fitted to a GMM to extract side chain interactions that undergo rearrangements during the simulation. (B) C_α-C_α interactions that sample a maximum distance of 15 Å during the simulation are used to calculate the C_α-C_α distance time series. A DCC matrix of all pairwise C_α-C_α distance time series are clustered and clusters with a minimum coefficient of 0.95 are extracted as domain motions of the protein. (C) Side chain rearrangements (blue nodes) and domain motions (green nodes) of the protein are considered dynamic components of the protein and are input into a DCC-based network to relate the two components to each other. Network connections are based on the correlation coefficients of pairwise dynamic components which are calculated as the average DCC coefficient of the pairwise time series belonging to each component.

$O_{\varepsilon 1}$, or $O_{\varepsilon 2}$ for E; atoms C_γ, $N_{\delta 1}$, $C_{\varepsilon 1}$, $N_{\varepsilon 2}$, or $C_{\delta 2}$ for H; atom N_ζ for K; atom O_γ for S; atom $O_{\gamma 1}$ for T; atom $N_{\varepsilon 1}$ for W; atom O_η for Y). All distance time series probability density functions are fit with a GMM to identify the number of substates that each interaction is sampling.

Distance time series with unimodal GMMs are considered to be stable during the simulations, contributing to the structural stability (robustness) of the protein. On the other hand, multi-modal GMMs are amongst the dynamic components of the protein and contribute to the protein's conformational transitions between different functional states.

2.2. Detection of Global Domain Motions Through DCCM

Global domain motions in proteins involve the collective motion of backbone atoms and aid in the transitioning of the protein between different functional states. This part of the computational framework entails the detection of these motions as a collection of highly correlated inter-C_α distance time series.

All alpha carbon interactions (at least three residues apart in sequence) within 15 Å at any point of the simulation are extracted, and all distance time series representing theses interactions are calculated. Pairwise dynamic cross-correlation of all distance time series are clustered based on their correlation coefficient and clusters with at least 0.95 correlation coefficient are extracted (Fig. 3A, B). Each cluster is a set of highly correlated time series that are localized to distinct protein sectors that exhibit different dynamic behaviors (Fig. 3C). The algorithm for hierarchical clustering used is provided in the *SciPy* library (scipy.cluster.hierarchy.linkage), and is performed on a condensed distance matrix using the Nearest Point Algorithm [23]. The condensed distance matrix is defined as a pairwise correlation coefficients matrix between extracted time series and is returned by the scipy.spatial.distance.pdist function.

The use of distance time series (rather than Cartesian coordinates) presents various advantages in molecular dynamics simulation analysis. Apart from reducing the dimensionality of the time series used (from three-dimensional Cartesian coordinates to one-dimensional distance

A

B

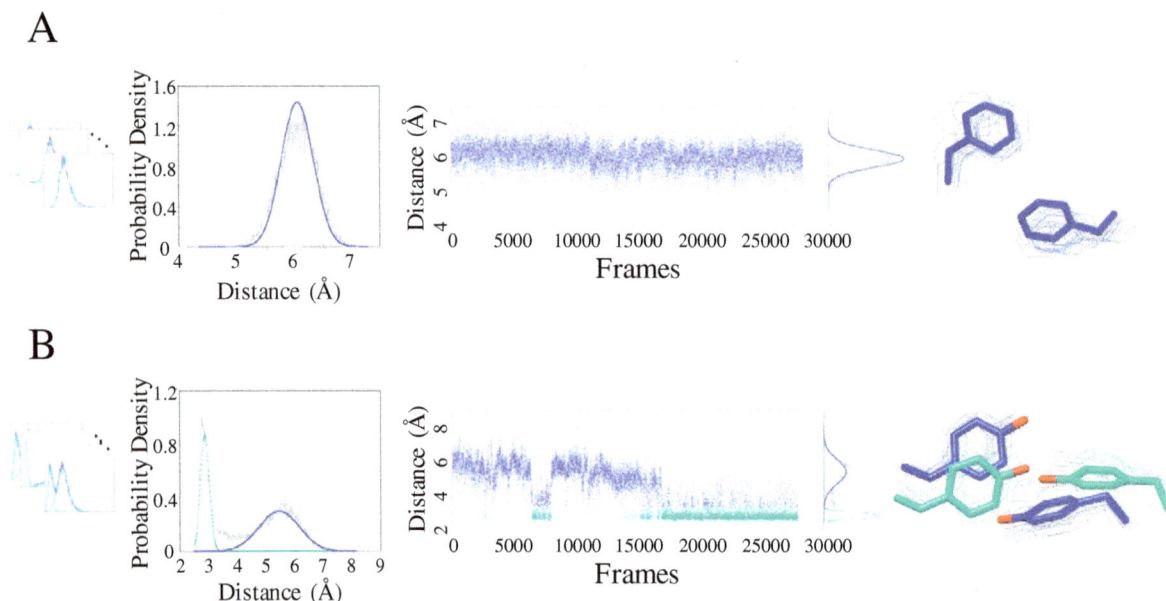

Fig. 2. Examples of side chain distance probability densities fitted using GMM. (A) Side chain distance probability densities fitted by unimodal distributions show a stable inter-residue interaction through the majority of the simulation. (B) Side chain distance probability densities fitted by multimodal distributions represent inter-residue interactions that undergo rearrangements during the simulation. The cyan and blue colors represent the Gaussian distribution sampled around 2.7 Å and 5.5 Å, respectively.

time series), the translation and rotation of the whole protein during the MD simulations can be ignored and, therefore, structure superimposition (alignment) can be omitted. These improvements allow us to accentuate the changes in the global structure of the protein and attenuate the effects of atomic fluctuations seen when using the Cartesian coordinates. Thus, clusters with high DCC coefficient better portray the global domain dynamic behavior of the protein.

2.3. Network of the Protein's Dynamic Components

To assess coupling between side chain rearrangements and global domain motions, these dynamic components of the protein are projected into a static network and classified into communities, using *igraph* [24]. We create a DCC-based network connecting the dynamic components of the protein (Fig. 4), extracted in the previous sections

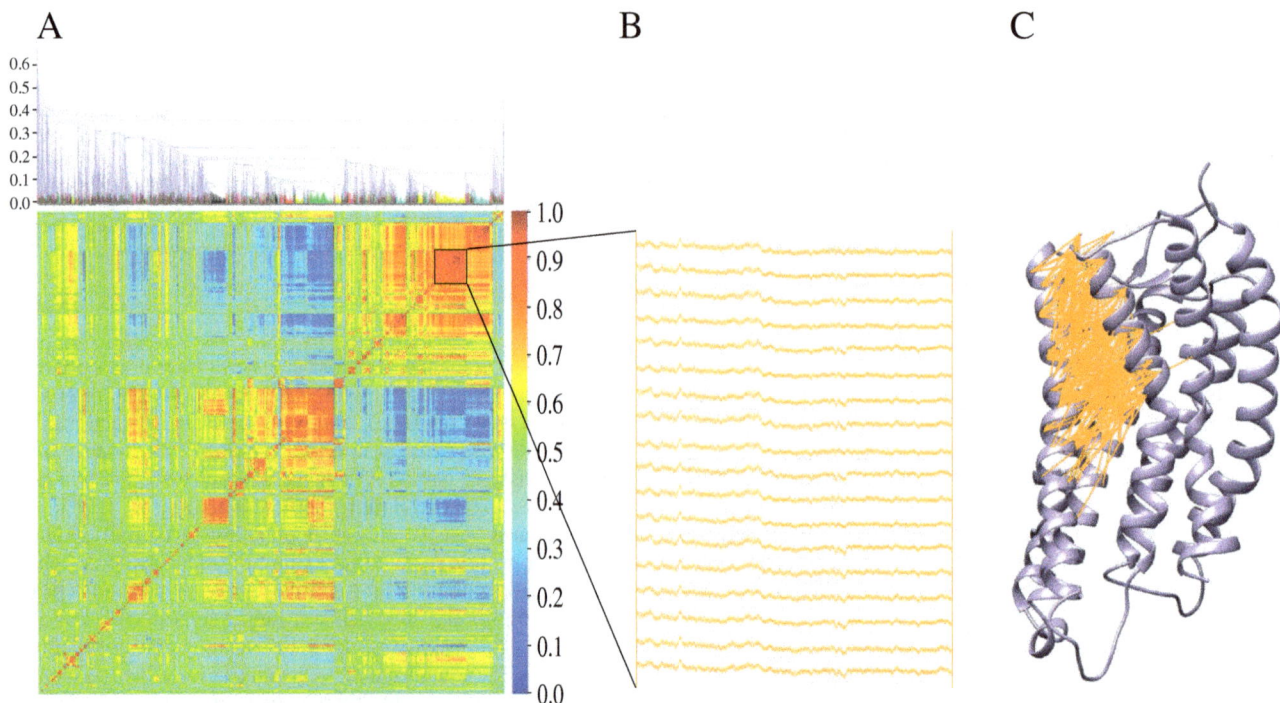

A   B   C

Fig. 3. DCC heat map of pairwise C_α-C_α distance time series are clustered using hierarchical clustering. (A) The clustering dendrogram is reported above the DCC heat map. The DCC coefficient is used as the distance calculated between two clusters and shown as the y-axis of the dendrogram. Each color of the dendrogram represents a different cluster of time series that are correlated at a cutoff DCC coefficient of 0.95. Due to the large number of C_α-C_α distance time series, only time series within the extracted clusters are shown in the DCC heat map. (B) An illustration of the time series within the highlighted cluster in (A). (C) An example of molecular graphics demonstrating the interacting residues involved in the domain motions between TM5 and TM6 illustrated in the highlighted cluster in (A). Each connection involves two C_α whose distance time series is within the highlighted cluster in (A).

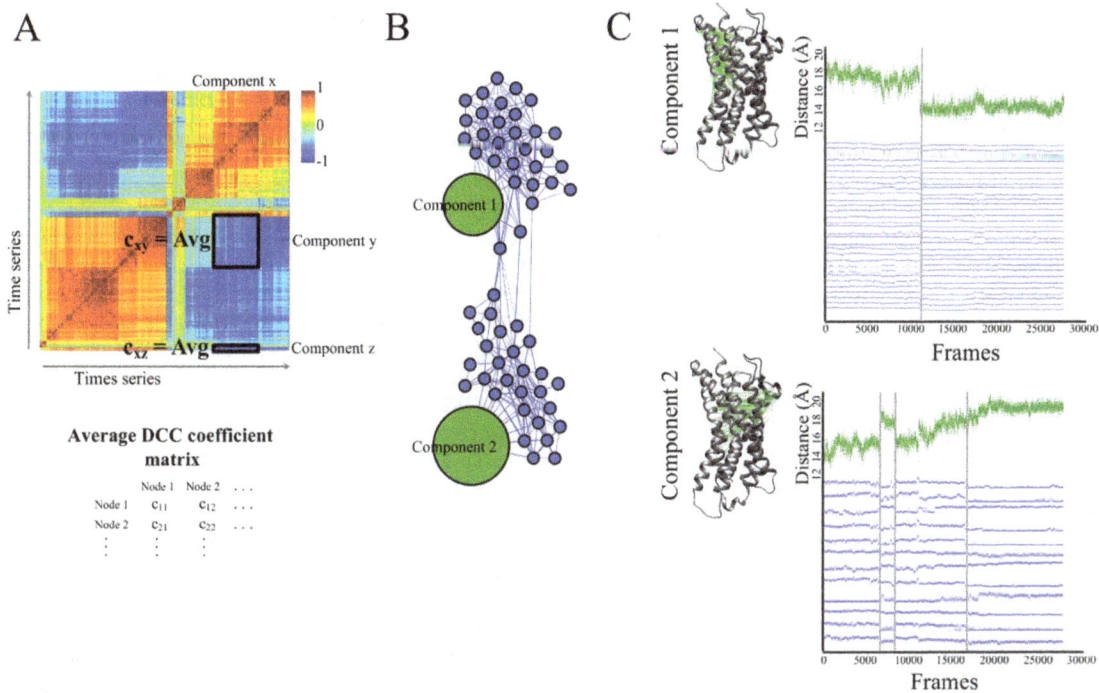

Fig. 4. DCC-based network illustration of the protein's dynamic components. (A) Correlation coefficients of pairwise dynamic components are calculated as the absolute average DCC coefficient of the pairwise time series belonging to each component as illustrated on a sample DCC heat map. Average correlations are calculated between pairwise domain motions (components x and y), between pairwise side chain rearrangement time series (component z) and across both components (components x and z). Average DCC coefficient matrix is generated for all pairwise dynamic components. (B) The network is built from a subset of the time series extracted from the MD simulation of CCL21-bound CCR7. The network is composed of two communities that are centered around domain motions labeled as component 1 and component 2. Network nodes represent the dynamic components extracted from the subset time series data and are colored blue for side chain rearrangements and green for domain motions. The size of each node is proportional to the number of time series the node represents. Edges connecting the dynamic components are based on the absolute average pairwise DCC coefficient of the time series involved in each of the components. Edges are drawn between dynamic components of a minimum coefficient of 0.75. C_α-C_α distance. (C) Time series that comprise each of components 1 and 2 are projected into the molecular graphics of CCR7 and labeled accordingly. Components 1 and 2 represent domain motions in a protein and are constituted of several highly correlated C_α-C_α distance time series. A sample time series from each of the domain motion components is shown in green. Blue time series are side chain time series for each of the blue nodes within each of the communities centered around components 1 and 2. All time series show coupled abrupt changes within each of the domain movements highlighted in grey. The network was built using Gephi [27].

and illustrated in Figs. 2 and 3. In the network (Fig. 4B), the blue and green nodes represent side chain and backbone interactions, respectively; and edges connect correlated components with a minimum absolute average correlation defined by the user (The network in our CCR7 case was constructed using a minimum average absolute correlation of 0.75).

Average correlation coefficients of pairwise dynamic components are calculated as the absolute value of the average DCC coefficient of the pairwise time series belonging to each component (Fig. 4A). The absolute average DCC coefficient matrix is generated for all the pairwise dynamic components and is then projected into the network where components are connected based on an average DCC coefficient cutoff (Fig. 4B). The absolute average correlation coefficient cutoff represents the degree of coupling between dynamic components, and can be adjusted to account for weak couplings between the fast side chain rearrangements (picosecond and nanosecond timescales) and the slow global motions of protein domains (microsecond and nanosecond timescales) (Fig. 4C, upper panel). In addition, while rearrangements in side chain interaction are manifested as abrupt changes in the distance time series, the global domain motion experiences more incremental changes that span hundreds of nanoseconds and can still be accounted for by also adjusting the cutoff absolute average DCC coefficient (Fig. 4C, lower panel). The size of each node is proportional to the number of distance time series each dynamic component (node) represents, where global dynamic components involve many time series, while side chain rearrangements are characterized by one distance time series (Fig. 4). Network communities are detected based on edge betweenness using the community_edge_betweenness method in *igraph* [24]. Each community is composed of side chain and backbone dynamics that are coupled to each other and represents the dynamic behavior of a protein subdomain

that is encoded in its side chain rearrangements and global domain motions (Fig. 4C).

2.4. Network Community Visualization Using Molecular Graphics Visualization Tools

MD simulations provide an insurmountable amount of dynamic information due to the high fluctuating and complex nature of protein dynamics. Here, the extracted communities reveal to be useful in reducing the MD data to its functional dynamic behavior, where each community is composed of coupled side chain rearrangements and global domain motions. These communities can be output into a protein data bank (PDB) file format to visualize the residues that make up the dynamic components of the community and each time series belonging to the dynamic components can be output as a pseudobond connecting two representative atoms of the time series' corresponding residues (Fig. 3C). This allows for better visualization and further analysis of the residues involved in mediating the allosteric communication within the protein.

3. Results and Discussion

3.1. Application to Molecular Dynamics Simulation Data

We apply our computational framework to previously published 7 μs-MD trajectory where we analyzed the simulations to understand the mechanism by which information is transmitted in CCR7 when bound to its agonist ligand, CCL21 [11]. We have determined key conformational changes that act as molecular switches and facilitate the transitioning of the receptor between its different states by inducing global motions of its transmembrane domain (TMD) helices [11]. The simulation dataset of CCR7 was originally analyzed through manual and visual inspection of a large set of distance time series and generic summary quantification, such as root mean square deviation (RMSD), principal component analysis (PCA), and comparison of the inter-residue mean distances between different time segments. Such non-systematic measures are very labor intensive and may not provide a complete analysis due to the overwhelming amount of the data output by the MD simulations. Nonetheless, we were able to detect a series of molecular switches that are mediated by various ligand-induced allosteric events. These molecular switches involve three tyrosine residues (Y112$^{3.32}$, Y255$^{6.51}$, and Y288$^{7.39}$), three phenylalanine residues (F116$^{3.36}$, F208$^{5.47}$, and F248$^{6.44}$), and a polar interaction between Q252$^{6.48}$ and R294$^{7.45}$ in the TMD of CCR7 [11]. Molecular events within these switches are coupled with global movements in the receptor's TM helices and contribute to the transitioning of the receptor to distinct states.

In our test case here, we apply our computational framework to the CCL21-bound CCR7 MD simulation data [11]. Using a distance cutoff of 5 Å, a total of ~1200 inter-residue side chain distance time series were imported and fit to a GMM in order to systematically extract all multi-modal distance probability densities. The selected contacts reduced our data set to ~600 time series. However, the majority of these contacts comprises of independent side chain rearrangements that do not

contribute to the protein's major motions, and only a fraction of these multi-modal contacts will remain in the final network of coupled dynamic components. The second part of our computational framework focused on extracting the receptor's global domain motions using inter-residue C$_\alpha$ distance time series with a cutoff of 15 Å. A pairwise DCC matrix was generated for ~6000 distance time series, and then clustered at a DCC coefficient cutoff of 0.95. The high DCC cutoff generated clusters with highly correlated distance time series that involve structurally adjacent amino acids. This part of the computational framework generated ~1000 clusters which included multiple clusters of more than one hundred time series (clusters containing a large number of time series represent a large number of residues involved in a global domain motion). After calculating all pairwise absolute average DCC coefficients between all dynamic components and projecting our data onto a DCC-based network, all dynamic components were then reduced to six communities with different dynamic behaviors that make up the orchestrated complex motions involved in transitioning CCR7 between two different states [11].

Using our computational framework, we systematically decomposed the protein dynamics into different sectors (subdomains) that show varying dynamic behaviors (Fig. 5). Our method reduces the protein dynamic interaction space of ~8000 time series into a network of 280 nodes representing side chain dynamic components and 127 nodes representing global domain dynamic components (node sizes is proportional to the number of time series representing the dynamic

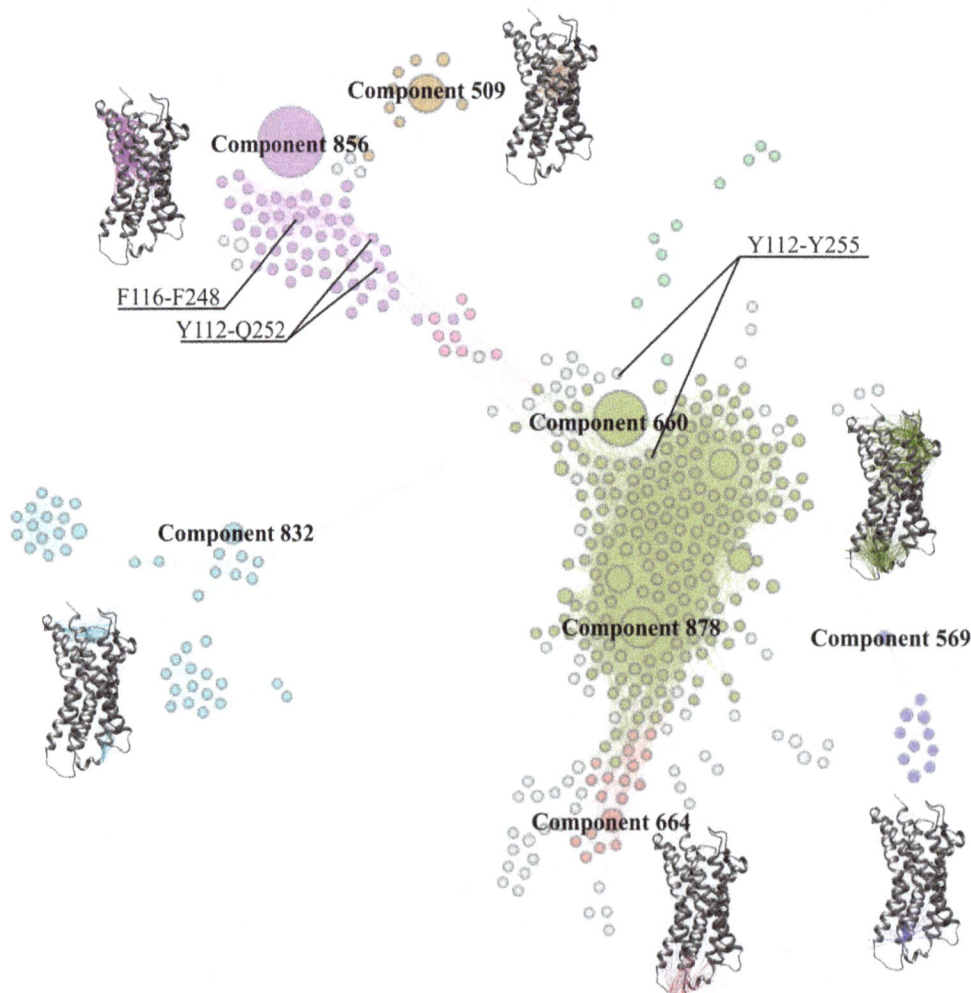

Fig. 5. A DCC-based network of the full CCL21-bound CCR7 MD simulation dataset. Network communities are colored differently and dynamic components representing domain motions are projected into a molecular graphics in which connections are colored according to the community they belong to. Previously determined molecular switches (F116-Q252, Y112-Q252, Y112-Y255) are labeled accordingly in the network [19].

components). Each community is composed of a few nodes that represent the main global motions of CCR7 and several nodes that represent side chain rearrangements. The network is decomposed into six communities that present a unified picture of the complex behavior of the protein's helices and loops. Each community contains all coupled dynamic components of the simulation and can be further analyzed to extract critical molecular switches that coordinate protein dynamics. Molecular switches consist of side chain rearrangements that switch controllably between two or more stable states in response to perturbations, and can be challenging to isolate from MD simulation data due to the complex dynamics of protein. Thus, summarizing and categorizing all dynamics data into a network will provide a clear picture of the large MD data sets of GPCRs that can be further analyzed using the extracted small set of communities. Analysis can be performed through manual and visual comparison between conformational states of each community, as performed by Vanatta et al. [25]. Each protein sector can also be clustered into states to extract control variables that could select for one or more of the conformational states using the Jensen-Shannon divergence statistics, as described by Fenley et al. [26]. Within our test case network, previously determined molecular switches highlighted in Fig. 5 (F116-F248, Y112-Y255, and Y112-Q252) were detected through visual and manual comparison of CCR7 conformational states [11]. These molecular switches belong to different communities centered around global motions of the receptor helices, which demonstrates their coupling to different global dynamic components of the receptor.

4. Concluding Remarks

This computational framework focuses on reducing the MD simulation data into a more manageable dynamic interaction space by mapping the GPCR dynamics into an intramolecular network of dynamic components composed of coupled side chain rearrangements and global conformational changes. This is done through the detection of side chain contacts with multi-modal probability density function and global domain motions manifested as clusters of highly correlated inter-residue C_α distance time series. Community detection in a DCC-based network of all extracted components correlate the side chain contacts to the domain motions in order to map all the different dynamic components of the protein into various communities of different dynamic behaviors.

As a proof of concept, this method was applied to a MD simulation of CCR7 to systematically detect the different protein sectors responsible for mediating the complex motions of its helices. Ultimately, our computational framework reduces the overall behavior of the protein to a set of communities composed of coupled side chain and global dynamic components. This method provides a reduced and more manageable dataset, where each community representing a separate protein sector can be further analyzed separately.

Acknowledgements

Anton computer time was provided by the National Center for Multiscale Modeling of Biological Systems (MMBioS) through Grant P41GM103712-S1 from the National Institutes of Health and the Pittsburgh Supercomputing Center (PSC). The Anton machine at PSC was generously made available by D.E. Shaw Research.

References

[1] Motlagh HN, Wrabl JO, Li J, Hilser VJ. The ensemble nature of allostery. Nature 2014; 508:331–9.
[2] Henzler-Wildman K, Kern D. Dynamic personalities of proteins. Nature 2007;450: 964–72.
[3] Grant BJ, Gorfe AA, McCammon JA. Large conformational changes in proteins: signaling and other functions. Curr Opin Struct Biol 2010;20:142–7.
[4] Popovych N, Sun S, Ebright RH, Kalodimos CG. Dynamically driven protein allostery. Nat Struct Mol Biol 2006;13:831–8.
[5] Salomon-Ferrer R, Götz AW, Poole D, Le Grand S, Walker RC. Routine microsecond molecular dynamics simulations with AMBER on GPUs. 2. Explicit solvent particle mesh ewald. J Chem Theory Comput 2013;9:3878–88.
[6] Shaw DE, Dror RO, Salmon JK, Grossman JP, Mackenzie KM, Bank JA, et al. Millisecond-scale molecular dynamics simulations on Anton. Proceedings of the conference on high performance computing networking, storage and analysis; 2009. p. 1–11.
[7] Miao Y, Feixas F, Eun C, McCammon JA. Accelerated molecular dynamics simulations of protein folding. J Comput Chem 2015;36:1536–49.
[8] Lindorff-Larsen K, Piana S, Dror RO, Shaw DE. How fast-folding proteins fold. Science 2011;334:517–20.
[9] Miao Y, Nichols SE, Gasper PM, Metzger VT, McCammon JA. Activation and dynamic network of the M2 muscarinic receptor. Proc Natl Acad Sci 2013;110:10982–7.
[10] Dror RO, Arlow DH, Maragakis P, Mildorf TJ, Pan AC, Xu H, et al. Activation mechanism of the β2-adrenergic receptor. Proc Natl Acad Sci 2011;108:18684–9.
[11] Gaieb Z, Lo DD, Morikis D. Molecular mechanism of biased ligand conformational changes in CC chemokine receptor 7. J Chem Inf Model 2016;56:1808–22.
[12] Kruse AC, Hu J, Pan AC, Arlow DH, Rosenbaum DM, Rosemond E, et al. Structure and dynamics of the M3 muscarinic acetylcholine receptor. Nature 2012;482:552–6.
[13] Huang W, Manglik A, Venkatakrishnan AJ, Laeremans T, Feinberg EN, Sanborn AL, et al. Structural insights into μ-opioid receptor activation. Nature 2015;524:315–21.
[14] Nygaard R, Zou Y, Dror RO, Mildorf TJ, Arlow DH, Manglik A, et al. The dynamic process of β2-adrenergic receptor activation. Cell 2013;152:532–42.
[15] Grant BJ, Rodrigues APC, ElSawy KM, McCammon JA, Caves LSD. Bio3d: an R package for the comparative analysis of protein structures. Bioinformatics 2006;22:2695–6.
[16] Sethi A, Eargle J, Black AA, Luthey-Schulten Z. Dynamical networks in tRNA:protein complexes. Proc Natl Acad Sci 2009;106:6620–5.
[17] Stolzenberg S, Michino M, LeVine MV, Weinstein H, Shi L. Computational approaches to detect allosteric pathways in transmembrane molecular machines. Biochim Biophys Acta Biomembr 2016;1858:1652–62.
[18] Ensign DL, Pande VS. Bayesian detection of intensity changes in single molecule and molecular dynamics trajectories. J Phys Chem B 2010;114:280–92.
[19] Wriggers W, Stafford KA, Shan Y, Piana S, Maragakis P, Lindorff-Larsen K, et al. Automated event detection and activity monitoring in long molecular dynamics simulations. J Chem Theory Comput 2009;5:2595–605.
[20] Fan Z, Dror RO, Mildorf TJ, Piana S, Shaw DE. Identifying localized changes in large systems: change-point detection for biomolecular simulations. Proc Natl Acad Sci 2015;112:7454–9.
[21] Shaw DE, Maragakis P, Lindorff-Larsen K, Piana S, Dror RO, Eastwood MP, et al. Atomic-level characterization of the structural dynamics of proteins. Science 2010; 330:341–6.
[22] Pedregosa F, Varoquaux G, Gramfort A, Michel V, Thirion B, Grisel O, et al. Scikit-learn: machine learning in python. J Mach Learn Res 2011;12:2825–30.
[23] van der Walt S, Colbert SC, Varoquaux G. The NumPy Array: a structure for efficient numerical computation. Comput Sci Eng 2011;13:22–30.
[24] Csardi G, Nepusz T. The igraph software package for complex network research. InterJournal Complex Syst 2006;1695:1–9.
[25] Vanatta DK, Shukla D, Lawrenz M, Pande VS. A network of molecular switches controls the activation of the two-component response regulator NtrC. Nat Commun 2015;6:7283.
[26] Fenley AT, Muddana HS, Gilson MK. Entropy–enthalpy transduction caused by conformational shifts can obscure the forces driving protein–ligand binding. Proc Natl Acad Sci 2012;109:20006–11.
[27] Bastian M, Heymann S, Jacomy M. Gephi: an open source software for exploring and manipulating networks. Third international AAAI conference on weblogs and social media; 2009.

7

Sequence comparison, molecular modeling, and network analysis predict structural diversity in cysteine proteases from the Cape sundew, *Drosera capensis*

Carter T. Butts [a,b,c,*], Xuhong Zhang [c], John E. Kelly [e], Kyle W. Roskamp [e], Megha H. Unhelkar [e], J. Alfredo Freites [e], Seemal Tahir [e], Rachel W. Martin [e,f,**]

[a] *Department of Sociology, UC Irvine, USA*
[b] *Department of Statistics, UC Irvine, USA*
[c] *Department of Electrical Engineering and Computer Science, UC Irvine, USA*
[e] *Department of Chemistry, UC Irvine, USA*
[f] *Department of Molecular Biology & Biochemistry, UC Irvine, Irvine, CA, 92697 USA*

ARTICLE INFO

Keywords:
Cysteine protease
Carnivorous plant
Protein sequence analysis
Protein structure prediction
Protein structure network
Rosetta
Molecular dynamics
Digestive enzyme
In silico maturation

ABSTRACT

Carnivorous plants represent a so far underexploited reservoir of novel proteases with potentially useful activities. Here we investigate 44 cysteine proteases from the Cape sundew, *Drosera capensis*, predicted from genomic DNA sequences. *D. capensis* has a large number of cysteine protease genes; analysis of their sequences reveals homologs of known plant proteases, some of which are predicted to have novel properties. Many functionally significant sequence and structural features are observed, including targeting signals and occluding loops. Several of the proteases contain a new type of granulin domain. Although active site residues are conserved, the sequence identity of these proteases to known proteins is moderate to low; therefore, comparative modeling with all-atom refinement and subsequent atomistic MD-simulation is used to predict their 3D structures. The structure prediction data, as well as analysis of protein structure networks, suggest multifarious variations on the papain-like cysteine protease structural theme. This *in silico* methodology provides a general framework for investigating a large pool of sequences that are potentially useful for biotechnology applications, enabling informed choices about which proteins to investigate in the laboratory.

1. Introduction

The proteases of carnivorous plants present attractive targets for exploitation in chemical biology and biotechnology contexts. Carnivorous plants, such as *Drosera capensis*, whose prey capture functions take place in the open have been rigorously selected by evolution for the ability to digest large prey over long time periods, without assistance from physical disruption of prey tissue, and in competition with ubiquitous fungi and bacteria. These evolutionary constraints have selected for highly stable enzymes with a different profile of substate specificities and cleavage patterns from those found in animal digestive enzymes. Carnivorous plant digestive enzymes function at pH values ranging from 2–6, depending on the species [1,2]. They also function over a

wide range of temperatures; *Drosera* are endemic to every continent except Antarctica and both tropical and temperate species exist. In particular, the pH of *D. capensis* mucilage is around 5 [3], and temperatures in the Western Cape region of South Africa where these plants are found typically range from 5–30 °C.

Characterization of carnivorous plant digestive enzymes could lead to their use in a variety of laboratory and applications contexts, including analytical use in proteomics studies as well as preventing fouling on the surface of medical devices that cannot be treated under harsh conditions. New proteases may also prove useful for cleaving amyloid fibrils, such as those responsible for the transmission of prion diseases or the formation of biofilms by pathogenic bacteria. The characterization of aspartic proteases from the tropical pitcher plants (*Nepenthes* sp.) [4–6], has already led to useful advances in mass spectrometry-based proteomics applications, where the ability to digest proteins using a variety of cut sites is essential for identifying proteins and peptides from complex mixtures. Proteases from plant and animal sources are also important components of pharmaceutical preparations for gluten

* Correspondence to: C. Butts, Calit2, UC Irvine, CA, 92697, UCI.
** Correspondence to: R. Martin, Department of Molecular Biology & Biochemistry, UC Irvine, Irvine, CA, 92697, USA.
E-mail addresses: buttsc@uci.edu (C.T. Butts), rwmartin@uci.edu (R.W. Martin).

intolerance, arthritis, and pancreatic disease [7]. Characterizing proteases from carnivorous plants has the potential to diversify the toolbox of proteases with different functional properties that are available for these and other applications.

Plant cysteine proteases form a large and diverse family of proteins that perform cellular housekeeping tasks, fulfill defensive functions, and, in carnivorous plants, digest proteins from prey. It is typical for plants to contain many different cysteine protease isoforms; for instance, in the case of tobacco (*Nicotiana tabacum*), more than 60 cysteine protease genes have been identified [8]. Many of the cysteine proteases of interest are classified by the MEROPS database as family C1 [9], a broad class of enzymes including cathepsins and viral proteases as well as plant enzymes that function to deter herbivory. C1 proteases can operate as endopeptidases, dipeptidyl peptidases, and aminopeptidases [10]. In plants, many C1 enzymes are used to degrade proteins in the vacuole, playing many of the same roles as their lysosomal counterparts in animals [11]. They are also found in fruits, particularly unripe ones; this protease activity impedes insect feeding and also serves to cleave endogenous proteins during fruit ripening. Some families of cysteine proteases in plants have been subject to diversifying selection due to a molecular arms race between these plants and their pathogens; as plants produce proteases that suppress fungal growth, fungi evolve inhibitors specific to these proteases, driving the diversification of plant proteases involved in the immune response [12].

The plethora of paralogs found in a typical plant is indicative of the need for a range of different substrate specificities; this is particularly important in the case of carnivorous plants, which must digest prey proteins to their component amino acids. Aspartic proteases have long been implicated in *Nepenthes* pitcher plant digestion [4,13], and more recently the cysteine protease dionain 1 has been confirmed as a major digestive enzyme in the Venus flytrap (*Dionaea muscipula*) [14]. In *D. capensis*, proteins from prey consititute the major nitrogen source for producing new plant tissue [15]. Given that plant carnivory appears to have evolved from defensive systems in general [16], and that the feeding responses are triggered by the same signaling pathway as is implicated in response to wounding [17], one would expect cysteine proteases to play a major role; here we investigate some of the many cysteine protease genes in *D. capensis* with the objective of adding to the portfolio of cleavage activities available for chemical biology applications. The *D. capensis* enzymes are particularly appealing for mass spectrometry-based proteomics applications, due to their ability to operate under relatively mild conditions, i.e. at room temperature and pH 5.

This study focuses on the C1 cysteine proteases from the Cape sundew (*D. capensis*), whose genome we have recently sequenced [18]. Here we use sequence analysis, comparative modeling with all-atom refinement and atomistic molecular dynamics (MD) simulation, and investigation of protein structure networks to identify structurally distinct subgroups of proteins for subsequent expression and biochemical characterization.

C1 cysteine proteases share a common papain-like fold, a property also predicted for the proteins studied here. Despite this conservation of the papain fold and critical active and structural residues, sequence analysis of the *D. capensis* cysteine proteases indicates that they represent a highly diverse group of proteins, some of which appear to be specific to the Droseraceae. In particular, a large cluster of proteases containing dionains 1 and 3 as well as many homologs from *D. capensis* has particular sequence features not seen in papain or other reference enzymes. Finally, a new class of granulin domain-containing cysteine proteases is identified, based on clustering of the granulin domains themselves.

Molecular modeling was performed in order to translate this sequence diversity into predicted structural diversity, which is more informative for guiding future experimental studies. Examination of the predicted enzyme structures potentially suggests diversity that may imply a variety of substrate preferences and cleavage patterns. The relationships between the shape of the substrate-recognition pockets and

variation in substrate cleavage activity have been established for other plant cysteine proteases, including the ervatamins [19], the KDEL-tailed CysEP protease from the castor bean [20], and in dionain 1 [21]. The sequence-structure relationships outlined here suggest hypotheses that can be tested in the laboratory, providing a starting point for discovering novel enzymes for use in biotechnology applications. In most cases, the sequences have only weak identity to known plant proteases, making traditional homology modeling of dubious utility. Instead, we use Rosetta [22,23] to perform comparative modeling with all-atom refinement, combining local homology modeling based on short fragments with de novo structure prediction. We then employ atomistic MD simulation of these initial structures in explicit solvent to produce equilibrated structures with corrected active site protonation states; these equilibrated structures serve as the starting point for further analysis.

Quality control was performed using both sequence alignment and inspection of the Rosetta structures; proteins that are missing one of the critical active residues (C 158 or H 292, papain numbering) were discarded, as were some lacking critical disulfide bonds or other structural features necessary for stability. After winnowing out sequences that are unlikely to produce active proteases, 44 potentially active proteases were chosen for further analysis. This methodology allows the development of hypotheses based on predicted 3D structure and activity, in contrast to focusing on the first discovered or most abundantly produced enzymes, enabling selection of the most promising targets for structural and biochemical characterization based on the priorities of technological utility rather than relative importance in the biological context.

2. Methods

2.1. Sequence Alignment and Prediction of Putative Protein Structures

Sequence alignments were performed using ClustalOmega [24], with settings for gap open penalty = 10.0 and gap extension penalty = 0.05, hydrophilic residues = GPSNDQERK, and the BLOSUM weight matrix. The presence and position of a signal sequence flagging the protein for secretion was predicted using the program SignalP 4.1 [25], while other localization sequences were identified using TargetP [26]. Structures were predicted using a three-stage process. First, an initial model was created for each complete sequence using the Robetta server [22]; the Robetta implementation of the Rosetta [23] system generates predictions from sequence information using a combination of comparative modeling and all-atom refinement based on a simplified forcefield. Second, any residues not present in each mature protein were removed, disulfide bonds were identified by homology to known homologs, and the protonation states of active site residues were fixed to their literature values. Finally, in the third phase, each corrected structure was equilibrated in explicit solvent under periodic boundary conditions in NAMD [27] using the CHARMM22 forcefield [28] with the CMAP correction [29] and the TIP3P model for water [30]; following minimization, each structure was simulated at 293 K for 500 ps, with the final conformation retained for subsequent analysis. This process was performed for the 44 protease sequences from *D. capensis*, as well as 10 reference sequences from other organisms (see below); where published structures were available, these were used as the initial starting model (following removal of heteroatoms and protonation using REDUCE [31] as required). For the 5 *D. capensis* sequences with granulin domains (as well as the two references with such a domain), steps (2) and (3) of the above were repeated after removal of the domain and any linking residues. This process provides predicted structures both with and without the domain in question. The PDB files corresponding to the equilibrated structures for all the proteins discussed in this manuscript are available in the Supplementary Information.

2.2. Network Modeling and Analysis

Each equilibrated protein structure was mapped to the network representation of Benson and Daggett [32] using custom scripts employing both VMD [33] and the statnet toolkit [34,35] within the R statistical computing system [36]. Each vertex within the resulting protein structure network (PSN) represents a chemical group, with edges representing potential interaction as determined by proximity within the protein structure. PSNs were then compared using the structural distance technique of Butts and Carley [37], which provides a uniform way to compare the underlying structures of networks (i.e. graphs) with different vertex sets; this involves mapping both graphs onto a common vertex set (adding isolated vertices to the smaller graph as needed) such that the differences between the two mapped networks are minimized with respect to an underlying metric. The value of this metric after mapping is the structural distance. Here, distances were computed between unlabeled graphs based on an underlying Hamming metric, and can be interpreted as the minimum number of edge changes required to transform a member of the isomorphism class of the first graph (i.e., the set of all graphs having the same underlying typology) into a member of the isomorphism class of the second (or vice versa). The raw structural distance between each pair of PSNs was then normalized by graph order, yielding a metric corresponding to edge changes per vertex. Normalized structural distances between PSNs were analyzed via metric multidimensional scaling and hierarchical clustering using R. Additional network visualization and analysis was performed using the sna library [38] within statnet.

3. Results and Discussion

3.1. D. capensis Cysteine Proteases Cluster Into Distinct Families Based on Resemblance to Known Homologs

All *D. capensis* sequences previously annotated as coding for MEROPS C1 cysteine proteases using the MAKER-P (v2.31.8) pipeline [39] and a BLAST search against SwissProt (downloaded 8/30/15) and InterProScan [40] were clustered by sequence similarity. Several previously-characterized cysteine proteases that have been identified from other plants are also included as reference sequences. Clustering of the *D. capensis* cysteine protease sequences reveals a broad range of cysteine protease types, some of which are homologous to known plant proteases (Fig. 1). Three of the six clusters contain only proteins from *D. capensis* or the related Venus flytrap *D. muscipula*, while many of the reference sequences cluster together despite coming from a variety of different plant species from diverse orders including both monocots and eudicots (Supplementary Table S1). The general types of plant protease features found correlate well with previous surveys of cysteine proteases in *Arabidopsis thaliana* [41], *Populus* sp. [42], and more recently, soybeans [43] and a broader group of plant proteases from a variety of species [44].

3.2. Residues Conserved in D. capensis Cysteine Proteases Include Active Sites and Important Sequence Features

A defining feature of C1A cysteine proteases is the Cys-His catalytic dyad, which is often accompanied by an Asn residue that stabilizes the protonated catalytic His [46,47]. The mechanism of these enzymes requires using the thiolate group on the deprotonated cysteine as a nucleophile to attack a carbonyl carbon in the backbone of the substrate. Preliminary sequence alignments comparing putative cysteine proteases from *D. capensis* were used to discard sequences lacking the conserved Cys and His residues of the catalytic dyad due to either substitution or truncation. Other conserved features were observed in many of the sequences, but were not treated as necessarily essential for activity. Reference sequences used include zingipain 1 from *Zingiber officianale* (UniProt P82473), pineapple fruit bromelain (*Ananas*

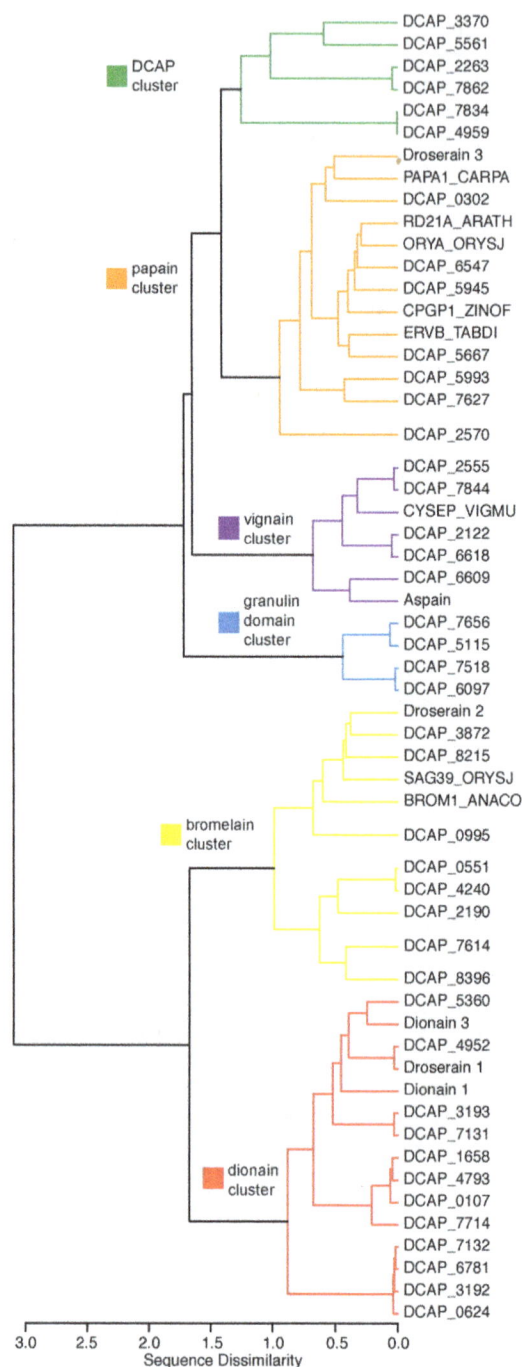

Fig. 1. Clustering of cysteine protease sequences identified from the *D. capensis* genome. Many are homologous to known plant cysteine proteases, including dionain 1 and dionain 3 from the Venus flytrap, *D. muscipula*. Dissimilarity between clusters is defined by the e-distance metric of Székely and Rizzo [45] (with $\alpha = 1$), which is a weighted function of within-cluster similarities and between-cluster differences with respect to a user-specified reference metric. The underlying input metric employed here is the raw sequence dissimilarity $(1 - (\%identity)/100)$.

comosus, UniProt O23791), RD21 from *A. thaliana* (UniProt P43297), oryzain alpha chain (UniProt P25776) and SAG39 (UniProt Q7XWK5) from *Oryza sativa* subsp. *japonica*, ervatamin b from *Tabernaemontana divaricata* (UniProt P60994), and dionains 1 and 3 from the related *D. muscipula* (UniProt A0A0E3GLN3, and A0A0E3M338, respectively). Several of the reference sequences, e.g. zingipain-1 [48], were characterized by mass spectrometric analysis of the mature enzyme; these sequences therefore lack the signal peptide and pro-sequence found in the initially transcribed sequence (see below).

Sequence alignments for the individual clusters (Supplementary Figs. S1–S7) are annotated to highlight individual amino acid properties, residues conserved within the cluster and/or shared with papain, as well as functional sequence features, as described in detail in the S.I. In addition to the cluster-specific reference sequences, all clusters include papain (*Carica papaya*, UniProt P00784) in order to have a common reference for all the C1A proteases discussed in this work.

Most of the clusters are named after a reference sequence or a distinguishing feature of its members. The DCAP cluster is highly diverse, yet it contains only sequences from *D. capensis*. The papain cluster contains many of the reference sequences, as well as several *D. capensis* proteases, some of which have granulin domains (Figs. S2 and S3), a feature that is peculiar to plant cysteine proteases. The vignain cluster (Fig. S4) contains vignain from *Vigna mungo* (UniProt P12412) as well as *D. capensis* homologs. Many of the proteins in the vignain cluster have C-terminal KDEL tags, indicating retention in the ER lumen, suggesting that they are involved in germination and/or senescence. In the granulin domain cluster (Fig. S5), every sequence but one contains a granulin domain connected to the catalytic domain by a proline-rich linker of about 40 residues; the one exception is truncated after the proline-rich region. Several sequences in the papain cluster also contain granulin domains, however the Pro-rich linkers in those sequences contain only about 16 residues and the sequence identity between the two types of granulin domains themselves is not high. The bromelain cluster (Fig. S6) contains homologs of both defensive and senescence-related enzymes. Every sequence in the dionain cluster (Fig. S7) contains an extra Cys residue immediately prior to the active site Cys. This CCWAF structural motif has been previously observed in the Arabidobsis protein SAG12 and homologs [44]; however, the function of the double Cys in unknown. It may have cataytic relevance, perhaps providing a second nucleophilic thiolate or operating as a redox switch.

Like many other proteases, the papain-family enzymes are expressed with an N-terminal pro-sequence blocking the active site. This sequence is cleaved during enzyme maturation, often upon the protein's entering a low-pH environment. This pro-sequence was found in most of the C1A proteases from *D. capensis* (highlighted with pink boxes in Figs. S1–S7 in the SI). Plant C1A protease pro-sequences are often bioactive in their own right, acting as inhibitors of exogenous cysteine proteases. This enables them to deter herbivory by insects [49], nematodes [50], and spider mites [51], protecting the plants from damage. This can be technologically exploited by producing transgenic crop varieties with protective cysteine proteases they would otherwise lack [52]. This approach has proven useful in protecting crops from Bt-resistant pests [53]. Despite some variation in the lengths of the C-terminal and N-terminal regions, all the cysteine proteases investigated here show substantial similarity in the pro-sequences; in particular, the ERFNIN motif (EX $_3$RX $_3$FX $_2$NX $_3$IX $_3$N) often found in the pro-sequence of C1A proteases [54] is conserved in many sequences spanning all the clusters. Interestingly, the alternative sequence EX $_3$RX $_3$FX $_2$NX $_3$AX $_3$Q, which is characteristic of the RD19 family of plant cysteine proteases, is found in only one of the *D. capensis* proteases, DCAP_3370 in the DCAP cluster. For all previously uncharacterized sequences, SignalP 4.1 [25] was used to predict the location of the signal sequences, if any, while the pro-sequences were predicted by sequence similarity and structural homology to papain. These sequence annotations were then used as the basis for further structure prediction and functional analysis.

In addition to the common sequence features in the N-terminal pro-region, other variations are observed, such as the presence of C-terminal granulin domains in some sequences and extra insertions that may be responsible for specific activities in others. Examples of organelle-specific targeting sequences are observed; several sequences have a C-terminal KDEL sequence targeting them for retention in the endoplasmic reticulum, while others have targeting sequences indicating their destination in the cells, including signals indicating transport to the vacuole (NPIR, but not FAEAI or LVAE) or the peroxisome (SSM at the C-terminus). The level of sequence conservation among the members of each cluster varies dramatically, as can be seen in Fig. S8, where sequence conservation is mapped onto the structure of a representative member of each cluster. The sequences in the DCAP cluster are less closely related to each other than the members of any of the other clusters, and some are homologous to reference sequences used by Richau et al. [44].

DCAP_2263 and DCAP_7862 belong to the Richau aleurain (cathepsin H) cluster. In humans, cathepsin H is an aminopeptidase that processes neuropeptides in the brain [55], as well as acting as a lysozomal protein in other tissues. Its barley (*Hordeum vulgare*) homolog, aleurain, has both aminopeptidase and endopeptidase activity [56], suggesting that DCAP_2263 and DCAP_7862 may have both types of activity as well. This hypothesis is supported by the presence of the Cathepsin H minichain sequence in its plant orthologs, as discussed in the section devoted to these proteins. DCAP_3370 is related to the Richau RD19 (cathepsin F) cluster, and is the only protease in this set that contains the characteristic pro-sequence motif (EX $_3$RX $_3$FX $_2$NX $_3$AX $_3$Q), of the RD19 (cathepsin F) family. Human cathepsin F is distinguished by its unusually long pro-domain, which is approximately 100 residues longer than that of other cysteine proteases and adopts a cystatin fold [57]. In contrast, the pro-sequence of DCAP_3370 is about 140 residues, typical for a plant cysteine protease. The last enzyme in the DCAP cluster, DCAP_5561 is not closely related to anything in either reference set. A BLAST search yields numerous matches to uncharacterized predicted cysteine proteases from a variety of plant genomes, however, the specific function of this enzyme remains enigmatic.

3.3. Molecular Modeling Predicts Many Variations on the Papain Structural Theme

Carnivorous plants require a variety of proteases with different substrate affinities and cleavage sites to effectively digest the proteins from their prey, in addition to the standard spectrum of protease activities required by all plants. Cysteine protease activity has previously been inferred from biochemical activity assays of the digestive fluids of *D. indica* [58], and dionain 1 from *D. muscipula* has been structurally and biochemically characterized [21]. However, with the exception of the nepethesins and dionain 1, these enzymes have yet to be extensively investigated. In this study, 44 cysteine proteases with moderate sequence homology to papain-like enzymes of known structure have been identified from the genome of *D. capensis*. For each putative functional protease, the structure of the full-length sequence, including the signal peptide, the pro-domain(s), and the granulin domain if present, was predicted using Rosetta. The resulting sequence was then subjected to *in silico* maturation, where known features of these enzymes were corrected, including addition of disulfide bonds and removal of pro-sequences and granulin domains, followed by equilibration using MD.

The *in silico* maturation and equilibration process allows for refinement of the initial Rosetta structure predictions. The Rosetta structure for a representative full-length protease (DCAP_7714) is shown in Fig. 2a. The full-length sequence consists of the active region, a secretion signal peptide (light orange), and an N-terminal pro-sequence (pink). The core sequence making up the mature form of this enzyme (dark blue) is structurally similar to papain, with two domains of approximately equal size, one primarily α-helical and the other mostly composed of β-strands, with the active site cleft between them. The inset in Fig. 2b shows the active Cys (yellow)/His (purple) dyad as well as the stabilizing Asn residue (magenta). In general, the structures predicted by Rosetta provide reasonable estimates for the overall folds of these enzymes, given their homology to papain. However, some details such as side chain rotamers are not perfectly consistent with known structures of papain-like enzymes. In particular, in the Rosetta structure, the S of the active cysteine is rotated up and away from the active histidine, and the side chains of cysteine residues predicted to be involved in disulfide bonds are not in the correct orientations (Fig. 2c). In order to generate more realistic structures for network analysis, *in silico* maturation and

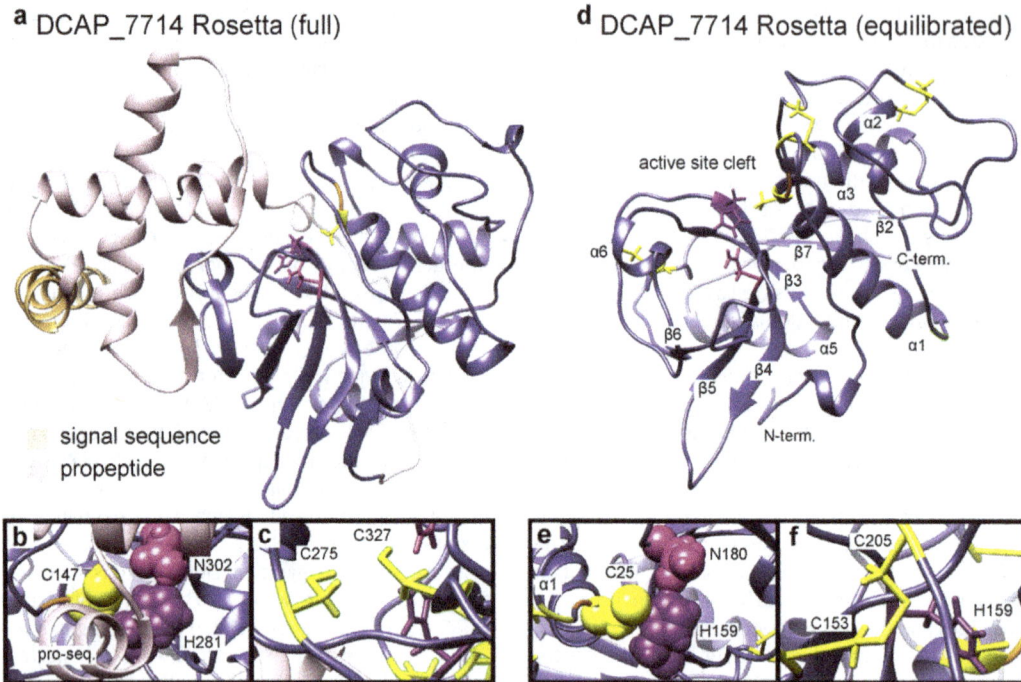

a DCAP_7714 Rosetta (full)

signal sequence
propeptide

d DCAP_7714 Rosetta (equilibrated)

Fig. 2. Predicted structures of DCAP_7714 before (a) and after (d) *in silico* maturation. (b) The active site residues (shown as space-filling models and with zymogen numbering) are in an unfavorable conformation prior to adjustment of their protonation states and equilibration in explicit solvent, whereas after equilibration. (e) the conformation is more consistent with that of an active cysteine protease (the same active site residues are shown but with mature sequence numbering). (c) In the initial Rosetta structure, the Cys rotamers (shown here for residues C275 and C327, zymogen numbering) are not generally in the ideal conformation for disulfide bonding, even in cases where it is expected. (f) Disulfide bonds (positions determined using sequence homology to papain) were added before equilibration. The residues shown here are the same as in panel (c), but with mature sequence numbering .

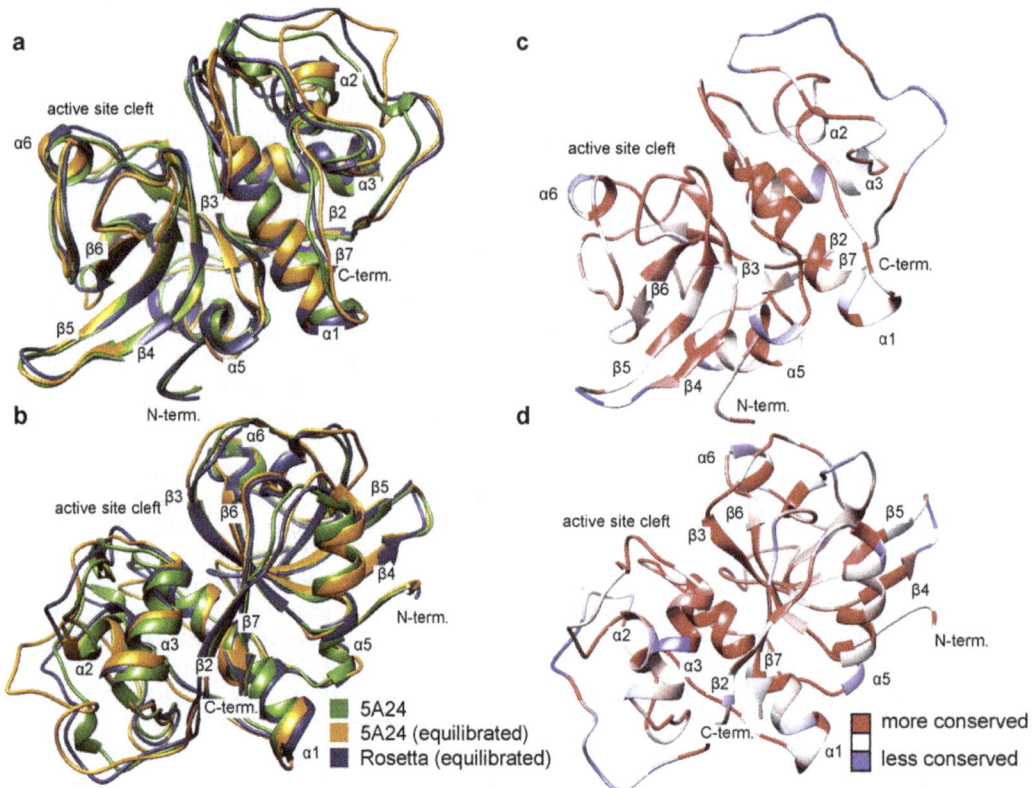

Fig. 3. a. and b. Structural comparison of the X-ray crystal structure of dionain 1 (PDBID 5A24) [21] (green) with the same structure after equilibration in solvent (orange) and the structure predicted by Rosetta after equilibration (blue), two different views. The Rosetta structure predicts all important secondary structure features observed in the crystal structure. Equilibration of the crystal structure in solvent prior to docking studies results in conformational changes to flexible loops as well as repositioning of side chains. c. and d. The percent conservation for each residue in the consensus sequence of the entire dionain cluster is plotted on the structure of dionain 1. Highly conserved residues tend to cluster in sequence regions where the predicted structure coincides with the observed structure, consistent with the idea that structurally important residues are strongly conserved.

equilibration were performed. The equilibrated structure of the mature form of DCAP_7714 is shown in Fig. 2d. Secondary structure elements are numbered according to the structure of Than et al. for the homologous *Ricinus communus* CysEP enzyme [20]. The protonation states of the active Cys and His were modified to reflect their expected states in the mature enzyme, resulting in more realistic side chain conformations in the equilibrated structure (Fig. 2e). The disulfide bonds were added to the structure before equilibration (Fig. 2d) based on homology to papain (and RD21A_ARATH in the case of enzymes containing a granulin domain).

As a validity check on our approach, we provide a comparison between our structural predictions (initial and refined) and an out-of-sample observation. The recently solved x-ray crystal structure of dionain 1 was published after our initial prediction and equilibration of this protein was performed, and hence this structure could not have been in the Rosetta training set. In Fig. 3a and b, the crystal structure (5A24, green) [21] is shown overlaid with its MD equilibrated counterpart (orange), and the structure predicted by Rosetta, after *in silico* maturation (blue). The Rosetta structure shows excellent agreement with the experimentally determined crystal structure, with nearly complete overlap of the major secondary structure elements, e.g. helices 1, 3, and 5, as well as the β-sheet formed by β-strands 3, 4, 5, and 6. As expected, substantially less agreement is observed in the loop regions, such as the flexible linkers between helices 1 and 2 and between helix 3 and strand 4. Molecular dynamics equilibration of the crystal structure in TIP3P solvent also results in movement of the loop regions, as is evident from comparison of the green and orange structures. Examination of the sequence conservation map of the dionain cluster plotted on the dionain 1 structure (Fig. 3 c and d) reveals that the most strongly conserved residues coincide with sequence regions predicted well by Rosetta, e.g. helix 1 and strands 3, 4, 5, and 6. In contrast helix 2, the loop regions, and the N-terminus have a higher RMSD between the crystal structure and the predicted structure, and display lower sequence conservation. This is consistent with the hypothesis that one reason for conservation of particular residues is that they are important for maintaining the structure. Overall, the close match between the predicted and observed dionain 1 structures indicates that our approach can provide excellent structural predictions within this class of proteins.

3.4. Some Cysteine Proteases Are Targeted to Specific Locations

Several of the cysteine proteases identified from *D. capensis* contain known targeting signals that mark the protein for delivery to specific cellular locations. The most common such signal is the N-terminal signal peptide targeting the protein for secretion. As expected, the majority of proteins in this set contain such a secretion signal. In plants, the secretory pathway delivers proteins to the vacuole, the vacuolar membrane, the cell wall, and the plasma membrane. In *D. capensis*, digestive enzymes are also expected to be secreted into the mucilage. In addition to the N-terminal signal sequences, tri- or tetrapeptides indicating that the protein is destined for a particular subcellular compartment are also found in many cases. Fig. 4 shows the structures predicted by Rosetta for three full-length cysteine proteases containing targeting signals, DCAP_2263, DCAP_5667, and DCAP_2122. Ribbon diagrams are shown for all three enzymes; a surface is also shown for DCAP_2122 in order to assist with visualization of the relationship of the pro-sequence, N-terminal signal peptide, and C-terminal localization sequence to the rest of the protein. The positioning of the pro-sequences (pink) and signal peptides (light orange) is highly variable, although in each example the pro-sequence blocks the active site and the signal sequences and other localization tags (light purple) are in highly exposed positions as expected based on their function.

In plants, the subsequence NPIR in the N-terminal region of a protein indicates targeting to the vacuole, a large acidic compartment that is specific to plant cells and serves the same function as the lysozome in animal cells. These compartments, which often occupy most of the

volume of the cell, contain a variety of hydrolases, including both aspartic and cysteine proteases, which normally act to recycle damaged or unneeded cellular components. Upon infection by viruses or fungal pathogens, the vacuole can also fuse with the plasma membrane to release defensive proteases into the extracellular space. Two putative vacuolar proteases, (DCAP_2263 and DCAP_7862) are found in the DCAP cluster. The NPIR tag is located in an exposed position between the secretion signal and the beginning of the N-terminal prosequence, as shown for DCAP_7862 in Fig. 1a. These proteases display sequence homology to mammalian cathepsin H, a lysozomal protein that is important in development and also implicated in cancer proliferation [59,60].

In human cathepsin H, aminopeptidase activity is modulated by the minichain sequence (EPQNCSAT). DCAP_2263 and DCAP_7862 (and aleurain, but no others in this set) contain the sequence AAQNCSAT, which may have a similar function. The hypothesis that this plant-specific minichain serves a similar role in modulating the substrate specificity is supported by comparing the predicted structures with the crystal structure of porcine cathepsin H (PDBID: 8PCH) [61]. Fig. 5 shows the predicted structures of mature DCAP_2263 (blue) and DCAP_7862 (green) overlaid with the crystal structure of porcine cathepsin H (gray). The predicted structures of the plant proteins coincide with the porcine protein in the major secondary structure elements, albeit with substantial variation in loops and linkers. The minichain sequence (EPQNCSAT in the porcine protein and AAQNCSAT in the *D. capensis* proteins) occupies a similar position in all three structures, allowing substrate approach to the active site cleft from one side (Fig. 5a), but not the other (Fig. 5b). Biochemical characterization of human cathepsin H has shown that deletion of the minichain abolishes aminopeptidase activity [62], making this protein a standard endopeptidase. Based on sequence homology and examination of the predicted structures, we hypothesize that this sequence plays a similar role in modulating the substrate specificity and activity patterns of DCAP_2263 and DCAP_7862.

Other proteases are targeted to the peroxisomes, organelles that bud from the ER membrane and primarily break down long-chain fatty acids, but are also involved in the synthesis of functional small molecules, such as isoprenoids, polyamines, and benzoic acid [63]. Some proteases in the peroxisome are involved in the maturation of other enzymes imported to this organelle, as well as disposal of oxidized proteins that build up in this challenging redox environment [64]. Others are active during different developmental stages, such as differentiation of seed glyoxysomes to mature leaf peroxisomes [65]. The most common type of targeting signal for transport to the peroxisome is one of several C-terminal tripeptides. The canonical example is SKL, but others have been discovered in a variety of plant proteins [66]. DCAP_5667, which is in the papain cluster (Fig. S3), has the tripeptide SSM at its extreme C-terminal end, indicating targeting to the peroxisome. DCAP_7656, which is in the granulin domain cluster (Fig. S5), contains the SKL sequence not at the C-terminal end, but at a highly exposed position near the C-terminus, suggesting possible peroxisome targeting for this protein also. DCAP_7656 contains the proline-rich linker common to this cluster, but its granulin domain is truncated. Another possibility is that the short sequence region following the SKL tripeptide may be cleaved under some circumstances, acting as a switch that determines whether this enzyme is sent to the peroxisome or elsewhere. Peroxisome-targeted proteases represent attractive targets for biotechnological studies, because they are optimized to remain stable and maintain their activity under harshly oxidizing conditions.

Proteins with the sequence KDEL at the C-terminus are retained in the lumen of the endoplasmic reticulum, enabling them to be stored in specialized vesicles as zymogens and released to mediate programmed cell death in response to a stressor or during a particular developmental phase. KDEL-tailed proteases such as vignain from *V. mungo* and CysEP from *R. communis* play an important role during

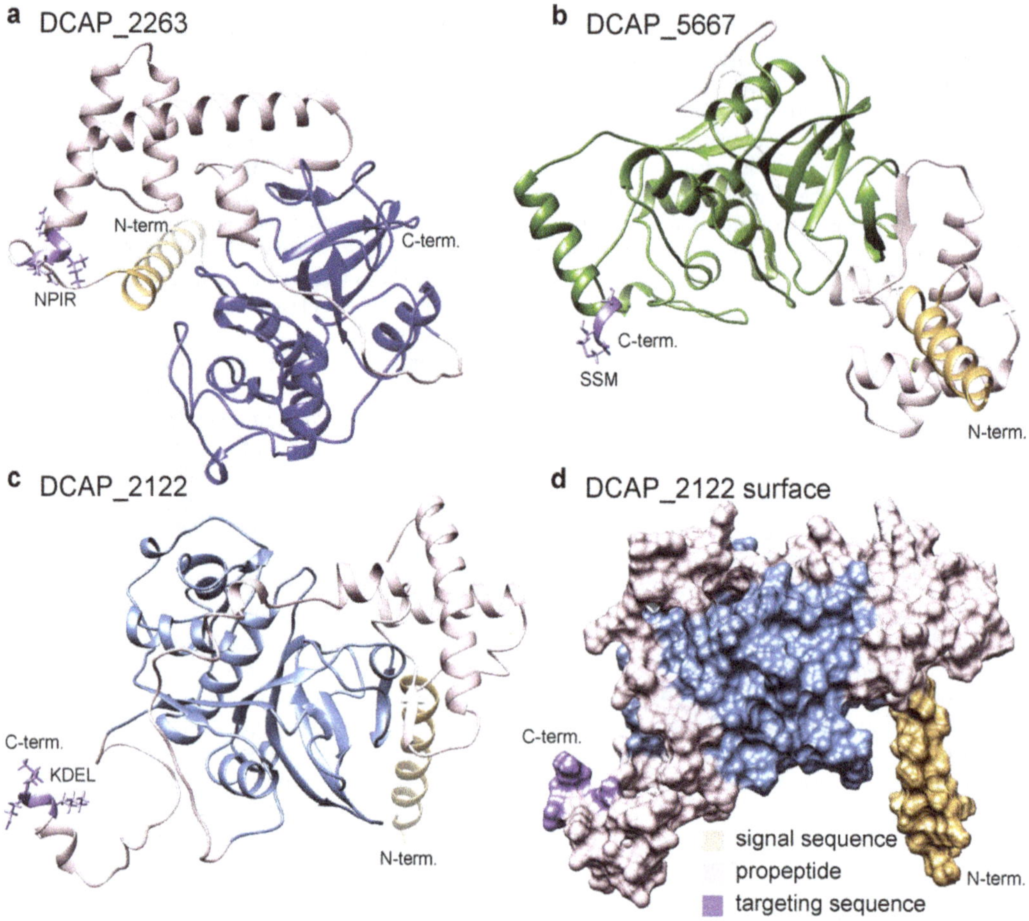

Fig. 4. Predicted structures for three full-length cysteine proteases. The secretion signals are highlighted in light orange, the pro-sequences in pink, and the localization tags in light purple. a. DCAP_2263 contains the target sequence NPIR, indicating localization to the vacuole. b. DCAP_5667 ends in the tripeptide SSM at the extreme C-terminus, indicating transport to the peroxisome c. and d. DCAP_2122 ribbon diagram and surface model, respectively. DCAP_2122 ends in the ER-retention signal KDEL, indicating that it is retained in the ER lumen.

germination, when proteins stored in endosperm tissue are degraded for use as the cotelydons develop. A C-terminal pro-peptide including the KDEL tag is removed along with the N-terminal pro-sequence during maturation, to yield the soluble, active enzyme [67]. The crystal structure and biochemical characterization of a homologous KDEL-tailed protein from the castor bean indicates that this enzyme has a strong preference for large, neutral amino acids in the substrate peptides, and has an unusually large and possibly flexible substrate-binding pocket that can accommodate a variety of sidechains, including proline [20].

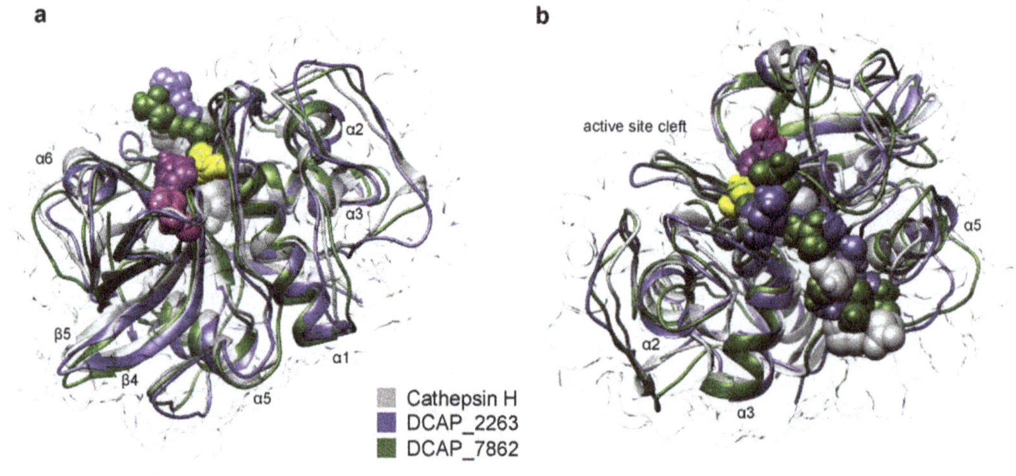

Fig. 5. Predicted structures for two vacuolar cysteine proteases (DCAP_2263, blue and DCAP_7862, green) with sequence homology to cathepsin H (PDBID: 8PCH gray). The active site residues and the minichain are shown as space-filling models. a. One side of the active site cleft is open and accessible to substrate. b. The other side of the active site cleft is blocked by the minichain. In cathepsin H, this partial occlusion of the active site confers aminopeptidase specificity.

3.5. Several Discovered Proteases Possess Novel Granulin Domains

Cysteine proteases with a C-terminal granulin domain are specific to plants, where they are involved in response to dessication or infection by pathogenic fungi [68]. This type of domain is found in two of the *D. capensis* protease clusters, the papain cluster Fig. S3) and the granulin domain cluster (Fig. S5). The reference sequences RD21A (RD21A_ ARATH) from arabidopsis and oryzain (ORYA_ORYSJ) from rice both contain granulin domains, as do three proteins in the papain cluster (Fig. S3) and three in the granulin domain cluster (Fig. S5). An additional two sequences in the papain cluster and one in the granulin domain cluster contain truncated versions that do not contain all four cysteine residues necessary to form the two disulfide bonds stabilizing the granulin domains. The granulin domain is separated from the catalytic domain by a proline-rich linker region. In RD21A, which is found in both the vacuole and the ER bodies [69], the granulin domain is removed from the mature enzyme. Maturation within the vacuole is relatively slow and involves accumulation of an intermediate where the N-terminal pro-sequence is removed and the C-terminal granulin domain remains attached [70]. This intermediate species forms aggregates that slowly release active enzyme following cleavage of the granulin domain, which is performed by RD21 itself [71]. This suggests that aggregation mediated by the granulin domain provides a mechanism for regulating protease activity during leaf senescence.

The granulin domain is attached to the catalytic domain by a proline-rich linker of variable length, as illustrated in Fig. 6. Granulins in animals act as growth factors, and contain distinct sequence and structural features: the characteristic sequence motif consists of four pairs of cysteine residues, with single conserved cysteines on both sides, and the resulting fold consists of β hairpins held together by disulfide bonds [72]. In plants, the granulin domain has two additional cysteines and an insertion of 6 residues between the first two Cys pairs, slightly modifying the structure (Fig. 7a). Clustering of the granulin domains themselves, separately from the catalytic domains, yields three clusters (Fig. 7b), two of which contain proteins from the *D. capensis* papain cluster and one of which is made up entirely of proteins from the *D. capensis* granulin domain cluster. The cluster analysis of Richau et al. [44] identified two subfamilies of granulin domain-containing cysteine proteases; comparison with those results places DCAP_0302 in their XBCP3 cluster, while DCAP_5945 and DCAP_6547 are in their RD21A cluster. Notably, the *D. capensis* granulin domain cluster represents a new subfamily of plant cysteine proteases that is not closely related to either of the previously described subfamilies.

The key sequence region of the canonical animal granulin motif is shown above the sequence alignment for comparison (Fig. 7c). The plant granulin sequences have two distinguishing features; an additional conserved Cys residue is present immediately after the first conserved CC pair in the animal sequence, and a 6-residue insertion containing another conserved C is present between the first and second CC pairs. In the granulin domain cluster, there is also a one-residue deletion between the first two conserved Cys residues. The first conserved glycine in the animal sequence is not conserved in the plant granulin domains, and in fact all of the examples shown here contain a bulky residue (F, Y, or L) at that position.

3.6. Protein Structure Networks Reveal a Tripartite Pattern of Structural Differentiation

In addition to the presence or absence of specific features, identifying broader patterns of structural differentiation can be helpful when selecting putative proteins for expression and characterization: proteins within different structural subgroups may differ with respect to other biophysically important properties such as thermal stability, substrate affinity pattern, overall activity, or aggregation propensity, and choosing a structurally diverse sample thus has the potential to maximize the chance of identifying proteins with functionally significant variation. Protein structure networks (PSNs) are a useful tool for such exploration, as they directly represent patterns of potential interaction among chemical groups rather than e.g. side chain dihedral angles or other properties that may vary substantially without inducing significant changes in protein function. Here, we employ the PSN representation of Benson and Daggett [32], which associates a vertex with each functionally distinct chemical group in the protein and assigns edges between vertices on the basis of their potential for direct interaction (as determined by a combination of inter-atomic distances and the identity of the groups in question). Using the structural distance approach of Butts and Carley [37], we can then directly compare protein structures via their PSNs.

Fig. 8a shows a metric multidimensional scaling (MDS) representation of the structural distances among PSNs in our sample. The MDS solution reveals a striking tripartite pattern of differentiation among the cysteine proteases, with protein structures exhibiting continuous and unilateral variation along three nearly orthogonal axes. A four-group hierarchical clustering solution (using Ward's method) on the underlying distance data is consonant with this pattern, yielding one cluster for each "spoke" of the tripartite structure (red, green, and blue points) and one cluster associated with the central "hub" (black points). The structure with the smallest median distance to all other structures (also the smallest maximum distance) is Aspain (orange point); notable reference structures within the central hub include papain and zingipain, which are in this sense among the most "typical" structures in the set. Oryzain characterizes the extreme end of the red spoke, which includes most of the proteins possessing a granulin domain (the remaining cases extending into the red sector of the central hub).

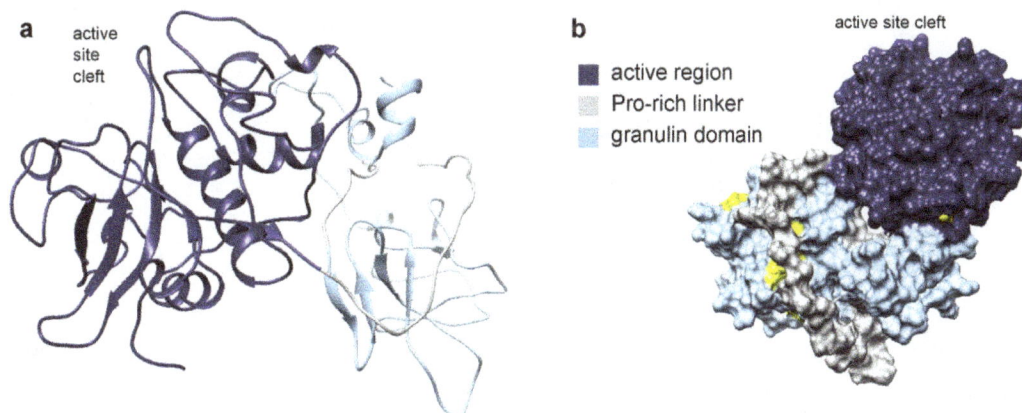

Fig. 6. a. Ribbon diagram for the predicted structure for a representative member of the granulin domain cluster (DCAP_5115), showing the catalytic domain (dark blue), the proline-rich linker (gray) and the granulin domain (light blue). b. Surface representation of the same structure rotated to show how the proline-rich linker interacts with the granulin domain.

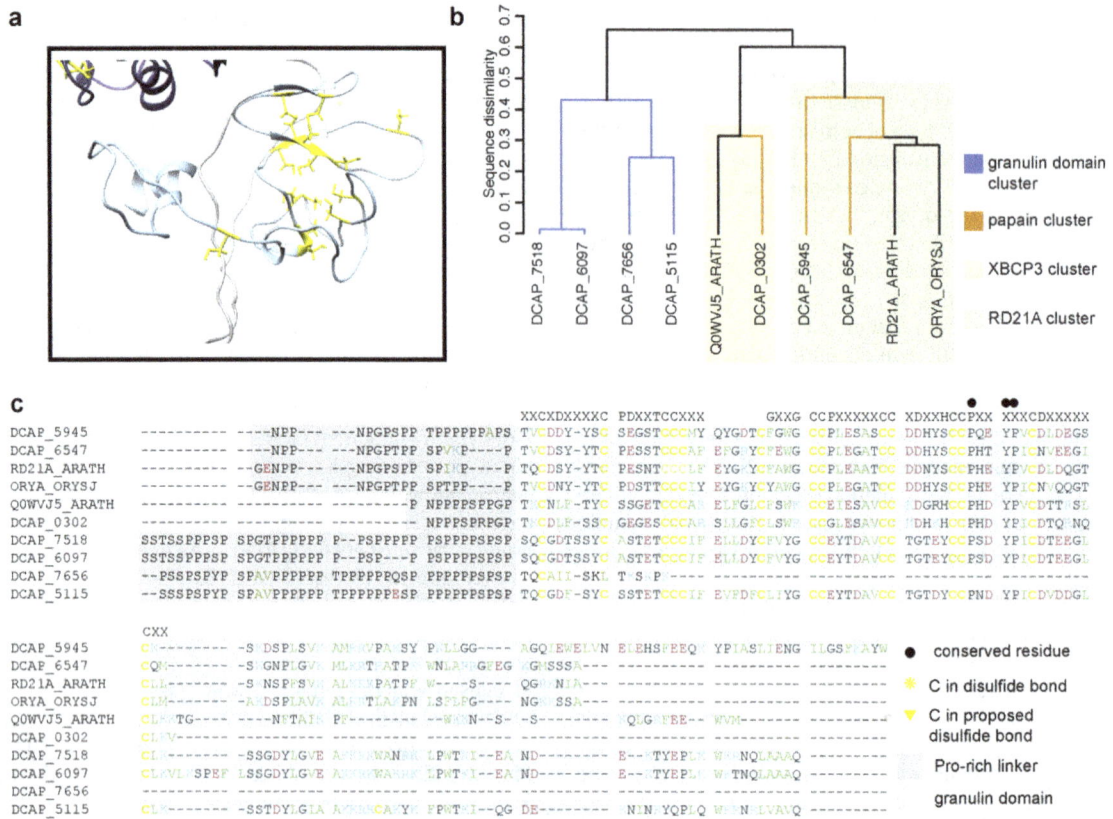

Fig. 7. a. Ribbon diagram of the DCAP_5115 granulin domain, with cysteine residues highlighted in yellow. b. Cluster analysis of granulin domains from *D. capensis* cysteine proteases and reference sequences. Solid colors denote membership in the clusters of Fig. 1, while the transparent boxes correspond to the clusters previously identified by Richau et al. [44]. Notably, the *D. capensis* granulin domain cluster appears to represent a new type of plant cysteine protease granulin domain. c. Sequence alignment of all the granulin domains found in the *D. capensis* cysteine proteases with reference sequences.

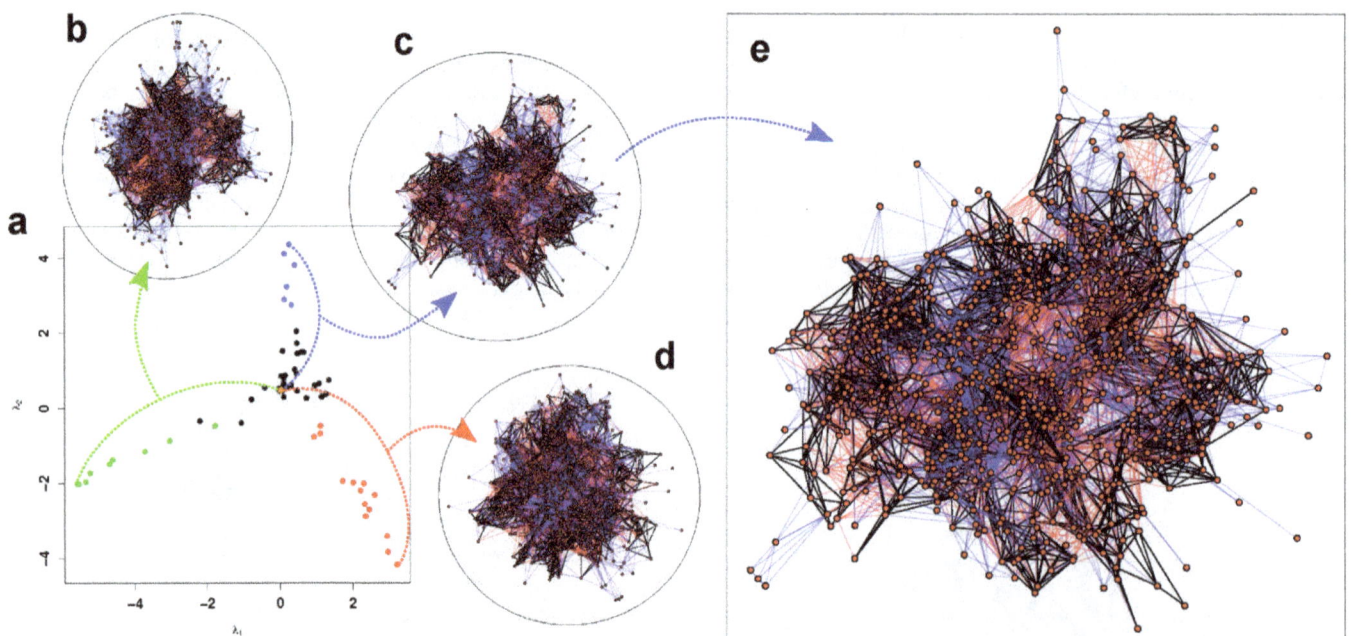

Fig. 8. a. Two-dimensional metric MDS solution for cysteine protease PSNs, based on structural distance; Euclidean distances between points approximate the structural distances between their corresponding PSNs. Protease PSNs show three distinct patterns of continuous variation from a central group of structures (black dots); the two-dimensional MDS solution is corroborated by the results of a four-group hierarchical clustering solution (indicated by point color), which also finds one central and three elongated peripheral groups. b., c., d., show minimum-distance mappings between the most central PSN (i.e., the PSN with the smallest median structural distance to all other PSNs, shown in orange) and the most extreme PSN on each axis of variation (see dotted lines). Black edges in each mapping show edges present in both PSNs, blue edges indicate edges present in the center but not the extreme PSN, and red edges indicate edges present in the extreme PSN but not the center. Differences between central and extreme PSNs are not localized to any particular location, but broadly diffused throughout the each graph. e. Detail of extreme/central PSN mapping for the blue axis, showing concentrated regions of edge addition (blue) and subtraction (red) as one approaches the center.

The most extreme structure in the blue spoke is DCAP_2555 with other members including references SAG39 and Dionain 3. Lastly, the green extreme corresponds to DCAP_4793; while this spoke contains almost exclusively *D. capensis* proteins, Ervatamin B lies at the interface of this spoke group with the hub cluster.

Given that each "spoke" in the MDS solution represents a distinct mode of variation, it is natural to suspect that the corresponding differences are structurally well-localized. Insets (b)–(d) of Fig. 8 show that this is not the case. Each inset shows the minimum distance mapping between the most extreme PSN on one MDS spoke and the central PSN (Aspain), with vertices placed using a standard (Fruchterman–Reingold) layout algorithm. Black edges within each network correspond to vertex interactions that are consistent between both structures, while red and blue edges reflect respectively interactions that are found only at the spoke extreme or central PSN. If structural differences along each spoke were well-localized, then red and blue regions within each inset would be concentrated within a particular part of the network structure; instead, we see a large number of "clouds" of red and blue edges in each network, indicating that systematic differences are found at numerous locations within the protein structure. Fig. 8e provides a detailed view of one such comparison. As can be seen, concentrations of red and blue edges tend to be separated by vertex sets that are well-connected by common (black) edges, indicating that the spoke differences typically correspond to broad shifts in the neighborhoods of large numbers of adjoining chemical groups (e.g., as produced by wholesale rotation or translation of large secondary structure elements). Each spoke contains several such rearrangements, extending throughout the entire protein structure, rather than either a very large number of idiosyncratic local changes or the reorganization of one portion of the protein.

This PSN-based analysis complements our findings regarding specific local features by suggesting that global differences in the structure of interaction among chemical groups fall along a small number of axes, from which it is then straightforward to select candidates for subsequent expression and biophysical characterization. By turns, the analysis also helps identify proteases with more "conventional" structures (e.g., papain, Aspain) that may serve as points of comparison vis a vis those from other groups.

4. Conclusion

In summary, 44 cysteine proteases were identified directly from the genomic DNA of *D. capensis*, and sorted into clusters based on sequence homology to known plant cysteine proteases. Molecular modeling and network analysis indicate that these proteases have distinct structural properties suggesting potential diversity in functional characteristics (e.g., thermal stability, substrate affinity). These diverse properties make this class of proteins an attractive target for further characterization studies, with rich potential for biotechnology applications.

One particularly attractive potential application for these proteases is in mass spectrometry-based proteomics. In bottom-up or shotgun proteomics, the proteins under investigation are digested into fragments using one or more proteases, followed by LC–MS/MS analysis [73]. In the most commonly performed experiments, the digestion is performed using trypsin, which is chosen because its propensity for producing fragments containing at least one basic (positively charged) residue makes it convenient for use with collisional activated tandem MS [74]. However, for investigating complex protein mixtures, using a combination of proteases with different specificities improves proteome sequence coverage; this is particularly true for proteins that are present at low abundance [75]. On the other hand, electron transfer or electron capture dissociation methodology does not depend on the presence of basic residues in the individual peptides, allowing more flexibility in the choice of proteases for digestion [76]. The use of multiple proteases also enables better characterization of post-translational modifications [77]. Identification and characterization of new proteases

from diverse sources, including carnivorous plants, adds to the repertoire of cleavage patterns that can be used in proteomics research.

Furthermore, the proteases from *D. capensis* are particularly attractive for mass spectrometry applications because they are optimized to function at pH 5. The optimal pH range of trypsin and other commonly used proteases, such as chymotrypsin and LysC, is neutral to mildly basic (7.8–8). This can be problematic for proteomic studies because at this pH, spontaneous deamidation of N and Q residues often occurs via formation of a succinimide intermediate [78,79]. This type of modification during sample preparation for mass spectrometry is relatively common [80–82]. Worse, it is not randomly distributed but depends on the neighboring residues [83], leading to artifactual modifications that can provide misleading results about protein aging [84,85] and the sites of N-linked glycosylation [86]. It has recently been shown that deamidation can be avoided by preparing the sample under slightly acidic conditions [87], a process potentially facilitated by the availability of *D. capensis* proteases. Disulfide scrambling can also be minimized by working at lower pH [88]. Future studies will include simulation of enzyme activity as a function of pH, as recently demonstrated for RNase A [89], as well as experimental characterization.

Author contributions

C.T.B. performed the cluster analysis, molecular dynamics simulations, and network visualization and analysis. X.Z., J.A.F., and C.T.B. created the protein structure networks. R.W.M. chose the protein set, generated the Rosetta structures, and performed sequence comparisons and structure analysis. J.E.K., K.W.R., M.H.U., and S.T. performed sequence annotation and comparisons. R.W.M. and C.T.B. designed the study and wrote the manuscript.

Acknowledgments

This work was made possible, in part, through access to the Genomic High Throughput Facility Shared Resource of the Cancer Center Support Grant (CA-62203) at the University of California, Irvine and NIH shared instrumentation grants 1S10RR025496-01 and 1S10OD010794-01; this research was also supported by NSF awards DMS-1361425 and OIA-1028394 and ARO award W911NF-14-1-0552. S.T. and R.W.M. acknowledge the California State Summer School for Math & Science (COSMOS) and NSF grant CHE-1308231. The authors acknowledge the Martin Lab for assistance with molecular modeling.

References

[1] Takeuchi Y, Chaffron S, Salcher MM, Shimizu-Inatsugi R, Kobayashi MJ, Diway B, et al. In situ enzyme activity in the dissolved and particulate fraction of the fluid from four pitcher plant species of the genus Nepenthes. PLoS One 2011;6, e25144.

[2] Adlassnig W, Peroutka M, Lendl T. Traps of carnivorous pitcher plants as a habitat: composition of the fluid, biodiversity and mutualistic activities. Ann Bot 2011;107: 181–94.

[3] Rost K, Schauer R. Physical and chemical properties of the mucin secreted by Drosera capensis. Phytochemistry 1977;16:1365–8.

[4] An CL, Fukusaki E, Kobayashi A. Aspartic proteinases are expressed in pitchers of the carnivorous plant Nepenthes alata Blanco. Planta 2002;214:661–7.

[5] Athauda S, Matsumoto K, Rajapakshe S, Kuribayashi M, Kojima N, Kubomura-Yoshida M, et al. Enzymic and structural characterization of nepenthesin, a unique member of a novel subfamily of aspartic proteinases. Biochem J 2004;381:295–306.

[6] Takahashi K, Athauda S, Matsumoto K, Rajapakshe S, Kuribayashi M, Kojima M, et al. Nepenthesin, a unique member of a novel subfamily of aspartic proteinases: enzymatic and structural characteristics. Curr Protein Pept Sci 2005;6:513–25.

[7] Lorkowski G. Gastrointestinal absorption and biological activities of serine and cysteine proteases of animal and plant origin: review on absorption of serine and cysteine proteases. Int J Physiol Pathophysiol Pharmacol 2012;4:10–27.

[8] Duwadi K, Chen L, Menassa R, Dhaubhadel S. Identification, characterization and down-regulation of cysteine protease genes in tobacco for use in recombinant protein production. PLoS One 2015;10, e0130556.

[9] Rawlings N, Waller M, Barrett A, Bateman A. MEROPS: the database of proteolytic enzymes, their substrates and inhibitors. Nucleic Acids Res 2014;42:D503–9.

[10] Novinec M, Lenarčič B. Papain-like peptidases: structure, function, and evolution. Biomol Concepts 2013;4:287–308.

[11] Turk V, Stoka V, Vasiljeva O, Renko M, Sun T, Turk B, et al. Cysteine cathepsins: from structure, function and regulation to new frontiers. Biochim Biophys Acta 2012; 1824:68–88.

[12] Kaschani F, Shabab M, Bozkurt T, Shindo T, Schornack S, Gu C, et al. An effector-targeted protease contributes to defense against *Phytophthora infestans* and is under diversifying selection in natural hosts. Plant Physiol 2010;154:1794–804.

[13] Buch F, Kaman WE, Bikker FJ, Yilamujiang A, Mithöfe A. Nepenthesin protease activity indicates digestive fluid dynamics in carnivorous *Nepenthes* plants. PLoS One 2015;10, e0118853.

[14] Libiaková M, Floková K, Novák O, Slováková L, Pavlovič A. Abundance of cysteine endopeptidase dionain in digestive fluid of Venus flytrap (*Dionaea muscipula* ellis) is regulated by different stimuli from prey through jasmonates. PLoS One 2014;9, e104424.

[15] Pavlovic A, Krausko M, Adamec L. A carnivorous sundew plant prefers protein over chitin as a source of nitrogen from its traps. Plant Physiol Biochem 2016;104:11–6.

[16] Ellison AM, Gotelli NJ. Energetics and the evolution of carnivorous plants—Darwin's 'most wonderful plants in the world'. J Exp Bot 2009;60:19–42.

[17] Nakamura Y, Reichelt M, Mayer VE, Mithöfer A. Jasmonates trigger prey-induced formation of outer stomach in carnivorous sundew plants. Proc R Soc B 2013;280.

[18] Butts CT, Bierma JC, Martin RW. Novel proteases from the genome of the carnivorous plant Drosera capensis: Structural prediction and comparative analysis. Proteins: Structure, Function, and Bioinformatics 2016. http://dx.doi.org/10.1002/prot.25095.

[19] Ghosh R, Chakraborty S, Chakrabarti C, Dattagupta JK, Biswas S. Structural insights into the substrate specificity and activity of ervatamins, the papain-like cysteine proteases from a tropical plant, *Ervatamia coronaria*. FEBS J 2008;275.

[20] Than ME, Helm M, Simpson DJ, Lottspeich F, Huber R, Gietl C. The 2.0 ácrystal structure and substrate specificity of the KDEL-tailed cysteine endopeptidase functioning in programmed cell death of *Ricinus communis* endosperm. J Mol Biol 2004;336: 1103–16.

[21] Risør MW, Thomsen LR, Sanggaard KW, Nielsen TA, Thøgersen IB, Lukassen MV, et al. Enzymatic and structural characterization of the major endopeptidase in the Venus flytrap digestion fluid. J Biol Chem 2016;291:2271–87.

[22] Kim D, Chivian D, Baker D. Protein structure prediction and analysis using the Robetta server. Nucleic Acids Res 2004;32:W526–31.

[23] Raman S, Vernon R, Thompson J, Tyka M, Sadreyev R, Pei J, et al. Structure prediction for CASP8 with all-atom refinement using Rosetta. Proteins 2009;77:89–99.

[24] Sievers F, Wilm A, Dineen D, Gibson TJ, Karplus K, Li W, et al. Fast, scalable generation of high-quality protein multiple sequence alignments using Clustal Omega. Mol Syst Biol 2011;7:539-539.

[25] Petersen T, Brunak S, von Heijne G, Henrik Nielsen H. SignalP 4.0: discriminating signal peptides from transmembrane regions. Nat Methods 2011;8:785–6.

[26] Emanuelsson O, Brunak S, von Heijne G, Nielsen H. Locating proteins in the cell using TargetP, SignalP and related tools. Nat Protoc 2007;2:953–71.

[27] Phillips JC, Braun R, Wang W, Gumbart J, Tajkhorshid E, Villa E, et al. Scalable molecular dynamics with NAMD. J Comput Chem 2005;26:1781–802.

[28] MacKerell AD, Bashford D, Bellott M, Dunbrack RL, Evanseck JD, Field MJ, et al. All-atom empirical potential for molecular modeling and dynamics studies of proteins. J Phys Chem B 1998;102:3586–616.

[29] Mackerell AD, Feig M, Brooks CL. Extending the treatment of backbone energetics in protein force fields: limitations of gas-phase quantum mechanics in reproducing protein conformational distributions in molecular dynamics simulations. J Comput Chem 2004;25:1400–15.

[30] Jorgensen WL, Chandrasekhar J, Madura JD, Impey RW, Klein ML. Comparison of simple potential functions for simulating liquid water. J Chem Phys 1983;79: 926–35.

[31] Word JM, Lovell SC, Richardson JS, Richardson DC. Asparagine and glutamine: using hydrogen atom contacts in the choice of side-chain amide orientation. J Mol Biol 1999;285:1735–47.

[32] Benson NC, Daggett V. A chemical group graph representation for efficient high-throughput analysis of atomistic protein simulations. J Bioinform Comput Biol 2012;10:1250008.

[33] Humphrey W, Dalke A, Schulten K. VMD: visual molecular dynamics. J Mol Graph 1996;14:33–8 [27–28].

[34] Handcock MS, Hunter DR, Butts CT, Goodreau SM, Morris M. statnet: Software tools for the representation, visualization, analysis and simulation of network data. J Stat Softw 2008;24:1–11.

[35] Butts CT. Network: a package for managing relational data in R. J Stat Softw 2008;24.

[36] R Core Team. R: a language and environment for statistical computing. Vienna, Austria: R Foundation for Statistical Computing; 2015.

[37] Butts CT, Carley KM. Some simple algorithms for structural comparison. Comput Math Organ Theory 2005;11:291–305.

[38] Butts CT. Social network analysis with sna. J Stat Softw 2008;24.

[39] Campbell M, Law M, Holt C, Stein J, Moghe G, Hufnagel D, et al. MAKER-P: a tool-kit for the rapid creation, management, and quality control of plant genome annotations. Plant Physiol 2013;164:513–24.

[40] Quevillon E, Silventoinen V, Pillai S, Harte N, Mulder N, Apweiler R, et al. InterProScan: protein domains identifier. Nucleic Acids Res 2005;33:W116–20.

[41] Beers EP, Jones AM, Dickerman AW. The S8 serine, C1A cysteine and A1 aspartic protease families in Arabidopsis. Phytochemistry 2004;65:43–58.

[42] García-Lorenzo M, Sjödin A, Jansson S, Funk C. Protease gene families in Populus and Arabidopsis. BMC Plant Biol 2006;6:30.

[43] van Wyk SG, Du Plessis M, Cullis CA, Kunert KJ, Vorster BJ. Cysteine protease and cystatin expression and activity during soybean nodule development and senescence. BMC Plant Biol 2014;14:294.

[44] Richau KH, Kaschani F, Verdoes M, Pansuriya TC, Niessen S, Stüber K, et al. Subclassification and biochemical analysis of plant papain-like cysteine proteases displays subfamily-specific characteristics. Plant Physiol 2012;158:1583–99.

[45] Székely GJ, Rizzo ML. Hierarchical clustering via joint between-within distances: Extending Ward's minimum variance method. J Classif 2005;22:151–83.

[46] Vernet T, Tessier DC, Chatellier J, Plouffe C, Lee TS, Thomas D, et al. Structural and functional roles of asparagine 175 in the cysteine protease papain. J Biol Chem 1995;270:16645–52.

[47] Ménard R, Plouffe C, Laflamme P, Vemet T, Tessier DC, Thomas DY, et al. Modification of the electrostatic environment is tolerated in the oxyanion hole of the cysteine protease papain. Biochemistry 1995;34:464–71.

[48] Choi K, Laursen R. Amino-acid sequence and glycan structures of cysteine proteases with proline specificity from ginger rhizome *Zingiber officinale*. Eur J Biochem 2000; 267:1516–26.

[49] Visal S, Taylor M, Michaud D. The proregion of papaya proteinase IV inhibits Colorado potato beetle digestive cysteine proteinases. FEBS Lett 1998;434:401–5 [PMID: 9742962].

[50] Silva F, Batista J, Marra B, Fragoo R, Monteiro A, Figueira E, et al. Prodomain peptide of HGCP-Iv cysteine proteinase inhibits nematode cysteine proteinases. Genet Mol Res 2004;3:342–55.

[51] Santamaria ME, Arnaiz A, Diaz-Mendoza M, Martinez I, Diaz M. Inhibitory properties of cysteine protease pro-peptides from barley confer resistance to spider mite feeding. PLoS One 2015;10:e0128323.

[52] Marra B, Souza D, Aguiar J, Firmino A, Sarto R, Silva F, et al. Protective effects of a cysteine proteinase propeptide expressed in transgenic soybean roots. Peptides 2009; 30:825–31.

[53] Rovenska G, Zemek R, Schmidt J, Hilbeck A. Altered host plant preference of *Tetranychus urticae* and prey preference of its predator *Phytoseiulus persimilis* (Acari: Tetranychidae, Phytoseiidae) on transgenic Cry3Bb-eggplants. Biol Control 2005;33:293–300 [PMID: 15781137].

[54] Karrer K, Peiffer S, DiTomas M. Two distinct gene subfamilies within the family of cysteine protease genes. Proc Natl Acad Sci U S A 1993;90:3063–7.

[55] Lu WD, Funkelstein L, Toneff T, Reinheckel T, Peters C, Hook V. Cathepsin H functions as an aminopeptidase in secretory vesicles for production of enkephalin and galanin peptide neurotransmitters. J Neurochem 2012;122:512–22.

[56] Holwerda BC, Rogers JC. Purification and characterization of Aleurain 1: a plant thiol protease functionally homologous to mammalian Cathepsin H. Plant Physiol 1992; 99:848–55.

[57] Nägler DK, Sulea T, Ménard R. Full-length cDNA of human cathepsin F predicts the presence of a cystatin domain at the N-terminus of the cysteine protease zymogen. Biochem Biophys Res Commun 1999;257:313–8.

[58] Takahashi K, Nishii W, Shibata C. The digestive fluid of drosera indica contains a cysteine endopeptidase ("droserain") similar to dionain from *Dionaea muscipula*. Carnivorous Plant Newsl 2012;41:132–4.

[59] Waghray A, Keppler D, Sloane BF, Schuger L, Chen YQ. Analysis of a truncated form of cathepsin h in human prostate tumor cells. J Biol Chem 2002;277: 11533–8.

[60] Jevnikar Z, Rojnik M, Jamnik P, Doljak B, Fonovi/'c UP, Kos J. Cathepsin H mediates the processing of talin and regulates migration of prostate cancer cells. J Biol Chem 2013;288:2201–9.

[61] Guncar G, Podobnik M, Pungercar J, Strukelj B, Turk V, Turk D. Crystal structure of porcine cathepsin H determined at 2.1 Å resolution: location of the mini-chain C-terminal carboxyl group defines cathepsin H aminopeptidase function. Structure 1998;6:51–61.

[62] Dodt J, Reichwein J. Human cathepsin H: deletion of the mini-chain switches substrate specificity from aminopeptidase to endopeptidaseh: deletion of the mini-chain switches substrate specificity from aminopeptidase to endopeptidase. Biol Chem 2003;384:1327–32.

[63] Baker A, Paudyal R. The life of the peroxisome: from birth to death. Curr Opin Plant Biol 2014;22:39–47.

[64] van Wijk KJ. Protein maturation and proteolysis in plant plastids, mitochondria, and peroxisomes. Annu Rev Plant Biol 2015;66:75–111.

[65] Palma JM, Sandalio LM, Corpas FJ, Romero-Puertas MC, McCarthy I, del Ro LA. Plant proteases, protein degradation, and oxidative stress: role of peroxisomes. Plant Physiol Biochem 2002;40:521–30.

[66] Lingner T, Kataya AR, Antonicelli GE, Benichou A, Nilssen K, Chen X-Y, et al. Identification of novel plant peroxisomal targeting signals by a combination of machine learning methods and *in vivo* subcellular targeting analyses. Plant Cell 2011;23:1556–72.

[67] Okamoto T, Nakayama H, Seta K, Isobe T, Minamikawa T. Posttranslational processing of a carboxy-terminal propeptide containing a KDEL sequence of plant vacuolar cysteine endopeptidase (SH-EP). FEBS Lett 1994;351:31–4.

[68] Shindo T, Misas-Villamil JC, Hörger AC, Song J, van der Hoorn RAL. A role in immunity for arabidopsis cysteine protease RD21, the ortholog of the tomato immune protease C14. PLoS One 2012;7, e29317.

[69] Hayashi Y, Yamada K, Shimada T, Matsushima R, Nishizawa NK, Nishimura M, et al. A proteinase-storing body that prepares for cell death or stresses in the epidermal cells of *Arabidopsis*. Plant Cell Physiol 2001;42:894–9.

[70] Yamada K, Matsushima R, Nishimura M, Hara-Nishimura I. A slow maturation of a cysteine protease with a granulin domain in the vacuoles of senescing *Arabidopsis* leaves. Plant Physiol 2001;127:1626–34.

[71] Gu C, Shabab M, Strasser R, Wolters PJ, Shindo T, Niemer M, et al. Post-translational regulation and trafficking of the granulin-containing protease RD21 of *Arabidopsis thaliana*. PLoS One 2012;7, e32422.

[72] Bateman A, Bennett HPJ. Granulins: the structure and function of an emerging family of growth factors. J Endocrinol 1998;158:145–51.

[73] Zhang Y, Fonslow BR, Shan B, Baek M, John JR, Yates R. Protein analysis by shotgun/bottom-up proteomics. Chem Rev 2013;113:2343–94.

[74] Dongre A, Jones J, Somogyi A, Wysocki V. Influence of peptide composition, gas-phase basicity, and chemical modification on fragmentation efficiency: evidence for the mobile proton model. J Am Chem Soc 1996;118:8365–74.

[75] Swaney DL, Wenger CD, Coon JJ. The value of using multiple proteases for large-scale mass spectrometry-based proteomics. J Proteome Res 2010;9:1323–9.

[76] Swaney D, McAlister G, Coon J. Decision tree-driven tandem mass spectrometry for shotgun proteomics. Nat Methods 2008;5:959–64.

[77] Gauci S, Helbig A, Slijper M, Krijgsveld J, Heck A, Mohammed S. Lys-N and trypsin cover complementary parts of the phosphoproteome in a refined SCX-based approach. Anal Chem 2009;81:4493–501.

[78] Reissner KJ, Aswad DW. Deamidation and isoaspartate formation in proteins: unwanted alterations or surreptitious signals? Cell Mol Life Sci 2003;60:1281–95.

[79] Klaene J, Ni W, Alfaro J, Zhou Z. Detection and quantitation of succinimide in intact protein via hydrazine trapping and chemical derivatizationuantitation of succinimide in intact protein via hydrazine trapping and chemical derivatization. J Pharm Sci 2014;103:3033–42.

[80] Krokhin OV, Antonovici M, Ens W, Wilkins JA, Standing KG. Deamidation of -Asn-Gly- sequences during sample preparation for proteomics: consequences for MALDI and HPLC-MALDI analysis. Anal Chem 2006;78:6645–50.

[81] Du Y, Wang F, May K, Wei X, Liu H. Determination of deamidation artifacts introduced by sample preparation using ^{18}O-labeling and tandem mass spectrometry analysis. Anal Chem 2012;84, 63556360.

[82] Hao Y, Ren P, Datta A, Tam JP, Sze SK. Evaluation of the effect of trypsin digestion buffers on artificial deamidation. J Proteome Res 2015;14, 13081314.

[83] Radkiewicz JL, Zipse H, Clarke S, Houk KN. Neighboring side chain effects on asparaginyl and aspartyl degradation: an ab initio study of the relationship between peptide conformation and backbone NH acidity. J Am Chem Soc 2001;123.

[84] Sharma KK, Santhoshkumar P. Lens aging: effects of crystallins. Biochim Biophys Acta 2009;1790:1095–108.

[85] Lampi KJ, Wilmarth PA, Murray MR, David LL. Lens β crystallins: the role of deamidation and related modifications in aging and cataract. Prog Biophys Mol Biol 2014;115:21–31.

[86] Palmisano G, Melo-Braga MN, Engholm-Keller K, Parker BL, Larsen MR. Chemical deamidation: a common pitfall in large-scale N-linked glycoproteomic mass spectrometry-based analyses. J Proteome Res 2012;11, 19491957.

[87] Liu S, Moulton KR, Auclair JR, Zhou ZS. Mildly acidic conditions eliminate deamidation artifact during proteolysis: digestion with endoprotease GluC at pH 4.5. Amino Acids 2016;48:1059–67.

[88] Pompach P, Man P, Kavan D, Hofbauerová K, Kumar V, Bezouška K, et al. Modified electrophoretic and digestion conditions allow a simplified mass spectrometric evaluation of disulfide bonds. J Mass Spectrom 2009;44:1571–8.

[89] Dissanayake T, Swails JM, Harris ME, Roitberg AE, York DM. Interpretation of pH–activity profiles for acid–base catalysis from molecular simulations. Biochemistry 2015;54:1307–13.

8

Metabolomic Profiling of Post-Mortem Brain Reveals Changes in Amino Acid and Glucose Metabolism in Mental Illness Compared with Controls

Rong Zhang [a,b], Tong Zhang [a], Ali Muhsen Ali [a,c], Mohammed Al Washih [a,d], Benjamin Pickard [a], David G. Watson [a,*]

[a] *Strathclyde Institute of Pharmacy and Biomedical Sciences, 161, Cathedral Street, Glasgow G4 0RE, Scotland, UK*
[b] *Institute of Clinical Pharmacology, Guangzhou University of Chinese Medicine, No. 12 Jichang Road, Guangzhou 510405, China*
[c] *Department of Clinical Biochemistry/Diabetes and Endocrinology Centre, Thi-Qar Health Office, Thi-Qar, Nassiriya, Iraq*
[d] *General Directorate of Medical Services, Ministry of Interior, Riyadh 13321, KSA*

ARTICLE INFO

ABSTRACT

Keywords:
Metabolomics
Schizophrenia
Depression
Bipolar disorder
Diabetes
Brain tissue
Branched chain amino acids
Sorbitol

Metabolomic profiling was carried out on 53 post-mortem brain samples from subjects diagnosed with schizophrenia, depression, bipolar disorder (SDB), diabetes, and controls. Chromatography on a ZICpHILIC column was used with detection by Orbitrap mass spectrometry. Data extraction was carried out with m/z Mine 2.14 with metabolite searching against an in-house database. There was no clear discrimination between the controls and the SDB samples on the basis of a principal components analysis (PCA) model of 755 identified or putatively identified metabolites. Orthogonal partial least square discriminant analysis (OPLSDA) produced clear separation between 17 of the controls and 19 of the SDB samples (R2CUM 0.976, Q2 0.671, p-value of the cross-validated ANOVA score 0.0024). The most important metabolites producing discrimination were the lipophilic amino acids leucine/isoleucine, proline, methionine, phenylalanine, and tyrosine; the neurotransmitters GABA and NAAG and sugar metabolites sorbitol, gluconic acid, xylitol, ribitol, arabinotol, and erythritol. Eight samples from diabetic brains were analysed, six of which grouped with the SDB samples without compromising the model (R2 CUM 0.850, Q2 CUM 0.534, p-value for cross-validated ANOVA score 0.00087). There appears on the basis of this small sample set to be some commonality between metabolic perturbations resulting from diabetes and from SDB.

1. Introduction

Mental illness, most commonly schizophrenia, depression, and bipolar disorder ('SDB') is common: schizophrenia has a European prevalence of 0.2–2.6%, depression 3.1–10.1%, and bipolar disorder 0.2–1.1% [1]. These conditions are a major burden on the health-care systems and on the relatives of affected people. Clinically, such illnesses are heterogeneous and present with psychosis or mood state features that vary over time and across individuals. Thus, it would be of great value to have an objective means to assist in diagnosis and categorisation of such illnesses and also give an insight into the best way to manage them [2]. Much diagnosis of mental illness remains subjective due to complex and poorly defined mechanisms underlying these diseases; there are no biomarkers and mental illness might be better viewed as a continuum rather than using absolute labelling [3].The significance of low-molecular-weight metabolites in driving or reflecting the aetiology of psychiatric disease has been researched for many years using serum samples that are a pragmatic choice for diagnostic testing and, additionally, brain tissue to investigate the central pathologies [4,5]. In the past 10 years, mass spectrometry-based metabolomics has evolved as a method for profiling a wide range of low-molecular-weight metabolites [6,7]. Metabolomics is a natural fit with metabolite profiling in mental illness where, for many years, targeted analysis was carried out in order to profile, for instance, biogenic amines in order to determine whether or not abnormalities in their levels might be causative [8,9]. There have been several studies which have carried out metabolomic profiling in mental illness [10–17], but these have not been as extensive as those into other diseases such as cancer.

There have been no untargeted metabolomics studies of human post-mortem brain samples although there was a study which examined disturbed glucose metabolism in post-mortem brains from psychotic patients [18]. In the current study, the availability of a unique library of post-mortem brain samples with extensive associated medical information allowed us to investigate whether or not these samples might reveal any underlying pathology which could be related to metabolic differences. Thus, we applied our established LC-MS-based metabolomic profiling methods [19,20] to determine if it was possible

* Corresponding author. Tel.: +44 1415482651.
  E-mail address: d.g.watson@strath.ac.uk (D.G. Watson).

to individually classify healthy control, depressive, schizophrenic, and bipolar brains. The observation of 'metabolic syndrome'-like features in those diagnosed with mental illness [21,22] prompted us to determine whether or not there was an overlap between metabolic perturbations in mental illness and diabetes. If such a link between mental illness and diabetes could be established then this might give some rationale for the evaluation of medicines used in the treatment of diabetes in the treatment of mental illness.

2. Materials and Methods

2.1. Chemicals

HPLC-grade acetonitrile was obtained from Fisher Scientific, UK. Ammonium carbonate, ammonium hydroxide solution (28–30%), acetic anhydride, pyridine, and methanol were purchased from Sigma–Aldrich, UK. HPLC grade water was produced by a Direct-Q 3 Ultrapure Water System from Millipore, UK. The mixtures of metabolite authentic standards were prepared as previously described [19,23] from standards obtained from Sigma–Aldrich, UK.

2.2. Post-Mortem Brain Samples

Post-mortem brain samples were obtained from the Sudden Death Bank collection held in the MRC Edinburgh Brain and Tissue Banks. Psychiatric diagnosis annotations for each sample were made by detailed study of donor case notes by qualified psychiatrists. Full ethics permission has been granted to the Banks for collection of samples and distribution to approved researchers (LREC 2003/8/37). The University of Strathclyde Ethics Committee also approved the local study of this material (UEC101123). Details of the brain samples are given in Table A1 in the Appendix. The information regarding the brain samples is summarised in Table 1.

2.3. Sample Extraction

The brain samples were thawed and then a sample of brain tissue (50 mg) was homogenised in ice cold methanol/water (1:1, 1.5 ml) using a handheld Lab Gen 7B homogeniser. The samples were then centrifuged at 16,000g, 15 min 4 °C and the supernatant was removed and the pellet reserved for further extraction to remove lipids. Lipids were extracted from the pellet with chloroform/methanol (3:1, 1.6 ml). The methanol/water extract was dried under a stream of nitrogen at 37 °C and redissolved in acetonitrile/water (80:20, 200 μl), the sample was centrifuged 16,000g, 15 min 4 °C to remove any insoluble material and then analysed by ZICHILIC and ZICpHILIC chromatography. The chloroform/methanol extract was dried under a stream of nitrogen at 37 °C and re-dissolved in either methanol/water (1:1, 200 μl) or methanol/chloroform (1:1, 200 μl) prior to chromatography on either C18 column or silica gel, respectively.

2.4. HILIC–HRMS Analysis

Sample analysis was carried out on an Accela 600 HPLC system combined with an Exactive (Orbitrap) mass spectrometer (Thermo Fisher Scientific, UK). An aliquot of each sample solution (10 μl) was injected onto a ZIC-pHILIC column (150 × 4.6 mm, 5 μm; HiChrom, Reading UK) with mobile phase A: 20 mM ammonium carbonate in HPLC grade water (pH 9.2), and B: HPLC grade acetonitrile. The LC and the MS conditions were as described previously [19,20]. Samples were submitted in random order for LC-MS analysis, and pooled quality control samples were injected at the beginning, middle, and end of the experiment to monitor the stability of the instrumentation. Standard mixtures containing authentic standards for 220 compounds were run in order to calibrate the column. Further analysis of the polar extract and of the lipophilic extracts were carried out on a ZICHILIC column (150 × 4.6 mm, 5 μm), ACE C18 column (150 × 3 mm, 3 μm), and an ACE silica gel column according to our previously described methods [23,24].

2.5. Analysis of Sugar Acids and Polyols by GC-MS

The individual standards for the polyols (100 μg) were treated with acetic anhydride/pyridine (1:1, 100 μl) for 30 min at 70 °C. The reagent was removed under a stream of nitrogen and the sample was re-dissolved in ethyl acetate 1 ml. The individual standards for the polyol acids were treated with methanol containing 1% HCl for 30 min at 70 °C, the reagent was removed under a stream of nitrogen and the sample was then treated as for the polyols. Brain tissue (200 mg) was extracted with acetonitrile/water (1:1, 1 ml) containing 2 μg/ml of pinitol internal standard, centrifuged and the supernatant was removed and evaporated to dryness with a stream of nitrogen at 70 °C and treated as for the polyol acids except that the residue was re-dissolved in 0.2 ml of ethyl acetate. GC-MS analysis was carried out on a DSQ GC-MS system (Thermo Fisher Scientific, UK) fitted with a GL Sciences Inert Cap 1 MS column from Hichrom, Reading UK(30 m × 0.25 mm × 0.25 μm film). The oven was programmed from 100 °C to 320 °C at 5 °C/min. The MS was operated in EI mode at 70 eV. For quantification of the sugars in brains selected, ion monitoring was carried out for ions at m/z 217, 200, 187, 145, 142, and 140, which are typical fragments of alditol acetates [25].

2.6. Data Extraction and Metabolite Identification

MZMine 2.14 [26] was used for peak extraction and alignment, as previously described [19,20]. Putative identification of metabolites was also conducted in MZMine by searching the accurate mass against our in-house database [18,19,23]. Background peaks present in the blank were removed in MZmine before transferring the data to an Excel file. Manual editing of the data was carried out in order to remove idiosyncratic peaks such as metabolites identified as drugs which were presumably from patient treatments and also nicotine metabolites which were particularly abundant in the brains of schizophrenic patients because of their well-established tendency to smoke much more than the general population [27] and ethyl sulphate which is from alcohol metabolism. The GC-MS data were extracted by using Sieve 1.3 (ThermoFisher Scientific UK), and the ions corresponding to the retention time of the sugar standards were extracted in order to build the OPLS-DA model.

Table 1
Summary information for the different groups of brain samples.

Group	Number	Male	Age range	Mean age ± RSD	Female	Age range	Mean age ± RSD
Control	21	18	26–74	47.4 ± 29.5	3	42–60	50.7 ± 17.8
Schizophrenic	11	10	25–69	44.4 ± 34.7	1	40	–
Bipolar	6	1	48	–	5	39–57	45.6 ± 16.2
Depressive	7	4	24–74	47.5 ± 49.1	3	20–57	60.3 ± 33.7
Diabetic	8	8	20–69	44.9 ± 35.2	0	–	–

2.7. Multivariate and Univariate Analysis

All data processing, including data visualisation, biomarker identification, diagnostics, and validation was implemented using SIMCA software v.14 (Umetrics AB, Umeå, Sweden). Prior to multivariate analysis, data were pareto scaled where the responses for each variable are centred by subtracting its mean value and then dividing by the square root of its standard deviation [28,29]. Principal component analysis (PCA) was used to provide an unsupervised model in order to explore how variables clustered regardless Y class [30]. Orthogonal projections to latent structures (OPLS) provides a supervised model that can predict Y from X and can separate variation in X that correlates to Y (predictive) and variation in X that is uncorrelated to Y (orthogonal/systemic). OPLS-DA is a discriminant analysis based on OPLS and employed to examine the difference between groups while neglecting the systemic variation [30]. The p-values of the biomarkers were evaluated for their significance applying the false discovery rate statistic (FDR) [31]. Variable importance in the projection (VIP) was employed in order to indicate the contribution of each variable in the in a given model compared to the rest of variables [32], the average VIP is equal to 1, based on that a variable larger than 1 has more contribution in explaining y than the average [33].

2.8. Diagnostics and Validation of Models

R^2 and Q^2 are diagnostic tools for supervised and unsupervised models; R^2 represents the percentage of variation explained by the model (the goodness of fit), Q^2 indicates the predictive ability of the model [34–36], a large discrepancy between between R^2 and Q^2 indicates overfitting of the model. A permutations test can be applied to supervised models to evaluate whether the specific grouping of the observations in the two designated classes is significantly better than any other random grouping in two arbitrary classes [34–36], and in Simca P, this is carried out by repeatedly leaving out 1/7th of the data an refitting the model, all the Q^2 values for the refitted models should be lower than the original Q^2 value. The criteria for validity for OPLSDA models tested via cross-validation are that all blue Q^2-values to the left are lower than the original points to the right or the blue regression line of the Q2-points intersects the vertical axis (on the left) at, or below zero. The R^2 values always show some degree of optimism. However, when all green R^2-values to the left are lower than the original point to the right, this is also an indication for the validity of the original model although this is not essential for the model to be valid. Model validity is also assessed using cross-validated ANOVA (CV-ANOVA) which corresponds to H_0 hypothesis of equal cross-validated predictive residuals of the supervised model in comparison with the variation around the mean [37]. Univariate comparisons were carried out in Excel.

3. Results

3.1. The Effect of the HPLC Column Used on the Results

The data produced from the analysis of the polar extracts on the ZICHILIC column were less satisfactory for producing separation in the sample sets than those produced on ZICpHILIC. There were similar trends in some of the metabolites but the clear-cut differences described below were not observed. This again supports our choice of ZICpHILIC as the best method for analysis of polar metabolites in metabolomics screens [13]. The ZICHILIC mobile phase produces a higher background which includes abundant sodium formate cluster ions and thus ion-suppression is potentially more of a problem. In addition, the chromatographic peaks for many metabolites are wide than om ZICpHILIC and the retention times from run to run are less stable which produces a greater challenge for the peak extraction software. The lipid fractions were analysed on silica gel and C18 columns and no major differences in

lipid profiles were observed between the controls and the SDB brains. This may be due in part to the fact that the initial methanol/water extraction also extracted many of the more polar lipids. The chromatography of lipids on the ZICpHILIC column is satisfactory but they are only weakly retained on this column so there is no separation of isomeric species.

3.2. Comparison of Control and Schizophrenic/Depressive/Bipolar/Diabetic Brains using PCA

Metabolites were identified to MSI levels 2 or 3 [38] according to either exact mass (<3 ppm deviation) or exact mass plus retention time matching to a standard. After data filtering, 755 metabolites from positive and negative ion modes were combined and used to build multivariate models. The sample set was selected by our collaborators at the sudden death brain and tissue bank to give us the best sample set available from samples in storage for making a comparison between controls, mental illness, and diabetes. Since the uncontrolled factors are highly variable in both control and affected samples, the expected result might be that variation in the data would preclude statistical separation unless the disease signature was very strong. In order to obtain a reasonable sample size, we treated schizophrenic, depression, bipolar (SDB), and diabetic (DI) samples as one group to compare against controls. Comparison of the data from schizophrenic, depression, bipolar (SDB), and diabetic (DI) samples and controls using PCA did not yield a clear separation of these diagnostic categories (Fig. 1). In order to rule out variation in level of technical precision across the ca 55 h required to complete the analysis, a pooled sample: (P1–6) was prepared by combining $5 \times 40\mu l$ of extract randomly selected from each sample type. Replicates were run as follows: P1 and 2 near the beginning of the sequence after running three blanks and four standard mixtures, P3 after ca 20 h, P5 after ca 39 h, P4 and 6 at the end of the run after ca 55 h. As can be seen in Fig. 1, the pooled samples all lie towards the centre of the PCA plot and individual sample points are close to each other. This indicates that there is only a small amount of instrumental drift and thus the results reflect biological, rather than technical, differences.

3.3. Comparison of SDB and SDBDI Samples Against Controls Using OPLSDA

When the DI samples were omitted, it was found that 36 of the 44 available SDB and control samples (see footnote to Table A1) could be combined where 19 SDB samples were compared against 17 control samples to produce a strong OPLS-DA model (Fig. 2) (R^2CUM 0.976, Q^2CUM 0.671) explaining 96.7% of the variation in the samples with six components. Q2 > 0.5 is generally accepted as being indicative of a robust model [35,36] and the model gave a permutations plot where all the permutated Q^2 values (n = 999) on the left are lower than the points on the right (Figure A 1) and the line plot intercepts the y-axis below 0 [34–36]. This preliminary model was used to inform the selection of the samples for univariate statistical comparison by excluding 8 samples that did not fit the model, four controls and four SDB samples. Despite variations arising from complex medical histories, length of sample storage and exact cause of death there appeared to be a strong metabolic signature associated with mental illness overriding these confounding factors which apply to both control and affected samples. Table 1 also shows the univariate statistical comparisons for the metabolites with VIP scores > 1 in the preliminary OPLSDA model. All of which are significantly different according to a two-tailed t-test and FDR statistics [31] based on 755 metabolites indicate all P values <0.05 are significant. A complete list of significantly different metabolites based on univariate comparison of the 17 controls and 19 SDB samples is given in Table A2. The initial application of OPLSDA based on 755 metabolites allowed us to focus on more limited list of metabolites than those listed in table A2.

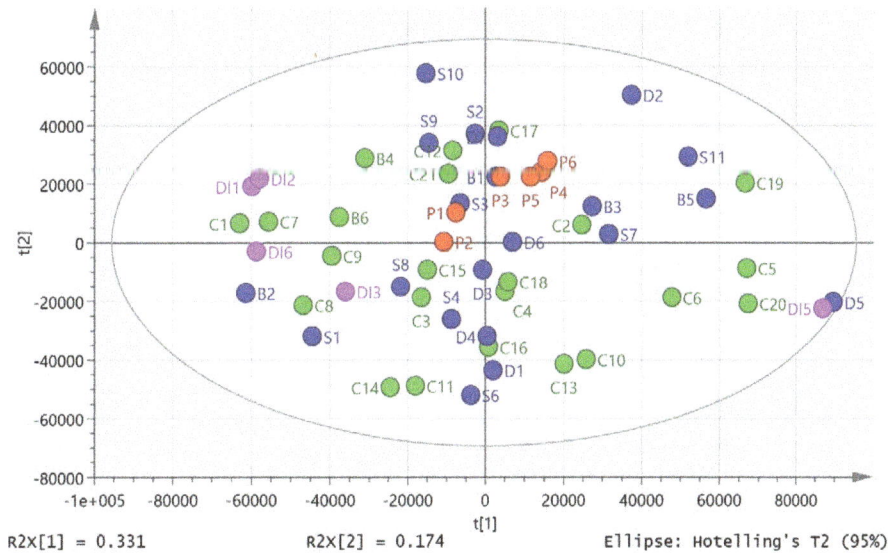

Fig. 1. PCA plot for control, SDB and DI samples (R^2X cum 0.61, Q^2 (cum) 0.464, 3 components) based on 755 metabolites from positive and negative ion modes. Three of the DI samples (DI4, 7 and 8) lay outside of the ellipse and were omitted from the model. P = pooled sample used to check instrument stability over time.

3.4. SDB Samples Show Differences in Branched Amino Acid, Neurotransmitter and Sugar Metabolism Compared With Controls.

Leucine/isoleucine have the highest VIP (8.1) this a very strong variable along with valine which has a VIP of 5.1. Thus, branched chain amino acids are highly correlated in the brains of SDB subjects and are also present in significantly higher levels than in the controls. The other neutral lipophilic amino acids methionine, phenylalanine, tyrosine, tryptophan, and proline also have high VIP values (Table 2 and table A2), are elevated in SDB brains and are important in the model. The important metabolites that are significantly lower in the SDB subjects than in the controls include GABA, its metabolite guanidino amino butyric acid, and the neuromodulator N-acetyl aspartyl glutamate (NAAG). In addition, there are higher levels of sugar metabolites, putatively identified according to the LC-MS

analysis as sorbitol, gluconic acid, and erythritol, in the SDB samples.

3.5. The Effect of Age on Metabolite Profiles of Brain Tissue

The brains were from subjects with a wide age range and the mean age of the control group at death was 45.9 and mean age for the SDB group was 46.4. An OPLS model (R^2X (cum) 0.706), R^2Y (cum) 0.979, Q^2 (cum) 0.476)) gave a very good correlation between age and metabolites (Figure A2) and there was no overlap between the metabolites used to discriminate the control and SDB brains and those which discriminated age (Table A3). The major changes with age were related to decreases in unsaturated fatty acids in the brain such as eicosatetraenoic, docosahexaenoic, and linoleic acid and increases in glycerol metabolites such as phosphoethanolamine and phosphocholine.

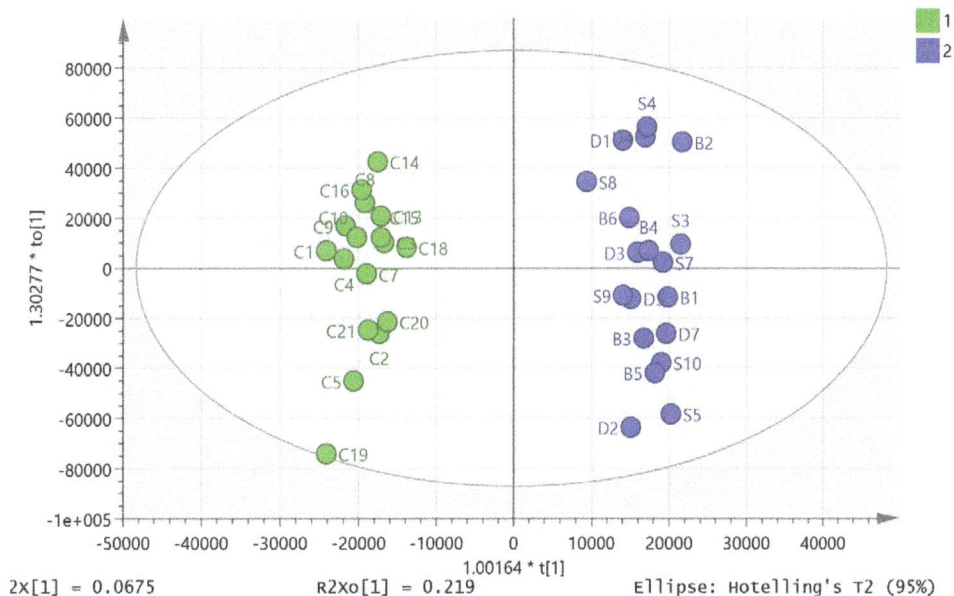

Fig. 2. OPLS-DA model (R^2CUM 0.976, Q^2CUM 0.671, 6 components) of control (n = 17) compared to SDB (n = 19) brain samples based on 755 metabolites from positive and negative ion modes.

Table 2
Metabolites with high impact on the model separating controls from SDB brains (18 control/18 SDB). *Matches retention time of standard. **Application of the Benjamini–Hochberg procedure [31] with a Q value of 0.1 indicates that the critical threshold for a regarding a P value as being significant is >0.05. ***Retention time does not match that of the standard. N = negative ion P = positive ion.

m/z	Rt min.	Metabolite	VIP	P value	Ratio SDB/C
N 130.087	11.2	*Leucine/isoleucine	8.3	0.0041	1.34
N 96.9698	13.6	**Orthophosphate (carbonic acid adduct of chloride)	7.5	0.0050	1.17
N 116.072	12.9	*Valine	5.1	0.0008	1.36
P 116.071	13.2	*Proline	4.2	0.0021	1.34
N 135.03	10.6	**Threonic acid isomer	4.3	0.0656	1.09
N 164.072	10.4	*Phenylalanine	4.0	0.0050	1.31
N 102.056	15.9	*GABA	3.2	0.0050	0.854
N 88.0404	15.2	*Sarcosine	3.1	0.0730	1.12
N 118.051	14.8	*Homoserine	3.1	0.0140	1.51
N 267.074	11.3	*Inosine	3.0	0.0340	0.81
N 148.044	11.8	*Methionine	2.8	0.0038	1.30
N 181.072	14.3	*Sorbitol/mannitol/iditol/dulcitol	2.6	0.0022	1.99
P 258.11	14.9	*sn-glycero-3-Phosphocholine	2.2	0.0415	1.61
N 273.039	15.7	Deoxy sedoheptulose phosphate	1.9	0.0020	0.5870
N 180.067	13.4	*Tyrosine	1.7	0.0170	1.23
N 121.051	12.1	*Erythritol/threitol	1.5	0.0011	1.57
N 241.012	17.4	D-myo-Inositol 1,2-cyclic phosphate	1.3	0.0060	0.580
N 239.115	16.6	**Homocarnosine isomer (anserine)	1.4	0.0037	0.622
N 303.084	17.2	*N-Acetyl-aspartyl-glutamate	1.3	0.0327	0.533
N 203.083	12.0	*Tryptophan	1.3	0.0250	1.21
N 195.051	14.4	*Gluconic acid	1.3	0.0011	2.20
N 215.033	13.7	Hexose (chloride adduct)	1.1	0.0019	2.19
N 209.067	14.4	*Sedoheptulose	1.0	0.0006	1.74
P 146.092	15.6	4-Guanidinobutanoate	1.0	0.00021	0.713

3.6. Inclusion of Diabetic Brains in the OPLSDA Model

There is evidence that there may be some shared pathology between diabetes and mental illness going back as far the pre-neuroleptic drug era and this provided the rationale for the insulin coma therapy which was used in the first half of the 20th century [39]. There were eight diabetes samples in the set of brain samples and these were subsequently added to the data set used to build the OPLSDA model described above. Two of the diabetic samples (DI1 and DI8) were extreme outliers and were excluded from the initial PCA plot (Fig. 1) since they were outside of the ellipse. They were also excluded from the combined OPLSDA model along with one of the SDB samples (S5) which was excluded

since it did not fit into the new OPLSDA model. Six of the diabetic samples could be classified with the SDB samples (Fig. 3, R^2 CUM 0.850, Q^2 CUM 0.534, p-value for cross-validated ANOVA score 0.00087) increasing the significance of the ANOVA score, the large decrease in the CVANOVA score implies considerable strengthening of the model since the score can be used as a guide to the optimal fitting of a model [36]. The permutations plot is shown in Figure A3 indicates a strong model. The addition of the diabetic samples to the model produced some change in the VIP values but basically most of the discriminating metabolites are the same (Table A3) which is perhaps not surprising since the model is strengthened by addition of these samples. However, when the univariate comparisons are examined most of the metabolites

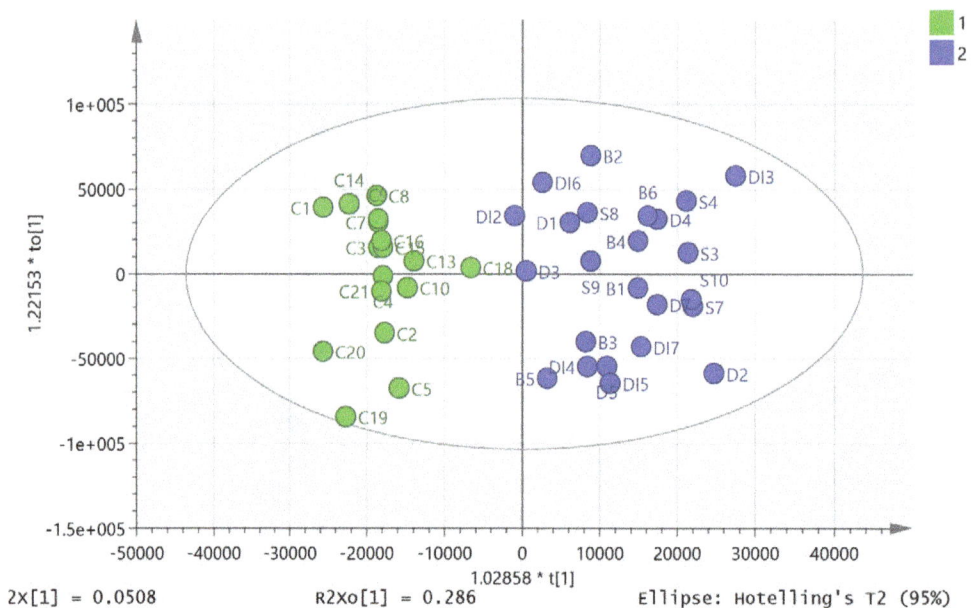

Fig. 3. OPLS-DA model (R^2 (cum) 0.850, Q^2 (cum) 0.534, 4 components) including six of the DI samples. Green control and blue SDB+ diabetic brain samples based on 755 metabolites from positive and negative ion modes.

with high VIP values in the model did not have significant p-values when the diabetic samples in the model are compared with controls. Thus, the similarities between diabetic and the SDB brains lie in the covariance of the set of important marker compounds shown in Table A3 rather than in the absolute levels. Leaving the diabetic samples out of the model and using them as a prediction set resulted in four of the samples being classified as SDB samples while two were unclassified but borderline to the SDB class.

3.7. Preparation of PCA Model with a Reduced Metabolite List

When the reduced list of 120 metabolites with low P values shown in Table A2 was used to prepare a PCA model, it was clear that the SDB samples contained subgroups. In particular, a group of nine SDB samples were quite different from the controls and the rest of the SDB samples as shown in Fig. 4 where HCA was used to define the groups (HCA tree shown in Figure A5). The metabolites defining the subgroup are shown in Table 3. This supports the proposal that there are similarities within the SDB group since the sub-group contains all three classes.

3.8. Refining the OPLSDA Models

The purpose of this study to try to better understand disease pathology in mental illness and thus the ideal outcome would be a list of related metabolites corresponding to the disease state in order to develop a hypothesis. Of lower priority was to provide a classification system in the current case since sampling brain tissue is not going to be a diagnostic test. With a high number of variables, there is the danger of overfitting and although the OPLSDA models shown in Figs. 2 and 3 performed well in cross-validation tests, there might still be some doubts with regard to their validity. Thus, the control/SDB OPLSDA model was refined by removing 600 of the lowest priority variables and then systematically removing variables one at a time from the remaining set while retaining variables that caused a reduction in the Q cum score of >0.05 when removed. This resulted in the model shown in Fig. 5 which had a CVANOVA score of 0.0006 and which could accommodate 38 out of the original 44 samples based on the six metabolites shown in Table 4. The cross-validation model is shown in Figure A5. Removal of samples belonging to each sub-group in order to create prediction sets gave the results shown in Table 5. This resulted in two out of 21

subjects being misclassified. The sample size is relatively small so removing in each case around 15% of the samples will considerably weaken the model. In reduced model, the branched chain amino acid valine and the neurotransmitter GABA retain their high importance. The same process of variable reduction was applied to the combined diabetic/SDB model which included six of the diabetic samples and resulted in a model based on six metabolites into which 45 out of 53 samples could be fitted and included seven of the diabetic samples (Fig. 6) and which had a CVANOVA score of 0.000013. The metabolites included in the model were the same as those used in the model shown in Fig. 5 except that the VIP values for each metabolite were different (Table 6). Removing the 7 diabetic samples and using them as a prediction set resulted in six of the samples being classified with the SDB group and one of the samples being classified with the controls (details shown in Table 5).

3.9. GC-MS Analysis to Quantify and Identify Polyols and Polyol Acids in Brain

In the process of analysing the data, it became apparent that sugar metabolism had an important role in distinguishing the control and SDB brains. The commercially available sugar alcohol standards were run on the ZICpHILIC column and isomeric compounds were found to co-elute or elute closely and thus were not distinguishable from each other. A GC-MS method was developed for the analysis of the sugar alcohols which were converted into their acetates after initial treatment with methanolic HCl to esterify the acidic groups in gluconic and gulonic acid. The retention times for the available standards are shown in Table A5. Fig. 7 shows the separation of some the polyols present in brain tissue in comparison with a mixture of standards. The major polyols present were erythritol plus an additional unknown tetritol, ribitol, arabinotol, xylitol, gluconic acid, and sorbitol (Fig. 7). Figure A7 shows OPLS-DA separation of control and SDB + diabetic samples based on the ions monitored for the polyol standards; there was not sufficient tissue to repeat analysis of all the samples and the model is based on 21 SDBDI samples compared to 15 control samples. Figure A8 shows the cross-validation for the model indicating the there was a robust discrimination. Calibration curves were prepared in the range 1–16 μg for all the sugar alcohols against 2 μg of pinitol which was used as an internal standard. The data for the calibration curves are shown in Table S4. The quantitative data for the sugar alcohols are shown in Table 7.

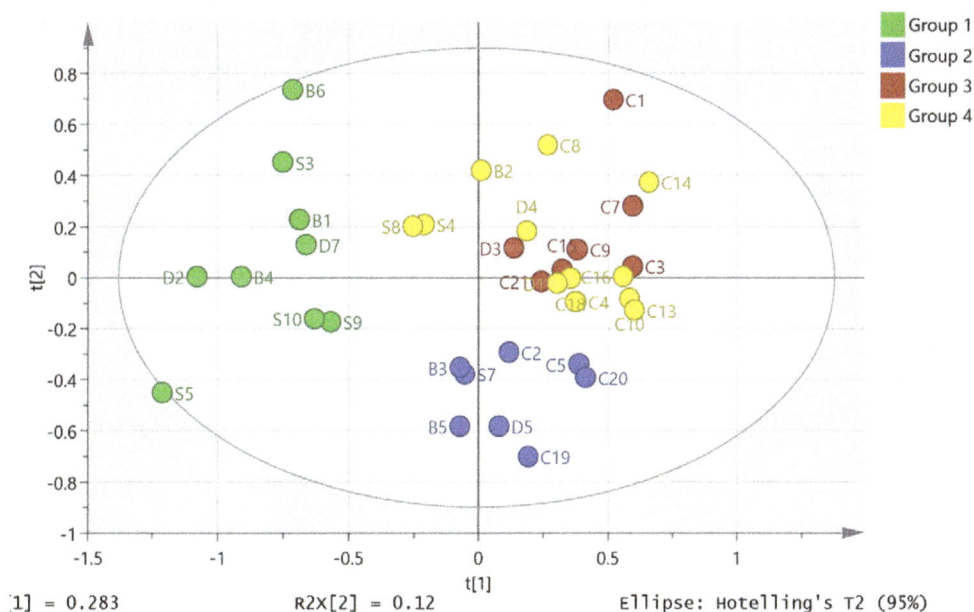

Fig. 4. PCA model (R²X = 0.68, Q²X = 0.283, 5 components) based on the metabolites with P values <0.05 when the control and SDB samples used to prepare the OPLS-DA model shown in Fig. 2 are compared and hierarchical cluster analysis is used to define subgroups. The analysis reveals a clear subgroup (group 1) containing depressive/bipolar and schizophrenic samples which is distinctly different from the rest of the samples.

Table 3
Important metabolites defining the sub-group of nine SDB brains shown in Fig. 4. *Matches retention time of standard. **Does not match the retention time of the standard therefore is an isomer of the named compound.

m/z	Rt min	Metabolite	P value	Ratio	VIP
P 227.114	10.3	**Carnosine isomer	0.0020	8.34	1.70
N 145.014	6.3	**Oxoglutarate isomer	0.0310	3.91	1.35
N 195.051	14.4	*Gluconic acid	<0.001	3.30	2.32
N 159.103	4.9	Ethyl-hydroxyhexanoate	0.0020	2.95	1.35
N 181.072	14.3	*Sorbitol/mannitol/iditol/dulcitol	<0.001	2.94	2.22
N 151.061	13.2	*Xylitol/ribitol/arabinotol	<0.001	2.38	1.56
N 209.067	14.4	*Sedoheptulose	<0.001	2.34	2.20
N 252.088	8.1	N-Acetylvanilalanine	<0.001	2.06	2.15
N 121.051	12.1	*Erythritol/threitol	<0.001	2.00	1.65
N 164.072	10.4	*Phenylalanine	<0.001	1.88	1.80
N 178.072	12.9	*Glucosamine	0.0010	1.88	1.48
N 130.087	11.2	*Leucine	<0.001	1.66	1.54
N 202.109	5.5	**O-Acetylcarnitine isomer	<0.001	1.60	2.13
P 116.072	12.9	*Valine	<0.001	1.58	1.62
N 103.004	16.1	*Malonate	0.0010	1.56	1.68
P 161.107	10.4	Tryptamine	0.0010	1.55	1.39
P 169.097	11.3	*Pyridoxamine	0.0010	1.53	1.34
N 114.056	13.2	*Proline	0.002	1.52	1.99
N 180.067	13.4	*Tyrosine	0.001	1.44	1.97
P 230.151	24.2	Gamma-Aminobutyryl-lysine	0.0280	1.42	1.62
N 220.083	12.3	*N-Acetyl-D-glucosamine	<0.001	1.40	1.99
P 104.071	15.9	*4-Aminobutanoate	<0.001	0.78	1.40
P 284.099	13.0	*Guanosine	<0.001	0.58	1.63
N 273.039	15.7	1-Deoxy-D-altro-heptulose 7-phosphate	0.0020	0.49	1.47
N 171.007	15.4	*Glycerol 3-phosphate	0.0040	0.38	1.88
N 231.099	16.8	N2-Succinyl-L-ornithine	0.0110	0.36	2.07
P 248.024	10.8	Norepinephrine sulfate	<0.001	0.32	1.48
P 305.098	17.2	*N-Acetyl-aspartyl-glutamate	0.0030	0.27	2.52
P 277.031	12.6	**Phospho-gluconate isomer	0.0030	0.06	1.92

Clearly, both the diabetic samples and SDB have high levels of sorbitol, gluconic acid, ribitol, and erythritol in comparison to the controls. Elevated levels of sorbitol in schizophrenic and bipolar brains have been reported before [18], but the addition of the other sugar metabolites re-enforces the importance of this pathway in the illness.

4. Discussion

4.1. High Levels of BCAs and Other Liphilic Amino Acids in SDB Samples

The OPLSDA models based on the larger number of variables (Figs. 2 and 3) and the univariate differences will be used to develop some hypotheses based on the underlying metabolite differences. The highest VIPs in the OPLS-DA model (Fig. 2) of the SDB samples against the controls are the branched chain amino acids leucine/isoleucine and valine (BCAs) which are elevated above the levels found in the controls. The importance of these metabolites in schizophrenia and bipolar disorder has recently been highlighted [40]. There have been a number of recent metabolomics studies of obesity and insulin resistance and it has been observed that there is a distinct metabolic signature linked to metabolic syndrome where the plasma levels of branched chain amino acids (BCAs) leucine, isoleucine, and valine were elevated together with methionine, glutamine, phenylalanine, tyrosine, asparagine, and arginine [41,42]. A study which was carried out on a cohort of 1872 individuals

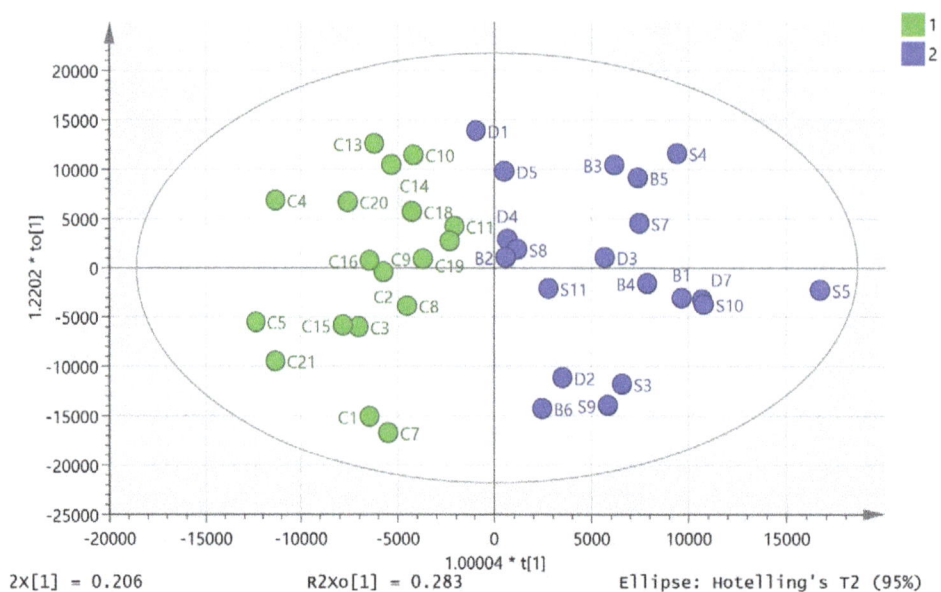

Fig. 5. OPLSDA (R^2Y cum 0.725, Q^2 cum 0.638, five components) model based on the six metabolites shown in Table 4 classifying 17 controls (1) and 21 SDB samples (2).

Table 4
Marker compounds used in the OPLSDA model shown in Fig. 5.

m/z	Rt (min)	Metabolite	P value*	Ratio *SDB/Control	VIP
116.072	12.9	L-Valine	0.00026	1.39	1.55
104.071	15.9	4-Aminobutanoate	0.0069	0.87	1.43
162.112	13.7	L-Carnitine	0.680	0.96	0.82
204.123	11.4	O-Acetylcarnitine	0.081	1.26	0.75
87.0087	8.3	Pyruvate	0.063	1.26	0.41
175.025	14.6	Ascorbate	0.85	1.03	0.37

* For the 38 samples in this model.

who were subdivided in lean, overweight, and obese groups proposed that BCA levels can provide a better signature of metabolic wellness than BMI [43]. Another group found that elevated levels of BCAs in plasma could be linked to obesity and potentially to the development of insulin resistance in children and adolescents [44]. BCAs are known to promote production of muscle protein [45,46] and an elevation in BCA levels may indicate that the uptake of BCAs into muscle tissues is reduced. Metabolomic profiling of plasma from schizophrenics, even before medication, has been found to indicate that they are at risk of developing metabolic syndrome [47]. Antipsychotic medications are known to significantly increase metabolic complications and induce weight gain and although the medication history of the relating to current samples is unknown it unlikely that variations in medication alone would be sufficiently systematic to account for the differences observed. In addition to BCAs, the neutral amino acids proline, methionine, tyrosine, and tryptophan are all elevated in the SDB/diabetic group and have high VIP values. There is an early report of marginal differences in the levels of lipophilic amino acids in plasma from schizophrenics with valine, phenylalanine, alanine, leucine, isoleucine, methionine, and tyrosine all being elevated [48]. However, the current data do not support the theory proposed by that paper, which was that elevation in lipophilic plasma acids might produce competition for lipophilic amino acid transporters into the CNS, resulting in reduced uptake of the amino acids tyrosine and tryptophan which are required for neurotransmitter biosynthesis.

Proline is a potential precursor of glutamate which is a neurotransmitter in brain and it has previously been shown to be increased in individuals diagnosed with schizophrenia. There is an extensive literature indicating that proline dehydrogenase (PRODH) activity may be up-regulated in schizophrenia [49,50]. However, this would be expected to lead to a fall in proline levels which does not fit with the current observation.

4.2. Alterations in Sugar Metabolism.

It was not possible to clearly identify the different sugar alcohols using LC-MS since the isomeric compounds have almost identical retention times and their MS/MS spectra are very similar. In order to get a clearer identification of the sugar alcohols in brain, standards and a brain extracts were derivatised and analysed by GC-MS. The high chromatographic resolving power of a capillary GC column was able to separate the isomers. The levels of glucose in these brains appeared to be very low and the major sugar in the brain was myo-inositol. As can be seen in Fig. 7, there are several sugar alcohols in the brain. The presence of these compounds has been observed before in human CSF

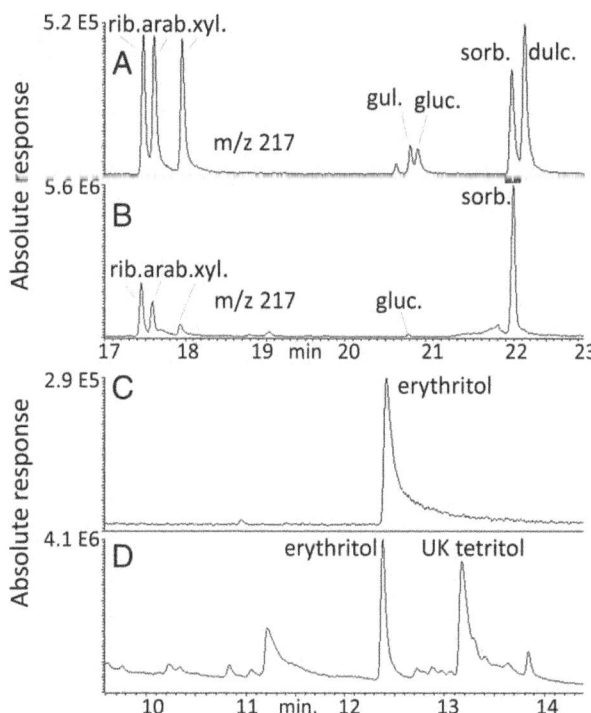

Fig. 6. OPLSDA model (R^2Y cum 0.798, Q^2 cum 0.691, 4 components) based on the six metabolites shown in Table 4 classifying 18 controls (1) and 27 SDBDI samples (2).

[51] where has been proposed that the likely source of the polyols was from the metabolic activity of the brain. The major hexitol in the brains is sorbitol but the pentitol peak observed in LC-MS as a single peak is due to the presence of three compounds, ribitol, arabinotol, and xylitol. In addition, there are two tetritols, erythritol, and an unknown isomer which are also elevated. It has been observed that the levels of these polyols in brain are elevated in response to osmotic stress [52]. In the current case, the levels of sorbitol, gluconic acid, ribitol, and erythritol are higher in the SDB/diabetic samples in comparison with the controls as judged from both the LC-MS and the GC-MS data (Table 3).

4.3. Elevation of Polyols and Oxidative Stress

Since the sugar alcohols are not closely linked within a particular pathway and several are elevated this suggests that the higher levels might be due to an upregulation of aldose reductase which has a wide substrate specificity [53] and is able reduce many different aldoses. Formation of sugar alcohols via aldose reductase activity is responsible for some of the complications of diabetes [53] and also generates oxidative stress since NADPH is consumed in carrying out the reduction. A previous paper observed that altered glucose metabolism is the brains of those diagnosed with depression and schizophrenia, where sorbitol was increased by a factor of 2.2 [18], similar to the elevations in the SDB brains in the current study. A recent metabolomic study observed altered glucose metabolism in peripheral blood mononuclear cells in

Table 5
Summary of the results obtained by removing the subsets B, S, D, and DI and using them as prediction sets.

Samples removed	R^2Y (cum) /Q^2 (cum) for new model	Correctly classified	Incorrectly classified
B1–B6	0.721/0.596	B2–B6	B1 as control
S3–S5, S7–S11	0.706/0.601	All	
D1_D5, D6	0.623/0.526	D2–5, D7	D1 as control
DI1–DI7	0.739/0.619	DI2-DI7	DI1 as control

Table 6
Marker compounds used in the OPLSDA model shown in Fig. 6.

m/z	Rt (min)	Metabolite	P value*	Ratio *SDBDI /Control	VIP
104.071	15.9	4-Aminobutanoate	0.023	0.83	1.64
116.072	12.9	L-Valine	0.0022	1.30	1.32
162.112	13.7	L-Carnitine	0.84	1.01	0.76
204.123	11.4	O-Acetylcarnitine	0.023	1.33	0.71
175.025	14.6	Ascorbate	0.834	1.04	0.50
87.0087	8.3	Pyruvate	0.063	1.26	0.48

* For the 45 samples used in the model.

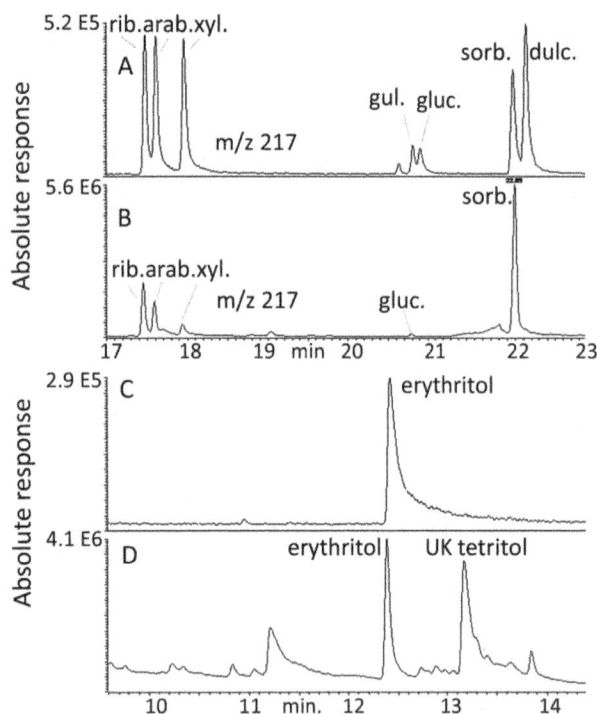

Fig. 7. GC-MS analysis of polyol standards (A and C 0.8 µg/0.2 ml) in comparison with polyols in brain (B and D).

schizophrenia with alterations in several glycolysis and Krebs cycle metabolites [11]. Glucaric acid, which is increased in SDB, also contributes to the model and has a high correlation to the model. Glucaric acid is of interest since it also relates to sorbitol and gluconic acid, being only a single oxidation step away from gluconic acid. Glucaric acid has been frequently monitored in urine as a marker of xenobiotic stress and urinary levels have been observed to rise in response to treatments with phenothiazines (such as the antipsychotic, chlorpromazine) [54]. Sedoheptulose is considerably elevated in SDB brains while a compound putatively identified as deoxysedoheptulose phosphate is depressed. Our published metabolomics study of brain tissue from a mouse model of psychiatric disorder, the *Npas3* knockout, also showed elevated levels of sedoheptulose (2.65-fold increase) [55]. This suggests that changes in glycolysis or nucleotide metabolism might be altering flux through the pentose phosphate pathway which could correlate with an increased requirement for NADPH as a co-factor for aldose reductase since NADP is converted to NADPH with formation of phosphogluconate at the entry to the pentose phosphate pathway. In addition, there is a deficit in the neuromodulator NAAG in the SDB brains; NAAG has been found to protect against neuronal death induced by exposure to glucose in a cell-culture model of diabetic neuropathy [56].

4.4. GABA Deficiency

GABA and its metabolite guanidino butyrate, which is formed directly from GABA via arginine–glycine amidinotransferase [57] are important variables in the OPLS-DA model separating control and SDB

brains. It is well established that there is a deficit in GABAergic transmission in schizophrenia [58–60]. The GABA receptor governs the entry of the chloride ion into cells [61] and one of the highest VIP values in Table 1 is for an adduct formed between chloride and carbonate which is present in the mobile phase which is strongly correlated with the SDB group. Initially, this peak was assigned to orthophosphate according to the library search but it runs earlier than the standard for orthophosphate. Inspection of the peak revealed a chlorine isotope pattern and the elemental composition matches the carbonic acid/chloride adduct. Chloride itself is below the lower mass range cut off for the instrument.

4.5. Altered Purine Metabolism

From the univariate comparisons, the purines guanine and guanosine were found to be lower in SDB brains and this can be correlated with elevated levels of uric acid in SDB brains. In a recent publication, it was observed that the severity of schizophrenic symptoms could be predicted from a high ratio of uric acid to guanine and the current data indicate that in SDB brains purine oxidation seems to be more active [62]. It has been suggested that that elevated uric acid is indicative of high levels of oxidative stress. Allopurinol, which inhibits purine oxidation, has been used as an experimental treatment for schizophrenia [63].

4.6. Elevation in the Level of a Homocarnosine Isomer

A compound, present in high amount in the brains, putatively identified as anserine since it is isomeric with homocarnosine but has a different retention time, is an important component is the OPLS-DA model separating control and SDB brains and it is significantly lower in the SDB samples according to the univariate data. Brain tissue accounts for around 20% of the oxygen consumption by the body and thus is a major site of oxidative stress. Carnosine, homocarnosine, and anserine are important antioxidants in brain and skeletal muscle [64] and lower levels of anserine might indicate increased oxidative stress in the SDB samples. Of the three commonly occurring histidine dipeptides, anserine has been observed to be the most effective anti-oxidant [64].

4.7. High Levels of Pyridoxine

The SDB brains contain higher levels of pyridoxine which has been used for many years as an experimental treatment for schizophrenia when given in conjunction with nicotinic acid [65].

4.8. Alterations in Biogenic Amine Metabolism

A number of neurochemically important compounds are significantly changed in the univariate data. Tryptamine has long been associated with mental illness particularly schizophrenia [8] and it is clearly slightly elevated in the SDB brain samples. In the SDB samples, there is a depression of norepinephrine sulphate levels; there are two isomers of this compound in brain both of which are depressed. Glucuronide and sulphate conjugates of dopamine and serotonin have been measured in brain dialysate previously [66]; there was no evidence in the current

Table 7
The amounts of sugar alcohols and gluconic acid in post-mortem brain samples.

Sugar	SDB + DI (RSD) µg/g	SDB (RSD) µg/g	Control (RSD) µg/g	SDB + DI/Control ratio (P value)	SDB/Control ratio (P value)
Sorbitol	22.7 (±56.4)	22.3 (±56.4)	13.5 (±57.0)	1.67 (0.0079)	1.65 (0.0015)
Gluconic acid	3.96 (±55.8)	4.2 (55.6)	1.96 (±40.7)	1.96 (0.0006)	2.06 (0.0012)
Ribitol	9.8 (±21.8)	10.3 (20.2)	9.3 (±25.9)	1.06 (0.18)	1.11 (0.06)
Arabinotol	7.9 (±26.6)	8.0 (26.6)	8.5 (±34.1)	0.93 (0.37)	0.94 (0.49)
Xylitol	4.1 (±28.3)	4.3 (28.3)	7.1 (±57.8)	0.58 (0.0042)	0.61 (0.006)
Erythritol	15.1 (±23.8)	15.1 (23.8)	12.9 (±34.9)	1.17 (0.028)	1.17 (0.056)

case for the presence norephinephrine glucuronide conjugates in brain. Although not significant in the OPLS-DA model, in the univariate analysis, N-acetylvanilalanine is significantly elevated in bipolar brains. This metabolite is of great interest since it is a marker for a deficiency of aromatic amino decarboxylase (AADC/DOPA Decarboxylase) deficiency which can lead to a deficit in the levels of several neurotransmitters [67]. This can be correlated with elevated levels of all the aromatic amino acids in the SDB brains. There are also elevated levels of kynurenine and kynurenamine which are metabolites of tryptophan which have neuropathological effects [68].

4.9. Differences in a Sub-group of SDB Samples

There are many other differences in the univariate data and it is difficult to rationalise them all. In order to determine if there are subgroups within the samples, a PCA model was fitted to the 36 samples used to produce the OPLS-DA model using only the metabolites shown in Table S1 which were significantly different according to univariate analysis. Hierarchical cluster analysis clearly highlighted a group of 9 SDB samples which were far away from the rest of the samples that did not clearly separate in the PCA plot (Fig. 4). The metabolites which were most significant in separating this sub-group from the controls in the PCA plot are listed in Table 3 along with P values and ratios derived from univariate comparison of these nine samples with the control samples. The brains in this subgroup contain much lower levels of NAAG in comparison with the rest of the SDB group and sorbitol, gluconic acid, and xylitol/ribitol/arabinotol are also higher than in the general cohort. In addition, N-acetylvanilalanine is higher in these samples along with tyrosine and phenylalanine than in the rest of the SDB samples which might indicate a greater degree of aromatic amino acid decarboxylase deficiency. However, tryptophan is not significantly different in this subgroup compared with the rest of the SDB samples although its metabolite tryptamine is elevated. In addition norepinephrine sulphate and guanosine are significantly lower in this group compared with the rest of the samples.

4.10. OPLSDA Models Based on Six Metabolites

Figures A10–14 show extracted ion traces for the six marker compounds used to produce the OPLSDA models shown in Figs. 5 and 6. GABA and valine are important components in the models shown in Figs. 2 and 3 and have been discussed above. Ascorbic acid does not show up strongly with regard to univariate statistics having a P value of 0.5 when comparing the samples modelled in Fig. 2. There have been reports of increased ascorbic acid requirements in schizophrenia with reduced urinary excretion being observed [69]. Carnitine and its acyl derivatives have been reported to have potential in the treatment of neurochemical disorders [70]. Finally, it was observed in a previous study that pyruvate levels were lowered in the thalmus of the post-mortem brains of schizophrenics in comparison with controls [71]. In the current case, pyruvate is slightly elevated in the SDB group but the part of the brain analysed in the current case was different.

5. Conclusion

In conclusion, many differences were observed in SDB versus control brains which have been observed by previous papers such as lower levels of NAAG and GABA in the SDB brains, elevated levels of sorbitol, and the importance of branched chain amino acids. Our strategy of treating the three psychiatric disorders as a single disease entity (SDB) may reduce the ability to detect specific aetiological biomarkers, but it increased sample size, diluted disease-specific medication effects, and most importantly, allowed the identification of a metabolic profile reflecting a shared pathological state. Since there are no biomarkers for mental illnesses and these diseases are multifaceted, diagnosis is never absolute and indeed this can be seen in scatter plots where each

individual is different. However, there is enough in common in the SDB group for them to be classified as more similar to each other than to the controls. Certain key metabolites highlighted as being more important in the pathology and it seems that abnormal sugar and branched chain amino acid metabolism might be a key element in SDB as reflected in the metabolic similarity between SDB and diabetes, and thus anti-diabetic treatments might have a role in the management of SDB. There are no previous metabolomics studies of post-mortem brain tissue in mental illness. Although this is only a small study, the findings are in agreement with several previous studies looking at specific markers and in respect of some markers with studies going back many years. The study has highlighted readily available markers which could be quantified in physiological fluids for the purpose of diagnosis or the monitoring of treatment.

Acknowledgements

The authors would like to thank Colin Smith, Robert Walker and Chris-Anne McKenzie of the MRC Sudden Death Brain and Tissue Bank for their assistance in sample provision. We would also like to thank the Scottish Life Sciences Alliance for funding mass spectrometry instruments.

References

[1] Wittchen H-U, Jacobi F. Size and burden of mental disorders in Europe—a critical review and appraisal of 27 studies. Eur Neuropsychopharmacol 2005;15:357–76.
[2] Weickert CS, Weickert TW, Pillai A, Buckley PF. Biomarkers in schizophrenia: a brief conceptual consideration. Dis Markers 2013;35:3–9.
[3] Corrigan PW. How clinical diagnosis might exacerbate the stigma of mental illness. Soc Work 2007;52:31–9.
[4] Pickard BS. Schizophrenia biomarkers: translating the descriptive into the diagnostic. J Psychopharmacol 2015;29:138–43.
[5] Perez VB, Swerdlow NR, Braff DL, Näätänen R, Light GA. Using biomarkers to inform diagnosis, guide treatments and track response to interventions in psychotic illnesses. Biomark Med 2013;8:9–14.
[6] Gika HG, Theodoridis GA, Plumb RS, Wilson ID. Current practice of liquid chromatography–mass spectrometry in metabolomics and metabonomics. J Pharm Biomed Anal 2014;87:12–25.
[7] Zhang T, Watson DG. A short review of applications of liquid chromatography mass spectrometry based metabolomics techniques to the analysis of human urine. Analyst 2015;140:2907–15.
[8] Burchett SA, Hicks TP. The mysterious trace amines: protean neuromodulators of synaptic transmission in mammalian brain. Prog Neurobiol 2006;79:223–46.
[9] Lindemann L, Hoener MC. A renaissance in trace amines inspired by a novel GPCR family. Trends Pharmacol Sci 2005;26:274–81.
[10] Koike S, Bundo M, Iwamoto K, Suga M, Kuwabara H, et al. A snapshot of plasma metabolites in first-episode schizophrenia: a capillary electrophoresis time-of-flight mass spectrometry study. Transcult Psychiatry 2014;4, e379.
[11] Liu M-L, Zhang X-T, Du X-Y, Fang Z, Liu Z, et al. Severe disturbance of glucose metabolism in peripheral blood mononuclear cells of schizophrenia patients: a targeted metabolomic study. J Transl Med 2015;13:226.
[12] Kaddurah-Daouk R, Krishnan KRR. Metabolomics: a global biochemical approach to the study of central nervous system diseases. Neuropsychopharmacol 2009;34:173–86.
[13] Yang J, Chen T, Sun L, Zhao Z, Qi X, et al. Potential metabolite markers of schizophrenia. Mol Psychiatry 2013;18:67–78.
[14] Yao J, Dougherty GG, Reddy R, Keshavan M, Montrose D, et al. Altered interactions of tryptophan metabolites in first-episode neuroleptic-naive patients with schizophrenia. Mol Psychiatry 2010;15:938–53.
[15] Xuan J, Pan G, Qiu Y, Yang L, Su M, et al. Metabolomic profiling to identify potential serum biomarkers for schizophrenia and risperidone action. J Prot Res 2011;10:5433–43.
[16] He Y, Yu Z, Giegling I, Xie L, Hartmann A, et al. Schizophrenia shows a unique metabolomics signature in plasma. Transcult Psychiatry 2012;2, e149.
[17] Orešič M, Tang J, Seppänen-Laakso T, Mattila I, Saarni SE, et al. Metabolome in schizophrenia and other psychotic disorders: a general population-based study. Genome Med 2011;3:19.
[18] Regenold W, Phatak P, Kling M, Hauser P. Post-mortem evidence from human brain tissue of disturbed glucose metabolism in mood and psychotic disorders. Mol Psychiatry 2004;9:731–3.
[19] Zhang R, Watson DG, Wang L, Westrop GD, Coombs GH, et al. Evaluation of mobile phase characteristics on three zwitterionic columns in hydrophilic interaction liquid chromatography mode for liquid chromatography-high resolution mass spectrometry based untargeted metabolite profiling of Leishmania parasites. J Chromatogr A 2014;1362:168–79.

[20] Zhang T, Watson DG, Wang L, Abbas M, Murdoch L, et al. Application of holistic liquid chromatography-high resolution mass spectrometry based urinary metabolomics for prostate cancer detection and biomarker discovery. PLoS One 2013;8, e65880.

[21] McEvoy JP, Meyer JM, Goff DC, Nasrallah HA, Davis SM, et al. Prevalence of the metabolic syndrome in patients with schizophrenia: baseline results from the Clinical Antipsychotic Trials of Intervention Effectiveness (CATIE) schizophrenia trial and comparison with national estimates from NHANES III. Schizophr Res 2005;80: 19–32.

[22] Newcomer JW. Metabolic syndrome and mental illness. Am J Manag Care 2007;13: S170–7.

[23] Zhang T, Creek DJ, Barrett MP, Blackburn G, Watson DG. Evaluation of coupling reversed phase, aqueous normal phase, and hydrophilic interaction liquid chromatography with Orbitrap mass spectrometry for metabolomic studies of human urine. Anal Chem 2012;84:1994–2001.

[24] Zheng L, T'Kind R, Decuypere S, von Freyend SJ, Coombs GH, et al. Profiling of lipids in Leishmania donovani using hydrophilic interaction chromatography in combination with Fourier transform mass spectrometry. Rapid Commun Mass Spectrom 2010;24:2074–82.

[25] Pluskal T, Castillo S, Villar-Briones A, Orešič M. MZmine 2: modular framework for processing, visualizing, and analyzing mass spectrometry-based molecular profile data. BMC Bioinform 2010;11:395.

[26] Ruiz-Matute AI, Hernandez-Hernandez O, Rodríguez-Sánchez S, Sanz ML, Martínez-Castro I. Derivatization of carbohydrates for GC and GC–MS analyses. J Chromatogr B 2011;879:1226–40.

[27] Dalack GW, Healy DJ, Meador-Woodruff JH. Nicotine dependence in schizophrenia: clinical phenomena and laboratory findings. Am J Psychol 1998;155:1490–501.

[28] Yang J, Zhao X, Lu X, Lin X, Xu G. A data preprocessing strategy for metabolomics to reduce the mask effect in data analysis. Front Mol Biosci 2015;2(4).

[29] van den Berg RA, Hoefsloot HC, Westerhuis JA, Smilde AK, van der Werf MJ. Centering, scaling, and transformations: improving the biological information content of metabolomics data. BMC Genomics 2006;7:142.

[30] Kirwan GM, Johansson E, Kleemann R, Verheij ER, Wheelock AM, Goto S, et al. Building multivariate systems biology models. Anal Chem 2012;84:7064–71.

[31] Benjamini Y, Hochberg Y. Controlling the False Discovery Rate a Practical and powerful Approach to Multiple Testing. J R Stat Soc 1995;57:289–300.

[32] Chong I-G, Jun C-H. Performance of some variable selection methods when multicollinearity is present. Chemom Intell Lab Syst 2005;78:103–12.

[33] Eriksson L, Byrne T, Johansson E, Trygg J, Vikstrom C. Multi- and Megavariate Data Analysis: Basic Principles and Application. 3 ed. Sweden: MKS Umetrics AB; 2013 455–6.

[34] Westerhuis JA, Hoefsloot HCJ, Smit S, Vis DJ, Smilde AK, van Velzen EJJ, et al. Assessment of PLSDA cross validation. Metabolomics 2008;4:81–9.

[35] Triba MN, Le Moyec L, Amathieu R, Goossens C, Bouchemal N, et al. PLS/OPLS models in metabolomics: the impact of permutation of dataset rows on the K-fold cross-validation quality parameters. Mol Biosyst 2015;11:13–9.

[36] Wheelock ÅM, Wheelock CE. Trials and tribulations of 'omics data analysis: assessing quality of SIMCA-based multivariate models using examples from pulmonary medicine. Mol Biosyst 2013;9:2589–96.

[37] Eriksson L, Trygg J, Wold S. CV-ANOVA for significance testing of PLS and OPLS® models. J Chemometr 2008;22:594–600.

[38] Sumner LW, Amberg A, Barrett D, Beale MH, Beger R, et al. Proposed minimum reporting standards for chemical analysis. Metabolomics 2007;3:211–21.

[39] Kohen D. Diabetes mellitus and schizophrenia: historical perspective. British J Psych 2004;184:S64–6.

[40] Burghardt KJ, Evans SJ, Wiese KM, Ellingrod VL. An Untargeted Metabolomics Analysis of Antipsychotic Use in Bipolar Disorder. Clin Transl Sci 2015;8:432–40.

[41] Newgard CB, An J, Bain JR, Muehlbauer MJ, Stevens RD, et al. A branched-chain amino acid-related metabolic signature that differentiates obese and lean humans and contributes to insulin resistance. Cell Metab 2009;9:311–26.

[42] Huffman KM, Shah SH, Stevens RD, Bain JR, Muehlbauer M, et al. Relationships between circulating metabolic intermediates and insulin action in overweight to obese, inactive men and women. Diabetes Care 2009;32:1678–83.

[43] Batch BC, Shah SH, Newgard CB, Turer CB, Haynes C, Bain JR, et al. Branched chain amino acids are novel biomarkers for discrimination of metabolic wellness. Metab Clin Exp 2013;62:961–9.

[44] McCormack SE, Shaham O, McCarthy MA, Deik AA, Wang TJ, et al. Circulating branched-chain amino acid concentrations are associated with obesity and future insulin resistance in children and adolescents. Pediatr Obes 2013;8:52–61.

[45] Krebs M. Amino acid-dependent modulation of glucose metabolism in humans. Eur J Clin Invest 2005;35:351–4.

[46] Rennie MJ, Bohé J, Smith K, Wackerhage H, Greenhaff P. Branched-chain amino acids as fuels and anabolic signals in human muscle. J Nutr 2006;136:264S–8S.

[47] Paredes RM, Quinones M, Marballi K, Gao X, Valdez C, et al. Metabolomic profiling of schizophrenia patients at risk for metabolic syndrome. Int J Neuropsychopharmacol 2014;17:1139–48.

[48] Bjerkenstedt L, Edman G, Hagenfeldt L, Sedvall G, Wiesel F. Plasma amino acids in relation to cerebrospinal fluid monoamine metabolites in schizophrenic patients and healthy controls. British J Psych 1985;147:276–82.

[49] Kempf L, Nicodemus KK, Kolachana B, Vakkalanka R, Verchinski BA, et al. Functional polymorphisms in PRODH are associated with risk and protection for schizophrenia and fronto-striatal structure and function. PLoS Genet 2008;4, e1000252.

[50] Tunbridge E, Burnet PW, Sodhi MS, Harrison PJ. Catechol-o-methyltransferase (COMT) and proline dehydrogenase (PRODH) mRNAs in the dorsolateral prefrontal cortex in schizophrenia, bipolar disorder, and major depression. Synapse 2004;51: 112–8.

[51] Shetty H, Holloway H, Schapiro M. Cerebrospinal fluid and plasma distribution of myo-inositol and other polyols in Alzheimer disease. Clin Chem 1996;42:298–302.

[52] Lien Y, Shapiro J, Chan L. Effects of hypernatremia on organic brain osmoles. J Clin Invest 1990;85:1427–35.

[53] Yabe-Nishimura C. Aldose reductase in glucose toxicity: a potential target for the prevention of diabetic complications. Pharmacol Rev 1998;50:21–34.

[54] Wright JH, Whitaker SB, Welch CB, Teller DN. Hepatic enzyme induction patterns and phenothiazine side effects. Clin Pharmacol Ther 1983;34:533–8.

[55] Sha L, MacIntyre L, Machell J, Kelly M, Porteous D, et al. Transcriptional regulation of neurodevelopmental and metabolic pathways by NPAS3. Mol Psychiatry 2012;17: 267–79.

[56] Berent-Spillson A, Robinson AM, Golovoy D, Slusher B, Rojas C, et al. Protection against glucose-induced neuronal death by NAAG and GCP II inhibition is regulated by mGluR3. J Neurochem 2004;89:90–9.

[57] Rowland LM, Kontson K, West J, Edden RA, Zhu H, et al. In vivo measurements of glutamate, GABA, and NAAG in schizophrenia. Schizophr Bull 2012. http://dx.doi.org/10.1093/schbul/sbs092.

[58] Jansen EE, Verhoeven NM, Jakobs C, Schulze A, Senephansiri H, et al. Increased guanidino species in murine and human succinate semialdehyde dehydrogenase (SSADH) deficiency. Biochim Biophys Acta Mol Basis Dis 2006;1762:494–8.

[59] Hashimoto T, Volk DW, Eggan SM, Mirnics K, Pierri JN, et al. Gene expression deficits in a subclass of GABA neurons in the prefrontal cortex of subjects with schizophrenia. J Neurosci 2003;23:6315–26.

[60] Guidotti A, Auta J, Davis JM, Dong E, Grayson DR, et al. GABAergic dysfunction in schizophrenia: new treatment strategies on the horizon. Psychopharmacol 2005; 180:191–205.

[61] Suzdak P, Schwartz RD, Skolnick P, Paul SM. Ethanol stimulates gamma-aminobutyric acid receptor-mediated chloride transport in rat brain synaptoneurosomes. Proc Natl Acad Sci 1986;83:4071–5.

[62] Yao JK, Condray R, Dougherty Jr GG, Keshavan MS, Montrose DM, et al. Associations between purine metabolites and clinical symptoms in schizophrenia. PLoS One 2012;7, e42165.

[63] Weiser M, Gershon AA, Rubinstein K, Petcu C, Ladea M, et al. A randomized controlled trial of allopurinol vs. placebo added on to antipsychotics in patients with schizophrenia or schizoaffective disorder. Schizophr Res 2012;138:35–8.

[64] Kohen R, Yamamoto Y, Cundy KC, Ames BN. Antioxidant activity of carnosine, homocarnosine, and anserine present in muscle and brain. Proc Natl Acad Sci 1988;85:3175–9.

[65] Xu X, Jiang G. Niacin-respondent subset of schizophrenia–a therapeutic review. Eur Rev Med Pharmacol Sci 2015;19:988–97.

[66] Suominen T, Uutela P, Ketola RA, Bergquist J, Hillered L, et al. Determination of serotonin and dopamine metabolites in human brain microdialysis and cerebrospinal fluid samples by UPLC-MS/MS: discovery of intact glucuronide and sulfate conjugates. PLoS One 2013;8, e68007.

[67] Abdenur JE, Abeling N, Specola N, Jorge L, Schenone AB, et al. Aromatic l-aminoacid decarboxylase deficiency: unusual neonatal presentation and additional findings in organic acid analysis. Mol Genet Metab 2006;87:48–53.

[68] Tan L, Yu J-T, Tan L. The kynurenine pathway in neurodegenerative diseases: mechanistic and therapeutic considerations. J Neurol Sci 2012;323:1–8.

[69] Yao JK, Reddy RD, Van Kammen DP. Oxidative damage and schizophrenia. CNS Drugs 2001;15:287–310.

[70] Malaguarnera M. Carnitine derivatives: clinical usefulness. Curr Opin Gastroenterol 2012;28:166–76.

[71] Martins-de-Souza D, Maccarrone G, Wobrock T, Zerr I, Gormanns P, et al. Proteome analysis of the thalamus and cerebrospinal fluid reveals glycolysis dysfunction and potential biomarkers candidates for schizophrenia. J Psychiatr Res 2010;44: 1176–89.

9

A guide to the identification of metabolites in NMR-based metabonomics/metabolomics experiments

Anthony C. Dona [a,1], Michael Kyriakides [a], Flora Scott [b], Elizabeth A. Shephard [b], Dorsa Varshavi [c], Kirill Veselkov [a], Jeremy R. Everett [c,*]

[a] *Department of Surgery and Cancer, Faculty of Medicine, Imperial College London, SW7 2AZ, United Kingdom*
[b] *Institute of Structural and Molecular Biology, University College London, London WC1E 6BT, United Kingdom*
[c] *Medway Metabonomics Research Group, University of Greenwich, Chatham Maritime, Kent ME4 4TB, United Kingdom*

ARTICLE INFO

ABSTRACT

Metabonomics/metabolomics is an important science for the understanding of biological systems and the prediction of their behaviour, through the profiling of metabolites. Two technologies are routinely used in order to analyse metabolite profiles in biological fluids: nuclear magnetic resonance (NMR) spectroscopy and mass spectrometry (MS), the latter typically with hyphenation to a chromatography system such as liquid chromatography (LC), in a configuration known as LC–MS. With both NMR and MS-based detection technologies, the identification of the metabolites in the biological sample remains a significant obstacle and bottleneck. This article provides guidance on methods for metabolite identification in biological fluids using NMR spectroscopy, and is illustrated with examples from recent studies on mice.

Keywords:
Nuclear magnetic resonance (NMR) spectroscopy
Metabolite identification
Molecular structure
Metabonomics
Metabolomics

1. Introduction

Metabonomics is defined as 'the quantitative measurement of the multiparametric metabolic response of living systems to pathophysiological stimuli or genetic modification' and is concerned with the study of the metabolic response of organisms to disease, environmental change or genetic modification [1]. The similar term metabolomics [2] was defined later and is now used interchangeably. In contrast to the interventional definition of metabonomics, metabolomics has an observational definition which is difficult if not impossible to achieve: 'a comprehensive analysis in which all the metabolites of a biological system are identified and quantified' [2]. In this work we will use the original term throughout. Metabonomics has many areas of application including biology and medicine [3] with new developments such as pharmacometabonomics (the ability to predict drug responses prior to drug dosing) and the more general area of predictive metabonomics, emerging recently [4–7].

There are many stages to a well-designed metabonomics experiment including: 1) definition of study aims and experimental design, 2) ethical approval of the study, 3) sample collection and storage, 4) sample preparation, 5) data acquisition, 6) data quality control, 7) spectroscopic data pre-processing (for NMR data this would include zero-filling, apodisation, Fourier transform, phasing, baseline correction and referencing), 8) statistical data pre-processing including peak alignment, scaling and normalisation, 9) statistical analysis of the data to interrogate e.g. differences in metabolite profiles due to a drug treatment, 10) identification of the metabolites that are responsible for the metabolite profile differences, 11) biological/biochemical interpretation of the role of those metabolites, including pathway analysis and 12) reporting of results and deposition of the data.

Many of the metabonomics study elements above have excellent literature reviews and references available to assist effective study execution [8–20]. However, the identification of the key biomarkers or metabolites that are responsible for discriminating between different groups in a study (Stage 10 above) is non-trivial for both NMR [15,21–28] and MS-based [28–34] metabonomics experiments. This guide aims to provide an insight into the methodologies that can be used for NMR-based metabolite identification in the course of a metabonomics project. It is assumed that the reader is familiar with the basics of NMR spectroscopy: many excellent books on the topic

* Corresponding author.
  E-mail address: j.r.everett@greenwich.ac.uk (J.R. Everett).
[1] Current address: Kolling Institute of Medical Sciences, Sydney Medical School, The University of Sydney, NSW 2065, Australia.

are available [35–38]. The focus of this guide is on the *use* of ^{1}H NMR, or ^{1}H NMR-detected heteronuclear 2D NMR experiments, for metabolite identification in metabonomics experiments on biological fluids.

2. Molecular structure information from 1D NMR spectra of metabolites

A surprising amount of information is available from a one-dimensional (1D) ^{1}H NMR spectrum, including: 1) chemical shifts, 2) signal multiplicities, 3) homonuclear ([1]H - [1]H) coupling constants, 4) heteronuclear coupling constants (typically [14]N–[1]H or [31]P–[1]H), 5) the first order or second-order nature of the signal, 6) the half bandwidth of the signal, 7) the integral of the signal and 8) the stability of the signal (changes in the integral with time). We will not cover: 9) spin-lattice relaxation times (T_1s) or 10) spin–spin relaxation times (T_2s). Whilst an appreciation of both these latter features is critical for the conduct of all NMR experiments, and differentiation of short from long T_2s is fundamental in the Carr–Purcell–Meiboom–Gill (CPMG) spin-echo pulse sequence for plasma analysis, these features are of minor importance *per se* for metabolite identification. We will deal with each of the first 8 features in turn and see how they can be used to assist metabolite identification.

2.1. ^{1}H NMR chemical shifts

Each chemically distinct hydrogen nucleus in each metabolite in a biological sample, such as a biofluid, will exhibit an NMR signal at a characteristic resonance frequency, which is measured as a chemical shift relative to a standard compound. For example, in metabonomics studies of urine, it is common to add a reference material such as 3-(trimethylsilyl)-2,2′,3,3′-tetradeuteropropionic acid (usually abbreviated to TSP) or deuterated forms of 4,4-dimethyl-4-silapentane-1-sulfonic acid (DSS) or its sodium salt, and define the chemical shift of the TSP or DSS methyl resonances as 0 ppm. Our preference is to use TSP as the reference material in biofluids without significant protein concentrations. The normal reference material for NMR spectroscopy in organic chemistry, tetramethylsilane (TMS) is rarely used in metabonomics studies, as it is insoluble in aqueous solutions.

The exact chemical shift of the NMR signal of a hydrogen nucleus in a metabolite is independent of the applied field strength, is highly reproducible and precisely characteristic of that nucleus, in that metabolite, in the particular matrix conditions. For biofluids such as blood plasma or serum, where DSS or TSP may become bound to macromolecule components, it is common to reference the spectra to the H1' anomeric proton of the alpha anomer of glucose at 5.233 ppm, to avoid variation in reference intensity and position due to binding [39]. However, care must be taken with temperature control as this signal has high temperature sensitivity and indeed, has been used as an NMR thermometer [40].

When comparing the experimental ^{1}H NMR chemical shifts of hydrogens in metabolites in intact biofluids with those of the corresponding pure reference standards in aqueous solution, it is usual for values to agree within 0.03 ppm. One of the strengths of NMR spectroscopy is that the chemical shifts are exquisitely sensitive to structural and environmental change. Indeed, sensitivity of chemical shifts to pH change can be used to distinguish or identify metabolites, especially those containing ionisable functional groups [41]. Whilst this is an excellent feature in terms of decreasing the likelihood of two similar molecules having identical ^{1}H NMR spectra, it does mean that for some metabolites, environmental change can have a significant effect on the spectra, including the ^{1}H NMR chemical shifts. A classic case of this sensitivity to the environment occurs for the diastereotopic methylene hydrogens in citric acid. Changes in pH between samples will alter the ionisation of the carboxylate groups in citric acid and thus affect the chemical shifts of the methylene hydrogens. In addition, it is well known [3] that citric acid can chelate metal ions such as calcium, magnesium and sodium. Thus, even if biofluid samples are buffered effectively to a constant pH,

changes in metal ion concentrations between samples, which are not readily apparent by ^{1}H NMR, may have a significant effect on the chemical shifts and the half bandwidths of the signals of the methylene hydrogens of citric acid and also any other metabolites with similar properties. This effect is observed in Fig. 1.

Many general resources are available which correlate the relationships between chemical structure and NMR chemical shifts [36,42], including web resources [43], whilst more specific metabonomics-focused databases are covered in Section 4.3 below. As for ^{13}C NMR chemical shifts (see Section 2.2 below), it is also possible to calculate ^{1}H NMR shifts, especially in discrete series [43].

2.2. ^{13}C NMR chemical shifts

Most metabonomics experiments are conducted with ^{1}H NMR detection. However, the 2D ^{13}C, ^{1}H HSQC NMR (Section 3.5 below) and 2D ^{13}C, ^{1}H HMBC NMR (Section 3.6) experiments which correlate ^{1}H NMR chemical shifts with ^{13}C NMR chemical shifts over 1-bond (HSQC) or 2 to 3 bonds (HMBC) are very important for metabolite identification, as they enable the determination of the ^{13}C NMR chemical shifts of metabolites, so an appreciation of the nature of ^{13}C NMR chemical shifts is required. One key feature of ^{13}C NMR chemical shifts is their much larger range of values compared with ^{1}H NMR chemical shifts. For common metabolites ^{13}C NMR chemical shifts occupy a huge range of values from ca 10 ppm for methyl carbons such as C4 in butanone to ca 222 ppm for the ketone (C2) carbon in the same molecule. Thus the range of ^{13}C NMR chemical shifts is ca 20 times that of ^{1}H NMR and this is the reason that their measurement is so important in metabolite identification: they are much more sensitive to small changes, or more remote changes in molecular structure, including stereoisomerism, than ^{1}H NMR chemical shifts.

For simple molecules ^{13}C NMR chemical shifts can be calculated by hand using simple additivity tables.[44] For example in simple substituted benzenes such as para-cresol (1-hydroxy-4-methylbenzene), the ^{13}C NMR chemical shifts of all of the carbons can be calculated by adding the known substituent effects of hydroxyl and methyl groups [44] to the chemical shift of benzene in an additive fashion.

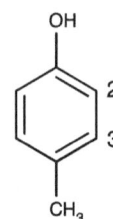

The substituent parameters for an hydroxyl group added to a benzene ring are $+26.9$ (ipso), -12.7 (ortho), $+1.4$ (meta) and -7.3 ppm (para position). For a methyl group the corresponding parameters are $+9.3$ (ipso), $+0.8$ (ortho), 0.0 (meta) and -2.9 ppm (para) [44]. The accepted ^{13}C NMR chemical shift of benzene is 128.5 ppm. Even if the molecule para-cresol was not in a metabolite database, we could calculate the ^{13}C NMR shifts with some degree of precision. For C2 and C3 the calculated shifts would be as follows:

$$\delta_{C2} = 128.5 - 12.7 \, (\text{OH ortho}) + 0 \, (\text{methyl, meta}) = 115.8 \, \text{ppm}$$

$$\delta_{C3} = 128.5 + 1.4 \, (\text{OH meta}) + 0.8 \, (\text{methyl, ortho}) = 130.7 \, \text{ppm}$$

For comparison, the actual values in the HMDB [45] for para-cresol, HMDB01858, in water at pH 7.0 are 117.9 and 132.8 ppm for C2 and C3 respectively. Modern NMR data processing software such as MNova [46] possesses more sophisticated ^{1}H, ^{13}C, and multinuclear NMR chemical shift calculation and prediction algorithms. The

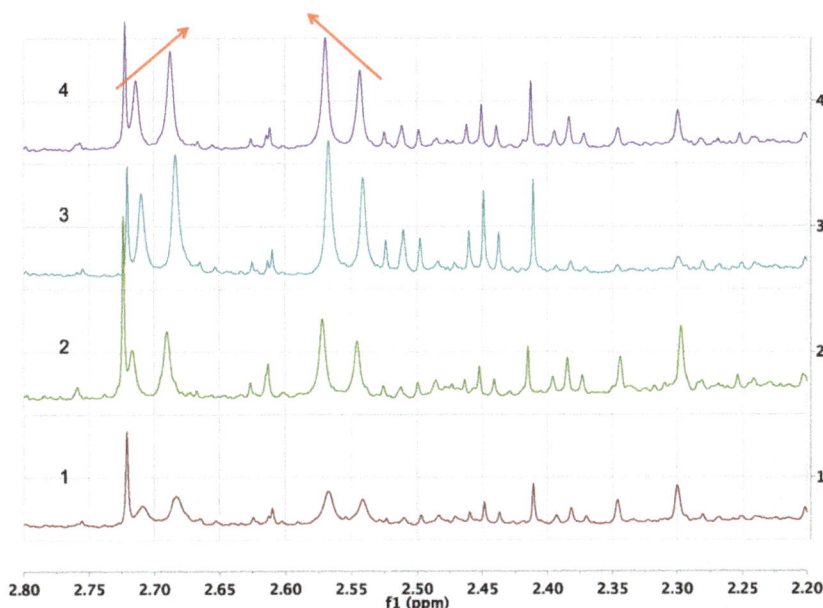

Fig. 1. an expansion of the 600 MHz ¹H NMR spectrum of the urine of four, 30-week-old, male C57BL/6 mice, in the region of the doublet signals of citrate at ca 2.70 and 2.56 ppm. Even though the urine is buffered to pH 7.4, there are differences in the chemical shifts of the citrate signals between the four urine samples and noticeable differences also in half bandwidth, with the signals of mouse 1 (bottom spectrum) being especially broadened. The 'roofing' of the doublet citrate signals towards one another is illustrated by the arrows above the citrate resonances of mouse 4. See Section 2.6 on 2nd order effects.

algorithm in MNova 10.0.0 predicted shifts of 117.6 and 130.0 ppm for C2 and C3 respectively, a very good fit to the real data for C2, but not quite as good as the simple hand calculation for C3. MarvinSketch v 6.1.1 from ChemAxon [47] also has ¹H and ¹³C NMR chemical shift calculation capabilities and it gave 115.3 and 130.0 ppm for C2 and C3 respectively. Calculations such as these can be useful when information about metabolites of interest is not in the existing databases: a common occurrence. However, users must be aware that these calculations are approximate, with precision varying according to the complexity of the metabolite and the relationship of the structure of the metabolite to the molecules in the prediction calculation database, or to those used to derive the substituent tables. In general a precision of better than $+/- 5$ ppm is usually achieved for ¹³C resonances.

It is beyond the scope of this guide to discuss factors that influence ¹³C NMR shifts in any detail. However, the key factors include: hybridisation of the carbon atom (sp3, sp2 or sp. hybridised), inductive substituent effects and mesomeric effects [43,48–52].

2.3. ¹H NMR multiplicities

The multiplicity is the pattern of peaks that is observed for a particular hydrogen signal in the ¹H NMR spectrum. In a first order ¹H NMR spectrum, the frequency difference between the resonances of coupled hydrogens is large ($\gg$ 10 times) relative to the value of the coupling constant between them. In those circumstances, the signals exhibit first order coupling patterns, which obey an $n + 1$ splitting rule, where n is the number of equivalent coupling partners. For instance, methyl groups such as those of lactic acid which couple with one hydrogen on an adjacent carbon via a homonuclear, 3-bond vicinal coupling, $^3J_{H,H}$, will be split into a doublet signal ($1 + 1 = 2$). Correspondingly, the signal of the lactate methyne CH proton will be split into a 4-line quartet due to the interaction with the 3 equivalent methyl hydrogens ($3 + 1 = 4$). The intensity ratios of these multiplet signals follow Pascal's triangle [35], being 1:1, 1:2:1, and 1:3:3:1 for a doublet, triplet and quartet respectively. An example of a 1:2:1 triplet from one of the methylene CH_2 groups in 2-oxoglutaric acid is clearly observed at 2.45 ppm in the ¹H NMR spectra of the urines of the mice in Fig. 1.

If a particular hydrogen is coupled to more than one group of hydrogens, then more complex coupling patterns or multiplicities are observed. For instance, the CH_2–3 methylene signal from the butyryl chain of N-butyrylglycine resonates as a triplet of quartets as it is coupled to both the terminal CH_3–4 protons and the CH_2–2 protons adjacent to the C1 amide carbon. If the coupling constants involved were non-equal then up to 12 lines could be observed (4×3). However, in this case, the CH_3–4 to CH_2–3 coupling constant (7.4 Hz) is almost equal to the CH_2–2 to CH_2–3 coupling constant (7.5 Hz) and the C3 methylene signal resonates as a pseudo-sextet due to signal overlap (Fig. 2).

The analysis of signal multiplicities, simple and complex [36,43], is important for the identification of metabolites. Multiplicity-editing spin-echo NMR experiments can also be used to distinguish between signals with different multiplicities and this can be helpful in metabolite identification. A good example is the use of spin-echo ¹H NMR in the identification of novel penicillin metabolites, where the characteristic singlet signals of the penicillin gem-dimethyl groups can be easily identified by Hahn spin-echo methods [53,54]. Many spectra do not obey first order requirements however and two main consequences arise from this; multiplicity intensities may be distorted, or in extreme cases additional lines may occur in the multiplets: see Section 2.6 below.

2.4. Homonuclear ¹H, ¹H coupling

Scalar coupling can occur between all non-equivalent hydrogen atoms in a metabolite. The key requirement here is magnetic non-equivalence. Hydrogens that are equivalent by molecular symmetry, such as the methyne hydrogens in tartaric acid, or equivalent by virtue of fast rotation, such as those of methyl hydrogens, will not show scalar coupling between themselves. Indeed the two methyne hydrogens of 2R, 3R-tartaric acid resonate as a characteristic, sharp singlet at 4.34 ppm in urine.

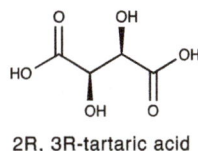

2R, 3R-tartaric acid

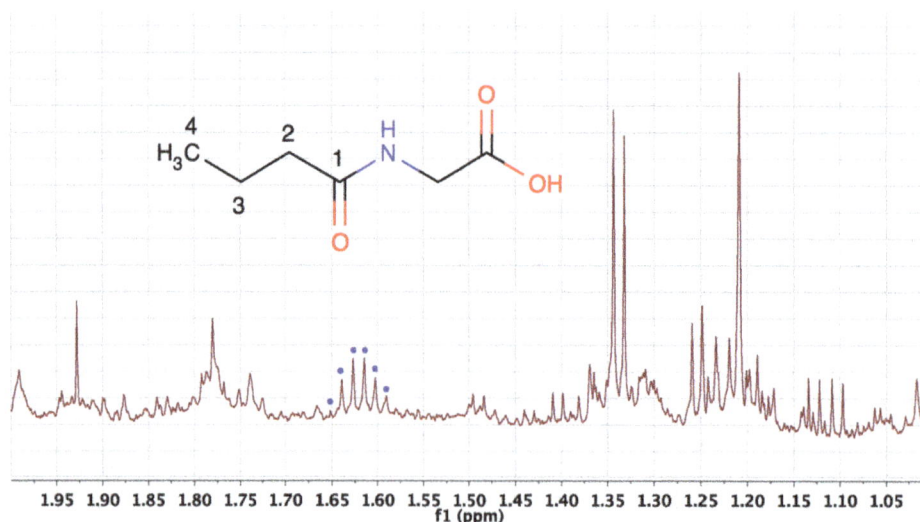

Fig. 2. An expansion of the 600 MHz ^{1}H NMR spectrum of the urine of a male, 15-week-old, C57BL/6 mouse, in the region of the signal from the CH$_2$–3 methylene protons of *N*-butyrylglycine at ca 1.62 ppm (dots). This pseudo-sextet signal is actually a triplet of quartets with the two ^{3}J$_{H,H}$ couplings being almost equal in magnitude (7.4 and 7.5 Hz) resulting in overlap of many of the lines. The molecular structure of the metabolite is superimposed.

The detection of the presence of scalar coupling between two hydrogens in a metabolite is very important in metabolite identification, as the magnitudes of the coupling constants are characteristic of the electronic pathway between the two hydrogens or groups of hydrogens. Scalar coupling is transmitted via the bonding electrons in the metabolites and drops off in magnitude as the number of bonds between the hydrogens increases. Most of the homonuclear scalar couplings observed in metabolites will be two-bond geminal couplings (^{2}J$_{H,H}$) between hydrogens on the same carbon, or three-bond, vicinal couplings (^{3}J$_{H,H}$) between hydrogens on adjacent carbons in a metabolite. In general, 2-bond geminal couplings are larger in magnitude than 3-bond, vicinal couplings. However, geminal couplings are affected by the hybridisation of the carbon atom and by the electronegativity of substituents, and in some alkenes, such as R$_1$R$_2$C $=$ $=$CH$_2$ the ^{2}J$_{H,H}$ value for the terminal $=$ $=$CH$_2$ will be close to 0 Hz. In passing, we should note that most geminal couplings are negative in sign and most vicinal couplings are positive, but this is not relevant for most analyses and we will ignore this feature henceforth. The large magnitude of geminal couplings in sp^{3}CH$_2$ groups is well illustrated by the spectrum of citrate shown in Fig. 1, where the geminal ^{2}J$_{H,H}$ coupling is ca 16.2 Hz. By contrast, the 3-bond, vicinal coupling between the C3–CH$_2$ group and its adjacent methyl and methylene group neighbours in *N*-butyrylglycine is ca 7.4 and 7.5 Hz respectively (Fig. 2). These ^{3}J$_{H,H}$ values are smaller and are typical of the values for free-rotating aliphatic moieties.

The values of vicinal couplings are particularly sensitive to stereochemistry in relatively rigid systems and this is well illustrated by metabolites such as D-glucose, which exists as two anomers in slow exchange with one another, so that separate signals are observed for each anomer.

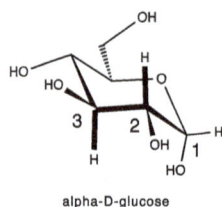

alpha-D-glucose

The anomeric proton at C1 in the alpha anomer is in an equatorial position on the 6-membered pyranose ring and has a modest ^{3}J$_{H,H}$ coupling of ca 3.7 Hz to the axial H-2 (equatorial–axial coupling). By

contrast, the coupling between H2 and H3 (both axial) has a value ^{3}J$_{H,H}$ ca 9.8 Hz because this is a favoured, di-axial coupling. Thus the magnitude of coupling constants can give information on the type of coupling and the stereochemistry of the interacting hydrogens. In addition to this, the values of coupling constants are affected by the electronegativity of groups in their vicinity, due to their impact on the electrons that transmit the coupling [43].

If ^{1}H NMR spectra are acquired with good spectral resolution, good digital resolution and good lineshape, it is possible to observe 4-bond, 5-bond and even 6-bond hydrogen-to-hydrogen couplings, ^{4}J$_{H,H}$, ^{5}J$_{H,H}$ and ^{6}J$_{H,H}$, in biofluids. For example, in cis-aconitic acid, it is usual to observe the olefinic proton at ca 5.74 as a triplet with ^{4}J$_{H,H}$ ca 1.4 Hz due to long-range, 4-bond coupling to the equivalent methylene CH$_2$ hydrogens across the double bond (Fig. 3).

In the trans-aconitic acid isomer, the olefinic proton at 6.60 ppm is a triplet with a smaller ^{4}J$_{H,H}$ ca 0.8 Hz coupling. Note the enormous sensitivity of the chemical shift to the geometry of the double bond: the olefinic proton shifts nearly 0.9 ppm just from the change of double-bond geometry, and the change in the coupling value for ^{4}J$_{H,H}$ is also diagnostic.

2.5. Heteronuclear ^{1}H, X coupling

These couplings are less common but will occur in phosphorous-containing metabolites such as adenosine monophosphate, where the presence of the NMR-active, 100% abundant, spin I $= 1/2$, ^{31}P isotope will give rise to additional 3-bond and 4-bond ^{3}J$_{P,H}$ and ^{4}J$_{P,H}$ couplings to the ribose ring protons, that are highly diagnostic [55]. Another less-commonly observed heteronuclear coupling in metabolites is due to the 99.6% natural abundance ^{14}N isotope which is NMR-active but quadrupolar, with spin quantum number I $= 1$. Due to quadrupolar relaxation, couplings to ^{14}N are not often observed, but in a symmetrical environment, the effects of quadrupolar relaxation are reduced and small couplings may be observed and these can also be critical for metabolite identification. For instance, in choline (HMDB00097), the almost symmetrical environment around the nitrogen allows the observation of a small ^{2}J$_{N,H}$ coupling of ca 0.6 Hz (1:1:1 triplet due to spin quantum number I $= 1$) to the methyl hydrogens due to 2-bond coupling to the ^{14}N. So, in this unusual case, the methyl signal is a narrow triplet instead of the expected singlet (Fig. 4).

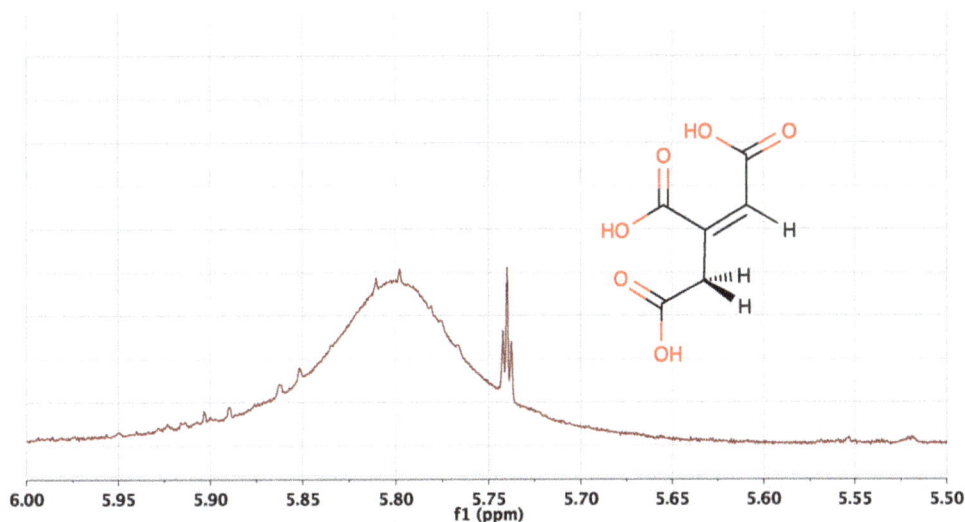

Fig. 3. An expansion of the 600 MHz ^{1}H NMR spectrum of the urine of pooled, male C57BL/6 mice at 15 weeks of age, in the region of the signal from the olefinic proton of *cis*-aconitic acid at ca 5.74 ppm. The signal is a triplet due to a long-range, 4-bond, $^4J_{H,H}$ coupling of ca 1.4 Hz to the two equivalent methylene hydrogens. The 1:2:1 nature of the triplet is clear, even though it is superimposed upon the very broad signal from urea at ca 5.80 ppm.

2.6. Second-order effects and strong coupling in ^{1}H NMR spectra

As the frequency separation in Hertz between coupled ^{1}H NMR signals decreases to less that ca 10 times the value of the coupling constant between them, distortions to expected multiplet peak intensities start to occur in the spectra. The spins are said to exhibit 'strong coupling', or to be in a second-order system. The spectra take on appearances that are different from those of systems that exhibit 'weak coupling' or are in first-order systems. Rather than being a problem, this is actually an aid to spectral interpretation and metabolite identification, as follows. In the simple case of two, non-equivalent hydrogen atoms coupling with one another, the intensity distortion is such that the doublets slope towards one another in an effect called 'roofing'. This is well illustrated in the spectra of the two, non-equivalent methylene protons in citric acid shown in

Fig. 1. The chemical shift difference between the resonances at 2.70 and 2.56 ppm is 0.14 ppm, which equates to ca 84 Hz at 600 MHz operating frequency. The $^2J_{H,H}$ coupling is ca 16.2 Hz and therefore the ratio of the frequency separation to the coupling constant is 84/16.2 = 5.2. This two-hydrogen spin system is formally designated AB: the two letters indicate that there are two distinct spins or hydrogen atoms involved in the coupling system; the closeness of the letters in the alphabet indicates that their chemical shifts are close in frequency. The roofing of the signals is clear to see in Fig. 1 and provides a way, without using 2D COSY NMR or any decoupling techniques, to determine that these hydrogens are coupled to one another; an important and often overlooked benefit of this feature. In a two-spin system that is first order, the nomenclature would be AX instead of AB to indicate that the two hydrogens are widely separated in chemical shifts, relative to the size of their mutual coupling.

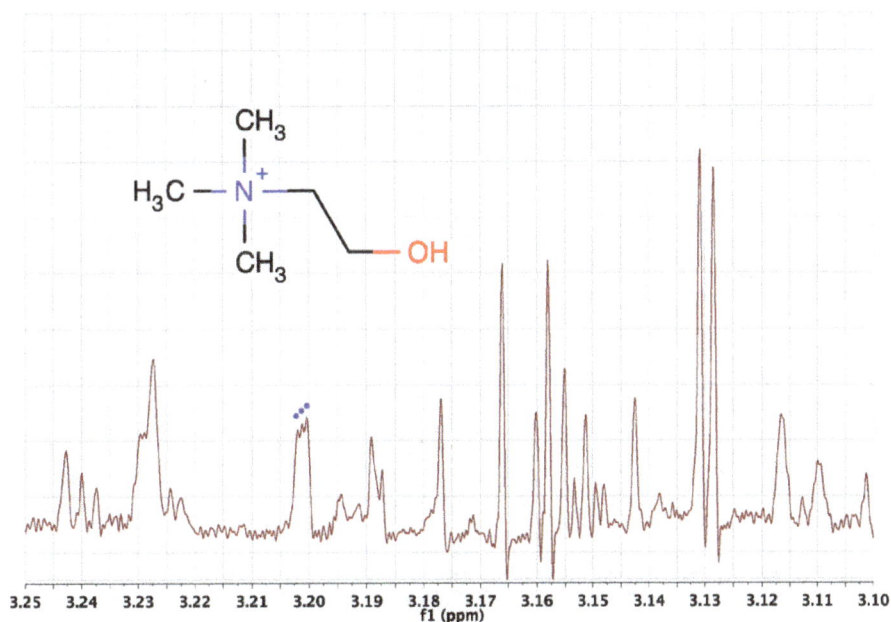

Fig. 4. An expansion of the 600 MHz ^{1}H NMR spectrum of the pooled urine of male C57BL/6 mice at 15 weeks of age, in the region of the signal from the methyl protons of choline (structure superimposed) at ca 3.20 ppm. The signal is a 1:1:1 triplet (dots) due to a 2-bond coupling of ca 0.6 Hz to the ^{14}N nucleus. Interestingly, the well-resolved doublet at ca 3.13 ppm is due to the methylene protons of cis-aconitic acid with $^4J_{H,H}$ coupling of ca 1.4 Hz (see also Fig. 3). The spectrum has been zero filled to 131,072 points and resolution enhanced by Gaussian multiplication, prior to Fourier transformation.

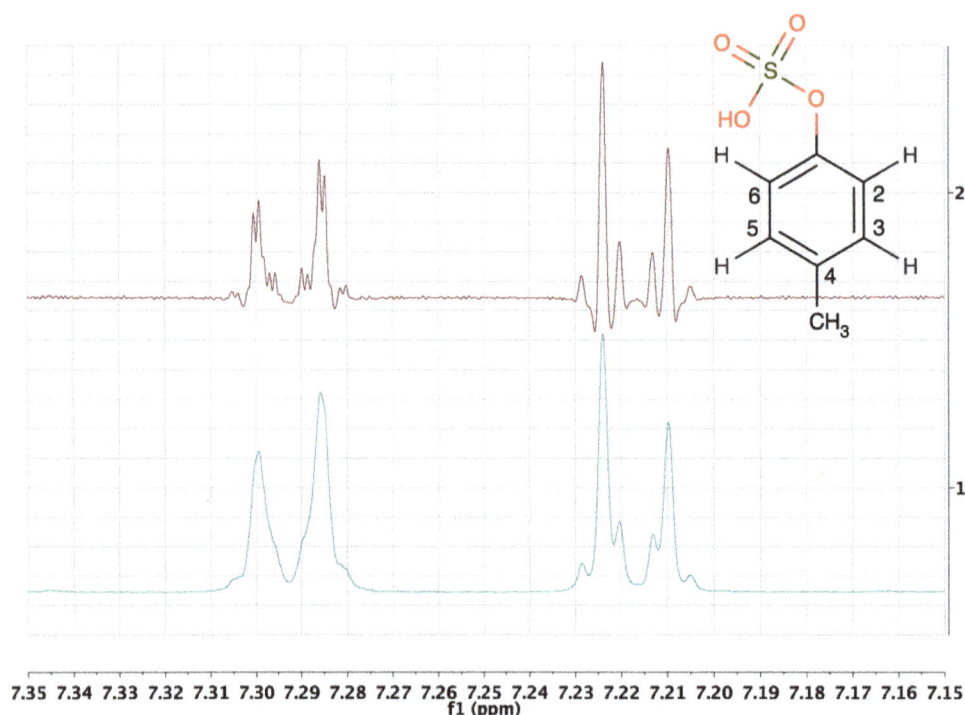

Fig. 5. Two versions of the 600 MHz ^{1}H NMR spectrum of an authentic sample of the metabolite *para*-cresol sulphate in deuterated phosphate buffer at pH 7.4, in the region of the signals from the aromatic hydrogens: 1) with a standard 0.3 Hz line-broadening and 2) resolution-enhanced using a Lorentzian to Gaussian transformation. The signal of the H2, H6 protons appears as a complex, second-order multiplet at ca 7.22 ppm, instead of a first-order doublet. The signal of the H3, H5 protons at ca 7.29 ppm displays additional complexity due to coupling to the methyl protons via a 4-bond coupling, in addition to the extra lines, clearly visible in this second-order system.

If the spin system is more complex, or the ratio of signal frequency separation to coupling constant becomes much smaller, the intensity distortions can become more significant and in extreme cases involving three spins or more, additional lines are seen in the resonances which are not always interpretable by first order analysis. This effect is commonly observed in the NMR signals for the aromatic hydrogens in symmetrically substituted benzene rings (Fig. 5).

For metabolites such as *para*-cresol sulphate, the phenomenon of magnetic non-equivalence appears [37]. The hydrogens on C2 and C6 are chemically equivalent by symmetry, as are those on C3 and C5.

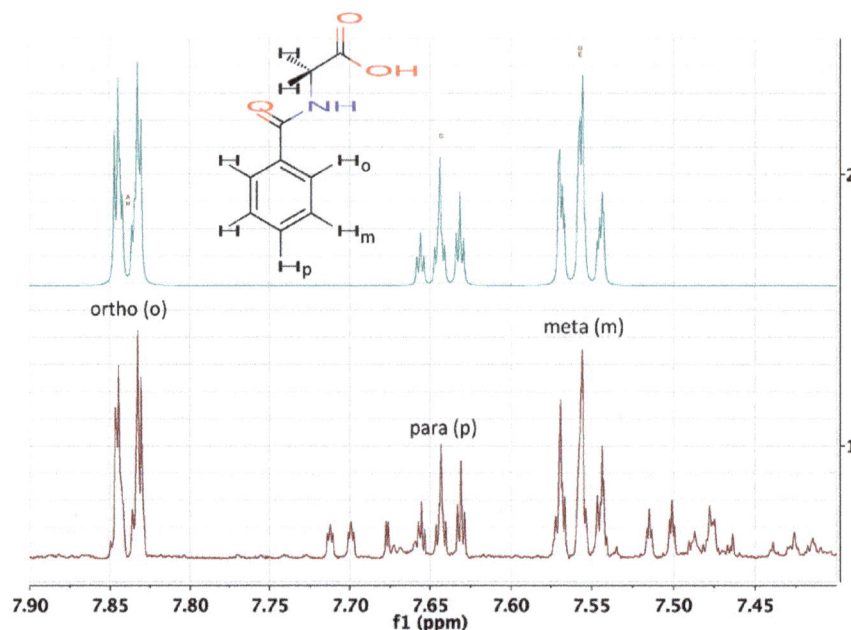

Fig. 6. 1) The 600 MHz ^{1}H NMR spectrum of the urine of a 30-week-old, male, flavin mono-oxygenase 5 (FMO5) knockout mouse [61] in the region of the aromatic signals from hippuric acid (structure inset). The spectrum is resolution enhanced by Gaussian multiplication. 2) A spin simulation of the aromatic signals from hippuric acid using the MNova spin simulation function. A good approximation to the complex, second-order signals was obtained. The complexity of the two ortho and two meta hydrogen signals is due to the fact that whilst these hydrogens are chemically equivalent (within each pair), they are magnetically non-equivalent and are part of a five hydrogen AA'BB'M spin system (see Section 2.6). Signals from 3-indoxyl sulphate and other metabolites are present in the real spectrum (1).

However, these pairs of hydrogens are NOT magnetically equivalent. The reason for this is as follows: H2 is *ortho* to H3 and has a 3-bond coupling to it. By contrast, the chemically equivalent proton H6 is *ortho* to H5 and *para* to H3. Thus, in terms of their nuclear magnetic interactions, these hydrogens are non-equivalent and this has consequences. The frequency difference between the signals of H2 and H6 is 0 Hz by definition and they are coupled by a favourable, long-range 'W' coupling over 4-bonds. Thus, the frequency separation to coupling ratio is 0, no matter what the value of the coupling constant and the resultant spectra are second-order [37]. This spin system is designated AA'BB', where A and A' represent H2 and H6 and the apostrophe signifies a chemically equivalent but magnetically non-equivalent nucleus. B and B' are H3 and H5 and the closeness of the letters in the alphabet is deliberate and signifies the closeness of the chemical shifts of these two groups of spins. In these extreme cases, additional lines appear in the spectra and the resonance patterns may not be readily interpretable by first order analysis. Instead of a simple pair of doublets, as might be expected, a complex pattern appears (Fig. 5). Typically, a computational, spin simulation program is used to calculate the spectrum and this is now a routine procedure. An important point to appreciate is that it may not be straightforward to extract chemical shifts or coupling constants from second-order spectra without spin simulation: see Section 2.10 and Fig. 6.

2.7. The half bandwidth of NMR signals

Another feature that provides information on the structure and the dynamics of metabolites is the half bandwidth of their signals. The half bandwidth, $\Delta\nu_{1/2}$ of a signal is related to the real spin–spin relaxation time of the hydrogen giving rise to that signal according to Eq. (1):

$$\Delta\nu_{1/2} = 1/\pi.T_{2}* \tag{1}$$

... where $T_{2}*$ is the real spin–spin relaxation time that takes into account underlying molecular relaxation processes, plus the effect of field inhomogeneities and the influence of factors such as the presence of paramagnetic species (including dissolved oxygen gas) in the sample. $T_{2}*$ can be shortened by interaction with quadrupolar spins, such as ^{14}N and by chemical exchange. In the case of chemical exchange between two forms of a metabolite, A and B, the lifetime of a spin in species A is necessarily limited to the lifetime of species A, as a maximum. Exchange-broadening of the signals will occur when the exchange rate in Hertz between forms A and B is of the same magnitude as the chemical shift difference in Hertz between the corresponding hydrogens in A and B. The broadening effects of exchange with water, quadrupolar relaxation and unresolved couplings to ^{14}N can be quite large, as can be seen in Fig. 3, where the hydrogen signal from urea has a half bandwidth of ca 50 Hz, in contrast to the much narrower linewidth of the olefinic proton in cis-aconitic acid, where all of these effects are absent and consequently the non-exchanging hydrogens have much larger $T_{2}*$ values.

2.8. The integral of NMR signals

When NMR experiments are run with sufficient delay times in between the acquisition of each successive free induction decay, the nuclei under observation will enjoy close to full spin–lattice relaxation. Under these conditions, the signals will not be partially saturated [35], and the area of a methyl (CH_3) signal in a metabolite in a biofluid will be precisely three times that of a methyne (CH) signal in the same metabolite in the same sample. NMR spectroscopy is thus an inherently quantitative technique and this is a huge advantage for the conduct of metabonomics experiments. It should be noted however that most NMR-based metabolic profiling experiments do not achieve full relaxation with the delay times typically used. Even so, the situation is in stark contrast to MS-based profiling, where the intensities of signals from metabolites may be significantly suppressed or enhanced by the presence of other

metabolites in the sample [56] and internal reference standards are required in order to achieve quantitation.

Quantifying the level of a metabolite in a biological fluid such as urine, by ^1H NMR spectroscopy, can be very difficult, because of spectral crowding and spectral overlap, and great care is required either with line fitting or direct integration quantification approaches. However, when careful approaches are taken, the analytical precision of the methodology is high [12,57,58] and this is critical for the statistical analysis of the data and the reliable discovery of discriminating biomarkers: see Section 4.1.

2.9. The stability of NMR signals

Generally, the metabolic profile of a biological fluid is stable over a significant period of time at room temperature, and certainly stable enough for the acquisition of routine 1D and 2D ^1H NMR data. However, there are exceptions. Some biological fluids are inherently unstable. A good example of this is human seminal fluid, where, post-ejaculation, enzymatic reactions take place that cause the biochemical transformation of some metabolites [59]. In addition, if a sample such as animal urine, has been in contact with animal faeces at any stage, it will be microbiologically contaminated and potentially unstable. Bacterial growth in a urine sample, for instance, will result in the transformation of certain metabolites into new products, as the bacteria scavenge the biofluid for fuel sources. It is common practice to add anti-bacterial agents such as sodium azide [8,16] to inhibit the growth of the bacteria. However, in our experience, even in the presence of sodium azide at 9 mM, bacterially-mediated metabolite transformations can still occur in mouse urine if kept at room temperature for extended periods, and hence, signals will be unstable over time: the signals of fermentation substrates will decrease, whereas those of products will increase. See Section 4.6, Biochemical Transformation and *In Vitro* Fermentation of Biofluids to Aid Metabolite Identification. A major improvement in this area has occurred with the development of cooled sample changers, such as the SampleJet system from Bruker Corporation (Billerica, Massachusetts, USA), that keeps queued samples at 4 C prior to their insertion into the NMR magnet, thus minimising sample instability.

2.10. Interpretation of 1D ^1H NMR spectra and metabolite identification

Metabolites that are present at relatively high concentrations or that have distinctive signals in relatively uncrowded spectral regions can be identified by inspection from a simple 1D ^1H NMR spectrum. This can be done manually by the spectroscopist interpreting the data, or with the assistance of software such as Chenomx NMR Suite (Chenomx, Edmonton, Canada), which has the advantage of a database of standard metabolite spectra at a variety of magnetic field strengths and a variety of pH values [60]. Obvious metabolites include citric acid (see Fig. 1) where the (somewhat variable) chemical shifts and large 'roofed' geminal couplings of the methylene protons are unmistakeable. Another easily identifiable metabolite is hippuric acid, whose second-order aromatic proton resonances between 7.9 and 7.4 ppm provide an unmistakable 'fingerprint' for identification (Fig. 6).

Certain other metabolites have distinctive singlet signals at characteristic chemical shifts, such as the methyl hydrogens of methylamine, dimethylamine and trimethylamine at ca 2.61, 2.73 and 2.88 ppm respectively. However, little information is present in the 1D ^1H NMR spectrum of these metabolites: just one singlet resonance. Hence, it is advisable to check the assignments of these types of resonances using a 2D ^{13}C, ^1H HSQC experiment to verify that the methyl carbons have the expected chemical shifts of ca 27.7, 37.6 and 47.6 respectively for methylamine, dimethylamine and trimethylamine. Note the uniform ca 10 ppm increase in methyl carbon chemical shift as each methyl group is added, due to the additive, two-bond or beta substituent effect.

The identification of metabolites present at relatively low levels, or that have signals that are partially or completely overlapped, will be difficult by 1D NMR methods and the use of two-dimensional NMR spectroscopic methods is required. In Section 4.7, we will review how much information is required in order to consider the identification of a known metabolite confident.

3. Molecular structure information from two-dimensional (2D) NMR spectroscopy

3.1. Introduction to 2D NMR spectroscopy

In a 1D NMR spectrum the NMR signals are acquired as a function of a single time variable (t2) in a free induction decay (FID), typically over 65,536 data points at a ^{1}H frequency of 600 MHz. This FID arises from the induction by an electric current in the receiver coils of the NMR probe by the excited nuclear magnetisations: there is no emission event detected in NMR. Fourier transformation of this FID gives rise to the conventional 1D ^{1}H NMR spectrum in which NMR signal intensity (y-axis) is plotted as a function of chemical shift (x-axis). By contrast in a 2D NMR experiment, a second time dimension (t1) is artificially created by the deliberate incrementing of a time delay, known as the evolution time, between two of the radiofrequency pulses in the pulse sequence used. An FID is collected for each of m values of the evolution time, such that at the end of the experiment, m x FIDs have been collected, each containing n data points. Double Fourier transformation of this data set over both t2 and t1 results in a single 2D NMR spectrum in which signal intensity (z-axis) is plotted as a function of two orthogonal signal frequency axes; f2 and f1 corresponding to t2 and t1 in the time domain (x and y respectively). The spectra are typically displayed as contour plots where signal intensity is represented by contour lines, in much the same way that the heights of mountains and hills are represented on maps.

We shall not go into the details of the design of the 2D NMR pulse sequences, nor the analysis of how those pulse sequences give the resulting spectra, as many excellent reference works are available in this area [35,37].

3.2. 2D ^{1}H J-resolved (JRES) NMR spectroscopy

The 2D ^{1}H J-Resolved NMR Spectroscopy (JRES) experiment is one of the simplest 2D NMR experiments and one of the most useful for the analysis of the complex ^{1}H NMR spectra of biological fluids [21,22,62]. The experimental radiofrequency pulse sequence is simply: RD - 90^0_H - t1/2 - 180^0_H - t1/2 - FID, where RD is a relaxation delay. The second proton pulse (180^0_H) occurs in the middle of the incremented evolution time (t1). In the resulting 2D ^{1}H NMR JRES spectrum, the chemical shifts run along the first frequency dimension, f2, as normal, and homonuclear coupling constants are modulated (spread out) across a second frequency dimension, f1. For simple, first order spin systems, no new signals are created: the existing signals are just spread out across two frequency dimensions instead of one. This has a tremendous effect in reducing signal overlap in crowded spectral regions. The spectra are typically tilted by 45^0 so that all the signals of a homonuclear multiplet appear at the exact same chemical shift. The projection of the 2D spectrum onto the chemical shift dimension, f2, is then effectively a broadband proton-decoupled proton NMR spectrum, in which each ^{1}H resonance is a singlet. It is important to note that heteronuclear couplings are unaffected by the [1]H 180^0 pulse in the 2D ^{1}H NMR JRES experiment and these are *not* modulated across the second dimension of the 2D spectrum [55].

The tremendous improvement in signal resolution by spreading the NMR signals out across a second dimension is clearly illustrated in Fig. 7. In the 1D ^{1}H NMR spectrum of the urine of an FMO5 knockout mouse [61], the triplet methyl signal for *N*-butyrylglycine (three blue circles at 0.926 ppm) is overlapped with the doublet methyl signal for isovaleric acid (two red squares at 0.916 ppm). By contrast in the 2D

^{1}H JRES NMR spectrum, these signals are completely resolved from one another.

The simple interpretation of 2D ^{1}H JRES NMR spectra only applies for first order systems in which there is weak coupling. If strong coupling exists (a second-order system) then artefacts can appear in the spectra [63]. This occurs because in a strongly coupled system the second ^{1}H pulse (a 180^0 or π pulse) will cause not just the modulation of the signals of a homonuclear-coupled spin across the second dimension, according to the size of its spin couplings, it will also cause the mixing of the transitions or signals between coupled spins, such as would normally occur in a chemical shift correlation experiment such as COSY (via the second 90^0 pulse). Thus, in a simple two hydrogen AB spin system such as citric acid, the two A transitions (doublet) become mixed with the two B transitions and in a tilted 2D ^{1}H JRES NMR spectrum, signals appear in the 2D spectrum at chemical shifts where there are no hydrogens! It is very important to recognise these 'artefacts' in order to avoid mis-assigning the spectra to non-existent metabolites with unreal J values! Fig. 8 shows an example of this feature for citric acid itself: the 2nd order signals in the 2D ^{1}H JRES NMR spectra are marked with stars.

Awareness of the origin of these signals allows chemical shift correlation information to be extracted from the 2D ^{1}H JRES NMR spectrum, so these artefacts can have real utility in spectral assignment and metabolite structure elucidation!

An important use of 2D ^{1}H JRES NMR spectra is to establish the magnitude of the coupling constants for the ^{1}H NMR signals of particular hydrogen atoms. This can readily be done even when the metabolites are at low levels and the signals are difficult to see in the 1D ^{1}H NMR spectra. For example, Fig. 9 shows an expansion from the 2D ^{1}H JRES NMR spectrum of the urine of an FMO5 KO mouse at 30 weeks of age. The signals at 2.003 and 1.845 are from the two methylene hydrogens at C3 in 2S-hydroxyglutaric acid (HMDB00694). The chemical shifts of the two hydrogens are close to the values reported in the HMDB (1.985 and 1.825 respectively) but the assignment of the two hydrogens is much more secure if the coupling constants can also be shown to match. In this case the 1D ^{1}H NMR FID of the authentic metabolite was downloaded from the HMDB and reprocessed. This showed that the line separations in the multiplets at 1.985 and 1.825 in the authentic metabolite were identical to those observed at 2.005 and 1.845 in the 2D ^{1}H JRES NMR spectrum of the urine of an FMO5 KO mouse at 30 weeks age, thus helping confirm this metabolite identification.

3.3. 2D ^{1}H chemical shift correlation spectroscopy (COSY)

The 2D ^{1}H chemical shift correlated spectroscopy (COSY) NMR experiment is a workhorse of metabonomics analyses for the identification of the metabolites in biological samples. Many variants of the 2D ^{1}H COSY NMR experiment exist [35] but all variants provide information on which hydrogens are spin–spin coupled together, and this is vital for metabolite structure identification. The basic pulse sequence is: RD - 90^0_H - t1 - 90^0_H - FID, where RD is a relaxation delay. The first 90^0 pulse excites all the nuclear spins: the second 90^0 pulse causes coherence transfer between the magnetisations of hydrogens which are spin-coupled to one another. The reason for the importance of the COSY experiment can be best illustrated with an example. If we observe a methyl doublet signal in a urine sample at 1.34 ppm and that doublet signal has a coupling constant of 6.9 Hz, we could infer that that signal originated from lactic acid. However, if a 2D ^{1}H COSY NMR spectrum of that urine sample indicated that the methyl doublet at 1.34 ppm was spin-coupled to a methyne proton at 4.13 ppm, that would be much stronger evidence that the methyl signal was indeed from lactic acid. The probability of known metabolite mis-identification decreases strongly with each successive connected spin matched to the corresponding signal in the spectrum of the authentic metabolite.

It is typical to run quick 2D ^{1}H COSY NMR spectra with low digital resolution and often low sensitivity. This can be appropriate for the rapid analysis of pure chemical compounds but is not appropriate for

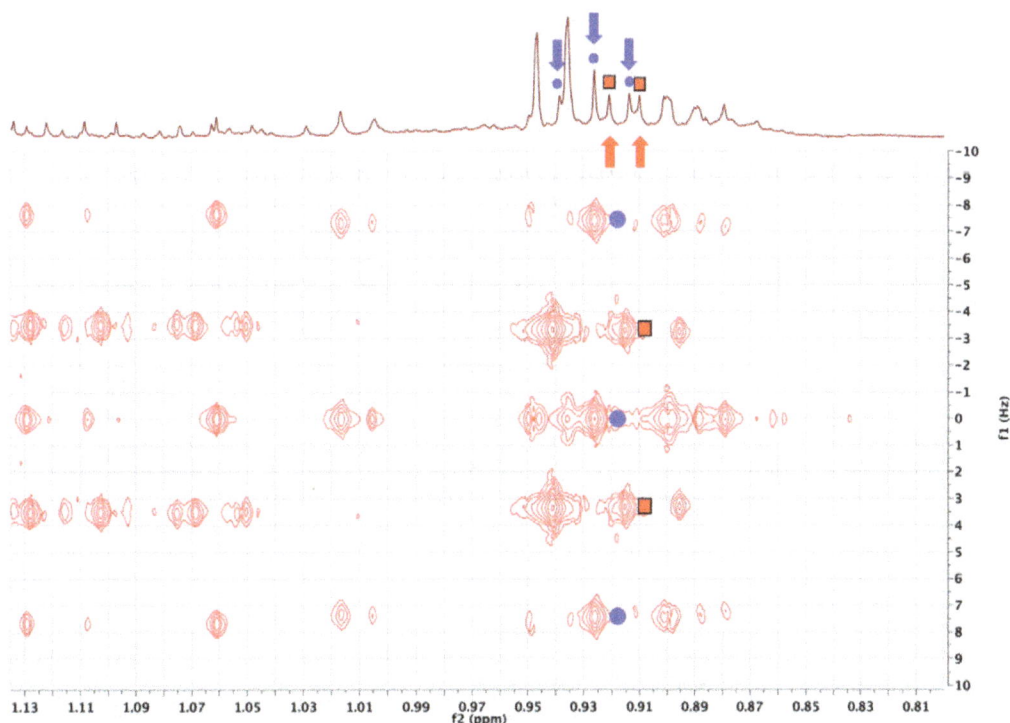

Fig. 7. The low frequency region of the 600 MHz 2D ^{1}H J-resolved NMR spectrum of the urine of a male, 30-week-old, FMO5 knockout mouse [61] displayed as a contour plot underneath the corresponding 1D ^{1}H NMR spectrum. The overlapping signals from the triplet methyl group of N-butyrylglycine (0.926 ppm, three blue circles and downward arrows) and the doublet methyl group of isovaleric acid (0.916 ppm, two red squares and upward arrows) are completely resolved in the 2D JRES NMR spectrum. The spectrum is tilted by 45^0, so that all the signals of each multiplet appear at the same chemical shift, and it is symmetrised.

metabonomics studies, as it results in the limited observation of hydrogen-to-hydrogen connectivities for major metabolites over 2-bonds or 3-bonds only. If the experiment is run at higher sensitivity and resolution, much more information can be gleaned, from a larger number of metabolites. Acquiring 2D COSY data at higher resolution can cost time, but this would not be done for every sample in a large metabonomics experiment. A high resolution COSY NMR spectrum would only be obtained on a handful of samples that are representative

of the different groups in the study, with the express purpose of aiding metabolite identification.

Using traditional methodology, a high resolution COSY spectrum might take several hours to acquire. For example, Fig. 10 shows an expansion of the 600 MHz 2D ^{1}H COSY NMR spectrum of the pooled urine from two FMO5 KO mice [61] at 60 weeks age. This experiment was acquired with spectral widths in f1 and f2 of 9578 Hz, and 4096 points in the FID (t2) for 512 values of the evolution time (t1): the

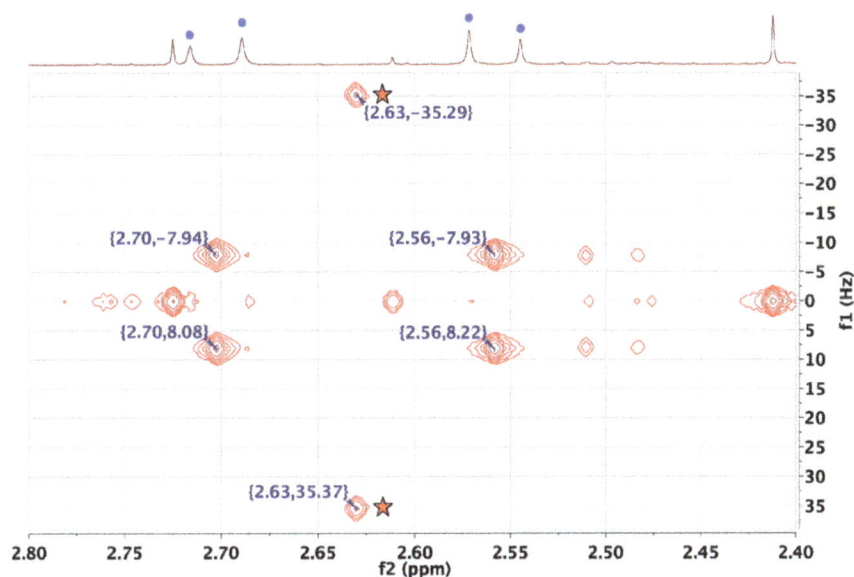

Fig. 8. An expansion of the 600 MHz 2D ^{1}H J-resolved NMR spectrum of the urine of a male, 30-week-old, FMO5 knockout mouse [61] in the region of the AB resonances from citric acid at ca 2.70 and ca 2.56 ppm (four dots in 1D spectrum), displayed as a contour plot underneath the corresponding 1D ^{1}H NMR spectrum. The spectrum is tilted by 45^0, so that all the signals of each multiplet appear at the same chemical shift, and symmetrised. The signals labelled with stars, appearing at ca 2.63 ppm, exactly in between the shifts of the two citrate signals are 2nd order effects caused by the mixing of transitions between the A and B spins by the 180^0 pulse, in the presence of strong coupling. As is clear from the 1D ^{1}H NMR spectrum, there are no real signals at 2.63 ppm!

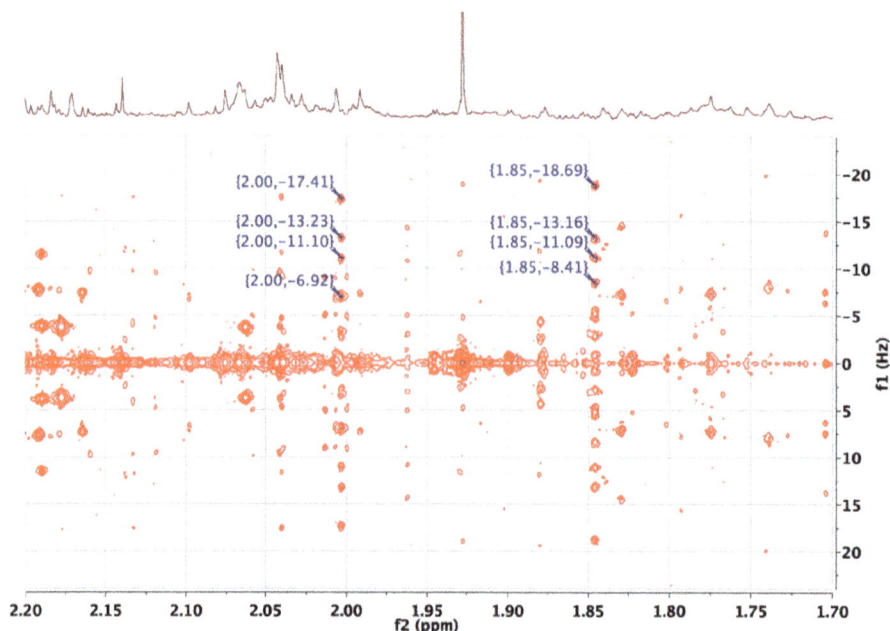

Fig. 9. An expansion of the 600 MHz 2D ^{1}H J-resolved NMR spectrum of the urine of a male, 30-week-old, FMO5 knockout mouse [61] in the region of the resonances from the C3 methylene hydrogens of 2S-hydroxyglutaric acid, displayed as a contour plot underneath the corresponding 1D ^{1}H NMR spectrum. The spectrum is tilted by 45^0, so that all the signals of each multiplet appear at the same chemical shift, and symmetrised. The peak picking allows a simple analysis of three of the four couplings that these hydrogens possess as 4.2, 6.3 and 10.5 Hz (2.003 ppm) and 5.5, 7.6 and 10.3 Hz (1.845 ppm). Note that these multiplets are invisible in the 1D ^{1}H NMR spectrum.

final spectrum was an 8192 by 2048 data matrix. The acquisition time was 0.428 s with a relaxation delay of 2 s, and 32 transients per increment of the evolution time, resulting in a total experiment time of just over 11 h, which is a significant investment of time. However, that additional time does allow correlations via small couplings over 4- to 6-bonds to be observed and these can be important for metabolite identification, as they enable connectivities to be established between parts of molecular structures isolated by so-called 'spectroscopically silent centres'. These silent centres are atoms with no hydrogens attached or no non-exchanging hydrogens. These spectroscopically silent centres

break up the chains of proton-to-proton connectivity in a metabolite that are important for metabolite identification by e.g. COSY NMR. In this case the silent centre is the pyridinium nitrogen, which has no hydrogens bonded to it.

When run at high resolution, the 2D ^{1}H COSY NMR spectrum can also be used to identify the multiplicity of signals that are completely buried in the 1D ^{1}H NMR spectrum, and even those that are buried in the 2D ^{1}H JRES NMR spectrum. For example, the signal for the C4H methyne proton of ketoleucine at 2.098 ppm was invisible in the 1D ^{1}H NMR spectrum (Fig. 11 top), or in the corresponding 2D ^{1}H

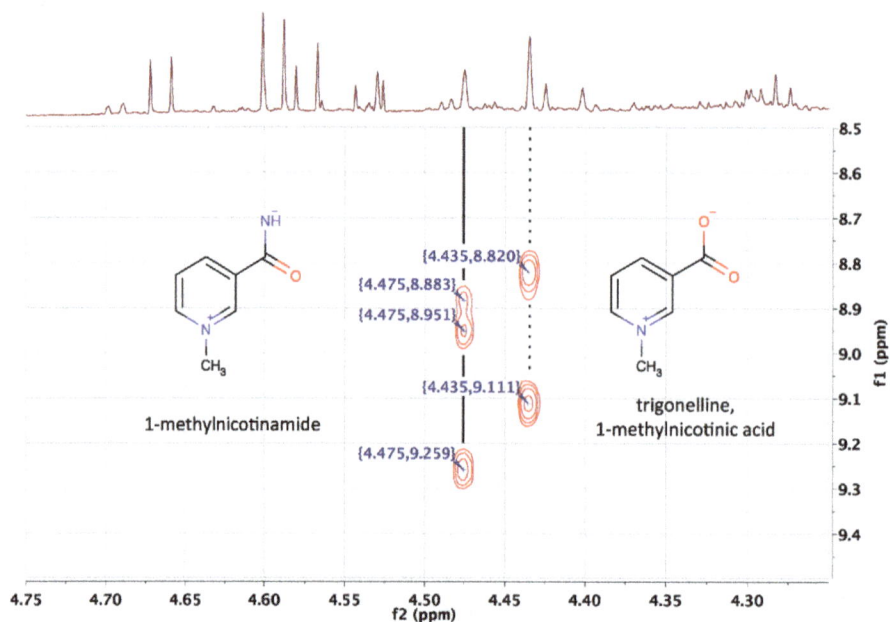

Fig. 10. An expansion of the 600 MHz 2D ^{1}H COSY NMR spectrum of the pooled urine of two male, 60-week-old, FMO5 knockout mice [61] in the region of the broad singlet methyl resonances from trigonelline at ca 4.435 and 1-methylnicotinamide ca 4.475 ppm, displayed as a contour plot underneath the corresponding resolution-enhanced 1D ^{1}H NMR spectrum. Trigonelline displays cross-peaks due to long-range, 4-bond coupling from the methyl protons to the H2 (9.111) and H6 (8.820) protons ortho to the pyridinium nitrogen. 1-methylnicotinamide displays the same cross-peaks to H2 (9.259) and H6 (8.951), but in addition, displays a clear and remarkable cross-peak via six-bond coupling to H4 (8.883). The ability to connect the methyl shift with the pyridinium proton shifts in this way can assist metabolite identification enormously.

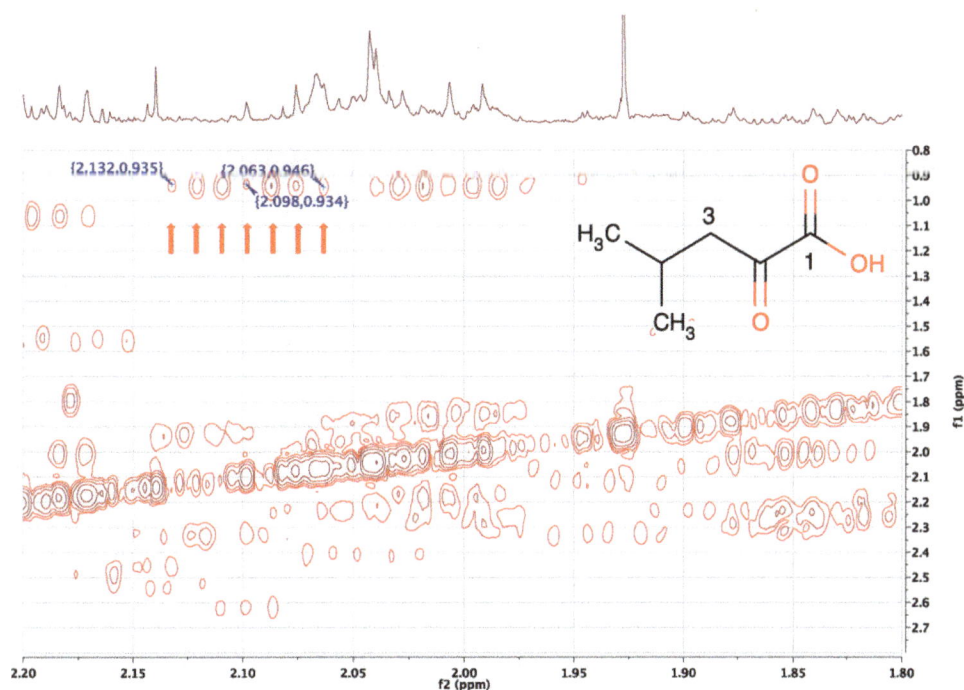

Fig. 11. An expansion of the 600 MHz 2D ^{1}H COSY NMR spectrum of the urine of a male, 30-week-old, FMO5 knockout mouse [61] highlighting with 7 arrows the cross-peak from the C4H proton of ketoleucine (structure inset) at 2.098 to the equivalent C5 and C6 methyl groups at 0.941 ppm. The signals from ketoleucine at 2.098 are not visible either in the 1D ^{1}H NMR spectrum (top), or in the 2D ^{1}H J-resolved NMR spectrum of the same sample (see Fig. 9) but the identification is confirmed from this high-resolution COSY spectrum. The seven cross peaks marked are the most intense peaks of the 9-line, pseudo-nonet, triplet of septets, the two outside lines of which are too weak to observe. See text for details.

J-resolved NMR spectrum of the same sample (Fig. 9) but its identification is confirmed from the high-resolution COSY spectrum (Fig. 11).

The C4H peak of ketoleucine is a triplet of septets which appears as a pseudo-nonet, as the coupling from C4H to the C3H$_2$ group (7.0 Hz) is very similar to the spin-coupling to the six equivalent methyl group protons (6.7 Hz). The two weak outside lines of the pseudo-nonet are weak and difficult to observe even in the authentic reference standard (BMRB, BMSE000383; HMDB00695 (caution the HMDB 1D ^{1}H NMR was run at pH 3! [accessed 12 September 2015]). The high resolution COSY spectrum shown in Fig. 11 allowed the measurement of the frequency separation of highest and lowest frequency lines observed in the multiplet at 2.098 as 41.4 Hz, which corresponded well (COSY digital resolution in f2 = 0.73 Hz) to the separation in authentic material: 40.8 Hz in BMSE000383, thus providing further confidence for the assignment of this cross-peak.

The COSY experiment should always be run with good resolution in the FID (t2) as that resolution is essentially 'free'. The increase in the acquisition time that this costs can be counterbalanced by a corresponding decrease in the relaxation delay between successive transients. Increasing the resolution across the second dimension, t1, does cost however, as does increasing the number of transients per value of the evolution time, and it is here that non-uniform sampling (NUS) methods and FAST NMR methods may lead to decreases in acquisition times in 2D NMR experiments for metabolite profiling in the future. Preliminary studies show promise and we await developments in this area with interest [64].

3.4. 2D ^{1}H total correlation spectroscopy (TOCSY)

The 2D ^{1}H TOCSY NMR experiment, sometimes called 'homonuclear Hartmann–Hahn spectroscopy' (HOHAHA), is a relatively simple NMR experiment often used in conjunction with the COSY experiment to elucidate further structural information on small molecules of interest [65]. TOCSY provides similar information to a COSY experiment with regards to directly coupled hydrogens, but provides further structural

information by identifying larger, interconnected groups of *indirectly* spin-coupled hydrogens.

In comparison to the COSY sequence, the second 90°$_H$ pulse is replaced by a spin-lock field, applied for 10 s of milliseconds, which can be considered to behave like a series of 180° $_H$ pulses. The spin-lock field eliminates chemical shifts during its application, but does not affect the scalar coupling. Due to the elimination of chemical shift differences in the spin-lock period, the spins are in a strong coupling regime, lose their individual identity and undergo magnetisation or coherence transfer. The magnetisation transfer that takes place is governed by the length of the spin-lock periods. Short spin-lock periods (20–100 ms) yield cross peaks for directly coupled spins. With longer spin-lock times (100–300 ms), coherence will be transferred more remotely down chains of spin-coupled hydrogens. Thus, if we have a spin system AMX, where A is coupled to M and M is coupled to X, but A is not coupled to X, two situations can arise in the TOCSY experiment. For short spin-lock periods, correlations will be seen between the chemical shifts of both A and M and of M and X. For longer spin-lock periods, cross-peaks will also be observed between A and X, even though they are not directly coupled.

A good example of this can be shown in *N*-butyrylglycine (HMDB00808) which has an alkyl chain three carbons long. In a COSY experiment, the protons from the terminal C4-methyl group (0.926 ppm) would only have cross peak correlations with the adjacent C3-methylene protons (1.617 ppm). However, further structural information for *N*-butyrylglycine is provided (Fig. 12, see also Figs. 2 and 7) when a cross peak is observed at the resonance of the remote C2-methylene protons (2.279 ppm).

A one-dimensional version of the TOCSY experiment is also available. The experiment involves the selective excitation of a signal followed immediately by the application of the spin-lock field to effect coherence transfer, essentially observing a slice of a 2D TOCSY. Chemical shift selective filter TOCSY (CSSF-TOCSY) uses excitation sculpting techniques with pulse field gradients to selectively excite overlapping proton signals with tiny chemical shift differences, enabling reliable extraction of coupling constants, important in metabolite identification.

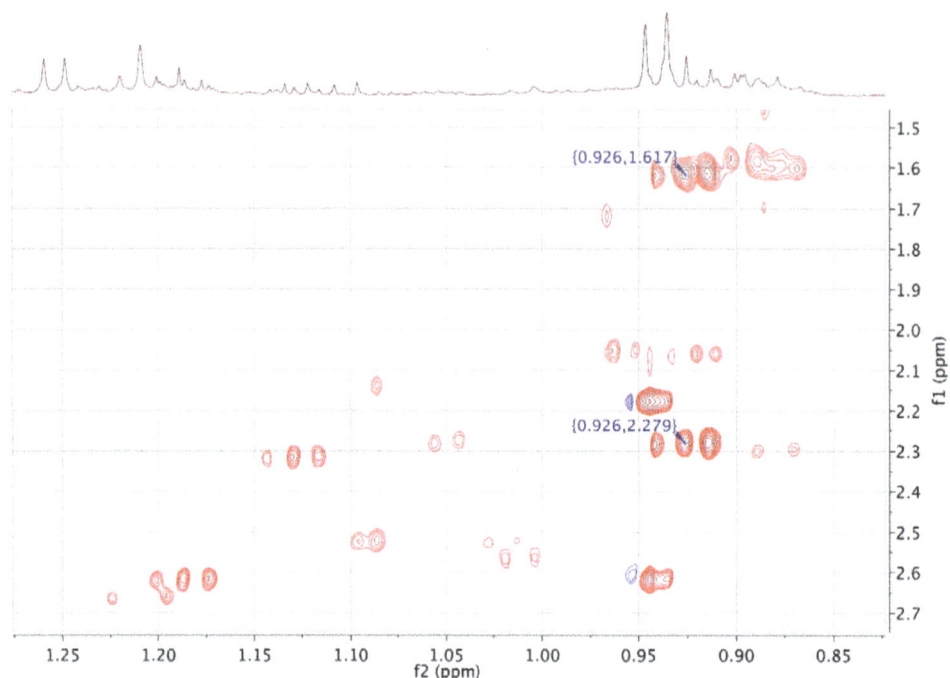

Fig. 12. An expansion of the 600 MHz 2D ^{1}H TOCSY NMR spectrum of the urine of a 30-week-old male C57BL/6 mouse. The cross peaks marked originate from the alkyl chain connectivities of *N*-butyrylglycine, from the terminal methyl group (C4). The cross peak marked at 0.926, 1.617 ppm represents a direct correlation from the C4 methyl protons to the adjacent C3 methylene group, equivalent to the cross-peak that would be observed in a 2D ^{1}H COSY experiment. Additional metabolite identification information is provided in this TOCSY experiment however, with the cross-peak at 0.926, 2.279 ppm establishing a connection between the C4 methyl protons and the C2 methylene group, even though there is no observable coupling between them.

3.5. 2D ^{13}C, ^{1}H Heteronuclear Single Quantum Correlation (HSQC) NMR spectroscopy

The 2D ^{13}C, ^{1}H Heteronuclear Single Quantum Correlation (HSQC) NMR Spectroscopy experiment is another fundamental experiment for metabolite identification. The experiment operates by correlating the chemical shifts of hydrogens with the chemical shifts of carbon-13 nuclei to which they are directly attached via $^1J_{C,H}$. The reason that this experiment is important is two-fold. Firstly, it introduces a completely new and orthogonal dimension beyond ^{1}H NMR to obtain information on the structure of metabolites: that available from the C-13 NMR chemical shift. Secondly, the chemical shifts of the carbon-13 nucleus extend over about 220 ppm for most metabolites: this is ca 20 times the range of proton NMR chemical shifts (ca 11 ppm), and these ^{13}C NMR shifts thus provide a much more sensitive response to minor changes in metabolite structure than does the ^{1}H NMR chemical shift: see Section 2.2.

Many variants of the 2D ^{13}C, ^{1}H HSQC NMR experiment are in current usage [35] and its successful execution does have some challenges. All variants of this experiment use ^{1}H detection for high sensitivity and thus, not only must the enormous proton signals from water be suppressed, but also all of the signals from hydrogen atoms that are bound to carbon-12 nuclei, which is 99% of the hydrogens in each metabolite. Fortunately, the availability of high performance digital NMR spectrometers and gradient pulses has made the experiment routine. Indeed, new variants suitable for metabolite profiling in biofluids are now available that even provide carbon multiplicity editing as well. In these experiments, the 2D ^{13}C, ^{1}H HSQC NMR spectrum not only displays the cross-peaks due to $^1J_{C,H}$ correlations, but also edits the cross-peaks in a phase-sensitive fashion so that the cross-peaks due to methyl (CH$_3$) and methyne (CH) moieties are of opposite phase to those of methylene groups (CH$_2$). This provides tremendous power for the assignment of signals in crowded regions of the ^{1}H NMR spectra of a biofluid: see Fig. 13.

The multiplicity-edited, 2D 13C, ^{1}H HSQC NMR spectrum in Fig. 13, readily distinguishes the red, positive cross-peaks (asterisked) arising from the methyl groups of creatinine (3.04, 32.98), creatine (3.04, 39.71) and dimethylamine (2.73, 37.36) from the blue, negative cross-peaks (no asterisks) due to the methylene groups in cis-aconitic acid (3.14, 46.34), 2-ketoglutaric acid (3.02, 38.63 and 2.45, 33.41) and succinic acid (2.42, 36.79). This experiment is a tremendous aid to the correct assignment of complex biofluid NMR spectra.

3.6. 2D ^{13}C, ^{1}H Heteronuclear Multiple Bond Correlation (HMBC) NMR spectroscopy

The 2D ^{13}C, ^{1}H Heteronuclear Multiple Correlation (HMBC) NMR Spectroscopy experiment [35] is another critical experiment in the identification of metabolites using NMR methods. The key reason for its importance is that it enables the establishing of connectivities between the parts of a metabolite's structure that are separated from one another by quaternary carbons or heteroatoms with no slow-exchanging, attached hydrogens. These are the so-called 'spectroscopically silent centres' mentioned earlier. The problem is that these silent centres interrupt the chains of proton-to-proton connectivity between regions of protonated carbons, resulting in isolated fragments of structure that may not be easy to piece together. To take a simple example, in the molecules cis- and trans-aconitic acid, the methylene moiety is separated from the olefinic proton by a quaternary carbon. In this case the HMBC experiment can help to connect the two fragments of protonated carbon structure together by establishing connectivities between hydrogens and carbon separated by two or three bonds (Fig. 14).

The HMBC experiment is critical for establishing connectivities between regions of protonated carbon structure when they are separated by quaternary carbons or heteroatoms. Although relatively insensitive, the HMBC experiment is sometimes the only way to obtain this information, if it is not available from alternatives such as high resolution COSY.

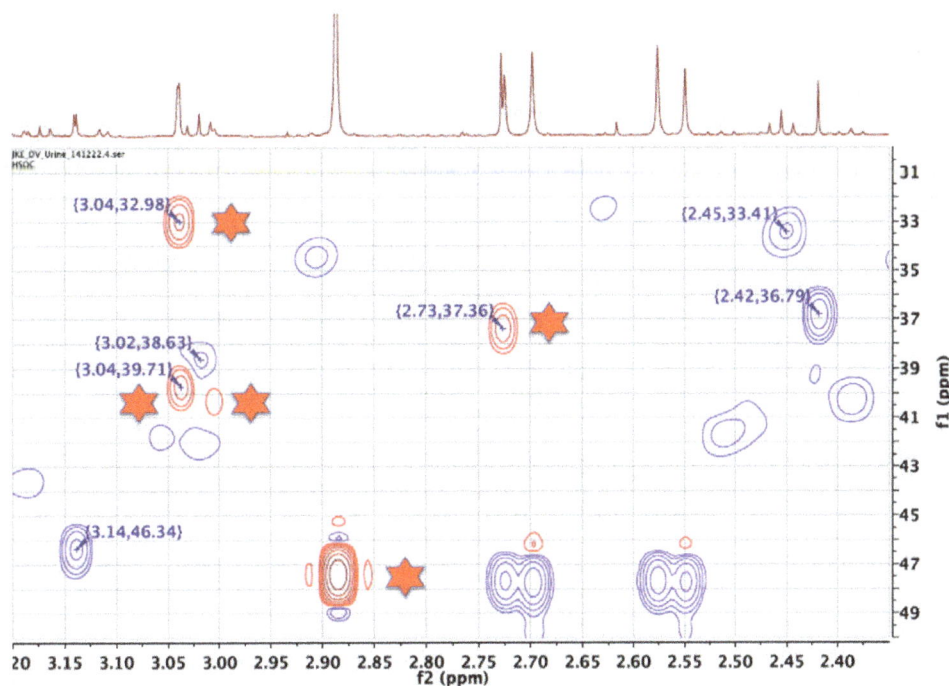

Fig. 13. An expansion of the 600 MHz, multiplicity-edited, 2D 13C, ¹H HSQC NMR spectrum of the pooled urine of 60-week-old, male, FMO5 knockout mice [61], displayed as a contour plot underneath the corresponding resolution-enhanced 1D ¹H NMR spectrum. In this phase-sensitive plot, positive peaks are represented by red contours (asterisked) and negative peaks by blue contours (no asterisks). See text for further explanation.

4. Metabolite identification

In this section we will bring together information obtained from 1D and 2D NMR experiments, together with information from metabolite databases and other sources to achieve metabolite identification and we will review methods for assessing the confidence in those metabolite identifications. There are essentially three strands to this activity:

1) the use of statistical methods to determine which NMR signals in a particular study are statistically significantly discriminating between groups of subjects in the study, or otherwise important, and therefore require identification and assignment, 2) the structure elucidation of novel metabolites, not previously described and 3) the structure confirmation of known metabolites. Some authors [28,66] have described novel metabolites as unknown unknowns and known metabolites as

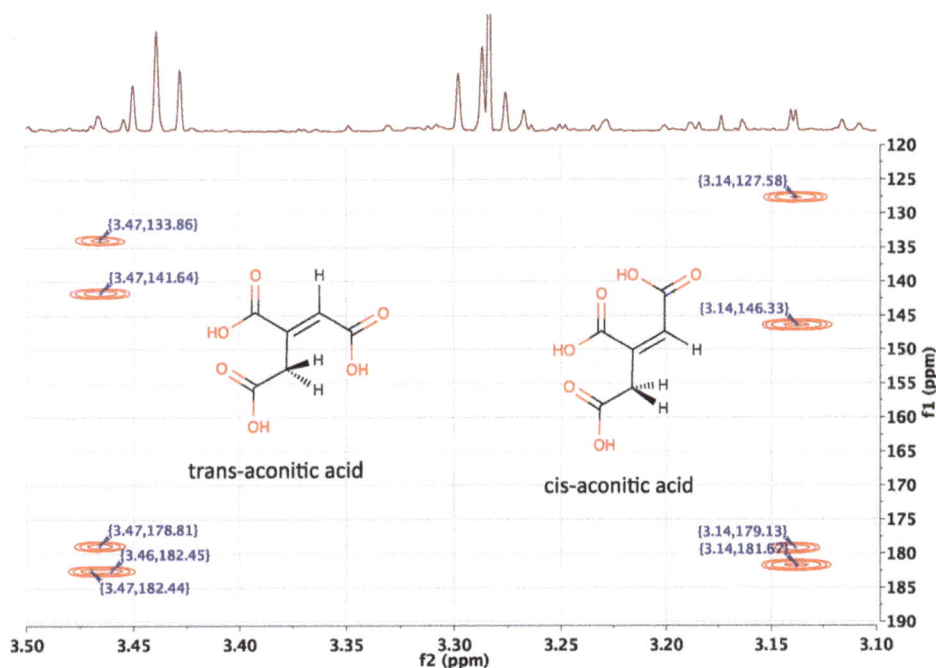

Fig. 14. An expansion of the 600 MHz 2D ¹³C, ¹H HMBC NMR spectrum of the pooled urine of male, 60-week-old, FMO5 knockout mice [61], displayed as a contour plot underneath the corresponding resolution-enhanced 1D ¹H NMR spectrum in the region of the signals from the methylene protons of cis-aconitic acid (3.14) and trans-aconitic acid (3.47 ppm). The methylene protons display all four possible 2- and 3-bond hydrogen-to-carbon connectivities, to both adjacent carboxylic acid carbons (178.8, 182.4 ppm, trans- and 179.1 and 181.7 ppm, cis-isomer) plus connections to the quaternary and protonated olefinic carbons at 141.6 and 133.9 (trans-) and 146.3 and 127.6 ppm (cis-isomer), respectively, thus establishing connectivities between the two regions of protonated carbon structure isolated from each other by the quaternary olefinic carbon.

known unknowns, but this language is confusing and unhelpful: we will retain the clear and simple distinction between novel metabolite structure *elucidation* and known metabolite structure *confirmation or identification*, that has been used in molecular structure studies by NMR spectroscopy for decades.

4.1. Identification of significant metabolites, or biomarkers, using multivariate statistics

The main objective in metabonomics is to extract relevant information from the large multivariate data sets. To this end pattern recognition (PR) and related multivariate statistical approaches can be used to discern meaningful patterns and identify metabolic signatures in the complex data sets that are of diagnostic or other classification value. A wide range of statistical methods is available today ranging from unsupervised methods, such as, principal component analysis (PCA), [67] or hierarchical clustering (HCA) [68], to supervised approaches like partial least squares (PLS) [69], partial least squares discriminant analysis (PLS-DA) and orthogonal partial least squares discriminant analysis (OPLS-DA) [70].

PCA is the most common technique in multivariate analysis that reduces the dimensionality of data and provides an unbiased overview of the variability in a dataset. In this approach samples are clustered based on their inherent similarity/dissimilarity with no prior knowledge of class membership. PCA represents most of the variance within a data set using a smaller set of variables, so-called principal components (PCs). Each PC is a weighted linear combination of the original variables, and each consecutive PC is orthogonal to the previous PC and describes the maximum additional variation in the data set that is not accounted for by the previous PCs. The results of a PCA are generally reported in terms of component scores, and loadings. In a scores plot, each point corresponds to a sample spectrum. Scores plots provide an overview of all samples and enable the visualisation of groupings, trends and outliers. A loadings plot illustrates which variables have the greatest contribution to the positioning of the samples on the scores plot and are therefore responsible for any observed clustering of samples. Since directions in the scores plot correspond to directions in the loadings plot, an examination of the loadings can explain spectral clustering observed on the scores plot [71–73]. Usually, PCA constitutes the first step in metabonomic data analysis and is commonly followed by supervised pattern recognition techniques. These methods use class information of the samples to maximise the separation between different groups of samples and detect the metabolic signatures that contribute to the classifications.

One commonly used supervised method is partial least squares, also known as projection to latent structures (PLS), which links a data matrix of predictors usually comprising spectral intensity values (an X matrix), to a matrix of responses containing quantitative values (a Y matrix). When the response matrix is categorical, i.e. the Y matrix contains sample class membership information, the application of PLS regression is called partial least squares-discriminant analysis (PLS-DA). PLS has also been used in combination with a pre-processing filter termed orthogonal signal correction (OSC), which excludes irrelevant parts of the data that are uncorrelated (orthogonal) with the response, often referred to as structured noise. This structured noise in the data set can be caused by analytical variation or by innate physiological variation (e.g. different diet, age, gender). Orthogonal partial least squares discriminant analysis (O-PLS-DA) has an advantage over the standard PLS because it filters the irrelevant variation and hence enhances the model interpretation and identification of important variables that are responsible for the observed classification [73–75]. Recently, a more advanced statistical technique, Statistical HOmogeneous Cluster SpectroscopY (SHOCSY), has been developed which can better address irrelevant variation in datasets and enhance the interpretation and predictive ability of the OPLS-DA model via the selection of 'truly' representative samples in each biological class [76].

In supervised techniques, loading weight, variable importance on projection (VIP) and regression coefficient plots are used to determine the most significant discriminating variables. Recently, a new approach has been introduced by Cloarec et al. that incorporates the back-transformed loading of an auto-scaled model with the respective weight of each variable in the same plot. The resulting loading plot created in this way has the same shape as that of a spectrum with colour-coded coefficients, according to statistical significance for each variable, which allows for easier interpretation of chemometric models [77].

Generally, supervised techniques are subject to overfitting, particularly in metabonomic studies where the number of variables is large and therefore the chance of false correlations is high. Proper model validation is therefore a key step to ensure model reliability and identification of true biomarkers. There are various validation methods including *k*-fold cross validation, permutation and test set validation. [78–81] Cross validation is performed in most cases, especially when the number of samples is low. Here, the *k* subset of samples is iteratively left out and predicted back into the model until all samples have been used once. However, truly robust model validation is achieved by dividing the data into a training set and a test set. The training set is used to construct a model and the test set is used to assess the model performance.

4.2. Statistical Correlation Spectroscopy (STOCSY) for metabolite identification

Statistical Correlation Spectroscopy (STOCSY) follows the concept of two dimensional correlation spectroscopy which had originally been implemented in other spectroscopic techniques including fluorescence and Raman spectroscopies [82]. The development and adaptation of STOCSY in NMR spectra was initially performed by Cloarec et al. and is traditionally applied to one dimensional ^{1}H NMR [74]. STOCSY takes advantage of the inherently linear relationship between intensity variables belonging to the same molecule in an NMR spectrum. It analyses the covariance of variables in a series of spectra and produces a correlation matrix, presented in the form of an NMR spectrum, which reveals the degree of correlation between each variable in the spectrum (either one-dimensional or two-dimensional; see Fig. 15). Depending on the strength of the correlation, correlated variables or resonances (consisting of many variables depending on the resolution) might belong to the same molecule (strong correlation) or molecules in the same metabolic pathway (weaker correlation). The correlation of each resonance, relative to the selected peak on which STOCSY is performed, is revealed by a colour scale which ranges from low correlation (typically 0) to high correlation (typically 1) [74]. In the field of metabonomics this technique is particularly useful in the analysis of complex mixtures, such as urine, where the identification of metabolites can be difficult due to the high density of resonances and potential overlapping [74].

It is important to clarify that the ability of STOCSY to detect the correct correlations is affected by the degree of overlap between resonances, as well as low concentrations. Significant overlapping with other peaks will distort the covariance of different resonances belonging to the same molecule in a spectrum, whilst resonances closer to the noise level are harder to analyse. Such deficiencies have led to the development of other techniques including SubseT Optimization by Reference Matching (STORM), which uses an iterative method to calculate the correlations and is better suited to dealing with potential overlaps or low concentrations [13].

4.3. Structure elucidation of novel metabolites

If a truly novel metabolite is identified in the course of a metabonomics study, then a full structure elucidation to the standard generally accepted for the identification of novel natural products [83] or novel drug degradation products [84] is required. This will usually entail the isolation and purification of the novel metabolite from

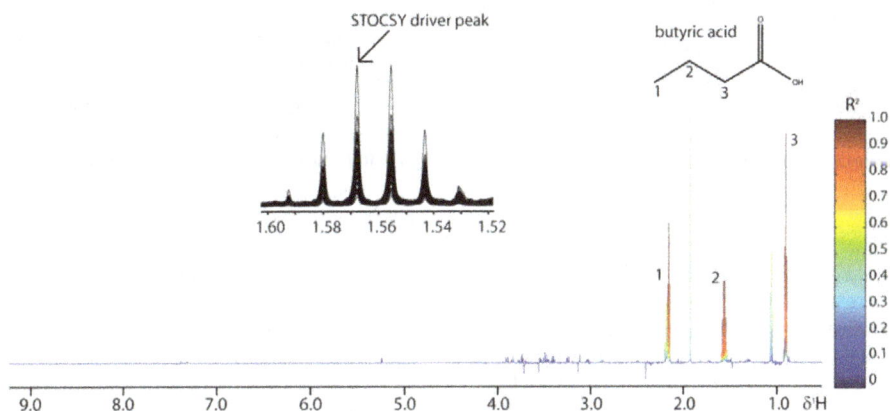

Fig. 15. NMR plot following a STOCSY analysis on a set of faecal water ^{1}H NMR spectra. The selected driver peak at 1.57 ppm was used to calculate the correlation matrix which reveals correlations ranging from 0 (low) to 1 (high). Two other resonances were revealed to have a positive correlation of 1, suggesting that they arise from the same molecule that was later identified as butyric acid.

the biofluid and a full structure elucidation, typically using NMR spectroscopy, MS, infrared spectroscopy and ultraviolet spectroscopy, and / or the synthesis of the metabolite for direct comparison with the data obtained from the biofluid.

4.4. Use of information from metabolite databases

Most metabolites observed by NMR spectroscopy in metabonomics studies will be known, and information on a proportion of these is available in various databases such as the Human Metabolome Database (HMDB) [45], the BioMagResBank (BMRB) [85] and the Birmingham Metabolite Library (BML) [86]. The HMDB is the largest repository of NMR and MS data on human metabolites that is currently available. As of September 4th 2015, the HMDB contained information on 41,993 metabolites. However, only 1381 of these metabolites have experimental NMR data, totalling 3186 NMR spectra. Thus, there are many metabolites for which it is not currently possible to access NMR data online. Databases such as the HMDB are valuable for four main reasons: 1) provide search facilities that allow the identification of known metabolites based on matches between user spectral data and database data on authentic metabolite samples, 2) provide interpreted 1D ^{1}H and 2D NMR spectra (particularly 2D ^{13}C, ^{1}H HSQC spectra) of metabolites; 3) provide access to the raw free induction decay data for authentic metabolites for downloading, processing and comparison with user data on metabolites from biofluids and 4) provide metadata on the metabolites and links to other databases.

The 2D ^{13}C, ^{1}H HSQC search facility in the HMDB is particularly useful, as it uses both ^{1}H and ^{13}C NMR chemical shift information, and searches for matches between HSQC cross-peak coordinates input by a user and those of authentic metabolites in the database. This is a good place to start a metabolite identification exercise. The user must input the tolerances for the chemical shift differences between the user input values and database values: metabolites whose cross-peak coordinates are inside those tolerances will be returned as 'candidate metabolites'. Chemical shifts will naturally be different between those of an authentic sample in water, D_2O or phosphate buffer and those of the same metabolite in a biofluid such as urine or plasma, but generally ^{1}H NMR chemical shifts should agree to $+/-0.03$ ppm and ^{13}C NMR chemical shifts to $+/-0.5$ ppm. These differences will increase for ^{1}H or ^{13}C NMR chemical shifts in metabolites which can undergo tautomerism [87] of any kind and the shift differences may also be larger for nuclei close to ionisable groups in metabolites: both these features will be sensitive to environment.

When reviewing the candidate structures returned by the database that have HSQC features matching the user query, other information about the metabolite of interest will be used to discriminate the candidates. This information could include the multiplicity in the ^{1}H NMR

spectrum of the hydrogen giving rise to the HSQC signal, or connectivity information linking further elements of the metabolite structure from COSY or HMBC spectra. Of course, if the database search is done on just one HSQC cross-peak observed in the spectrum of the biofluid, ALL remaining HSQC cross-peaks in that metabolite should also be observable in the biofluid HSQC spectrum, and the absence of any of the expected HSQC cross-peaks would put a question mark over the identification of the metabolite. On the other hand, as seen above, even databases as large as the HMDB are incomplete and searches will return no candidate structures for known metabolites if: (i) the metabolite is not entered into the database, (ii) the metabolite is in the database but the relevant NMR data is not, or (iii) the metabolite is in the database but the relevant NMR data is not correctly entered.

A further caveat to the use of metabolite databases is that they are only as good as the quality of the data entered into them. Users must beware that errors of several type are present at a low level in current databases such as the HMDB, including incorrect samples, incorrect structures for the metabolites, impure samples and incorrect assignments. A good approach is to always download the original data and check it against expectations, and/or check the values given across more than one database where possible.

4.5. Prediction of NMR spectra of metabolites for structure confirmation

An ideal situation for the confident identification of known or novel metabolites would be to be able to predict their NMR spectra computationally without the need for access to authentic, real samples. In Section 2.2, we saw that ^{13}C NMR chemical shifts could be predicted by hand for simple molecules. Accurate chemical shift prediction would allow the expansion of databases such as the HMDB to include all known metabolites and the confident identification of novel as well as known metabolites. At present, this approach is not generally possible. Software such as MNova [46] and Marvin [47] allows the prediction of ^{1}H and ^{13}C NMR spectra. In our experience, these approaches are useful and somewhat successful but may fail in cases where the metabolite structure is complex, or is complicated by tautomerism or multiple sites of ionisation, and the methodology cannot always compute these with confidence for the relevant biological matrix.

4.6. Biochemical transformation and in vitro fermentation of biofluids to aid metabolite identification

One successful approach to metabolite identification that is currently under-utilised is the biochemical transformation of unknown metabolite A in a biofluid to known metabolite B. This approach was used in the identification of para-cresol sulphate (PCS) as the key biomarker in human urine for the prediction of the metabolic fate of paracetamol

[5]. Incubation of samples of the human urine containing PCS with a sulphatase enzyme led to the transformation of PCS to the known metabolite para-cresol, which was then readily identified in this first human pharmacometabonomics study.

A more extreme and more random, but still useful, implementation of this approach can occur if biofluids are left at room temperature for extended periods of time. Biofluids such as mouse urine will quite likely have been in contact with faecal material and thereby be contaminated with bacteria from the animal's microbiome. It is standard practice in metabonomics studies to add a low concentration of an anti-bacterial agent such as sodium azide to animal urine samples to inhibit bacterial growth, but unless the concentration of azide is high, bacterial growth may still occur. This will cause in vitro fermentation in the urine and will transform large numbers of metabolites into different but related metabolite products. For instance, bacterial fermentation in a sample of urine from a male FMO5 KO mouse at age 30 weeks, led to the 100% conversion of hippuric acid (benzoylglycine) to benzoic acid and glycine (Fig. 16).

Compared with single enzymatic transformations, the in vitro fermentation approach is less specific. However, it is still a potentially useful tool to clarify metabolite identifications, by transforming unknown metabolites into known metabolites, or just to decrease crowding in a particular spectral region.

4.7. Confidence levels in known metabolite identification and confirmation of known metabolite identity

The Metabolomics Standards Initiative recognises 4 levels of known metabolite identification:

Level 1: Identified Compound: A minimum of two independent and orthogonal data (such as retention time and mass spectrum) compared directly relative to an authentic reference standard
Level 2: Putatively Annotated Compound: Compound identified by analysis of spectral data and/or similarity to data in a public database but without direct comparison to a reference standard as for Level 1
Level 3: Putatively Characterised Compound Class: unidentified per se but the data available allows the metabolite to be placed in a compound class
Level 4: Unknown Compound: unidentified or unclassified but characterised by spectral data

These categorisations are somewhat vague in terms of the degree of fit between the data on the metabolite and that on the reference standard it is being compared to. They have not been widely adopted since their publication in 2007 [88,89], and this has been commented upon recently [90]. Various modifications to the original categorisations have been suggested [91,92] in order to improve them but with no general agreement on the way forward. A call to the community was made for engagement with this problem [91]. Encouragingly, a new, quantitative Bayesian method for annotation of metabolites in LC–MS experiments has recently emerged. [93] New quantitative NMR spectroscopy-based proposals have also been published [94] that reject the notion that known metabolite identification (as opposed to putative annotation (Level 2)) must always be based on a direct comparison of the experimental data on the metabolite in a biofluid with that of an authentic reference standard (Level 1 above). The new methods are based on the matching of information obtained experimentally from NMR studies of biofluids with that contained on authentic metabolites in databases such as the HMDB. These methods analyse the amount of matching 1D and 2D ^{1}H NMR spectroscopic information obtained on each metabolite, relative to the number of carbon atoms or heavy atoms in the molecule. One promising new approach is called Metabolite Identification Carbon Efficiency (MICE) [94] and provides a logical, quantitative and systematic method for assessing confidence in known metabolite identification by NMR methods.

The use of metabolite database information, as opposed to information directly from the actual reference standards, to underpin metabolite identification is appropriate for NMR spectroscopy-based methods. In general, there is very good agreement between the chemical shifts of a metabolite in a buffered biological fluid such as urine and in a pure buffer solution of the same metabolite at the same pH. As mentioned in Section 4.4 above, generally, ^{1}H NMR chemical shifts should agree to $+/-0.03$ ppm and ^{13}C NMR chemical shifts to $+/-0.5$ ppm for most metabolites, although there will be cases of metabolites with

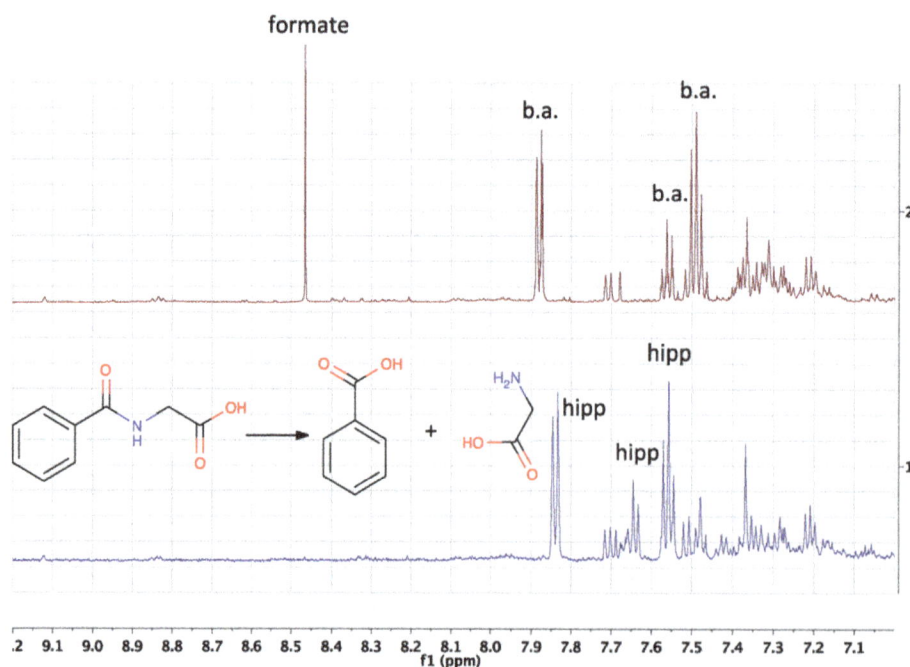

Fig. 16. An expansion of the 600 MHz ^{1}H NMR spectra of the urine of a male, 30-week-old, FMO5 knockout mouse [61]: 1) before bacterial fermentation and 2) after bacterial fermentation after leaving the sample at ambient temperature for several days. The bacterial fermentation caused many metabolic transformations including that of hippuric acid (hipp) to benzoic acid (b.a.) and glycine (3.57 ppm, not shown) and the formation of formate. The lower spectrum 1) prior to fermentation shows many signals including those from the ortho (7.84), para (7.64) and meta (7.56) protons of hippuric acid, whereas post-fermentation, spectrum 2) at top, shows corresponding signals from the ortho (7.88), para (7.56) and meta (7.49 ppm) protons of benzoic acid.

greater chemical shift sensitivity, due to the arrangement of ionisable groups in their molecular structures, for instance, citric acid. There will be an even closer agreement between the chemical shifts of a reference standard run in similar buffers between one laboratory and the next. Therefore access to the NMR spectral data on a metabolite from a database such as HMDB is, in most cases, equivalent to having run the NMR spectrum of that material under the same conditions in the user laboratory. It must be stressed however, that all database data should be checked for quality and for matching to the expected structure. Mistakes in databases do occur: users should be aware. On the other hand, for MS-based metabonomics approaches, such as LC–MS or UPLC–MS, the use of authentic reference standards is more important, due to variations in metabolite retention times and peak intensities that can occur in these experiments, although new methods are making the metabolite annotations more secure [93].

The MICE method mentioned above is one of many new variants that can be used for the assessment of known metabolite identification confidence. In its recommended HSQC-level implementation [94], MICE counts and sums the number of bits of spectroscopic identification information obtained from ^{1}H NMR chemical shifts, multiplicities, coupling constants, second-order flags (flag = 0 if metabolite signals are first order; flag = 1 if signals second-order [strict definition: additional lines present in the spectra]), 2D COSY cross-peaks and 2D HSQC cross peaks, for each metabolite, that *match* corresponding database values for the authentic metabolite. The MICE value is then obtained by dividing this information bit sum total by the number of carbon atoms in the metabolite. For example, the following signal features were observed for the metabolite ketoleucine, (4-methyl-2-oxopentanoic acid, HMDB00695, see Fig. 11 and structure below), in the 600 MHz ^{1}H NMR spectra of the urine of a male, 30-week-old FMO5 knockout mouse [61]: a doublet ($^3J_{H,H}$ ca 7.0 Hz) for the H3 protons at 2.618 with a COSY to 2.098 (triplet of septets, H4), itself with a COSY to the equivalent methyl groups H5, and H6 at 0.941 (doublet, 6.7 Hz) and these with an HSQC to 24.5 ppm (C5, C6). Thus for this metabolite, we observed 3 ^{1}H NMR chemical shifts, 3 multiplicities, two coupling constants, two COSY connectivities and one HSQC connectivity: a total of 11 pieces of information, all of which are a good match to the corresponding values in the HMDB. The guidelines for a good match are: within ± 0.03 ppm for ^{1}H, and ± 0.5 ppm for ^{13}C NMR shifts and ± 0.2 Hz for proton couplings. There are 6 carbon atoms in the molecule, so the Metabolite Identification Carbon Efficiency (MICE) = 11/6 = 1.8. MICE values of > 1 with a good match of spectral features to those of the standard in a database are considered confidently identified, as in this case.

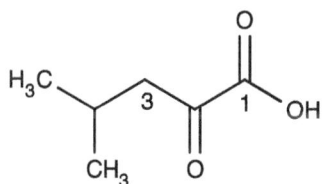

Even if a known metabolite is confidently identified by NMR spectroscopy using the MICE methodology, it can sometimes still be important to further confirm the identification, especially if the particular metabolite is an important biomarker. Three basic approaches are frequently used: 1) authentic metabolite spiking; 2) orthogonal analyses using MS-based approaches and 3) isolation or purification of the metabolite using chromatographic procedures. Metabolite spiking involves the addition of a small quantity of an authentic sample of the metabolite into the biofluid of interest and re-running the NMR spectrum. If the metabolite is present in the biofluid, then the signals of the spiked material should overlap exactly with those assigned to that metabolite in the original biofluid. For this experiment to work well, spectral resolution and lineshape must be optimal and it is best to spike in a quantity

equalling between 25% and 50% of the material in the biofluid: too little material spiked can lead to uncertainty as to whether the signals of interest have increased in intensity: too much material spiked may swamp the signals and lead to uncertainty as to whether the spike signals match those of the metabolite of interest. Liquid chromatography mass spectrometry (LC–MS) or ultra performance liquid chromatography MS (UPLC-MS) [29,30] is often used as an orthogonal confirmatory technique for metabolites identified by NMR spectroscopy. This joint NMR and MS approach will provide the maximal confidence in the identification of known biomarkers that are particularly important. Isolation or purification procedures may use liquid chromatography, solid-phase extraction or liquid–liquid extraction methods [95,96].

5. Conclusions and future thoughts

Metabonomics/metabolomics is undergoing a period of very rapid technology development and a huge increase in the number of applications, using mainly NMR spectroscopy or MS detection technologies. In this guide, we have focused upon the metabolite identification stage of a project using NMR spectroscopy-based detection of metabolites. Compared with MS, NMR spectroscopy is much less sensitive, but has the key advantages of better spectrometer stability, absence of spectrum quenching or enhancement phenomena, full quantitation of metabolites and the ability to use a huge range of the most powerful experiments for metabolite structure elucidation. NMR-detected metabonomics/metabolomics has been delivering answers to important questions in medicine, biology and other sciences for over 30 years and we confidently predict that it will continue to do so for decades more.

Many key advances in NMR spectroscopy-based metabonomics are emerging and these are expected have a significant impact on the utility of the technology. We can highlight the following: 1) the development of highly stable digital spectrometers producing spectra of unparalleled quality; 2) the development of probes with multiple receiver coils enabling the parallelisation of some data acquisition; [97] 3) the development of non-uniform sampling and spatially-encoded 'ultrafast' methods [64] of 2D NMR data acquisition, which hold out the prospect that in the future the default metabonomics experiment may be 2D COSY or 2D J-resolved rather than the current standard: 1D ^{1}H NMR; 4) huge advances in the computational analysis of NMR data in methods derived from STOCSY that hold out the prospect of a systems biology analysis directly from the NMR data [98] and finally 5) the use of reliable, chilled, NMR sample automation systems which mean that large-scale experiments on hundreds or thousands of samples are feasible, enabling the advent of large-scale phenome analyses [12]. We await this future with excitement and much anticipation.

Glossary of terms

Term	Meaning
1D	one-dimensional
2D	two-dimensional
90°_H	a 90 degree pulse to the ^{1}H channel
180°_H	a 180 degree pulse to the ^{1}H channel
BML	Birmingham Metabolite Library
BMRB	BioMagResBank
CAWG	Chemical Analysis Working Group
COSY	COrrelation SpectroscopY
CPMG	Carr–Purcell–Meiboom–Gill
CSSF-TOCSY	Chemical Shift Selective Filter TOCSY
δ_H	^{1}H, hydrogen-1 or proton NMR chemical shift
δ_C	^{13}C or carbon-13 NMR chemical shift
FID	Free Induction Decay
FMO5	Flavin Mono-Oxygenase 5

(continued on next page)

(*continued*)

Term	Meaning
GC–MS	Gas Chromatography–Mass Spectrometry
HCA	Hierarchical Cluster Analysis
HMBC	Heteronuclear Multiple Bond Correlation spectroscopy
HMDB	Human Metabolome DataBase
HOHAHA	HOmonuclear HArtman HAhn
HSQC	Heteronuclear Single Quantum Correlation spectroscopy
ID	identification
$^3J_{H,H}$	three-bond spin–spin coupling between two hydrogens
JRES	J-resolved spectroscopy
KO	gene Knock Out
LC–MS	Liquid Chromatography-Mass Spectrometry
MHz	MegaHertz = Hertz x 10^6
MICE	Metabolite Identification Carbon Efficiency
MS	Mass Spectrometry
MSI	Metabolomics Standards Initiative
NOESY	nuclear Overhauser enhancement spectroscopy
NMR	Nuclear Magnetic Resonance
O-PLS-DA	Orthogonal-Partial Least Squares-Discriminant Analysis
PC	Principal Component
PCA	Principal Components Analysis
PLS	Partial Least Squares (Projection to Latent Structures)
RD	Relaxation Delay
SHOCSY	Statistical HOmogeneous Cluster SpectroscopY
STOCSY	Statistical TOtal Correlation SpectroscopY
STORM	SubseT Optimization by Reference Matching
t1	evolution time in a 2D NMR experiment
t2	the acquisition time over which the FID is measured
T_1	spin–lattice relaxation time
T_2^*	real spin–spin relaxation time
TOCSY	TOtal Correlation SpectroscopY
TSP	sodium 3-(trimethylsilyl) propionate-2, 2, 3, 3-d4
UPLC–MS	Ultra-Performance Liquid Chromatography Mass Spectrometry
VIP	Variable Importance on Projection

Acknowledgements

We thank Professors Jeremy Nicholson, John Lindon and Elaine Holmes for access to the NMR facilities at Imperial College, and for enjoyable and fruitful metabonomics collaborations over more than 30 years.

References

[1] Lindon J, Nicholson J, Holmes E, Everett J. Metabonomics: metabolic processes studied by NMR spectroscopy of biofluids. Concepts Magn Reson 2000;12(5):289–320.

[2] Fiehn O. Metabolomics—the link between genotypes and phenotypes. Plant Mol Biol 2002;48(1–2):155–71.

[3] Lindon JC, Nicholson JK, Holmes E. The Handbook of Metabonomics and Metabolomics. Amsterdam, Oxford: Elsevier; 2007.

[4] Clayton T, Lindon J, Cloarec O, et al. Pharmaco-metabonomic phenotyping and personalized drug treatment. Nature 2006;440(7087):1073–7.

[5] Clayton TA, Baker D, Lindon JC, Everett JR, Nicholson JK. Pharmacometabonomic identification of a significant host–microbiome metabolic interaction affecting human drug metabolism. Proc Natl Acad Sci U S A 2009;106(34):14728–33.

[6] Everett JR, Loo RL, Pullen FS. Pharmacometabonomics and personalized medicine. Ann Clin Biochem 2013;50:523–45.

[7] Everett JR. Pharmacometabonomics in humans: a new tool for personalized medicine. Pharmacogenomics 2015;16(7):737–54.

[8] Beckonert O, Keun HC, Ebbels TMD, et al. Metabolic profiling, metabolomic and metabonomic procedures for NMR spectroscopy of urine, plasma, serum and tissue extracts. Nat Protoc 2007;2(11):2692–703.

[9] Dunn WB, Lin W, Broadhurst D, et al. Molecular phenotyping of a UK population: defining the human serum metabolome. Metabolomics 2015;11:9–26.

[10] Craig A, Cloarec O, Holmes E, Nicholson JK, Lindon JC. Scaling and normalization effects in NMR spectroscopic metabonomic data sets. Anal Chem 2006;78(7):2262–7.

[11] Bylesjo M, Rantalainen M, Cloarec O, Nicholson JK, Holmes E, Trygg J. OPLS discriminant analysis: combining the strengths of PLS-DA and SIMCA classification. J Chemometr 2006;20(8–10):341–51.

[12] Dona AC, Jimenez B, Schaefer H, et al. Precision high-throughput proton NMR spectroscopy of human urine, serum, and plasma for large-scale metabolic phenotyping. Anal Chem 2014;86(19):9887–94.

[13] Posma JM, Garcia-Perez I, De Iorio M, et al. Subset Optimization by Reference Matching (STORM): an optimized statistical approach for recovery of metabolic biomarker structural information from H-1 NMR spectra of biofluids. Anal Chem 2012; 84(24):10694–701.

[14] Bouatra S, Aziat F, Mandal R, et al. The human urine metabolome. PLoS One 2013; 8(9).

[15] Emwas A-HM, Salek RM, Griffin JL, Merzaban J. NMR-based metabolomics in human disease diagnosis: applications, limitations, and recommendations. Metabolomics 2013;9(5):1048–72.

[16] Emwas A-H, Luchinat C, Turano P, et al. Standardizing the experimental conditions for using urine in NMR-based metabolomic studies with a particular focus on diagnostic studies: a review. Metabolomics 2015;11(4):872–94.

[17] Teahan O, Gamble S, Holmes E, et al. Impact of analytical bias in metabonomic studies of human blood serum and plasma. Anal Chem 2006;78(13):4307–18.

[18] Pinto J, Domingues MRM, Galhano E, et al. Human plasma stability during handling and storage: impact on NMR metabolomics. Analyst 2014;139(5):1168–77.

[19] Izquierdo-Garcia JL, Villa P, Kyriazis A, et al. Descriptive review of current NMR-based metabolomic data analysis packages. Prog Nucl Magn Reson Spectrosc 2011;59(3):263–70.

[20] Want EJ, Wilson ID, Gika H, et al. Global metabolic profiling procedures for urine using UPLC–MS. Nat Protoc 2010;5(6):1005–18.

[21] Fonville JM, Maher AD, Coen M, Holmes E, Lindon JC, Nicholson JK. Evaluation of full-resolution J-resolved 1H NMR projections of biofluids for metabonomics information retrieval and biomarker identification. Anal Chem 2010;82(5):1811–21.

[22] Ludwig C, Viant MR. Two-dimensional J-resolved NMR Spectroscopy: review of a key methodology in the metabolomics toolbox. Phytochem Anal 2010;21(1).

[23] Cui Q, Lewis IA, Hegeman AD, et al. Metabolite identification via the Madison metabolomics consortium database. Nat Biotechnol 2008;26(2):162–4.

[24] Tulpan D, Leger S, Belliveau L, Culf A, Cuperlovic-Culf M. MetaboHunter: an automatic approach for identification of metabolites from H-1-NMR spectra of complex mixtures. BMC Bioinforma 2011;12.

[25] van der Hooft JJJ, de Vos RCH, Ridder L, Vervoort J, Bino RJ. Structural elucidation of low abundant metabolites in complex sample matrices. Metabolomics 2013;9(5): 1009–18.

[26] van der Hooft JJJ, Mihaleva V, de Vos RCH, Bino RJ, Vervoort J. A strategy for fast structural elucidation of metabolites in small volume plant extracts using automated MS-guided LC–MS-SPE-NMR. Magn Reson Chem 2011;49:S55–60.

[27] Jacob D, Deborde C, Moing A. An efficient spectra processing method for metabolite identification from H-1-NMR metabolomics data. Anal Bioanal Chem 2013;405(15): 5049–61.

[28] Wishart DS. Advances in metabolite identification. Bioanalysis 2011;3(15):1769–82.

[29] Dunn WB, Erban A, Weber RJM, et al. Mass appeal: metabolite identification in mass spectrometry-focused untargeted metabolomics. Metabolomics 2013;9(1):S44–66.

[30] Watson DG. A rough guide to metabolite identification using high resolution liquid chromatography mass spectrometry in metabolomic profiling in metazoans. Comput Struct Biotechnol J 2013;4:e201301005.

[31] Peirorcely JE, Rojas-Cherto M, Tas A, et al. Automated pipeline for de novo metabolite identification using mass-spectrometry-based metabolomics. Anal Chem 2013; 85(7):3576–83.

[32] Kertesz TM, Hill DW, Albaugh DR, Hall LH, Hall LM, Grant DF. Database searching for structural identification of metabolites in complex biofluids for mass spectrometry-based metabonomics. Bioanalysis 2009;1(9).

[33] Theodoridis G, Gika HG, Wilson ID. Mass spectrometry-based holistic analytical approaches for metabolite profiling in systems biology studies. Mass Spectrom Rev 2011;30(5):884–906.

[34] Kind T, Fiehn O. Advances in structure elucidation of small molecules using mass spectrometry. Bioanal Rev 2010;2(1–4):23–60.

[35] Claridge T. High-Resolution NMR Techniques in Organic Chemistry. Oxford, UK: Elsevier; 2009.

[36] Williams DH, Fleming I. Spectroscopic methods in organic chemistry. London: McGraw-Hill; 2008.

[37] Keeler J. Understanding NMR Spectroscopy. Oxford: Wiley; 2010.

[38] Gunther H. NMR Spectroscopy: Basic Principles, Concepts and Applications in Chemistry. 3rd ed. Wiley VCH; Sept. 25 2013.

[39] Pearce JTM, Athersuch TJ, Ebbels TMD, Lindon JC, Nicholson JK, Keun HC. Robust algorithms for automated chemical shift calibration of 1D H-1 NMR spectra of blood serum. Anal Chem 2008;80(18):7158–62.

[40] Farrant RD, Lindon JC, Nicholson JK. Internal temperature calibration for H-1-NMR spectroscopy studies of blood-plasma and other biofluids. NMR Biomed 1994; 7(5):243–7.

[41] Fan TWM, Lane AN. Structure-based profiling of metabolites and isotopomers by NMR. Prog Nucl Magn Reson Spectrosc 2008;52(2–3):69–117.

[42] Schaller RB, Arnold C, Pretsch E. New parameters for predicting H-1-NMR chemical-shifts of protons attached to carbon-atoms. Anal Chim Acta 1995;312(1):95–105.

[43] Reich HJ. Web-based NMR information, 2015, http://www.chem.wisc.edu/areas/reich/chem605/index.htm

[44] Brown DW. A short set of C-13 NMR correlation tables. J Chem Educ 1985;62(3): 209–12.

[45] Wishart DS, Jewison T, Guo AC, et al. HMDB 3.0—the human metabolome database in 2013. Nucleic Acids Res 2013;41(D1):D801–7.

[46] Mestrelab Research MNova NMR Software. v 10.0.0, 2015, http://mestrelab.com/software/mnova/nmr/

[47] ChemAxon's NMR Predictor in MarvinSketch, Marvin 6.1.1, 2013, ChemAxon (http://www.chemaxon.com).

[48] Bremser W. Expectation ranges Of C-13 NMR Chemical-Shifts. Magn Reson Chem 1985;23(4):271–5.

[49] Breitmaier E, Haas G, Voelter W. Atlas of Carbon-13 NMR data. London: Heyden; 1979.

[50] Breitmaier E, Voelter W. Carbon-13 NMR Spectroscopy: High-Resolution Methods And Applications In Organic Chemistry and Biochemistry. 3rd ed. Weinheim, Germany: Wiley-VCH; 1987.

[51] Wehrli FW, Marchand AP, Wehrli S. Interpretation of Carbon-13 NMR Spectra. Chichester: Wiley; 1988.

[52] Kalinowski H-O, Berger S, Braun S. Carbon-13 NMR Spectroscopy. Chichester: Wiley; 1988.

[53] Everett J, Jennings K, Woodnutt G, Buckingham M. Spin-Echo H-1-NMR spectroscopy — a new method for studying penicillin metabolism. J Chem Soc Chem Commun 1984;14:894 5.

[54] Connor S, Everett J, Jennings K, Nicholson J, Woodnutt G. High-resolution H-1-NMR spectroscopic studies of the metabolism and excretion of ampicillin in rats and amoxicillin in rats and man. J Pharm Pharmacol 1994;46(2):128–34.

[55] Everett J, Hughes D, Bain A, Bell R. Homonuclear and heteronuclear coupling in 5'-Amp as probed by 2-dimensional proton nuclear magnetic-resonance spectroscopy. J Am Chem Soc 1979;101(22):6776–7.

[56] Leverence R, Avery MJ, Kavetskaia O, Bi H, Hop CECA, Gusev AI. Signal suppression/enhancement in HPLC-ESI-MS/MS from concomitant medications. Biomed Chromatogr 2007;21(11):1143–50.

[57] Keun HC, Ebbels TMD, Antti H, et al. Analytical reproducibility in H-1 NMR-based metabonomic urinalysis. Chem Res Toxicol 2002;15(11):1380–6.

[58] Barton RH, Nicholson JK, Elliott P, Holmes E. High-throughput ^{1}H NMR-based metabolic analysis of human serum and urine for large-scale epidemiological studies: validation study. Int J Epidemiol 2008;37(Suppl. 1):i31–40.

[59] Tomlins A, Foxall P, Lynch M, Parkinson J, Everett J, Nicholson J. High resolution (1)H NMR spectroscopic studies on dynamic biochemical processes in incubated human seminal fluid samples. Biochim Biophys Acta Gen Subj 1998;1379(3):367–80.

[60] Mercier P, Lewis MJ, Chang D, Baker D, Wishart DS. Towards automatic metabolomic profiling of high-resolution one-dimensional proton NMR spectra. J Biomol NMR 2011;49(3–4):307–23.

[61] Malagon SGG, Melidoni AN, Hernandez D, et al. The phenotype of a knockout mouse identifies flavin-containing monooxygenase 5 (FMO5) as a regulator of metabolic ageing. Biochem Pharmacol 2015;96(3):267–77.

[62] Viant MR. Improved methods for the acquisition and interpretation of NMR metabolomic data. Biochem Biophys Res Commun 2003;310(3):943–8.

[63] Bain AD, Burton IW, Reynolds WF. Artifacts in 2-dimensional NMR. Prog Nucl Magn Reson Spectrosc 1994;26:59–89.

[64] Le Guennec A, Giraudeau P, Caldarelli S. Evaluation of fast 2D NMR for metabolomics. Anal Chem 2014;86(12):5946–54.

[65] Davis DG, Bax AD. Assignment of complex 'H NMR spectra via wo-dimensional homonuclear Hartmann–Hahn spectroscopy. J Am Chem Soc 1985;107:2820–1.

[66] Wishart DS. Computational strategies for metabolite identification in metabolomics. Bioanalysis 2009;1(9).

[67] Lindon JC, Holmes E, Nicholson JK. Pattern recognition methods and applications in biomedical magnetic resonance. Prog Nucl Magn Reson Spectrosc 2001;39(1):1–40.

[68] Wold S, Kettaneh N, Tjessem K. Hierarchical multiblock PLS and PC models for easier model interpretation and as an alternative to variable selection. J Chemometr 1996;10(5–6):463–82.

[69] Wold S, Sjöström M, Eriksson L. PLS-regression: a basic tool of chemometrics. Chemom Intell Lab Syst 2001;58(2):109–30.

[70] Trygg J, Wold S. Orthogonal projections to latent structures (O-PLS). J Chemometr 2002;16(3):119–28.

[71] Holmes E, Foxall PJ, Nicholson JK, et al. Automatic data reduction and pattern recognition methods for analysis of ^{1}H nuclear magnetic resonance spectra of human urine from normal and pathological states. Anal Biochem 1994;220(2):284–96.

[72] Holmes E. Chemometric models for toxicity classification based on NMR spectra of biofluids. Chem Res Toxicol 2000;13:471–8.

[73] Trygg J, Holmes E, Lundstedt T. Chemometrics in metabonomics. J Proteome Res 2007;6(2):469–79.

[74] Cloarec O, Dumas M-E, Craig A, et al. Statistical total correlation spectroscopy: an exploratory approach for latent biomarker identification from metabolic ^{1}H NMR data sets. Anal Chem 2005;77(5):1282–9.

[75] Fonville J, Richards S, Barton R, et al. The evolution of partial least squares models and related chemometric approaches in metabonomics and metabolic phenotyping. J Chemother 2010;24(11–12):636 49.

[76] Zou X, Holmes E, Nicholson JK, Loo RL. Statistical HOmogeneous cluster SpectroscopY (SHOCSY): an optimized statistical approach for clustering of ^{1}H NMR spectral data to reduce interference and enhance robust biomarkers selection. Anal Chem 2014;86(11):5308–15.

[77] Cloarec O, Dumas ME, Trygg J, et al. Evaluation of the orthogonal projection on latent structure model limitations caused by chemical shift variability and improved visualization of biomarker changes in ^{1}H NMR spectroscopic metabonomic studies. Anal Chem 2005;77(2):517–26.

[78] Anderssen E, Dyrstad K, Westad F, Martens H. Reducing over-optimism in variable selection by cross-model validation. Chemom Intell Lab Syst 2006;84(1–2):69–74.

[79] Rubingh C, Bijlsma S, Derks EPA, et al. Assessing the performance of statistical validation tools for megavariate metabolomics data. Metabolomics 2006;2(2):53–61.

[80] Szymańska E, Saccenti E, Smilde AK, Westerhuis JA. Double-check: validation of diagnostic statistics for PLS-DA models in metabolomics studies. Metabolomics 2012;8(Suppl. 1):3–16.

[81] Golland P, Fischl B. Permutation Tests for Classification: Towards Statistical Significance in Image-Based Studies. In: Taylor C, Noble JA, editors. Information Processing in Medical Imaging. Berlin Heidelberg: Springer; 2003. p. 330–41.

[82] Noda I. Generalized 2-dimensional correlation method applicable to infrared, Raman, and other types of spectroscopy. Appl Spectrosc 1993;47(9):1329–36.

[83] Baker G, Dorgan R, Everett J, Hood J, Poulton M. A novel series of milbemycin antibiotics from streptomyces strain-E225.2. Isolation, characterization, structure elucidation and solution conformations. J Antibiot 1990;43(9):1069–76.

[84] Ashline K, Attrill R, Chess E, et al. Isolation, structure elucidation, and synthesis of novel penicillin degradation products — Thietan-2-Ones. J Chem Soc Perkin Trans 1990;2(9):1559–66.

[85] Ulrich EL, Akutsu H, Doreleijers JF, et al. BioMagResBank. Nucleic Acids Res 2008;36: D402–8.

[86] Ludwig C, Easton JM, Lodi A, et al. Birmingham Metabolite Library: a publicly accessible database of 1-D H-1 and 2-D H-1 J-resolved NMR spectra of authentic metabolite standards (BML-NMR). Metabolomics 2012;8(1):8–18.

[87] Antonov L, editor. Tautomerism: Methods and Theories. Weinheim, Germany: Wiley-VCH; 2014.

[88] Sumner LW, Amberg A, Barrett D, et al. Proposed minimum reporting standards for chemical analysis. Metabolomics 2007;3(3):211–21.

[89] Fiehn O, Robertson D, Griffin J, et al. The metabolomics standards initiative (MSI). Metabolomics 2007;3(3):175–8.

[90] Salek RM, Steinbeck C, Viant MR, Goodacre R, Dunn WB. The role of reporting standards for metabolite annotation and identification in metabolomic studies. GigaScience 2013;2(1):13.

[91] Creek DJ, Dunn WB, Fiehn O, et al. Metabolite identification: are you sure? And how do your peers gauge your confidence? Metabolomics 2014;10(3):350–3.

[92] Sumner L, Lei Z, Nikolau B, Saito K, Roessner U, Trengove R. Proposed quantitative and alphanumeric metabolite identification metrics. Metabolomics 2014;10(6): 1047–9.

[93] Daly R, Rogers S, Wandy J, Jankevics A, Burgess KEV, Breitling R. MetAssign: probabilistic annotation of metabolites from LC–MS data using a Bayesian clustering approach. Bioinformatics 2014;30(19):2764–71.

[94] Everett JR. A new paradigm for known metabolite identification in metabonomics/metabolomics: metabolite identification efficiency. Comput Struct Biotechnol J 2015;13:131–44.

[95] Tang YQ, Weng N. Salting-out assisted liquid–liquid extraction for bioanalysis. Bioanalysis 2013;5(12):1583–98.

[96] Mushtaq MY, Choi YH, Verpoorte R, Wilson EG. Extraction for metabolomics: access to the metabolome. Phytochem Anal 2014;25(4):291–306.

[97] Gierth P, Codina A, Schumann F, Kovacs H, Kupče Ē. Fast experiments for structure elucidation of small molecules: Hadamard NMR with multiple receivers. Magn Reson Chem 2015.

[98] Robinette SL, Lindon JC, Nicholson JK. Statistical spectroscopic tools for biomarker discovery and systems medicine. Anal Chem 2013;85(11):5297–303.

10

Under-detection of endospore-forming *Firmicutes* in metagenomic data

Sevasti Filippidou [a], Thomas Junier [a,b], Tina Wunderlin [a,1], Chien-Chi Lo [c], Po-E Li [c], Patrick S. Chain [c], Pilar Junier [a,*]

[a] *Laboratory of Microbiology, Institute of Biology, University of Neuchatel, CH-2000, Neuchâtel, Switzerland*
[b] *Vital-IT group, Swiss Institute of Bioinformatics, CH-1015 Lausanne, Switzerland*
[c] *Bioscience Division, Los Alamos National Laboratory, Los Alamos, NM 87545, USA*

ARTICLE INFO

Keywords:
Endospores
gpr
Metagenomics
Profile analysis
spo0A

ABSTRACT

Microbial diversity studies based on metagenomic sequencing have greatly enhanced our knowledge of the microbial world. However, one caveat is the fact that not all microorganisms are equally well detected, questioning the universality of this approach. *Firmicutes* are known to be a dominant bacterial group. Several *Firmicutes* species are endospore formers and this property makes them hardy in potentially harsh conditions, and thus likely to be present in a wide variety of environments, even as residents and not functional players. While metagenomic libraries can be expected to contain endospore formers, endospores are known to be resilient to many traditional methods of DNA isolation and thus potentially undetectable. In this study we evaluated the representation of endospore-forming *Firmicutes* in 73 published metagenomic datasets using two molecular markers unique to this bacterial group (*spo0A* and *gpr*). Both markers were notably absent in well-known habitats of *Firmicutes* such as soil, with *spo0A* found only in three mammalian gut microbiomes. A tailored DNA extraction method resulted in the detection of a large diversity of endospore-formers in amplicon sequencing of the 16S rRNA and *spo0A* genes. However, shotgun classification was still poor with only a minor fraction of the community assigned to *Firmicutes*. Thus, removing a specific bias in a molecular workflow improves detection in amplicon sequencing, but it was insufficient to overcome the limitations for detecting endospore-forming *Firmicutes* in whole-genome metagenomics. In conclusion, this study highlights the importance of understanding the specific methodological biases that can contribute to improve the universality of metagenomic approaches.

1. Introduction

Metagenomic studies have emerged as promising methods for the collective study of microbial communities directly extracted from environmental samples [1–3]. These approaches have been successfully applied to a variety of environments and have helped to unveil new functional pathways and metabolic processes within the microbial world [4–8].

Biases, however, can occur at all the steps involved in a metagenomic workflow. They can be associated to the specific type of environment [9, 10], the DNA yields obtained [11], the DNA extraction method [12], the amplification (for example in amplicon sequencing), but also in the sequencing and the analysis of the sequences. These limitations have

been highlighted in the recent literature and result in problems such as low coverage of the less abundant taxa (the so-called "depth bias" for example in the detection of ribosomal genes [13]), low reproducibility of results [14] and underrepresentation of certain taxa, as discussed herein. In order to overcome these limitations, new approaches have been developed including single-cell genomics or culture-dependent methodologies such as culturomics [15,16] which, in their turn, have their own limitations.

Even though methodological bias of metagenomic diversity surveys associated to particular types of environments such as soil has been demonstrated experimentally [9,10], the specific coverage of individual microbial groups within the community is still unknown. One example of a bacterial group that can be used to test coverage bias in metagenomic datasets is endospore-forming *Firmicutes*. Even though, culturing of microorganisms is largely acknowledge to be biased, according to previous research based on culture collections as well as whole-genome sequencing, *Firmicutes* is the second most abundant bacterial phylum [17]. Endospore formers live in a wide range of environments on Earth's surface and subsurface [18,19]. The hardy

* Corresponding author
E-mail addresses: sevasti.filippidou@unine.ch (S. Filippidou), thomas.junier@unine.ch (T. Junier), tina.wunderlin@mq.edu.au (T. Wunderlin), chienchi@lanl.gov (C.-C. Lo), po-e@lanl.gov (P.-E. Li), pchain@lanl.gov (P.S. Chain), pilar.junier@unine.ch (P. Junier).
[1] Current address: Department of Biological Sciences, Macquarie University, Australia.

outer cortex of endospores and the small acid–soluble proteins stabilizing their DNA [20–22], allow these bacteria to be distributed into every habitat on Earth [23]. However, a phylogenetic assessment of the microbial communities in four metagenomic datasets has revealed surprisingly few endospore formers [24]. This might appear surprising considering their ubiquity, but endospores are known to withstand many traditional methods of DNA isolation and are thus potentially undetectable in a sample. Recently, a DNA extraction method for the extraction of resistant structures such as endospores has been developed by our group [12]. This DNA extraction method was combined with amplicon sequencing of the gene coding the master regulator for the initiation of sporulation (spo0A gene) to demonstrate an improved detection of endospore-forming Firmicutes in sediment samples [12]. Our group has developed further methods to separate endospores from vegetative cells, which has open the possibility to carry out genomic studies only focused on endospores [12,25]. These two studies demonstrate by amplicon sequencing that the diversity of endospore-forming Firmicutes is far from uncovered. However, the effectiveness of the improved DNA extraction method for whole-genome metagenomic studies is unknown.

The aim of this study was to measure the level of detection of endospore formers in metagenomic studies carried out so far, and to evaluate the effect of an improved DNA extraction method on the detectability of this group. To do this, we initially searched for functional gene markers of endospore formation in metagenomic datasets using profiles. We then applied a modified DNA extraction method that is tailored to release DNA from resistant structures such as endospores [12] in a selected environmental sample. Amplicon sequencing of the 16S rRNA and spo0A genes were performed on the sample in order to assess the relative abundance and phylogenetic diversity of Firmicutes. This was complemented by shotgun sequencing and classification of the metagenome reads. Our results indicate that endospore-forming Firmicutes are overlooked in environmental diversity surveys using traditional whole metagenomic approaches.

2. Materials and Methods

2.1. Genome Sequence Retrieval

Complete and draft genome sequences of endospore-forming Firmicutes were downloaded from the Comprehensive Microbial Resource (CMR, 24.0 data release, cmr.jcvi.org) and Integrated Microbial Genomes (IMG, 3.0, img.jgi.doe.gov) websites. Protein and nucleotide sequences of spore-related genes were obtained by search for role category/function sporulation and germination (CMR) and sporulating (IMG). Additional information on all retrieved genomes was obtained from the GenBank database (www.ncbi.nlm.nih.gov/genome).

2.2. Detection of Orthologous Sporulation Genes Common to All Endospore-Formers

Orthologous groups were delineated based on best reciprocal BLASTp hits [26]. BLASTp was used to align each sequence in the set against all sequences except those of the same species (thus avoiding paralogs). The best hit in each species was retained, and sequence pairs, that were each other's best match, were defined as best reciprocal hits (BRHs). Putative orthologous groups were defined using the algorithm used by OrthoDB [27]. OrthoDB has data on Fungi, Metazoa, and Bacteria. An early version of the BRHCLUS program (unpublished at the time) was obtained from its author, Dr. Tegenfeldt (pers. comm) and run according to the author's instructions. The program is now available from http://orthodb.org/. To our knowledge, its utility does not depend on the clade it is used for — OrthoDB uses the same clustering program for all data in its scope.

2.3. Profile Construction and Validation

The genomic sequences were filtered in such as way as to keep only one (randomly chosen) sequence per genus, thus reducing taxonomic sampling bias. Multiple alignments of Spo0A and Grp were produced with MAFFT [28]. Gribskov-style sequence profiles were constructed with EMBOSS's prophecy program [29]. The profiles' score cutoffs were determined by searching with EMBOSS's prophet program against the original Spo0A (resp. Gpr) sequence set as a positive control, and against shuffled versions of the same as negative set.

2.4. Metagenomic Datasets Retrieval

The metagenome datasets (supplementary Table 1) were downloaded from IMG, GOLD (genomesonline.org), or the metagenomes subset of the WGS section of EMBL (ebi.ac.uk/genomes/wgs.html). These datasets included all the metagenomic studies available at EMBL when the profile analysis was performed. Only sequences or contigs of >800 bp, which are slightly shorter than the full-length sporulation genes, were kept for analysis.

2.5. Environmental Sampling, DNA Extraction and Quantitative PCR

The sample was collected at Nea Apollonia (NAP) geothermal spring (N 40° 39,191′ E 22° 56,707′), Greece, in June 2011. Geothermal reservoir was reached through a 120 m drilling pipe, used mostly for pumping 80 °C water for bathing purposes. Biofilm from the pipe interior was collected and frozen within 2 h of collection. Upon arrival at the laboratory, a tailored DNA extraction method previously described [12] was applied to the sample. More precisely, DNA was extracted using the FastDNA Spin Kit for Soil (MP Biomedicals, California), using a modified protocol in order to ensure that DNA was not only extracted from vegetative cells but also from spores and other cells difficult to lyse. These modifications were (a) a separation of the biomass from the soil, using a Na-hexa-meta-phosphate solution and (b) a sequential bead-beating step (three times) to ensure mechanical disruption of cells. In total, 10ug of high molecular RNA-free DNA was obtained.

Moreover, 16S rRNA gene and spo0A gene copy numbers were calculated using a quantitative PCR assay, as previously described [30].

2.6. Amplicon Sequencing of the 16S rRNA and spo0A Genes

In order to verify the presence and relative abundance of endospore formers, 454 pyrosequencing of a fragment of the 16S rRNA and spo0A genes was firstly applied to the sample NAP. Sequencing was done using the services of Eurofins MWG Operon (Ebersberg, Germany). For 16S rRNA amplicon sequencing, fragments of approximately 500 bp were retrieved using primers Eub8f (5′-AGAGTTTGATCCTGGC TCAG-3′) and Eub519r (5′-GTATTACCGCGGCTGCTGG-3′), as previously described [31]. 16S rRNA gene raw sequence data was analyzed with QIIME [32], using the pipeline for de novo OTU picking. OTUs were identified using a threshold of 97% sequence similarity. The sequences were then clustered into putative OTUs with the pick_otus.py program from the QIIME package using the Uclust method [32]. The single sequence picked by the program as a representative of each OTU was used to build a phylogeny.

For the spo0A amplicon sequencing, a 602 bp sequence of the spo0A gene was amplified using the degenerated primer spo0A166f (5′-GATA THATYATGCCDCATYT-3′) and spo0A748r (5′-GCNACCATHGCRATR AAYTC-3′) [12]. 42′151 sequences were received from the sample. Sequences were then filtered according to Phred [33] quality score (minimum of 30) and sequences of length shorter than 600 bp were removed. Remaining sequences were translated to their amino acid sequence; resulting full-length ORFs were then matched against the spo0A profile, in order to confirm that the primers actually amplified the spo0A sequences.

Phylogenies were constructed from Phylip-formatted alignments with PhyML [34], using default parameters. The trees were re-rooted, condensed according to protocol, and displayed with the Newick Utilities [64]. Each branch represents a cluster of OTUs of > 97% sequence similarity. Identification of the closest relatives of the environmental sequences was done by protein BLAST [26] with the translated protein sequences using a reference database of 581 spo0A protein sequences from the InterPro site [35].

All metagenomic sequences were submitted to GenBank. The 16S rRNA amplicon sequencing data can be retrieved under the BioProject ID PRJNA267761 and BioSample ID SAMN03198953 and the spo0A amplicon sequencing data under the BioProject ID PRJNA276803 and Biosample ID SAMN03392534.

2.7. Metagenomic Sequencing

Once high prevalence of endospore formers was confirmed in the 16S rRNA pyrosequencing data (41% of total bacterial community), whole-metagenome sequencing of NAP was performed on a full plate of a GS FLX platform, followed by de novo assembly using the services of GATC- biotech (Konstanz, Germany). The metagenome dataset can be retrieved from GenBank under the BioProject ID PRJNA271123 and BioSample ID SAMN03273062.

2.8. Metagenome Data Annotation

Several tools were used to produce the read-based metagenomic analysis of NAP metagenome dataset. GOTTCHA [36] was run using BWA [37] against 4 databases consisting of Phylum, Genus, Species and Strain-level unique signatures. MetaPhlAn v1.7.7 [38] was run using BowTie2 [39] with default parameters against its clade-specific maker genes database. Kraken was run with its reduced taxonomic-specific 31-mer database (mini-database). BWA v0.7.4-r385 used as a stand-alone tool was run locally using BWA-backtrack algorithm to map reads against a custom database of bacterial, archaeal and viral complete genomes retrieved from NCBI RefSeq database [40]. The mapped reads were subsequently assigned to organisms by mapping the GI numbers of aligned references to NCBI taxonomic ID and rolled up to higher ranks. mOTUs v1.0 [41] was run with the database composed of 10 universal marker genes and LMAT v1.2.1 [42] was run with the pre-computed reference search database (kML.18mer.16bit. reduced.db) with default parameters. Since BWA (standalone), Kraken and LMAT only reported read counts of taxonomies, the relative abundances were represented by the portion of total classified reads in these tools. While each tool tries to identify similarities among the reads and the databases used, each tool is centered around a different algorithmic approach to solve this complex challenge, using either a unique search algorithm, a uniquely designed database, or both. The interpretation of the results from each tool should thus be taken within its own context. For example, mOTUs and MetaPhlAn use pre-selected marker genes to perform the analysis, however different marker genes are used and different methods are used to identify reads that are similar to these marker genes. Kraken and LMAT both use subsequences within reads (k-mers) and match k-mers observed within the reads with those observed within known reference genomes. Meanwhile BWA is a read-mapping tool that we use against the refseq database to report matching reads.

3. Results and Discussion

3.1. Selection of Functional Markers for Endospore-formation

We recently identified functional marker genes involved in endospore formation in endospore-forming Firmicutes [12]. Bidirectional

Table 1

Prevalence of Firmicutes in 16S rRNA gene amplicon sequencing and shotgun metagenomic sequencing applied to the NAP sample. Different prediction tools were used to establish the five most frequent Phyla in the samples. With the exception of the 16S rRNA gene amplicon sequencing, the relative percentage indicated corresponded to the fraction of the sequences that could be classified and not to the frequency of any of the groups for the total reads generated after sequencing.

Prediction tool		Top 5 Phyla	Frequency	Relative %
16S RNA gene amplicon pyrosequencing (QIIME)	1	Firmicutes	41.70	41.70%
	2	Proteobacteria	26.14	26.14%
	3	Bacteroidetes	10.55	10.55%
	4	Planctomycetes	5.35	5.35%
	5	Chlorobi	3.88	3.88%
Kraken (mini database)	1	Proteobacteria	16644	82.71%
	2	Actinobacteria	1744	8.67%
	3	Firmicutes	322	1.60%
	4	Bacteroidetes	298	1.48%
	5	Cyanobacteria	192	0.95%
MetaPhlAn	1	Proteobacteria	82.01061	82.01%
	2	Chloroflexi	9.24158	9.24%
	3	Actinobacteria	2.32449	2.32%
	4	Bacteroidetes	2.08071	2.08%
	5	Acidobacteria	1.54098	1.54%
BWA	1	Proteobacteria	452	75.21%
	2	Firmicutes	32	5.32%
	3	Thaumarchaeota	28	4.66%
	4	Actinobacteria	26	4.33%
	5	Bacteroidetes	17	2.83%
LMAT	1	Ascomycota	425	35.68%
	2	Cyanobacteria	385	32.33%
	3	Proteobacteria	190	15.95%
	4	Thaumarchaeota	145	12.17%
	5	Basidiomycota	20	1.68%

BLAST of the genes annotated as part of the cellular function of sporulation allowed to select six highly conserved orthologous genes as part of the endospore-forming Firmicutes proteome. Among those, spo0A and gpr, were selected for the construction of profiles based on their consistent phylogenetic reconstruction with the 16S rRNA gene phylogeny. These two genes represent significant stages of the endospore-formation process, namely the commitment to enter sporulation (spo0A) and the proteolytic activity on acid-soluble spore proteins (SASPs) during germination (gpr) [43]. In recent studies analyzing the minimal set of endospore-formation genes required by endospore-formers had indicated that spo0A is indeed one of the most conserved genes almost exclusively found among this bacterial group [44–46]. In the case of gpr, it has been shown that it belongs to a category of genes present in Bacillus and Clostridium without any known ortholog in Gram-negative Proteobacteria or Cyanobacteria [21].

3.2. Profile Analysis of Sporulation Genes in Metagenomes

Profiles of SpoOA and Gpr were constructed and compared to metagenomic datasets to find sequences of high similarity with spo0A and gpr. Profiles are models of conserved sequences built from an alignment and are more sensitive than BLAST or other pair-wise comparisons especially for protein searches [47]. The sequence profiles were generated based on 14 aligned sequences. They were validated on genomes of known endospore-forming and non-sporulating bacteria (Fig. 1A). A single positive hit was found in the genome of each endospore-forming bacterium, while no hits were found in the negative controls. This result also allowed determining a score cut-off for spo0ASpo0A (2000) and Gpr (2500) profiles to distinguish between positive and negative hits. Using this cut-off value one orthologous sequence of each of the two genes could be detected in a further 59 genomes of endospore-forming bacteria (Fig. 1B) reported in the genomic databases of the Comprehensive Microbial Resource (CMR) and Integrated Microbial Genomes (IMG) (Supplementary Table 1).

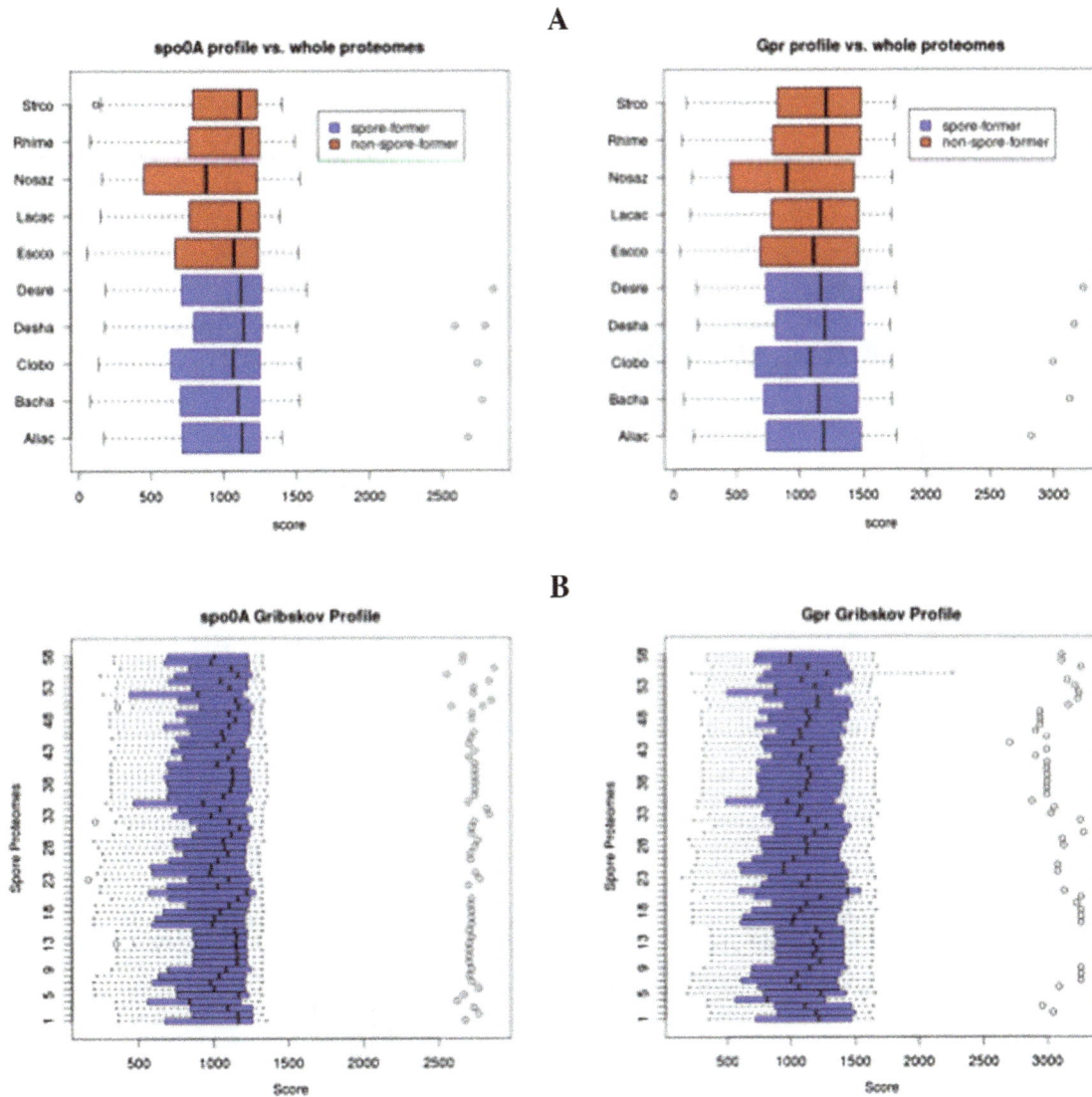

Fig. 1. A. Validation of the profiles created for the genes *spo0A* and *gpr* compared to a selection of genomes of endospore-forming *Firmicutes* (blue bars) and non spore-forming genomes (red bars). In endospore-forming *Firmicutes* a single hit with a score above 2000 (Spo0A) and 2500 (Gpr) distinguish between positive and negative hits. Strco = *Streptomyces coelicolor*; Rhime = *Rhizobium melliloti*; Nosaz: *Nostoc azollae*; Lacac = *Lactobacillus acidophilus*; Escco = *Escherichia coli*; Desre = *Desulfotomaculum reducens*; Desha = *Desulfitobacterium hafniense*; Clobo = *Clostridium botulinum*; Bacha = *Bacillus halodurans*; Aliac = *Alicyclobacillus acidocaldarius*. B. The same analysis was repeated using all 59 endospore-forming genomes retrieved from IMG and CMR databases (see supplementary Table 1).

The profile analysis was then used to detect Spo0A or Gpr in publicly available environmental metagenomes. For this, 73 microbial metagenomic datasets (Supplementary Table 2) from a total of 25 publications or direct submissions were retrieved. The datasets consisted of 6,220,494 sequences of average length of 957 bp and represented different environments, including marine, fresh- and ground-waters, acid mine drainage, compost, hypersaline environments, hot springs, soils, sludge, food and organism-associated environments (ant fungus garden, coral, fish and human gut).

The profile analysis revealed only three sequences with a score above the cutoff of the Spo0A profile in all metagenomic datasets (Fig. 2A). All three metagenomes (AAQL, BAAY, BAAZ) originated from human gut [48,49], in which *Firmicutes* are known to be one of the dominant bacterial groups [50,51]. For the *gpr* gene profile (Fig. 2B), no sequences were found with a similarity score above the cutoff value. These results are surprising considering that some of these metagenomes were sampled in environments with high abundance of endospore-forming *Firmicutes* (e.g. gut or soil; [52, 53]). These results showed that these two genes from endospore-

forming *Firmicutes* are underrepresented in metagenomes. This had been alluded to earlier by von Mering et al., [24], and is now confirmed here.

A methodological bias during the DNA extraction of resistant structures such as bacterial endospores has been suggested as the origin of an underrepresentation of microbial groups producing this structure [24]. Indeed, independently of the methodological approach taken (i.e. whole genome shotgun analysis, activity- or sequence-driven screening), the first and most crucial step in any metagenomic project is the extraction of nucleic acids. The isolated DNA should be representative of all cells in the sample and of sufficient quality and amount for subsequent sequencing [54]. Clearly, not all microbial species are equally amenable to the DNA extraction methods used today [9,10], especially considering the diversity of morphological and physiological states in which microbes can be found in environmental samples. Therefore, complementary information, in particular concerning the method used for DNA extraction of the metagenomes was thus considered. The described DNA extraction methods (Supplementary Table 2)

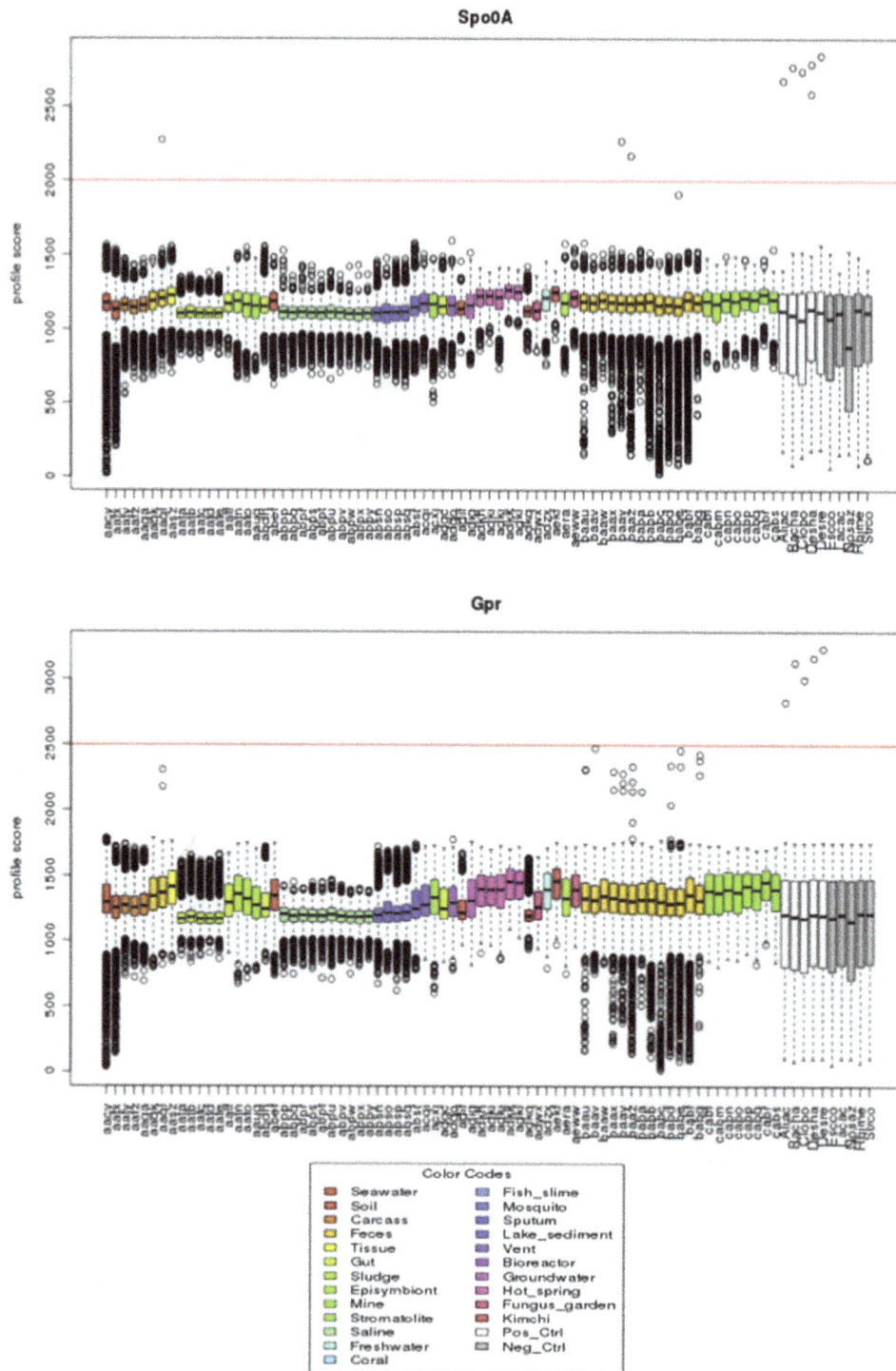

Fig. 2. Profile similarity hits for SpoOA and Gpr protein profiles in metagenomes from different origins. The color code identifying different environments is presented under the results. The genomes included in profile testing (see Fig. 1A) were also included in the analysis and are presented in white (endospore-formers) and gray (non-spore formers).

consisted of enzymatic or chemical protocols (18 datasets) or mechanical procedures of cell lysis (8 datasets). Sequences associated to *Firmicutes* are reported for some of the analyzed metagenome projects regardless of the DNA extraction protocol. For example, sequences of Clostridia (30%) and Bacilli (1%) were reported in the wallaby gut extracted enzymatically [55]. Also, in the compost metagenome extracted by bead beating, more than 13% of sequences were reported as members of endospore-formers *Bacillus* spp. or *Paenibacillus* spp. [56]. Our profile analyses however, do not show

positive hits for SpoOA and Gpr in either of these metagenomes. Whether this is due to the extraction method applied, to the depth of sequencing or to other specific bias is hard to establish.

We have developed a tailored DNA extraction method that allows a better assessment of the abundance and diversity of endospore-formers in environmental samples for amplicon sequencing [12,57]. Therefore, we next evaluated if using this extraction protocol in an environmental sample could improve the detection of endospore-formers in a metagenome.

3.3. Amplicon Sequencing of an Environmental Sample With High Prevalence of Endospore-forming Firmicutes

We performed amplicon sequencing from a sample in which high prevalence of endospore-forming Firmicutes was suspected from the ratio of 16S rRNA (bacterial) and spo0A (endospore-formers) gene numbers measured by quantitative PCR [58]. This ratio was obtained from DNA extracted using our modified protocol. Sequencing of the 16S rRNA and spo0A gene amplicons was conducted and revealed not only a high prevalence of endospore-forming Firmicutes, but also a high diversity of endospore formers (Fig. 3).

In the amplicon sequencing of the 16S rRNA gene, Firmicutes accounted for 41.70% of the total bacterial community. The abundance of 16S rRNA amplicons corresponding to Firmicutes was nearly double the amount of Proteobacteria, which was the second most abundant bacterial Phylum (26.14%). Among the endospore-formers observed in the pyrosequencing results, the genera Clostridium and Desulfosporosinus dominated the community in the sample, indicating a clear dominance of anaerobic endospore-formers [59] as could be expected considering the temperature and other environmental conditions at this geothermal spring. Amplicons affiliated to Clostridium and Desulfosporosinus were also dominant in the spo0A amplicon sequencing, which also showed the dominance of anaerobic endospore-formers. Even though spo0A sequences related to aerobic endospore-formers (e.g. Geobacillus and Bacillus) were also obtained, the classification of the spo0A from aerobic endospore-formers was ambiguous as shown by the existence of, for example, clades related to Anoxybacillus but placed at different positions in the phylogeny (Fig. 3C). In fact, only recently environmental spo0A sequences have started to be obtained [12], and the phylogenetic assignment needs to be refined.

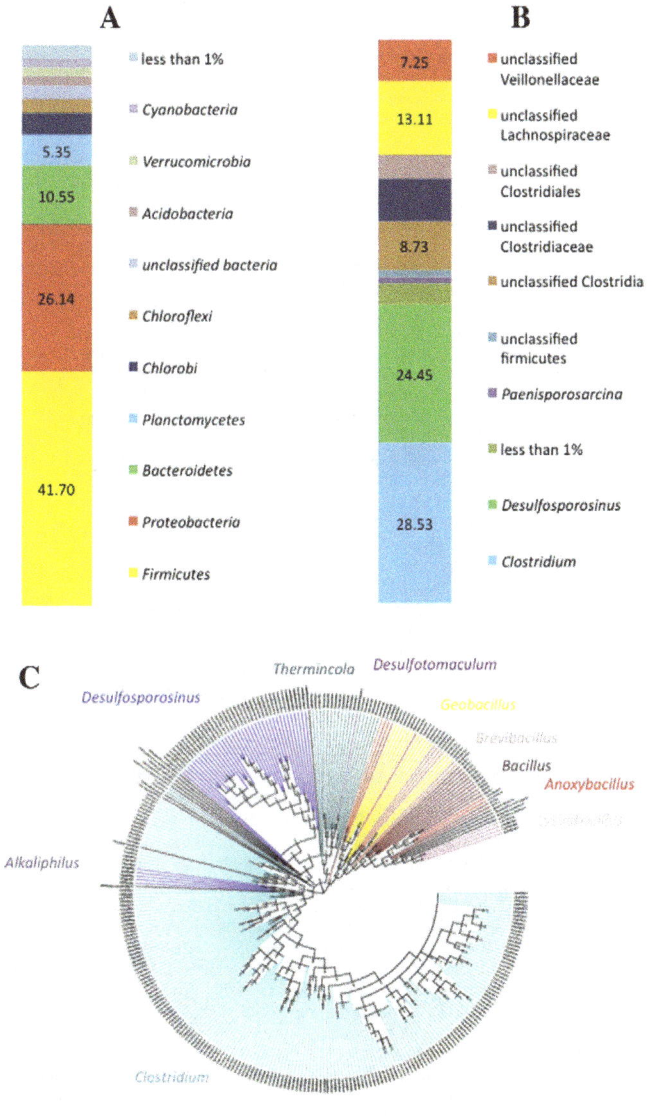

3.4. Metagenomic Sequencing

In addition to pyrosequencing, the same sample was also subjected to metagenomic sequencing. It is worth mentioning that in whole-genome metagenomics a PCR amplification bias does not apply and thus we did not necessarily expect to find the same groups or the same frequency detected in the amplicon sequencing. However, the results of the qPCR quantification and the amplicon sequencing were taken as an indication of the prevalence of Firmicutes in this specific environmental sample. The NAP dataset consisted of a total of 481,810 sequences of average length of 330 bp. When the Spo0A and Gpr profile analyses were conducted on this metagenome, none of the two genes were detected. However, looking only at two specific genes could be an issue, since those could be, for various reasons, underrepresented in the sequences. Therefore, an extended search for reads that could be assigned to Firmicutes using different prediction tools on the assembled metagenome was also carried out.

Relative abundances from classified reads were considered to establish the five most prevalent Phyla present in the sample (Table 1). Firmicutes appear in the top five Phyla only for two of the four prediction tools used. In the case of Kraken, Firmicutes reads corresponded to 1.60% of the classified data, being the third most abundant phylum (the most abundant one was Proteobacteria with 82.71%). BWA predicted 5.32% of the classified sequences as to belong to Firmicutes (second most abundant phylum after Proteobacteria with 75.21%). Firmicutes were not listed after classification with MetaPhlAN and LMAT. Likewise, when reconstruction of full bacterial genomes was attempted for the NAP metagenome using MetaPhlAn, none of the top 5 microorganisms was assigned to Firmicutes (data not shown).

Thus, even though amplicon sequencing revealed a large fraction of the community as belonging to Firmicutes, this was not observed in the shotgun metagenome. There are several possible explanations for these results. One of those is the fact that the ribosomal (rrn) operon is normally found in several copies and thus the representation of a microbial community based on 16S rRNA gene sequencing is skewed. Furthermore, the average number of rrn operon copies depends on the group of bacteria. An average value of 7.01 copies of 16S rRNA genes was found for the phylum Firmicutes in the rrnDB [60], which implies that this group can be overrepresented in 16S rRNA gene amplicon libraries. In addition, it should be noted that for all the tools used, classification was poor and only a very small fraction of the sequences could be actually assigned to a particular taxonomic group. Therefore, the lack of detection of Firmicutes could be due to the current limitations of the analysis tools. In fact, recent sequencing technologies generate such large quantities of data as to bring along a new set of challenges in data analysis, the so-called bioinformatics bottleneck [61]. On the level of interpretation of metagenomic data there is still an important amount of unexplored information available from the results, simply because the advances in sequencing technologies are greater than the complementary progress in annotation, data inventory and standardization of metadata [14].

Fig. 3. Analysis of pyrosequencing results obtained from 16S rRNA gene and spo0A amplicons, from an environmental sample with high prevalence of endospore-forming Firmicutes (Nea Apollonia, NAP). (A) Total 16S rRNA gene community composition to the phylum level. (B) Firmicute fraction of the total community (16S rRNA gene) to the genus level. (C). Cladogram representing the community composition of Firmicutes using the spo0A gene. Sequences color coded by genus.

4. Conclusions

Since Staley and Konopka introduced the "great plate count anomaly" [62,63], revealing that only a small fraction of the microbial community can be cultured in the laboratory, one of the great challenges in environmental microbiology is the understanding of the diversity and metabolic capabilities of microbes in a culture-independent manner. That bias was partly overcome by moving into the direction of directly extracting genetic material from environmental samples. However, our results reveal that for specific microbial groups, we are still in a phase in which, similar to a percentage of the community being *not culturable* in culture-based approaches, a fraction of the genomes of the community might be considered as *not detectable* for culture-independent approaches. Nonetheless, profiling of the taxonomic and phylogenetic composition of microbial communities is at the heart of many metagenomic studies, and it is an obligatory step to draw conclusions on the role of microorganisms in the environment based on metagenomics. Our results suggest that in the case of endospore-forming *Firmicutes*, classification by various methods still lags behind. However, starting from samples such as NAP, in which evidence for high frequency of this bacterial group exists, could be the first step towards developing improved methods of classification and phylogenetic assignment of metagenomic data.

Acknowledgments

This work was supported by the Swiss National Science Foundation grant Nos. 31003A-132358/1 and 31003A_152972, from Fondation Pierre Mercier pour la science and from REGARD for equality of women in science.

References

[1] Suenaga H. Targeted metagenomics: a high-resolution metagenomics approach for specific gene clusters in complex microbial communities. Environ Microbiol 2012; 14:13–22. http://dx.doi.org/10.1111/j.1462-2920.2011.02438.x.

[2] Warnecke F, Hugenholtz P. Building on basic metagenomics with complementary technologies. Genome Biol 2007;8:231. http://dx.doi.org/10.1186/gb-2007-8-12-231.

[3] Xu J. Microbial ecology in the age of genomics and metagenomics: concepts, tools, and recent advances. Mol Ecol 2006;15:1713–31. http://dx.doi.org/10.1111/j.1365-294X.2006.02882.x.

[4] Béjà O, Suzuki MT, Heidelberg JF, Nelson WC, Preston CM, Hamada T, et al. Unsuspected diversity among marine aerobic anoxygenic phototrophs. Nature 2002;415:630–3. http://dx.doi.org/10.1038/415630a.

[5] Béjà O, Aravind L, Koonin EV, Suzuki MT, Hadd A, Nguyen LP, et al. Bacterial rhodopsin: evidence for a new type of phototrophy in the sea. Science 2000;289:1902–6.

[6] Ram RJ, Verberkmoes NC, Thelen MP, Tyson GW, Baker BJ, Blake RC, et al. Community proteomics of a natural microbial biofilm. Science 2005;308:1915–20. http://dx.doi.org/10.1126/science. 1109070.

[7] Venter JC, Remington K, Heidelberg JF, Halpern AL, Rusch D, Eisen JA, et al. Environmental genome shotgun sequencing of the Sargasso Sea. Science 2004;304:66–74. http://dx.doi.org/10.1126/science.1093857.

[8] Voget S, Leggewie C, Uesbeck A, Raasch C, Jaeger K-E, Streit WR. Prospecting for novel biocatalysts in a soil metagenome. Appl Environ Microbiol 2003;69:6235–42.

[9] Delmont TO, Robe P, Cecillon S, Clark IM, Constancias F, Simonet P, et al. Accessing the soil metagenome for studies of microbial diversity. Appl Environ Microbiol 2011;77:1315–24. http://dx.doi.org/10.1128/AEM.01526-10.

[10] Lombard N, Prestat E, van Elsas JD, Simonet P. Soil-specific limitations for access and analysis of soil microbial communities by metagenomics. FEMS Microbiol Ecol 2011; 78:31–49. http://dx.doi.org/10.1111/j.1574-6941.2011.01140.x.

[11] Pinard R, de Winter A, Sarkis GJ, Gerstein MB, Tartaro KR, Plant RN, et al. Assessment of whole genome amplification-induced bias through high-throughput, massively parallel whole genome sequencing. BMC Genomics 2006;7:216. http://dx.doi.org/10.1186/1471-2164-7-216.

[12] Wunderlin T, Junier T, Roussel-Delif L, Jeanneret N, Junier P. Stage 0 sporulation gene A as a molecular marker to study diversity of endospore-forming Firmicutes. Environ Microbiol Rep 2013;5:911–24. http://dx.doi.org/10.1111/1758-2229.12094.

[13] Batmalle CS, Chiang H-I, Zhang K, Lomas MW, Martiny AC. Development and bias assessment of a method for targeted metagenomic sequencing of marine cyanobacteria. Appl Environ Microbiol 2014;80:1116–25. http://dx.doi.org/10.1128/AEM.02834-13.

[14] Yilmaz P, Gilbert JA, Knight R, Amaral-Zettler L, Karsch-Mizrachi I, Cochrane G, et al. The genomic standards consortium: bringing standards to life for microbial ecology. ISME J 2011;5:1565–7. http://dx.doi.org/10.1038/ismej.2011.39.

[15] Lagier J-C, Armougom F, Million M, Hugon P, Pagnier I, Robert C, et al. Microbial culturomics: paradigm shift in the human gut microbiome study. Clin Microbiol Infect 2012;18:1185–93. http://dx.doi.org/10.1111/1469-0691.12023.

[16] Lagier J-C, Hugon P, Khelaifia S, Fournier P-E, Scola BL, Raoult D. The rebirth of culture in microbiology through the example of culturomics to study human gut microbiota. Clin Microbiol Rev 2015;28:237–64. http://dx.doi.org/10.1128/CMR.00014-14.

[17] Hugenholtz P. Exploring prokaryotic diversity in the genomic era. Genome Biol 2002;3 [reviews0003.1–reviews0003.8].

[18] Nicholson WL, Munakata N, Horneck G, Melosh HJ, Setlow P. Resistance of *Bacillus* endospores to extreme terrestrial and extraterrestrial environments. Microbiol Mol Biol Rev 2000;64:548–72.

[19] Nicholson WL. Roles of *Bacillus* endospores in the environment. Cell Mol Life Sci 2002;59:410–6.

[20] Driks A. Overview: development in bacteria: spore formation in *Bacillus* subtilis. Cell Mol Life Sci 2002;59:389–91.

[21] Onyenwoke RU, Brill JA, Farahi K, Wiegel J. Sporulation genes in members of the low G + C Gram-type-positive phylogenetic branch (Firmicutes). Arch Microbiol 2004; 182:182–92. http://dx.doi.org/10.1007/s00203-004-0696-y.

[22] Yudkin MD, Clarkson J. Differential gene expression in genetically identical sister cells: the initiation of sporulation in *Bacillus* subtilis. Mol Microbiol 2005;56: 578–89. http://dx.doi.org/10.1111/j.1365-2958.2005.04594.x.

[23] Martiny JBH, Bohannan BJM, Brown JH, Colwell RK, Fuhrman JA, Green JL, et al. Microbial biogeography: putting microorganisms on the map. Nat Rev Microbiol 2006; 4:102–12. http://dx.doi.org/10.1038/nrmicro1341.

[24] Von Mering C, Hugenholtz P, Raes J, Tringe SG, Doerks T, Jensen LJ, et al. Quantitative phylogenetic assessment of microbial communities in diverse environments. Science 2007;315:1126–30. http://dx.doi.org/10.1126/science.1133420.

[25] Wunderlin T, Junier T, Roussel-Delif L, Jeanneret N, Junier P. Endospore-enriched sequencing approach reveals unprecedented diversity of Firmicutes in sediments. Environ Microbiol Rep 2014;6:631–9. http://dx.doi.org/10.1111/1758-2229.12179.

[26] Altschul SF, Madden TL, Schäffer AA, Zhang J, Zhang Z, Miller W, et al. Gapped BLAST and PSI-BLAST: a new generation of protein database search programs. Nucleic Acids Res 1997;25:3389–402.

[27] Kriventseva EV, Rahman N, Espinosa O, Zdobnov EM. OrthoDB: the hierarchical catalog of eukaryotic orthologs. Nucleic Acids Res 2008;36:D271–5. http://dx.doi.org/10.1093/nar/gkm845.

[28] Katoh K, Misawa K, Kuma K, Miyata T. MAFFT: a novel method for rapid multiple sequence alignment based on fast Fourier transform. Nucleic Acids Res 2002;30:3059–66.

[29] Rice P, Longden I, Bleasby A. EMBOSS: the European Molecular Biology Open Software Suite. Trends Genet 2000;16:276–7.

[30] Bueche M, Wunderlin T, Roussel-Delif L, Junier T, Sauvain L, Jeanneret N, et al. Quantification of endospore-forming firmicutes by quantitative PCR with the functional gene spo0A. Appl Environ Microbiol 2013;79:5302–12. http://dx.doi.org/10.1128/AEM.01376-13.

[31] Li H, Zhang Y, Li D, Xu H, Chen G, Zhang C. Comparisons of different hypervariable regions of rrs genes for fingerprinting of microbial communities in paddy soils. Soil Biol Biochem 2009;41:954–68. http://dx.doi.org/10.1016/j.soilbio.2008.10.030.

[32] Caporaso JG, Kuczynski J, Stombaugh J, Bittinger K, Bushman FD, Costello EK, et al. QIIME allows analysis of high-throughput community sequencing data. Nat Methods 2010;7:335–6. http://dx.doi.org/10.1038/nmeth.f.303.

[33] Ewing B, Green P. Base-calling of automated sequencer traces using phred. II. Error probabilities. Genome Res 1998;8:186–94.

[34] Guindon S, Gascuel O. A simple, fast, and accurate algorithm to estimate large phylogenies by maximum likelihood. Syst Biol 2003;52:696–704.

[35] Mulder NJ, Apweiler R, Attwood TK, Bairoch A, Bateman A, Binns D, et al. InterPro: an integrated documentation resource for protein families, domains and functional sites. Brief Bioinform 2002;3:225–35.

[36] Freitas TAK, Li P-E, Scholz MB, Chain PSG. Accurate read-based metagenome characterization using a hierarchical suite of unique signatures. Nucleic Acids Res 2015. http://dx.doi.org/10.1093/nar/gkv180 [gkv180].

[37] Li H, Durbin R. Fast and accurate short read alignment with Burrows–Wheeler transform. Bioinformatics 2009;25:1754–60. http://dx.doi.org/10.1093/bioinformatics/btp324.

[38] Segata N, Waldron L, Ballarini A, Narasimhan V, Jousson O, Huttenhower C. Metagenomic microbial community profiling using unique clade-specific marker genes. Nat Methods 2012;9:811–4. http://dx.doi.org/10.1038/nmeth.2066.

[39] Langmead B, Salzberg SL. Fast gapped-read alignment with Bowtie 2. Nat Methods 2012;9:357–9. http://dx.doi.org/10.1038/nmeth.1923.

[40] Pruitt KD, Brown GR, Hiatt SM, Thibaud-Nissen F, Astashyn A, Ermolaeva O, et al. RefSeq: an update on mammalian reference sequences. Nucleic Acids Res 2014; 42:D756–63. http://dx.doi.org/10.1093/nar/gkt1114.

[41] Sunagawa S, Mende DR, Zeller G, Izquierdo-Carrasco F, Berger SA, Kultima JR, et al. Metagenomic species profiling using universal phylogenetic marker genes. Nat Methods 2013;10:1196–9. http://dx.doi.org/10.1038/nmeth.2693.

[42] Ames SK, Hysom DA, Gardner SN, Lloyd GS, Gokhale MB, Allen JE. Scalable metagenomic taxonomy classification using a reference genome database. Bioinformatics 2013;29:2253–60. http://dx.doi.org/10.1093/bioinformatics/btt389.

[43] Stragier P, Losick R. Molecular genetics of sporulation in *Bacillus* subtilis. Annu Rev Genet 1996;30. http://dx.doi.org/10.1146/annurev.genet.30.1.297 [297–241].

[44] Abecasis AB, Serrano M, Alves R, Quintais L, Pereira-Leal JB, Henriques AO. A genomic signature and the identification of new sporulation genes. J Bacteriol 2013;195: 2101–15. http://dx.doi.org/10.1128/JB.02110-12.

[45] Traag BA, Pugliese A, Eisen JA, Losick R. Gene conservation among endospore-forming bacteria reveals additional sporulation genes in *Bacillus* subtilis. J Bacteriol 2013;195:253–60. http://dx.doi.org/10.1128/JB.01778-12.

[46] Galperin MY, Mekhedov SL, Puigbo P, Smirnov S, Wolf YI, Rigden DJ. Genomic determinants of sporulation in Bacilli and Clostridia: towards the minimal set of sporulation-specific genes. Environ Microbiol 2012;14:2870–90. http://dx.doi.org/10.1111/j.1462-2920.2012.02841.x

[47] Gribskov M, McLachlan AD, Eisenberg D. Profile analysis: detection of distantly related proteins. Proc Natl Acad Sci U S A 1987;84:4355–8.

[48] Gill SR, Pop M, Deboy RT, Eckburg PB, Turnbaugh PJ, Samuel BS, et al. Metagenomic analysis of the human distal gut microbiome. Science 2006;312:1355–9. http://dx.doi.org/10.1126/science.1124234.

[49] Kurokawa K, Itoh T, Kuwahara T, Oshima K, Toh H, Toyoda A, et al. Comparative metagenomics revealed commonly enriched gene sets in human gut microbiomes DNA Res 2007;14:169–81. http://dx.doi.org/10.1093/dnares/dsm018.

[50] Zoetendal EG, Vaughan EE, De Vos WM. A microbial world within us. Mol Microbiol 2006;59:1639–50. http://dx.doi.org/10.1111/j.1365-2958.2006.05056.x.

[51] Suzuki TA, Worobey M. Geographical variation of human gut microbial composition. Biol Lett 2014;10:20131037. http://dx.doi.org/10.1098/rsbl.2013.1037.

[52] Felske ADM, Tzeneva V, Heyrman J, Langeveld MA, Akkermans ADL, De Vos P. Isolation and biodiversity of hitherto undescribed soil bacteria related to Bacillus niacini. Microb Ecol 2004;48:111–9. http://dx.doi.org/10.1007/s00248-003-2025-4.

[53] Hoyles L, Honda H, Logan NA, Halket G, La Ragione RM, McCartney AL. Recognition of greater diversity of Bacillus species and related bacteria in human faeces. Res Microbiol 2012;163:3–13. http://dx.doi.org/10.1016/j.resmic.2011.10.004.

[54] Thomas T, Gilbert J, Meyer F. Metagenomics — a guide from sampling to data analysis. Microb Inf Exp 2012;2:3. http://dx.doi.org/10.1186/2042-5783-2-3.

[55] Pope PB, Denman SE, Jones M, Tringe SG, Barry K, Malfatti SA, et al. Adaptation to herbivory by the Tammar wallaby includes bacterial and glycoside hydrolase profiles different from other herbivores. Proc Natl Acad Sci U S A 2010;107:14793–8. http://dx.doi.org/10.1073/pnas.1005297107.

[56] Allgaier M, Reddy A, Park JI, Ivanova N, D'haeseleer P, Lowry S, et al. Targeted discovery of glycoside hydrolases from a switchgrass-adapted compost community. PLoS One 2010;5:e8812. http://dx.doi.org/10.1371/journal.pone.0008812.

[57] Wunderlin T, Junier T, Roussel-Delif L, Junier P. Profile analyses of sporulation genes reveal underrepresentation of endospore- forming bacteria in metagenomes. ISME14, Copenhagen, Denmark; 2012.

[58] S. Filippidou, M. Bueche, T. Wunderlin, T. Junier, L. Roussel-Delif, N. Jeanneret, et al. Survival strategy meets classic ecological theory: the case of diversity and abundance of endospore-forming Firmicutes in extreme environments. Prep n.d.

[59] Schleifer KH. Classification of bacteria and archaea: past, present and future. Syst Appl Microbiol 2009;32:533–42. http://dx.doi.org/10.1016/j.syapm.2009.09.002.

[60] Lee ZM-P, Bussema C, Schmidt TM. rrnDB: documenting the number of rRNA and tRNA genes in bacteria and archaea. Nucleic Acids Res 2009;37:D489–93. http://dx.doi.org/10.1093/nar/gkn689.

[61] Scholz MB, Lo C-C, Chain PSG. Next generation sequencing and bioinformatic bottlenecks: the current state of metagenomic data analysis. Curr Opin Biotechnol 2012;23:9–15. http://dx.doi.org/10.1016/j.copbio.2011.11.013.

[62] Staley JT, Konopka A. Measurement of in situ activities of nonphotosynthetic microorganisms in aquatic and terrestrial habitats. Annu Rev Microbiol 1985;39:321–46. http://dx.doi.org/10.1146/annurev.mi.39.100185.001541.

[63] Amann RI, Ludwig W, Schleifer KH. Phylogenetic identification and in situ detection of individual microbial cells without cultivation. Microbiol Rev 1995;59:143–69.

[64] Junier T, Zdobnov EM. The Newick Utilities: High-throughput Phylogenetic tree Processing in the UNIX Shell. Bioinformatics 2010;26:1669–70.

11

Identification of miRNAs contributing to neuroblastoma chemoresistance

Duncan Ayers [a,b,*], Pieter Mestdagh [c], Tom Van Maerken [c], Jo Vandesompele [c]

[a] Centre for Molecular Medicine and Biobanking, University of Malta, Msida, Malta
[b] Manchester Institute of Biotechnology, Faculty of Medical and Human Sciences, The University of Manchester, United Kingdom
[c] Center for Medical Genetics Ghent, Ghent University Hospital, Ghent, Belgium

ARTICLE INFO

Keywords:
miRNA
Drug
Resistance
Chemoresistance
Neuroblastoma

ABSTRACT

Background: The emergence of the role of microRNAs (miRNAs) in exacerbating drug resistance of tumours is recently being highlighted as a crucial research field for future clinical management of drug resistant tumours. The purpose of this study was to identify dys-regulations in expression of individual and/or networks of miRNAs that may have direct effect on neuroblastoma (NB) drug resistance.

Methods: Individual subcultures of chemosensitive SH-SY5Y and UKF-NB-3 cells were rendered chemoresistant to doxorubicin (SH-SY5Y, UKF-NB-3) or etoposide (SH-SY5Y). In each validated chemoresistance model, the parental and subcultured cell lines were analysed for miRNA expression profiling, using a high-throughput quantitative polymerase chain reaction (RT-qPCR) miRNA profiling platform for a total of 668 miRNAs.

Results: A unique expression signature of miRNAs was found to be differentially expressed (higher than 2-fold change) within all three NB chemoresistance models. Four miRNAs were upregulated in the subcultured chemoresistant cell line. Three miRNAs were found to be downregulated in the chemoresistant cell lines for all models.

Conclusions: Based on the initial miRNA findings, this study elucidates the dys-regulation of four miRNAs in three separate NB chemoresistant cell line models, spanning two cell lines (SH-SY5Y and UKF-NB-3) and two chemotherapeutic agents (doxorubicin and etoposide). These miRNAs may thus be possibly linked to chemoresistance induction in NB. Such miRNAs are good candidates to be novel drug targets for future miRNA based therapies against aggressive tumours that are not responding to conventional chemotherapy.

1. Background

Undoubtedly, one of the most frequently occurring malignancies in childhood is neuroblastoma (NB) [1], in most cases affecting children under the age of five [2]. Neuroblastoma is a tumour that develops in the embryonic stage [3] and derives from primordial cells of the sympathetic nervous system known as neural crest cells [4]. These progenitor cells cease to differentiate and mature, which is the natural course of events in neural crest cells en-route to the development of the sympathetic nervous system in the embryo [5]. The tumourigenesis from these progenitor cells is thought to be due to the formation of self-regenerating tumour stem cells that have the ability to produce a distinct range of different NB cell lineages according to their histology [6].

The clinical manifestations and degree of severity of NB may be highly varied [7]. The initial phase of the condition is the presence of a painless lump on the abdomen, neck or chest of the child [8]. However, the tumour usually undergoes metastasis and the child consequently presents symptoms due to the NB tumour acting as a space-occupying lesion in the areas of the body that are affected. In order to aid clinicians and oncologists to determine the severity of NB in the individual patient and consequently implement more bespoke treatment strategies, an International Neuroblastoma Staging System (INSS) [9] was devised, based on the degree of NB metastasis and microscopic analysis of the afflicted tissues of the patient. This staging system has proved to be of immense value and is still presently being applied for neuroblastoma diagnosis.

In overview, patients diagnosed with stage 1 or 2 NB conditions are mainly subjected to surgery for excising the tumour mass, without the use of chemotherapeutic cycles or other treatments [10]. However, patients with stage 3 or 4 NB are at high risk and thus tumour de-bulking/removal surgery is combined to other treatment strategies such as chemotherapy (with or without bone marrow transplantation) and low-dose radiotherapy [10].

Unfortunately, the emergence of chemoresistance within tumour cells of solid tissues is one of the main reasons for treatment failure and relapse in patients suffering from metastatic cancer conditions [11]. Resistance of the tumour cell to chemotherapeutic agent exposure may be innate, whereby the genetic characteristics of the tumour cells are naturally resistant to chemotherapeutic drug exposure [12].

* Corresponding author.

Alternatively, chemoresistance can be acquired through development of a drug resistant phenotype over a defined time period of exposure of the tumour cell to multiple chemotherapy combinations [11,12]. Apart from other biological and genetic factors influencing the chemoresistance properties of such tumours, the role of non-coding RNAs such as micro RNAs (miRNAs) [13,14] in the induction of such a phenotype is rapidly being recognised [15–19].

Within the context of NB, there already exist links between miRNA dysregulated expression patterns and NB clinical severity [20]. However, scientific literature provides limited studies relating to the identification of miRNAs directly contributing to NB chemoresistance, with one of them being *miR-17-5p* [21]. This study revealed that the exposure of chemotherapy resistant NB murine tumour models to antagomirs for miR-17-5p resulted in an increase in prognosis, due to enhanced apoptosis and cell cycle arrest effects [21]. In addition, miR-21 is a known oncomir and was found to be over-expressed in cisplatin resistant NB cell lines [22]. Another study has also identified miR-204 as a potentially important miRNA that conveys suppressive properties against the chemoresistance phenotype in NB [23].

2. Methodology

The purpose of this study was to identify and validate (if any) specific miRNAs that could be directly involved in the exacerbation of chemoresistance properties in NB cell lines. The utilisation of RT-qPCR based miRNA profiling can nowadays be employed for serving such research objectives.

2.1. Description of NB cell line chemoresistance models

Eight separate NB cell line models were implemented in this study. Each model consisted of two cell lines, a parental chemosensitive together with a sub-cultured cell line that was rendered chemoresistant to a single conventional chemotherapeutic agent through repeated exposure (see Table 1).

2.2. Validation of chemosensitivity status for NB cell lines

Prior to miRNA profiling of all NB cell lines, validation of their chemosensitivity status was paramount to ensure that the subcultured chemoresistant cell line in each model retained its drug resistance properties over several culturing passages without further drug exposure.

The method adopted for validating chemoresistance status in each model consisted of luminescence-based cell viability analysis, through the utilisation of the Cell Titer Glo assay [Promega, USA] on a Fluorostar Optima luminescence plate reader platform [BMG, Germany], according to the manufacturer's protocol. Due care was applied in maintaining a seeding cell population across all culture wells utilised in the functional assay, in order to avoid false positive viability readings.

2.2.1. Cell culturing conditions and harvesting

The growth medium utilised for culturing of all cell lines in this study consisted of RPMI 1640 [Gibco, USA] treated with 10% foetal calf serum [Gibco, USA] and 1% by volume of 1:1 penicillin/streptomycin mix 10,000 U/mL, kanamicin 10 mg/mL and L-glutamate 200 mM [Gibco, USA] respectively. Prior to a three-minute step within T25 cell culturing flasks [Corning, USA], versene solution [Gibco, USA] was used for washing all adherent cells within the culture flasks. The trypsin solution utilised consisted of 0.05% trypsin with EDTA [Gibco, USA]. Harvesting of cell suspensions for RNA extraction was performed by adding 0.7 mL Qiazol solution [Qiagen, Germany].

2.2.2. NB chemoresistance model cell viability assay

For each chemoresistance model investigated in this study, both the parental chemosensitive and subcultured chemoresistant NB cell lines were grown in culture and simultaneously harvested and transferred to a sterile 50 mL tube [BD Bioscience, USA]. The cell population density for each cell line was counted by extraction of a 20 µL sample (following vigorous pipette mixing) and analysed on a Cellometer Auto T4 cell counter platform [Nexcelom Bioscience, USA] according to manufacturer's protocol. Both cell cultures where then diluted as necessary with sufficient growth medium to produce a cell culture with a cell population density of approximately 10,000 cells/95 µL. Consequently, 95 µL aliquots from each cell line were transferred (following vigorous pipette mixing) to an opaque Nunclon 96 well microtitre plate [Nunc, USA].

For each cell line, 18 wells were utilised for consequent drug treatment, and nine wells were to serve as non-treated NB negative controls (NB cell culture only, with no drug exposure). A further six wells were treated with 195 µL growth medium only, in order to act as negative controls (drug exposure only). All remaining wells were treated with 200 µL growth medium. Finally, the microtitre plate was placed in an incubator at 37 °C and 5% CO_2 for 24 h, to allow cell adhesion on well surface.

Following the 24-hour incubation, the microtitre plate containing adherent cells was retrieved and treated with the appropriate chemotherapeutic agents for each chemoresistance cell line model. The varying drug dilutions (see Table 2) were prepared on the same day when used for treatment, using filter sterilised water [Sigma, USA] within the confines of a Class II laminar flow cabinet. All light sensitive drug dilutions were also covered in aluminium foil until required.

Once retrieved from the incubator, the plate was placed inside a laminar flow cabinet and all wells designated for treatment with the appropriate chemotherapeutic agent were exposed. A 5 µL aliquot of the appropriate drug and dose was pipetted, following vigorous pipette mixing of the drug solution, into the appropriate triplicate wells. An extra 5 µL aliquot of each drug dose was applied to a single well containing only growth medium, in order to obtain background luminescence data from a drug/growth medium solution for future use following cell viability analysis. The plate was then re-incubated for a further 36 h.

Following this time period, the plate was collected and a cell viability assay was performed on all wells, by using the Cell Titer Glo assay [Promega, USA] according to manufacturer's protocol.

Table 1
List of NB chemoresistance cell line models. Subcultures of each parental cell line were induced to acquire chemoresistance properties towards a single chemotherapeutic agent. The SH-SY5Y and Kelly cell lines were obtained from the Eggert group, Essen, Germany. The UKF-NB-3 cell lines were obtained from the Cinatl Jr group, Frankfurt, Germany.

NB cell line	Chemotherapeutic agent
SH-SY5Y	Cisplatin
SH-SY5Y	Doxorubicin
SH-SY5Y	Etoposide
KELLY	Cisplatin
KELLY	Doxorubicin
KELLY	Etoposide
UKF-NB-3	Doxorubicin
UKF-NB-3	Vincristine

Table 2
Chemotherapeutic drug ranges utilised for investigating chemosensitivity status in each cell line. All treatments were performed in triplicate for each drug dose.

Plate row location	Chemotherapeutic Drug dose range (ng/mL)			
	Cisplatin	Doxorubicin	Etoposide	Vincristine
B	50,000	28,999	1,000,000	4615
C	5000	2899.9	100,000	461.5
D	500	289.99	10,000	46.15
E	50	28.999	1000	4.615
F	5	2.9	100	0.465
G	0.5	0.29	10	0.046

2.2.3. Data analysis and statistical analysis

Following the assay, the Excel sheet containing all raw luminescence data was collected for analysis, with outliers being identified and discarded. The chemosensitivity profiles, based on the cell viability analysis data set results, were plotted by application of the Graph Pad Prism software package [GraphPad Software Inc., USA]. No detailed statistical analysis was performed on the resultant data sets as long as the inhibiting concentration for 50% cell viability (IC50) demonstrated distinct values, reflecting the chemosensitivity profiles.

2.3. NB cell line miRNA profiling

2.3.1. Cell line harvesting and lysis

All cell lines from each validated NB chemoresistance model were cultured and harvested accordingly. Following trypsin incubation step, each T25 flask was subjected to horizontal mechanical shock, necessary for dislodging all cells from the flask surface. The resultant cell suspension from each flask was transferred to an individually labelled, sterile 15 mL collection tube [BD Biosciences, USA] and centrifuged at 1500 rpm for five minutes to allow all cells to form a pellet at the base of the collection tube. The supernatant from each collection tube was aspirated by disposable, sterile glass pipette with due care to avoid inadvertent aspiration of cell pellet. Consequently, the remaining cell pellet was treated with 700 μL of Qiazol solution [Qiagen, Germany] in order to induce cell lysis. This step was performed within the confines of a fume cupboard and thorough pipette-mixing was applied for ensuring a resultant homogenous cell lysate suspension. Immediately after cell lysis induction, all cell lysate suspensions were snap frozen in liquid nitrogen solution and transferred to a −80 °C freezer until RNA extraction was performed.

2.3.2. miRNA extraction procedure

All NB cell line lysates were allowed to thaw to room temperature prior to miRNA extraction. Consequently, all lysates were treated with the miRNeasy [Qiagen, Germany] miRNA extraction protocol, within the confines of a fume cupboard. The resultant 30 μL volume of extracted miRNA from each NB cell line lysate was quantified through utilisation of the Nanodrop UV spectrophotometry platform [Nanodrop Technologies, USA]. The extracted miRNA samples were then stored at −80 °C until further use. The RNA quality was not tested following extraction, as previous experience from earlier studies repeatedly demonstrated that the RNA quality derived from NB cell lines was always of an excellent nature (with RIN values above 9.0), thus such a step was not performed.

2.3.3. RT-qPCR based miRNA profiling

All miRNA samples were allowed to thaw prior to further treatments. The protocol utilised in this step was a validated, high throughput RT-qPCR based miRNA profiling method [30]. In summary, miRNA samples were diluted to a standardised concentration and initially subjected to a megaplex reverse transcription protocol [30]. This was followed by a pre-amplification step and eventual qPCR analysis [30].

2.3.4. Data normalisation and analysis

All raw Cq values obtained from each individual run were corrected by inter-run calibrators (five individual small nucleic RNA assays – RNU24, RNU44, RNU48, RNU6B, U6 snRNA; two technical replicates/assay/plate) and consequently normalised against the average Cq value obtained from the total quantity of miRNAs assayed within the same run [27]. The normalised Cq expression data for each miRNA was compared between the constituent chemosensitive and chemoresistant cell lines for each chemoresistance model. All miRNAs found to be dysregulated within the chemoresistant cell line, following the adoption of a +/− 2× linear fold change cut-off value, were deemed to be putative chemoresistance miRNAs within the individual chemoresistance model. This nominal cut-off value is subjective and was deemed to be acceptable to attain a

legitimate short list of miRNAs having a tangible expression dysregulation profile of sufficient weight as to lead to an overall shift in chemosensitivity profile (in this case, see Discussion section).

Ultimately, all miRNAs dysregulated in all three chemoresistance models' miRNA shortlists were identified and selected for further investigation.

2.3.5. Multiplex RT-qPCR for putative chemoresistance miRNAs

The putative miRNAs identified from miRNA profiling to be dysregulated in the chemoresistant cell lines were further analysed by multiplex RT-qPCR.

Taqman miRNA RT kit [ABI, USA] was utilised for the reverse transcription step. Reverse transcriptase (RT) primers specific for the relevant target miRNAs [ABI, USA] were diluted (5 nmol) with 250 μL nuclease free water [Sigma, USA], and 10 μL aliquots from each individual primer solution were pooled in a 1.5 mL Eppendorf tube. The primer pool was then diluted with nuclease free water or concentrated by vacuum centrifugation in order to obtain a final volume equivalent to 20% of the total RT reaction volume utilised in this step. The RT reaction

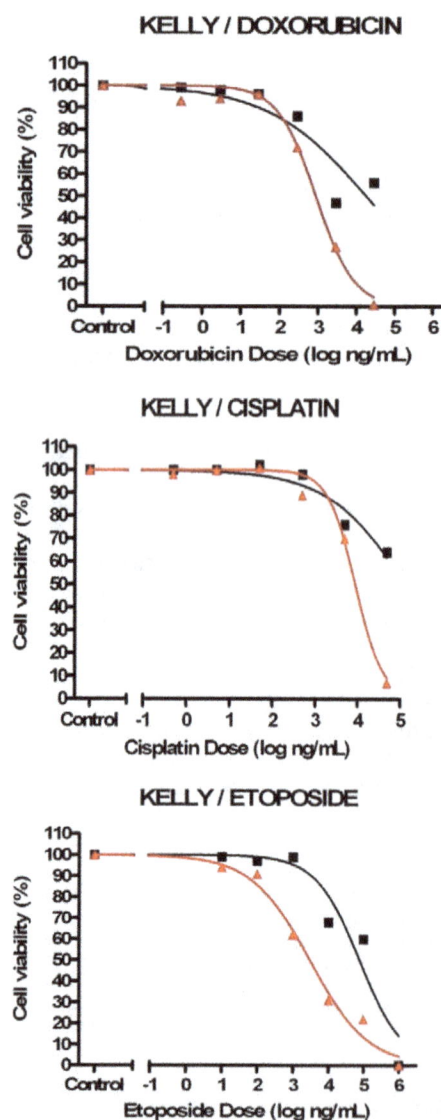

Fig. 1. Results for cell viability analysis of Kelly NB chemoresistance models (n = 1, three technical replicates/data point, SEM not illustrated). In all three assays the chemosensitivity status for each component NB cell line was not suitable for inclusion in the miRNA profiling study (black – chemosensitive parental NB cell line; red – chemoresistant NB cell line).

volume (per well) consisted of 4 µL stem-loop primer pool, 0.4 µL of 100 nM dNTPs, 4 µL MultiScribe reverse transcriptase [Applied Biosystems, USA], 2 µL of 10× RT buffer, 0.25 µL RNase inhibitor [Applied Biosystems, USA] and 0.65 µL nuclease free water. The total reaction volume was then placed in a thermal cycler and subjected to 30 min at a temperature of 16 °C, followed by 30 min at 42 °C, 1 s at 50 °C, 5 min at 85 °C and 5 min at 4 °C respectively. The finalised reaction volume was diluted by a factor of five, prior to RT-qPCR step.

The TaqMan® miRNA assay for RT-qPCR quantification was utilised in this step for all miRNA assays. Assays for hsa-mir-99b, hsa-mir-125a and hsa-mir-425 were also prepared as reference miRNAs for post-run data normalisation and analysis. All miRNA assays for each cell line sample cDNA were performed in triplicate, with water — containing negative controls. The reaction volume (per well) contained 2.5 µL TaqMan Master mix, 0.125 µL TaqMan miRNA probe and primers, 2 µL of sample cDNA and 0.375 µL nuclease free water. The cycling protocol was run on the ABI 7900HT qPCR platform [ABI, USA] and consisted of an initial holding stage of 95 °C for 10 min, followed by 40 cycles of 95 °C for 15 s/60 °C for 1 min, ending with a final holding stage of 37 °C for 5 min. Following the qPCR run, the data was exported and analysed, including statistical analysis, on the qBasePlus software package [Biogazelle, Belgium].

3. Results

3.1. Validation of chemosensitivity status for NB cell lines

Prior to performing miRNA profiling for each individual cell line within the varying NB chemoresistance models, it was essential to confirm the chemosensitivity profiles. This was implemented by performing a luminescence based cell viability assay, following a pre-determined exposure period by the cell lines to the relevant chemotherapeutic agent. However, the luminescence detection platform also required preliminary analysis to confirm optimum performance.

From the eight NB chemoresistance cell line models available, cell viability analyses only confirmed three models to still have the necessary chemosensitvity profiles required for investigating miRNA expression profiling (see Figs. 1–4). The three validated models (SH-SY5Y/ETOPO, SH-SY5Y/DOXO, UKF-NB-3/DOXO) were analysed by two individual cell viability assays to ensure chemosensitivity profiles.

Each NB chemoresistance cell line model was analysed to confirm chemosensitivity status of its constituent cell lines. The cell lines that confirmed chemosensitivity status were re-analysed with a second, identical cell viability analysis in order to ascertain such status prior to miRNA profiling.

Following the set of cell viability assays on each NB chemoresistance model, only three out of the initial eight cell line models were confirmed to have maintained their chemosensitivity profiles, namely the SH-SY5Y/DOXO, SH-SY5Y/ETOPO and UKF-NB-3/DOXO NB chemoresistance models. The extracted RNA from constituent cell lines, from each of the three cell line models, were consequently eligible for miRNA profiling.

3.2. NB cell line miRNA profiling

The expression profiling of 668 miRNAs for each constituent cell line of the three validated and selected NB chemoresistance cell line models demonstrated that approximately 50% of all miRNAs are expressed above the 35 Cq value (see Figs. 5–7). Additionally, the expression scatterplot profiles highlight that overall miRNA expression is cell line dependent.

The results of the miRNA expression profiling revealed that a putative chemoresistance signature expression of seven miRNAs was present in all three NB chemoresistance cell line models (see Fig. 8).

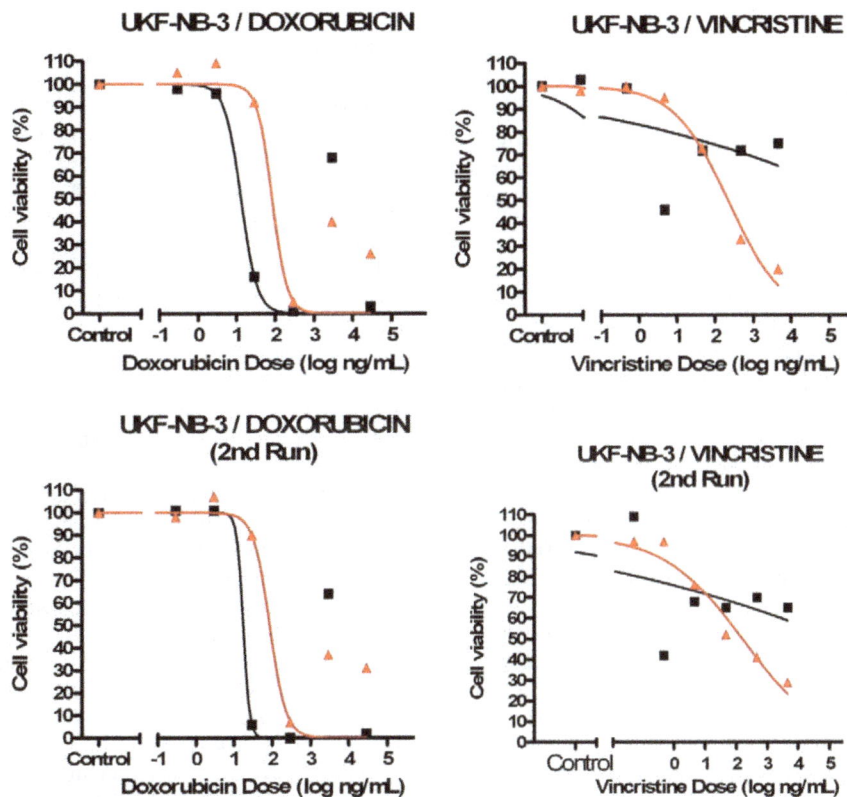

Fig. 2. Results for cell viability analysis of UKF-NB-3 NB chemoresistance models (two separate runs, n = 1, three technical replicates/data point, SEM not illustrated). The doxorubicin chemoresistance model was included in the miRNA profiling study due to a 10-fold linear change in the IC50 dose for both constitutive cell lines. However, at elevated doxorubicin doses a reproducible, artificial increase in cell viability was noticed within this NB chemoresistance model. The UKF-NB-3/vincristine NB chemoresistance model was not included in the study due to incongruent chemosensitivity profiles for each constituent NB cell line (black — chemosensitive parental NB cell line; red — chemoresistant NB cell line).

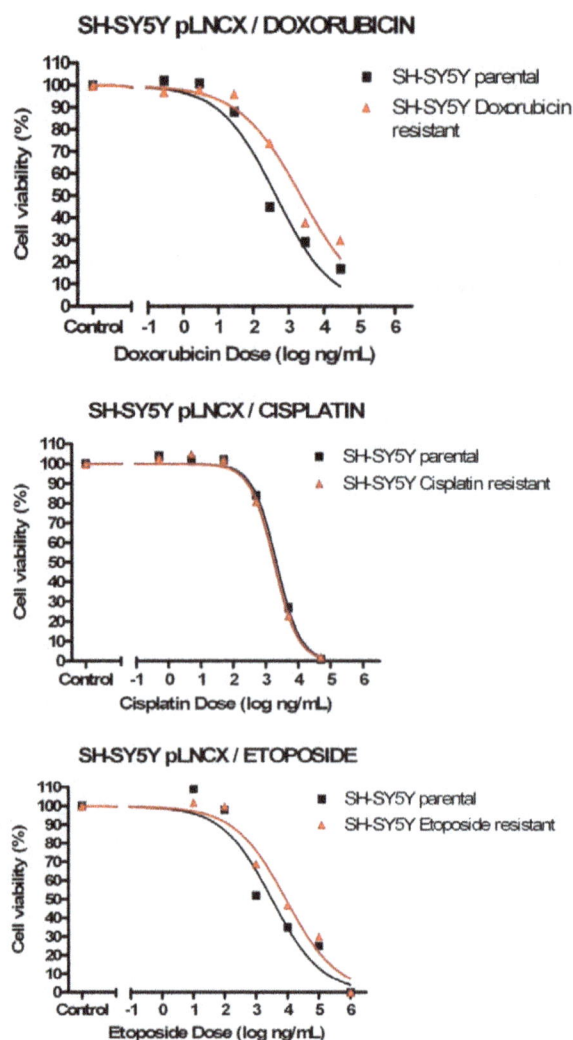

Fig. 3. Results for cell viability analysis of SH-SY5Y NB chemoresistance models (n = 1, three technical replicates/data point, SEM not illustrated). The cisplatin chemoresistance NB model was not included in the study due to lack of a suitable chemosensitivity profile by the predicted chemoresistant NB cell line component for this model.

All seven miRNAs were consequently selected for validation of expression by multiplex RT-qPCR miRNA assay, prior to proceeding to chemoresistance function validation assays.

All identified miRNAs deemed to be involved in NB chemoresistance were consequently eligible for downstream validation studies.

3.3. Multiplex RT-qPCR for putative chemoresistance miRNAs

Once the short list of putative chemoresistance miRNAs was elucidated, it was deemed necessary to validate such findings by re-confirming the degree of dysregulated expression of each individual putative miRNA.

Since the miRNA profiling step did not utilise technical replicates for each individual miRNA RT-qPCR assay (one replicate/miRNA), it was essential to re-confirm the dysregulated expression status for each individual putative NB chemoresistance miRNA within each constitutive NB cell line of each chemoresistance model utilised in this study. This was implemented by performing a secondary RT-qPCR assay, with three technical replicates for each miRNA assay, in order to ascertain the specific miRNA expression levels.

Following re-analysis of RNA from each of the three NB chemoresistance models' cell line constituents by multiplex RT-qPCR miRNA assay, only four out of the original seven-member putative

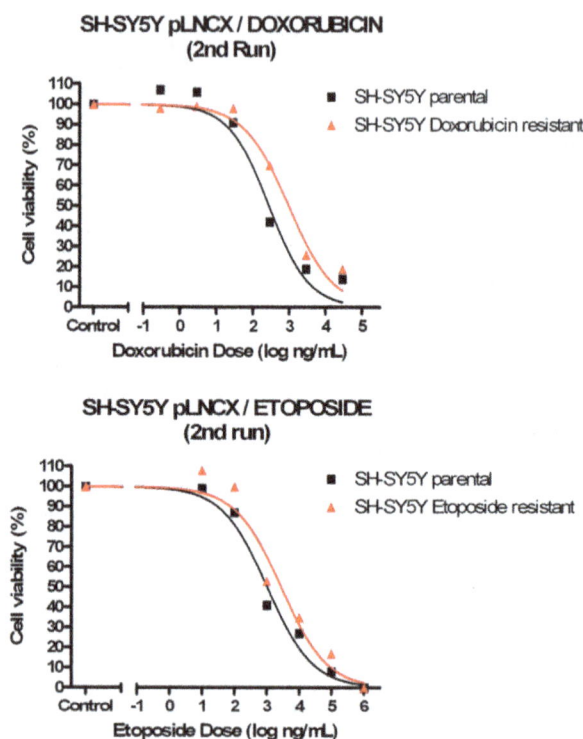

Fig. 4. Cell viability analysis results (2nd run, n = 1, three technical replicates/data point, SEM not illustrated) for re-confirmation of chemosensitivity profiles for each constituent NB cell line chemoresistance model.

chemoresistance miRNA signature were pursued for functional validation (see Figs. 9–15).

4. Discussion

Initial performance results of the luminescence reader platform proved to be successful in achieving specificity of luminescence analysis, with a standard deviation of approximately 1%. Additionally, the dynamic range of the luminescence platform was also deemed optimal, with the lower end of the platform sensitivity spectrum denoted at a cell suspension density of 12,500 cells/mL.

From the eight NB chemoresistance cell line models at disposal in-house, only three models (SH-SY5Y/ETOPO, SH-SY5Y/DOXO and UKF-NB-3/DOXO) demonstrated distinct chemosensitivity profiles for the constitutive cell lines.

Analysis of the three Kelly NB chemoresistance models highlighted incongruencies in the chemosensitivity profiles of the component cell lines and was therefore excluded from the study.

For the UKF-NB-3 chemoresistance models, the UKF-NB-3/vincristine model was also excluded due to inability to confirm the projected chemosensitivity profiles for the two component cell lines, over two separate cell viability assays post-vincristine exposure. However, the UKF-NB-3/DOXO cell line reproducibly demonstrated to confirm the predicted chemosensitivity profiles, with the doxorubicin-resistant component cell line proving to be resistant to the drug exposure dose on comparison with the chemosensitive counterpart cell line by a factor of ten, on comparing the doxorubicin doses required to induce 50% decrease in cell viability. A peculiar observation was also denoted within the UKF-NB-3/DOXO model, in that at the elevated doxorubicin doses the cell viability in both component cell lines seems enhanced. This observation was found to be reproducible across two separate cell viability assays and was solely noticed within this specific NB chemoresistance model (UKF-NB-3 cell line exposed to doxorubicin). Possible explanations for such an anomaly would be technical interference with the luminescence assay, though the actual source thought to

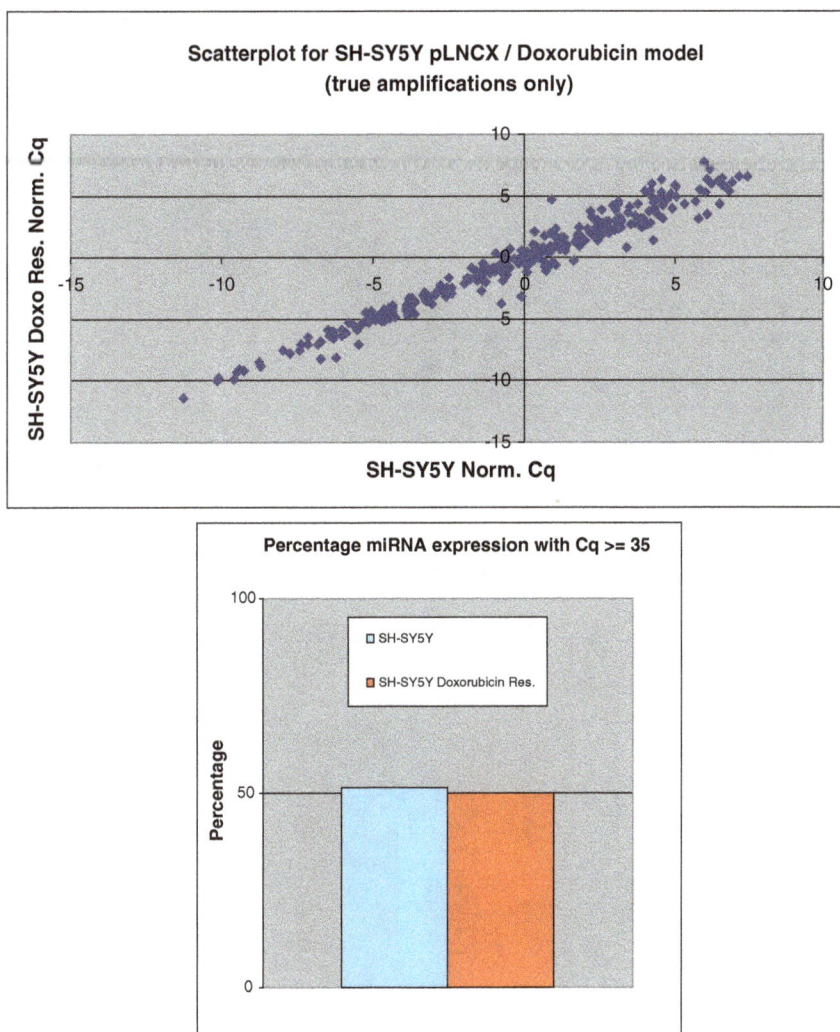

Fig. 5. Scatterplot of miRNA expression for SH-SY5Y/DOXO cell line model constituents. This highlights only the marginal variation in the percentage of miRNAs that are actually expressed within the individual cell lines.

cause such interference is still unclear. Consequently, for the following transient transfection validation assays concerning the UKF-NB-3/DOXO model, the doxorubicin dose range utilised had to be reduced accordingly.

Analysis of the SH-SY5Y chemoresistance models revealed that the SH-SY5Y/cisplatin NB model had lost its chemoresistance phenotype, due to the observation of near-identical chemosensitivity profiles by both components of the NB model and therefore was excluded from the study. For the SH-SY5Y/DOXO and SH-SY5Y/ETOPO chemoresistance models, both met the predicted chemosensitivity profiles required for inclusion in the study, although it was also denoted that the SH-SY5Y/DOXO chemoresistance model demonstrated to be more robust. This was due to the enhanced chemoresistance properties highlighted by the doxorubicin-resistant cell line model component, on comparison with its etoposide-resistant counterpart cell line. Such phenotypes were confirmed for both NB chemoresistance models, following secondary cell viability assays.

The utilisation of RT-qPCR technology in a high throughput manner for the purpose of quantitatively analysing the miRnome in clinical and scientific studies is, of late, a rapidly developing tool for investigating miRNA expression. The essential reason for such popularity lies in the fact that such a method combines the high-throughput efficiency of the platform to simultaneously analyse almost 700 miRNAs from a single biological sample, though retaining the robustness and quantitative analytical advantage of RT-qPCR techniques. Such a technique was

utilised in this study for identifying dysregulated expression of individual and/or networks of miRNAs within the selected NB chemoresistance cell line models.

Within the context of this study, pre-determined criteria were applied for the purpose of identifying putative chemoresistance miRNAs. All Cq values obtained for the investigated miRNAs, observed at a value above 35, would lead to the interpretation that such miRNAs were either not expressed at all, or expressed at very low levels in the corresponding biological sample. This specific Cq threshold value was thus utilised in order to eliminate false positive expression data inclusion due to quantification noise, which is normally introduced due to natural biological variances or also due to the technical setup of the RT-qPCR assay [24]. This is a fine balance that was applied to ensure that true amplification read-outs for miRNAs expressed at a low level (Cq value in the 30–35 range) would not be disregarded. Additionally, a defined threshold was applied for identifying which miRNAs in this study were deemed as putative chemoresistance miRNAs. Only the miRNAs that demonstrated linear fold change of above 2.0 (up-regulated) and below 0.5 (down-regulated) in the chemoresistant cell line component of each study model were included in the putative chemoresistance miRNA shortlist, on comparison of the normalised Cq values for each individual miRNA.

Such a fold change threshold range is deemed to be more restrictive as a selection strategy utilised for identifying relatively elevated levels of dysregulated expression by the individual miRNA. It is important to

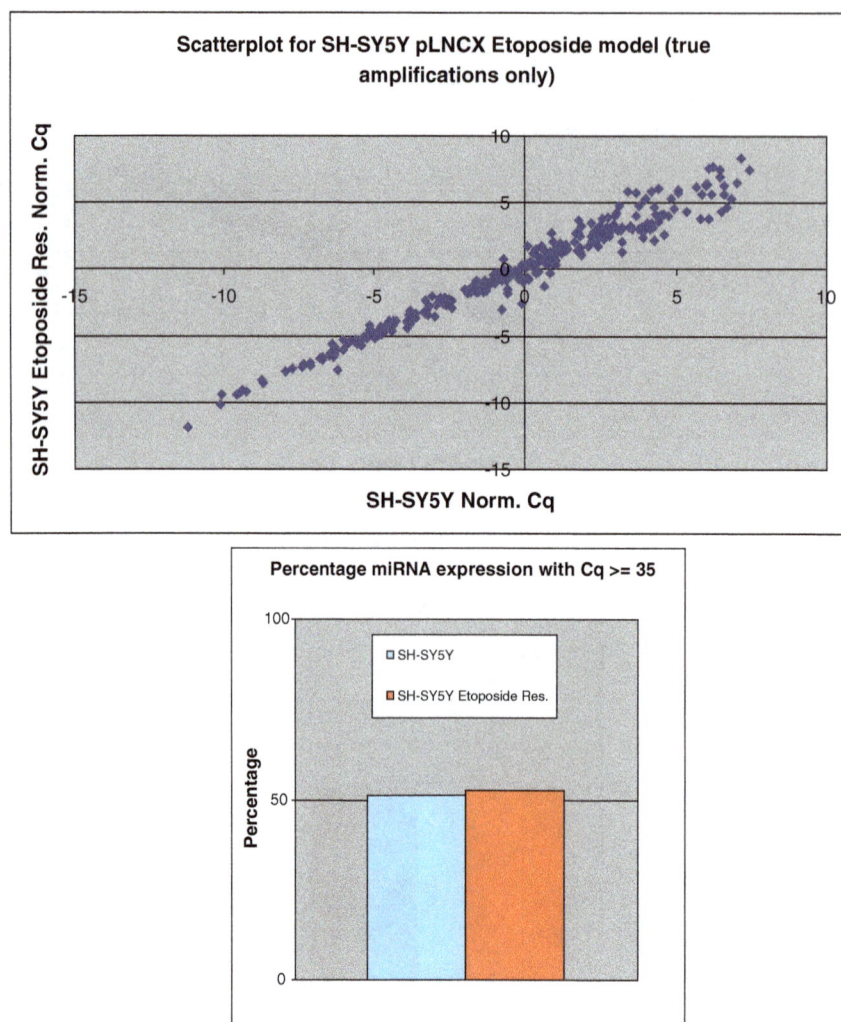

Fig. 6. Scatterplot of miRNA expression for SH-SY5Y/ETOPO cell line model constituents. This highlights only the marginal variation in the percentage of miRNAs that are actually expressed within the individual cell lines.

realise that such restrictive selection criteria might have excluded minor miRNA dysregulations (ie. less than two fold up-regulation or less than two-fold down-regulation) that may still prove to induce a tangible phenotypic influence (eg. fold-change × 1.8).

However, the fold change threshold was implemented in this study at such cut-off points for practical reasons, with the assumption that only those miRNAs falling within the dysregulation thresholds were deemed to have possible influence on NB cell line model chemoresistance properties.

Consequently, it must be emphasised that the threshold fold-change levels are entirely subjective and therefore a careful balance must be attained whereby reliable hits can be selected, though without elevating the stringency levels of the threshold fold-change values to the point at which identification of even a low quantity of hits would prove to be a challenge. Since the minimum miRNA fold change acceptable for scientific publications is a threshold of 1.5-fold, it would be recommended to set the least stringent miRNA fold-change selection criteria at this level (+/− 1.5 fold-change) for all studies of this nature [25–27].

The overview analysis of miRNA expression profiling of all three NB chemoresistance models demonstrated that approximately 50% of the miRNAs analysed in this study had a raw Cq value at/above a value of 35. Those miRNAs having a Cq value above the value of 35 were expressed at low/single molecule level and excluded from further analyses. The notion that half of all miRNAs are expressed at low levels (or not expressed) can be attributed to the fact that in most cancer conditions

there exists an overall down-regulation of miRNA expression [28], with the overall percentage of miRNAs being expressed in normal, tumour-free tissues being higher than 50% (though such tissues were not analysed in this study).

In addition, the scatterplot analysis of normalised Cq values for each individual miRNA, obtained from each constitutive NB model component cell line, revealed overall low levels of data spreading. This implies that only a minute fraction of the entire miRnome may be deemed to be grossly dysregulated in such investigated cell line models.

For the three NB chemoresistance models investigated in this study and following implementation of the above selection criteria, a total of seven putative chemoresistance miRNAs were observed to be dysregulated across all three NB models (see Fig. 8). From such a shortlist of putative chemoresistance miRNAs, miR-188-5p demonstrated to have reliable RT-qPCR, due to all miR-188-5p dysregulations across all three NB chemoresistance models having Cq values below the 35.0 threshold. For the remaining six putative miRNAs, such an observation was not denoted from the RT-qPCR miRNA profiling analysis. This implies that for such miRNAs, in one or more NB cell line model, the dysregulated effect was deemed to be an 'on/off' effect, with the Cq value for the individual miRNA from one of the two NB cell line model components being above the pre-defined 35.0 value.

Following miRNA expression profiling, it was necessary to confirm dysregulation of the selected putative chemoresistance miRNAs by multiplex RT-qPCR techniques. This step was of importance for

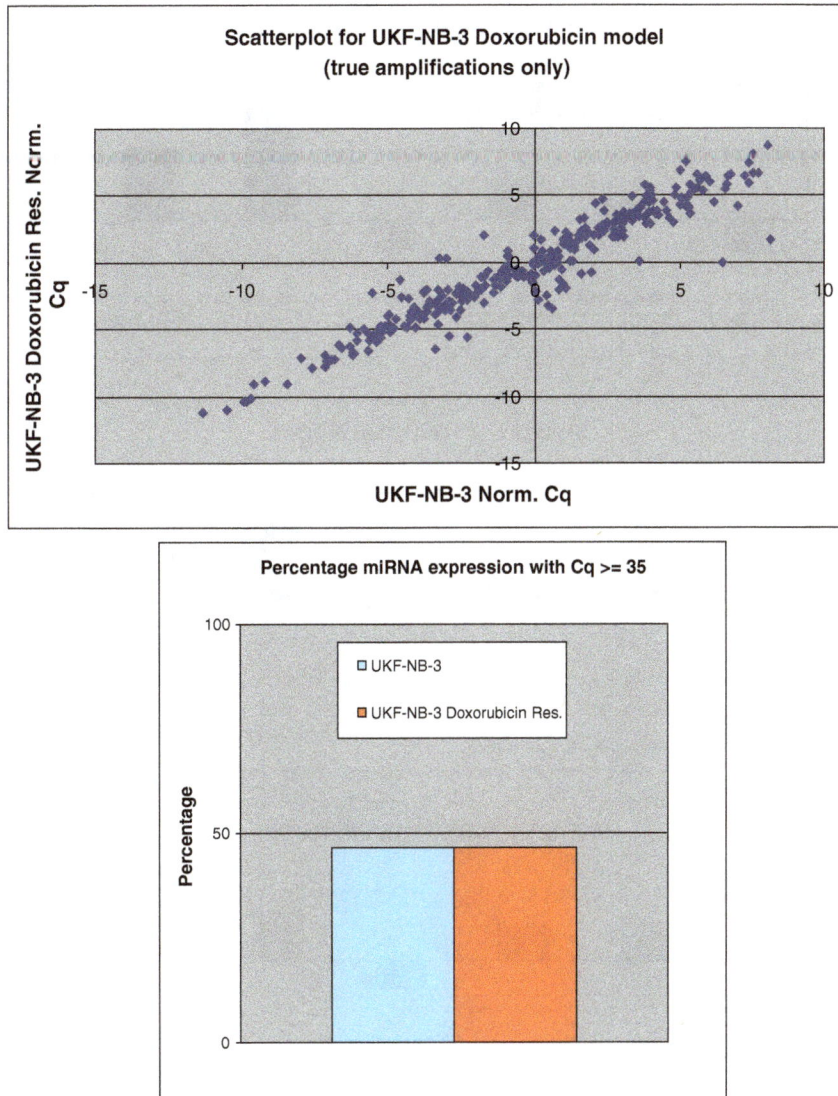

Fig. 7. Scatterplot of miRNA expression for UKF-NB-3/DOXO cell line model constituents. This highlights insignificant variation in the percentage of miRNAs that are actually expressed within the individual cell lines.

confirming the results obtained from miRNA profiling, since the latter RT-qPCR assays were performed using one sole technical replicate for each individual miRNA. Consequently, confirmation by multiplex RT-qPCR analysis utilising three technical replicates allowed for such validated putative chemoresistance miRNAs to be further validated by functional assay implementation.

Results from the multiplex RT-qPCR assays fully confirmed the up-regulated expression profile for miR-188-5p across all three NB chemoresistance models. Two additional putative miRNAs were also of major interest, namely miR-125b-1# and miR-501-5p, due to confirmation of dysregulated miRNA expression in at least two NB chemoresistance models. The fourth putative miRNA was miR-204. However, the down-regulated expression of miR-204 was only confirmed within the UKF-NB-3/DOXO chemoresistance model. This suggests that, due to the mutated p53 status present in UKF-NB-3, this dysregulation is cell line dependent and is totally absent in wild type p53 status NB cell lines such as SH-SY5Y [29].

The remaining miRNAs failed to confirm their dysregulated expression profile within the majority (or all) of the three NB chemoresistance models. The main reasons for such results stems from the RT-qPCR technique implemented for miRNA profiling. A typical step utilised for

sample preparation prior to miRNA profiling is the pre-amplification of RNA collected from the biological sample [30].

Therefore, any miRNAs identified as expressed following miRNA profiling with a raw Cq value between 30 and 35 might eventually not be equally identified on multiplex/singleplex RT-qPCR analysis, since the latter is applied on non-pre-amplified RNA samples solely. Ultimately, such low expressed miRNA profiling expression values must be investigated with due attention and awareness of such technical influences [30].

Following the results of this study, exact mechanistic links between miRNA dys-regulation levels and chemoresistance phenotype cannot be concluded. However, it certainly is of interest to highlight the possible involvement of such miRNAs in affecting the most common means of chemoresistance-inducing cellular pathways within such cell lines. A typical example would be the effect of such miRNAs on the ABC transporter system and its member genes that regulate drug efflux properties of the cell [36].

Such mechanistic links have been identified within other cancer models. The study conducted by Ma and colleagues [37] demonstrated the key roles played by miR-133a and miR-326 in conferring adriamycin chemoresistance properties to the HepG2 hepatocellular carcinoma cell line model, through their regulatory effects on the ABCC1 gene.

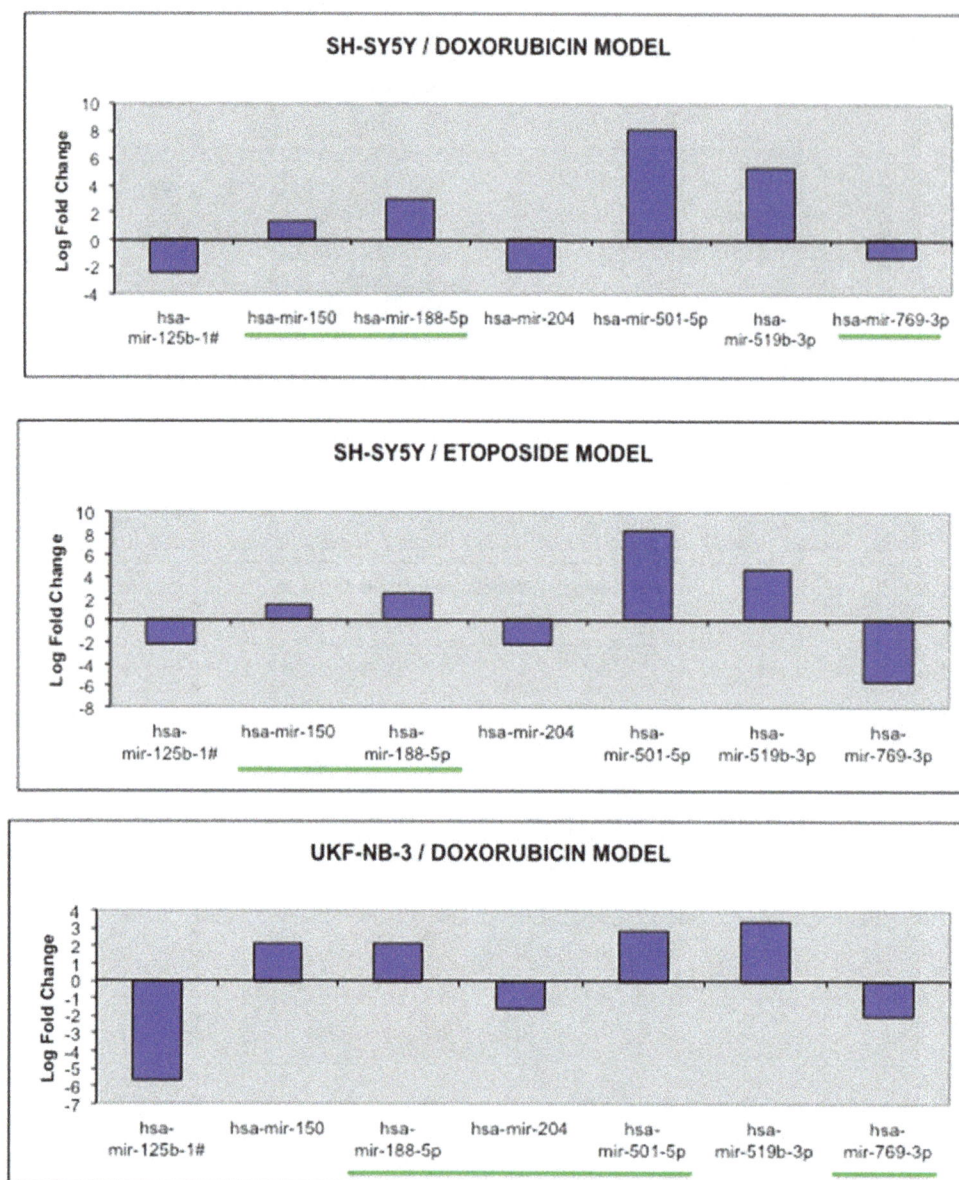

Fig. 8. Graphs demonstrating log-fold change in miRNA expression for all three selected NB chemoresistance models. This miRNA signature of seven miRNAs was identified as dysregulated in the chemoresistant NB cell line components of all three selected NB chemoresistance models. miRNAs highlighted in green denote dysregulations in which both constituent cell lines had a raw Cq value below the 35 cutoff value. This specific Cq threshold value was utilised in order to eliminate false positive expression data inclusion due to quantification noise.

Furthermore, another study has highlighted the involvement of miR-27b in sensitising liver and kidney carcinoma cells to conventional cancer chemotherapies [38]. The study also demonstrated that miR-27b exerts its drug sensitivity influences through the activation of p53-dependent apoptotic pathways [38]. In colorectal cancer, miR-22 was recognised to affect drug sensitivity properties to 5-fluorouracil treatments through its involvement in the regulation of autophagy mechanisms, mainly due to miR-22 direct action on B-cell translocation gene 1 (BTG1) — a key molecular player in autophagy [39]. In gastric carcinoma, miR-129-5p CpG island methylations were identified to affect multi-drug resistance properties through targeted influences on ABC transporter genes [40].

Future research focusing on the influence of miRNA effects on NB chemoresistance properties may be highly varied and of equal importance. One specific research option would be to monitor the miRNA expression profile of individual NB cell lines during prolonged exposure to specific conventional chemotherapeutic agents, until such cell lines develop sustainable chemoresistance properties. Such a study would provide great insight onto which miRNA networks are most affected by prolonged cytotoxic drug exposure, through evaluation of miRNA profiles for dysregulated miRNA expression across the varying phases of the drug exposure period. Consequently, the miRNA profiles across the entire timeline for rendering a specific NB cell line chemoresistant to individual or combinations of chemotherapeutic agents may provide a wealth of information regarding which miRNAs could possibly have key roles in orchestrating such a phenotypic shift at the cellular level.

Furthermore, the prospect of performing miRNA profiling of NB cell lines having combinations of differing NB oncogene expression, as described above, could also shed more light on the effect of NB oncogene expression on miRNA expression profiles. Downstream miRNA dysregulations correlated to exacerbated oncogene expression may consequently be investigated for the identification and validation of key miRNAs deemed to act as intermediate effectors of chemoresistance.

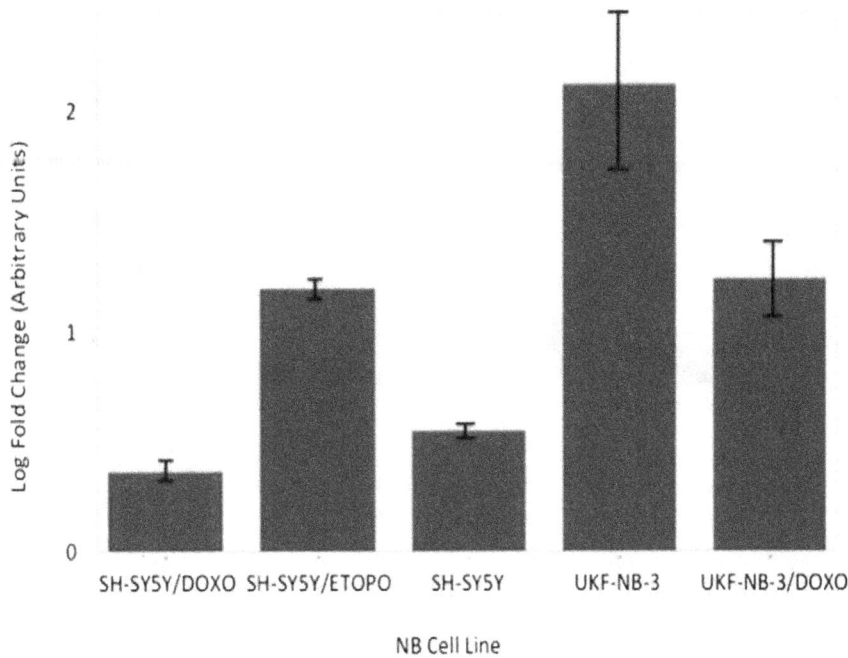

Fig. 9. Multiplex RT-qPCR assay results (log-fold change) for hsa-miR-125b-1# following normalisation with reference miRNAs. Expected down-regulation of putative miRNA was confirmed in UKF-NB-3/DOXO and SH-SY5Y/DOXO models. Error bars represent standard deviation. Assays for hsa-mir-99b, hsa-mir-125a and hsa-mir-425 were also prepared as reference miRNAs for post-run data normalisation and analysis.

The implementation of a systems biology approach might also be of valuable use for enhanced mapping and characterisation of miRNA networks with the capacity to interact and regulate complex gene networks directly affecting the various biological processes within the neuroblast that ultimately lead to the attainment of pronounced chemoresistance properties [31].

Ideally, should any miRNA have be successfuly validated within *in vitro* evaluations folowing our findings, further *in vivo* murine model research could be enacted, consisting of NB xenograft implants bearing constructs with individual putative chemoresistance miRNAs, which are actively expressed on oral administration of tetracycline by the individual mouse, thus minimising distress to the animal. Following

miRNA induced expression, the animals would be exposed to cytotoxic agents and tumour progression is monitored accordingly. Induced expression of the putative chemoresistance modulating miRNA would be expected to lead to NB tumours having resilience to chemotherapy.

Clinical relevance of the identification of individual/networks of miRNAs directly linked with NB chemoresistance properties is two-fold.

Firstly, the identification and proper validation of hallmark miRNAs which are proven to be key players in conferring individual and/or multi — drug chemoresistance properties to NB tumours may be utilised for the development and implementation of novel diagnostic miRNA expression profile and screens [32]. These screens may prove essential for early recognition of NB tumours suspected to possess chemoresistance

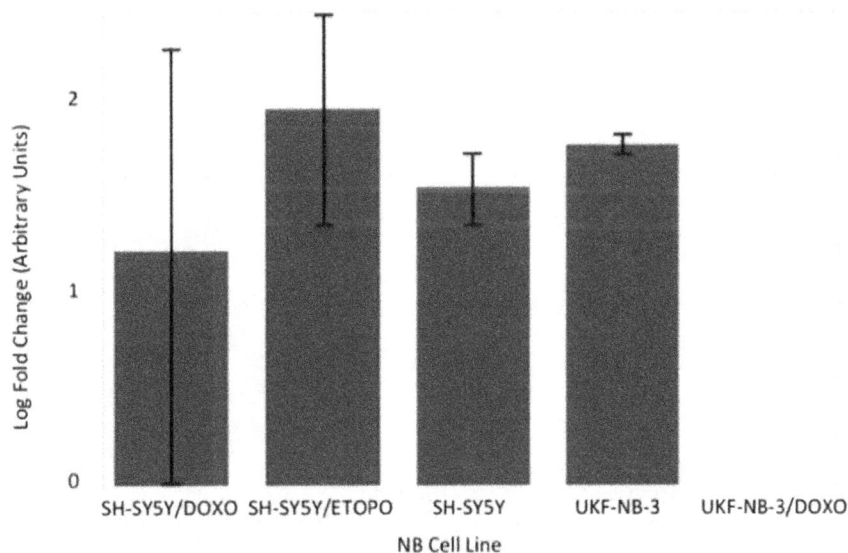

Fig. 10. Multiplex RT-qPCR assay results (log-fold change) for hsa-miR-150 following normalisation with reference miRNAs. Expected up-regulation of putative miRNA was not confirmed in all three chemoresistance models. Error bars represent standard deviation. Assays for hsa-mir-99b, hsa-mir-125a and hsa-mir-425 were also prepared as reference miRNAs for post-run data normalisation and analysis.

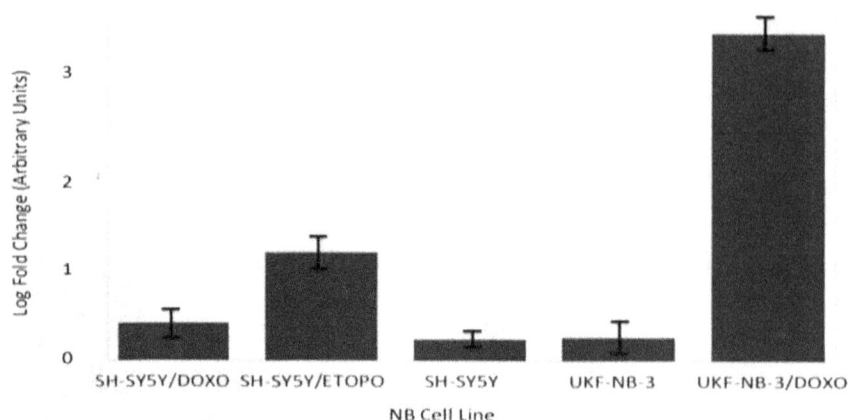

Fig. 11. Multiplex RT-qPCR assay results (log-fold change) for hsa-miR-188-5p following normalisation with reference miRNAs. Expected up-regulation of putative miRNA was confirmed in all three NB chemoresistance models. Error bars represent standard deviation. Assays for hsa-mir-99b, hsa-mir-125a and hsa-mir-425 were also prepared as reference miRNAs for õpost-run data normalisation and analysis.

properties, thus giving the clinician an accurate picture of the clinical manifestation in the individual child suffering from NB. Following miRNA profiling and consequent functional assay validation techniques described above, the results are still inconclusive regarding adopting the seven-miRNA signature as hallmark biomarkers for NB chemoresistance. The results from the RT-qPCR study of oncogene over-expression on MDR gene dysregulation also proved to be inconclusive. Therefore, future studies such as the experimental designs suggested for future research described above might help to identify other miRNAs with more reliable and robust validation assay results to confirm their involvement in the induction of NB chemoresistance properties. Such identified miRNAs would ultimately be utilised in the clinical setting for diagnostic purposes to recognise NB tumours having chemoresistance properties to conventional chemotherapeutic agents.

Similar diagnostic screens have already been designed for accurate diagnosis of other severe conditions such as human papilloma viral infectious strains, though such bespoke microarray technology could be implemented for the recognition of specific miRNA sequences present in any clinical sample [33]. For NB diagnostic purposes,

identification of novel biomarkers may also be attained through the profiling of other classes such as long non-coding RNA molecules (utilising RT-qPCR), whole genome microarray profiling and also exome next-generation sequencing.

However, the validation methods for confirming individual candidate biomarkers may be time consuming. This is a particular problem for exome sequencing approaches, due to the vast array of data that is commonly generated and thus requires thorough bioinformatic analysis prior to even identifying candidate exome sequence biomarkers for consequent validation assays. In distinct contrast, the possibility of identifying novel body fluid biomarkers, such as free-circulating miRNAs [34], ultimately provides an extremely rapid diagnostic/prediction test within the clinical setting due to the sole requirement of a blood sample for RT-qPCR analysis. However, biomarker validation techniques are very much dependant on the specific phenotype with which the individual candidate biomarker needs to be reliably linked to (*e.g.* chemoresistance, cell proliferation, cell differentiation).

For the validation of miRNAs associated with chemoresistance, transient transfection of miRNA antagonists remains the mainstay

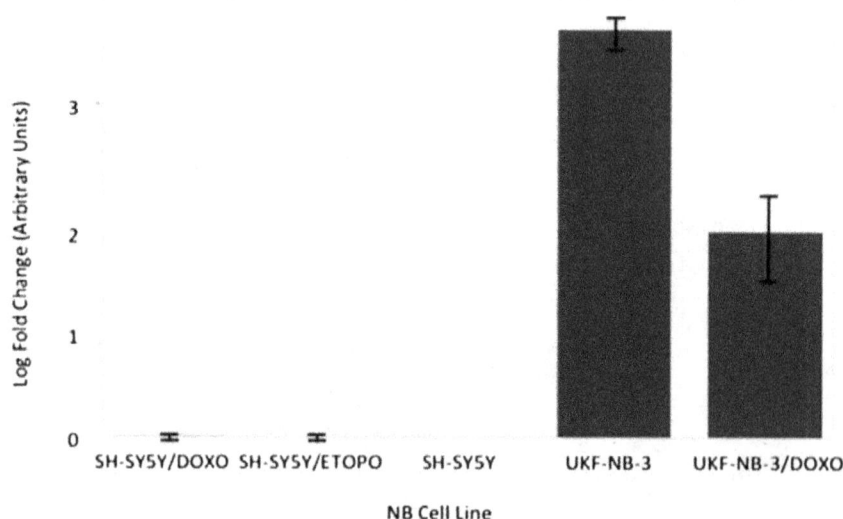

Fig. 12. Multiplex RT-qPCR assay results (log-fold change) for hsa-miR-204 following normalisation with reference miRNAs. Expected down-regulation of putative miRNA was solely confirmed in UKF-NB-3/DOXO NB model. Error bars represent standard deviation. Assays for hsa-mir-99b, hsa-mir-125a and hsa-mir-425 were also prepared as reference miRNAs for post-run data normalisation and analysis.

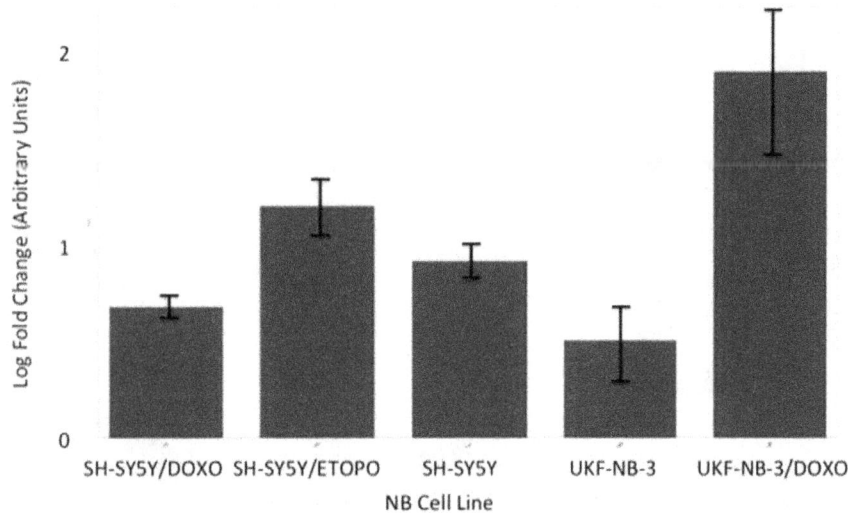

Fig. 13. Multiplex RT-qPCR assay results (log-fold change) for hsa-miR-501-5p following normalisation with reference miRNAs. Expected up-regulation of putative miRNA was confirmed in UKF-NB-3/DOXO and SH-SY5Y/ETOPO models. Error bars represent standard deviation. Assays for hsa-mir-99b, hsa-mir-125a and hsa-mir-425 were also prepared as reference miRNAs for post-run data normalisation and analysis.

method due to the rapidity of the assay (72–96 h). Unfortunately, this method also has the drawback that direct quantification of candidate biomarker miRNA by RT-qPCR cannot be performed, unless the miRNA in question is known to down-regulate specific target genes (which have been previously confirmed in other studies).

Following the acquisition of additional information from these novel biomarkers, the therapeutic avenues chosen by the clinician will be decided with the inclusion of pre-emptive strategies against chemoresistant tumours, based on the additional awareness of the chemosensitivity status of the individual NB tumour. The therapeutic options applied will thus maximise the chances of survival of the patient, with minimal suffering due to chemotherapy-induced adverse effects.

Secondly, such miRNAs are good candidates to be novel drug targets for future RNAi based therapies against aggressive tumours that are not responding to conventional chemotherapy [35]. These miRNA-directed therapies are still in their infancy, though may in future prove to be a novel molecular approach aimed at enhancing the efficacy of conventional chemotherapeutic treatments, with possible reduction in the dose and frequency of chemotherapy cycles through which the child has to undergo, ultimately reducing the level of distress and suffering from chemotherapy-induced adverse effects.

Acknowledgements

Ali Rihani[2], Alexander Schramm[3], Martin Michaelis[4], Jindrich Cinatl Jr.[4], Angelika Eggert[3], Philip JR Day[1], Frank Speleman[2].

[1]Manchester Institute of Biotechnology, The University of Manchester, UK.

[2]Center for Medical Genetics, Ghent University Hospital, Ghent, Belgium.

[3]Division of Hematology/Oncology, University Children's Hospital, Essen, Germany.

[4]Institut für Medizinische Virologie, Klinikum der J.W. Goethe Universität, Frankfurt am Main, Germany.

Lighthouse Laboratories (Perth, AUS) for funding the miRNA profiling.

Fig. 14. Multiplex RT-qPCR assay results (log-fold change) for hsa-miR-519b-3p following normalisation with reference miRNAs. Expected up-regulation of putative miRNA was not confirmed for all three chemoresistance models. Error bars represent standard deviation. Assays for hsa-mir-99b, hsa-mir-125a and hsa-mir-425 were also prepared as reference miRNAs for post-run data normalisation and analysis.

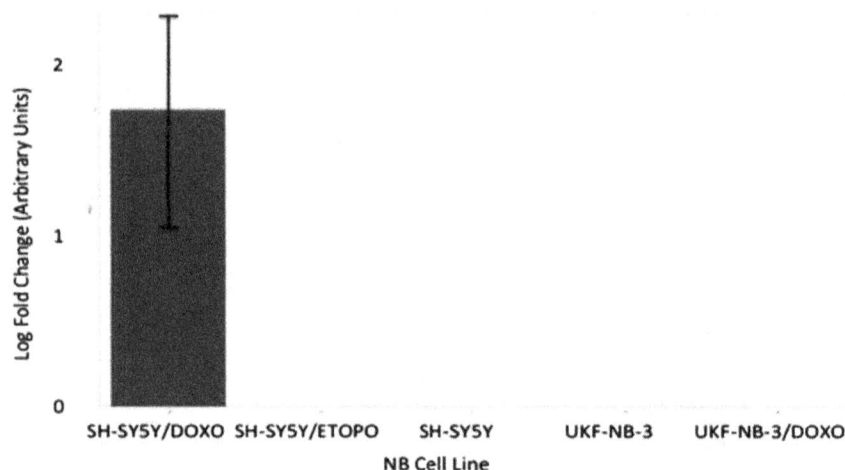

Fig. 15. Multiplex RT-qPCR assay results (log-fold change) for hsa-miR-769-3p following normalisation with reference miRNAs. Expected down-regulation of putative miRNA was not confirmed in all three chemoresistance models. Error bars represent standard deviation. Assays for hsa-mir-99b, hsa-mir-125a and hsa-mir-425 were also prepared as reference miRNAs for post-run data normalisation and analysis.

References

[1] Izbicka E, Izbicki T. Therapeutic strategies for the treatment of neuroblastoma. Curr Opin Investig Drugs Dec 2005;6(12):1200–14.

[2] American Cancer Society. What are the most common types of childhood cancers? [Internet]. 2010 [cited 2010 Nov 28]. Available from: http://www.cancer.org/Cancer/CancerinChildren/DetailedGuide/cancer-in-children-types-of-childhood-cancers.

[3] Kushner BH, Cheung N-KV. Neuroblastoma—from genetic profiles to clinical challenge. N Engl J Med Nov 24 2005;353(21):2215–7.

[4] Grau E, Oltra S, Orellana C, Hernández-Martí M, Castel V, Martínez F. There is no evidence that the SDHB gene is involved in neuroblastoma development. Oncol Res 2005;15(7-8):393–8.

[5] Magro G, Grasso S. The glial cell in the ontogenesis of the human peripheral sympathetic nervous system and in neuroblastoma. Pathologica Oct 2001;93(5):505–16.

[6] Walton JD, Kattan DR, Thomas SK, Spengler BA, Guo H-F, Biedler JL, et al. Characteristics of stem cells from human neuroblastoma cell lines and in tumors. Neoplasia Dec 2004;6(6):838–45.

[7] Tanaka T, Iehara T, Sugimoto T, Hamasaki M, Teramukai S, Tsuchida Y, et al. Diversity in neuroblastomas and discrimination of the risk to progress. Cancer Lett Oct 18 2005;228(1-2):267–70.

[8] American Cancer Society. How is neuroblastoma diagnosed? [Internet]. 2010 [cited 2010 Nov 28]. Available from: http://www.cancer.org/Cancer/Neuroblastoma/DetailedGuide/neuroblastoma-diagnosis

[9] Brodeur GM, Pritchard J, Berthold F, Carlsen NL, Castel V, Castleberry RP, et al. Revisions of the international criteria for neuroblastoma diagnosis, staging, and response to treatment. J Clin Oncol Aug 1993;11(8):1466–77.

[10] Saad DF, Gow KW, Milas Z, Wulkan ML. Laparoscopic adrenalectomy for neuroblastoma in children: a report of 6 cases. J Pediatr Surg Dec 2005;40(12):1948–50.

[11] Longley DB, Johnston PG. Molecular mechanisms of drug resistance. J Pathol Jan 2005;205(2):275–92.

[12] Kerbel RS, Kobayashi H, Graham CH. Intrinsic or acquired drug resistance and metastasis: are they linked phenotypes? J Cell Biochem Sep 1994;56(1):37–47.

[13] Garofalo M, Croce CM. microRNAs: Master Regulators as Potential Therapeutics in Cancer. Annu Rev Pharmacol Toxicol [Internet]. 2010 Jan 18 [cited 2010 Sep 26]; Available from: http://www.ncbi.nlm.nih.gov/pubmed/20809797

[14] Mestdagh P, Fredlund E, Pattyn F, Schulte JH, Muth D, Vermeulen J, et al. MYCN/c-MYC-induced microRNAs repress coding gene networks associated with poor outcome in MYCN/c-MYC-activated tumors. Oncogene Mar 4 2010;29(9):1394–404.

[15] Sorrentino A, Liu C-G, Addario A, Peschle C, Scambia G, Ferlini C. Role of microRNAs in drug-resistant ovarian cancer cells. Gynecol Oncol Dec 2008;111(3):478–86.

[16] Zhou M, Liu Z, Zhao Y, Ding Y, Liu H, Xi Y, et al. MicroRNA-125b confers resistance of breast cancer cells to paclitaxel through suppression of pro-apoptotic Bcl-2 antagonist killer 1 (Bak1) expression. J Biol Chem Jul 9 2010;285(28):21496–507.

[17] DeVere White RW, Vinall RL, Tepper CG, Shi X-B. MicroRNAs and their potential for translation in prostate cancer. Urol Oncol Jun 2009;27(3):307–11.

[18] Galluzzi L, Morselli E, Vitale I, Kepp O, Senovilla L, Criollo A, et al. miR-181a and miR-630 regulate cisplatin-induced cancer cell death. Cancer Res Mar 1 2010;70(5):1793–803.

[19] Miller TE, Ghoshal K, Ramaswamy B, Roy S, Datta J, Shapiro CL, et al. MicroRNA-221/222 confers tamoxifen resistance in breast cancer by targeting p27Kip1. J Biol Chem Oct 31 2008;283(44):29897–903.

[20] Stallings RL. MicroRNA involvement in the pathogenesis of neuroblastoma: potential for microRNA mediated therapeutics. Curr Pharm Des 2009;15(4):456–62.

[21] Fontana L, Fiori ME, Albini S, Cifaldi L, Giovinazzi S, Forloni M, et al. Antagomir-17-5p abolishes the growth of therapy-resistant neuroblastoma through p21 and BIM. PLoS One 2008;3(5):e2236.

[22] Chen Y, Tsai Y-H, Fang Y, Tseng S-H. Micro-RNA-21 regulates the sensitivity to cisplatin in human neuroblastoma cells. J Pediatr Surg Oct 2012;47(10):1797–805.

[23] Ryan J, Tivnan A, Fay J, Bryan K, Meehan M, Creevey L, et al. MicroRNA-204 increases sensitivity of neuroblastoma cells to cisplatin and is associated with a favourable clinical outcome. Br J Cancer Sep 4 2012;107(6):967–76.

[24] Reiter M, Kirchner B, Müller H, Holzhauer C, Mann W, Pfaffl MW. Quantification noise in single cell experiments. Nucleic Acids Res Oct 2011;39(18):e124.

[25] Dharap A, Vemuganti R. Ischemic pre-conditioning alters cerebral microRNAs that are upstream to neuroprotective signaling pathways. J Neurochem Jun 2010;113(6):1685–91.

[26] Wang L-L, Zhang Z, Li Q, Yang R, Pei X, Xu Y, et al. Ethanol exposure induces differential microRNA and target gene expression and teratogenic effects which can be suppressed by folic acid supplementation. Hum Reprod Mar 2009;24(3):562–79.

[27] Mestdagh P, Van Vlierberghe P, De Weer A, Muth D, Westermann F, Speleman F, et al. A novel and universal method for microRNA RT-qPCR data normalization. Genome Biol 2009;10(6):R64.

[28] Lee YS, Dutta A. MicroRNAs in cancer. Annu Rev Pathol 2009;4:199–227.

[29] Michaelis M, Cinatl J, Anand P, Rothweiler F, Kotchetkov R, von Deimling A, et al. Onconase induces caspase-independent cell death in chemoresistant neuroblastoma cells. Cancer Lett May 18 2007;250(1):107–16.

[30] Mestdagh P, Feys T, Bernard N, Guenther S, Chen C, Speleman F, et al. High-throughput stem-loop RT-qPCR miRNA expression profiling using minute amounts of input RNA. Nucleic Acids Res Dec 2008;36(21):e143.

[31] Logan JA, Kelly ME, Ayers D, Shipillis N, Baier G, Day PJR. Systems biology and modeling in neuroblastoma: practicalities and perspectives. Expert Rev Mol Diagn Mar 2010;10(2):131–45.

[32] Iguchi H, Kosaka N, Ochiya T. Versatile applications of microRNA in anti-cancer drug discovery: from therapeutics to biomarkers. Curr Drug Discov Technol Jun 1 2010;7(2):95–105.

[33] Ayers D, Platt M, Javad F, Day PJR. Human papilloma virus strain detection utilising custom-designed oligonucleotide microarrays. Methods Mol Biol 2011;688:75–95.

[34] Schwarzenbach H, Hoon DSB, Pantel K. Cell-free nucleic acids as biomarkers in cancer patients. Nat Rev Cancer Jun 2011;11(6):426–37.

[35] Ayers D, Day PJ. Unlocking the potential of RNA interference as a therapeutic tool. Malta Medical Journal Volume 21 (2009); Issue 3 September. MMJ. 2009 Sep;21(3): 13–9.

[36] Ayers D, Nasti A. Utilisation of nanoparticle technology in cancer chemoresistance. J Drug Deliv 2012;2012:265691.

[37] Ma J, Wang T, Guo R, Yang X, Yin J, Yu J, et al. Involvement of miR-133a and miR-326 in ADM resistance of HepG2 through modulating expression of ABCC1. J Drug Target Feb 2015;25:1–6.

[38] Mu W, Hu C, Zhang H, Qu Z, Cen J, Qiu Z, et al. miR-27b synergizes with anticancer drugs via p53 activation and CYP1B1 suppression. Cell Res Apr 2015;25(4):477–95.

[39] Zhang H, Tang J, Li C, Kong J, Wang J, Wu Y, et al. MiR-22 regulates 5-FU sensitivity by inhibiting autophagy and promoting apoptosis in colorectal cancer cells. Cancer Lett Jan 28 2015;356(2 Pt B):781–90.

[40] Wu Q, Yang Z, Xia L, Nie Y, Wu K, Shi Y, et al. Methylation of miR-129-5p CpG island modulates multi-drug resistance in gastric cancer by targeting ABC transporters. Oncotarget Nov 30 2014;5(22):11552–63.

12

Internal Transcribed Spacer 1 (ITS1) based sequence typing reveals phylogenetically distinct *Ascaris* population

Koushik Das [1], Punam Chowdhury [1], Sandipan Ganguly *

Division of Parasitology, National Institute of Cholera and Enteric Diseases, P-33, CIT Road, Scheme XM, Beliaghata, Kolkata 700010, India

ARTICLE INFO

Keywords:
Ascariasis
Ascaris sp
Genetic diversity
Phylogeny
Internal transcribed spacer 1 (ITS1)
Sequence typing

ABSTRACT

Taxonomic differentiation among morphologically identical *Ascaris* species is a debatable scientific issue in the context of Ascariasis epidemiology. To explain the disease epidemiology and also the taxonomic position of different *Ascaris* species, genome information of infecting strains from endemic areas throughout the world is certainly crucial. *Ascaris* population from human has been genetically characterized based on the widely used genetic marker, internal transcribed spacer1 (ITS1). Along with previously reported and prevalent genotype G1, 8 new sequence variants of ITS1 have been identified. Genotype G1 was significantly present among female patients aged between 10 to 15 years. Intragenic linkage disequilibrium (LD) analysis at target locus within our study population has identified an incomplete LD value with potential recombination events. A separate cluster of Indian isolates with high bootstrap value indicate their distinct phylogenetic position in comparison to the global *Ascaris* population. Genetic shuffling through recombination could be a possible reason for high population diversity and frequent emergence of new sequence variants, identified in present and other previous studies. This study explores the genetic organization of Indian *Ascaris* population for the first time which certainly includes some fundamental information on the molecular epidemiology of Ascariasis.

1. Introduction

Human Ascariasis caused by gastrointestinal nematode *Ascaris lumbricoides* (L) is one of the major Soil Transmitted Helminthiases (STHs).The disease has been included in World Health Organization (WHO) list of Neglected Tropical Diseases (NTD), infecting more than one billion people [1]. Transmission is normally through the ingestion of infective *Ascaris* sp. egg in sewage contaminated soil and vegetables. Majority of infections are asymptomatic, while some chronic infection develops symptoms like abdominal pain, nausea, lung inflammation, anemia, stunted growth, diminished physical fitness etc. [1]. This has a certain impact on socio-economic development of low-income countries [2]. Like human, pigs are also infected with closely related species of *A. lumbricoides* (L), *Ascaris suum* Goeze [3]. Taxonomic separation between *A. lumbricoides* and *A. suum* represents a debatable scientific issue in the context of Ascariasis

epidemiology due to the absence of distinguishing morphological characteristics among them [4]. Proper identification and genetic characterization of infecting strains from endemic areas throughout the world are certainly important to explain the disease epidemiology and also the taxonomic status of two *Ascaris* species. Several molecular epidemiological investigation based on polymorphic markers like-internal transcribed spacer 1 (ITS1), mitochondrial cytochrome c oxidase subunit 1 (cox1), NADH dehydrogenase subunit 1 (nad1) and microsatellite markers have been proposed to explain the origin of the two ascarid taxa in their respective hosts and their taxonomic status [5–8]. However, non-repetitive genomic regions have been

Table 1
List of gene specific primers, used in the study.

Gene name	PCR round	Primer name	Primer sequence (5′ to 3′)
Internal transcribed spacer 1 (ITS1)	Primary	ITS F 1	CGAGCAGAAAAAAAAAGTCTCC
		ITS R1	GGAATGAACCCGATGGCGCAAT
	Secondary	ITS F 2[a]	CGAGCAGAAAAAAAAAAAGTCTCC
		ITS R 2[a]	GCTGCGTTCTTCATCGAT

[a] Gene specific primer pairs used for sequencing of amplified PCR products.

* Corresponding author.
E-mail addresses: koushikdas55@gmail.com (K. Das), punam.bt07@gmail.com (P. Chowdhury), sandipanganguly@gmail.com (S. Ganguly).
[1] Authors contributed equally to this work.

preferred over repetitive regions as genotyping marker for their high genetic stability and evolutionary significance. Single nucleotide substitution occurred just once in the phylogenetic history of a species, unlikely to mutate again to either a novel or ancestral genotype [9]. Using the non-repetitive marker ITS1, 5 *Ascaris* genotypes (G1-G5) in human and 3 *Ascaris* genotypes (G1-G3) in pig have been identified. G1 frequently infects human, while G3 is predominant in pigs. The other three has been detected in lower frequencies in

their respective hosts [5]. Recently, a study from Brazil based on ITS1 marker reported a new *Ascaris* genotype G6 in human [10]. However, no such information regarding genetic pattern and diversity of *Ascaris* population from India are available still date. Hence, the present study was designed to generate an idea about genetic patterns and diversity of Indian *Ascaris* population and also to determine their phylogenetic relation with the global *Ascaris* population. The result revealed a considerable amount of polymorphism within

Table 2
List of polymorphic sites identified within internal transcribed spacer 1 (ITS1) region among various *Ascaris* isolates from India and Worldwide.

Sample ID	GenBank accession number	Genotypes/ haplotypes	Country	Host	Nucleotide variation at alignment position[a]	References
WP01	AJ554036	G1[b]	Ba[c]/Br[d]/Ja[e]/Ch[f]	Hu[g]/P[h]	120T, 122T, 123T, 124T, 125T, 127T, 128--[i], 129 --[i], 130G, 131C, 132G,133G, 134A, 135C, 139T, 142A, 148 --[i], 150T, 151T, 155A, 156T, 157--[i], 162T, 167A, 168A, 169G, 172T, 173G, 176T, 181T, 183T, 203G, 205C, 206G, 207C, 218T, 229T, 231C, 233T, 238A, 248T	[5,10,21]
WP40	AJ554037	G2[b]	Ch[f]	Hu[g]/P[h]	128--[i]/T, 133G/S[j]	[5]
WP17	AJ554038	G3[b]	Ja[e]/Ch[f]	Hu[g]/P[h]	128--[i]/T, 133G/C, 248T/A	[5,21]
WP03	AJ554039	G4[b]	Ch[f]	Hu[g]	173G/R[k]	[5]
WP37	AJ554040	G5[b]	Ch[f]	Hu[g]	133G/S[j], 248T/W[l]	[5]
DL04	EF153621	G6[b]	Br[d]	Hu[g]	120T/--[i]	[10,17]
AJ000895H	AJ000895	AI[m]	Au[n]	Hu[g]	205C/S[j]	[7]
AJ000896P	AJ000896	As[o]	UK[p]/De[q]	P[h]	128--[i]/T, 129--[i]/T, 133G/C, 205C/S[j], 248T/A	[7]
DL02Ec9	GQ339794	H1[b]	Br[d]	Hu[g]	203G/A	[17]
DL04Ec1	EU635686	H2[b]	Br[d]	Hu[g]	124T/--[i], 127T/--[i], 156T/C, 231C/T	[17]
DL04Ec2	EU635687	H3[b]	Br[d]	Hu[g]	127T/--[i], 183T/A	[17]
DL04Ec3	EU635688	H4[b]	Br[d]	Hu[g]	156T/C, 127T/--[i]	[17]
DL13Ec2	GQ339795	H5[b]	Br[d]	Hu[g]	167A/G, 120T/--[i]	[17]
DL13Ec4	GQ339796	H6[b]	Br[d]	Hu[g]	150T/C	[17]
DL15Ec1	GQ339797	H7[b]	Br[d]	Hu[g]	120T/--[i], 229T/C	[17]
DL16Ec1	GQ339798	H8[b]	Br[d]	Hu[g]	120T/C, 233T/C	[17]
DL16Ec3	GQ339799	H9[b]	Br[d]	Hu[g]	124T/--[i], 218T/C	[17]
DL17Ec2	EU635694	H10[b]	Br[d]	Hu[g]	127T/--[i], 238A/G	[17]
DL17Ec14	EU635695	H11[b]	Br[d]	Hu[g]	127T/--[i], 130G/T, 155A/G, 156T/C, 231C/T	[17]
041-1Ec4	GQ339800	H12[b]	Br[d]	Hu[g]	120T/--[i], 229T/A, 248T/C	[17]
104-5Ec1	GQ339801	H13[b]	Br[d]	Hu[g]	124T/C	[17]
I158	JN176638	G1[b]	IND[r]	Hu[g]	I[s]	[*][t]
I305	JN176639	G1[b]	IND[r]	Hu[g]	I[s]	[*][t]
I300	JN176640	G1[b]	IND[r]	Hu[g]	I[s]	[*][t]
I152	JN176641	G1[b]	IND[r]	Hu[g]	I[s]	[*][t]
I6	JN176642	G1[b]	IND[r]	Hu[g]	I[s]	[*][t]
I172	JN176643	G1[b]	IND[r]	Hu[g]	I[s]	[*][t]
I149	JN176644	G1[b]	IND[r]	Hu[g]	I[s]	[*][t]
I31	JN176645	G1[b]	IND[r]	Hu[g]	I[s]	[*][t]
I203	JN176646	G1[b]	IND[r]	Hu[g]	I[s]	[*][t]
I170	JN176647	G1[b]	IND[r]	Hu[g]	I[s]	[*][t]
I449	JN176648	G1[b]	IND[r]	Hu[g]	I[s]	[*][t]
I450	JN176649	G1[b]	IND[r]	Hu[g]	I[s]	[*][t]
I2878	JN176655	G1[b]	IND[r]	Hu[g]	I[s]	[*][t]
I198	JN176658	G1[b]	IND[r]	Hu[g]	I[s]	[*][t]
I189	JN176659	G1[b]	IND[r]	Hu[g]	I[s]	[*][t]
I33	JN176660	G1[b]	IND[r]	Hu[g]	I[s]	[*][t]
I2212	JN176663	G1[b]	IND[r]	Hu[g]	I[s]	[*][t]
I2946	JN176664	G1[b]	IND[r]	Hu[g]	I[s]	[*][t]
I236	JN176665	G1[b]	IND[r]	Hu[g]	I[s]	[*][t]
I2910	JN176666	G1[b]	IND[r]	Hu[g]	I[s]	[*][t]
I17	JN176667	G1[b]	IND[r]	Hu[g]	I[s]	[*][t]
I2864	JN176668	G1[b]	IND[r]	Hu[g]	I[s]	[*][t]
I2838	JN176669	G1[b]	IND[r]	Hu[g]	I[s]	[*][t]
I2942	JN176670	G1[b]	IND[r]	Hu[g]	I[s]	[*][t]
I611	JN176672	G1[b]	IND[r]	Hu[g]	I[s]	[*][t]
I75	JN176673	G1[b]	IND[r]	Hu[g]	I[s]	[*][t]
I161	JN176674	G1[b]	IND[r]	Hu[g]	I[s]	[*][t]
I451	JN176651	IND1[u]	IND[r]	Hu[g]	142A/C, 148--[i]/G, 157--[i]/T	[*][t]
I2990	JN176653	IND2[u]	IND[r]	Hu[g]	130G/A, 131C/G, 173G/C	[*][t]
I2788	JN176654	IND3[u]	IND[r]	Hu[g]	122T/A	[*][t]
I288	JN176656	IND4[u]	IND[r]	Hu[g]	162T/--[i]	[*][t]
I196	JN176657	IND5[u]	IND[r]	Hu[g]	120T/A	[*][t]
I2905	JN176661	IND6[u]	IND[r]	Hu[g]	122T/A, 123T/A, 124T/A, 125T/A, 131C/G, 132G/A, 134A/G, 135C/A, 151T/G, 168A/T, 169G/T, 172T/A, 181T/G	[*][t]
I567	JN176662	IND7[u]	IND[r]	Hu[g]	134A/G	[*][t]
I722	JN176671	IND8[u]	IND[r]	Hu[g]	139T/--[i]	[*][t]

our *Ascaris* population. Along with the previously reported and widely distributed genotype G1 [5,10], 8 new sequence variants of ITS1 (IND1-IND8) have been identified. Since, *Ascaris sp.* multiply through sexual reproduction and genetic recombination during meiosis is a natural phenomenon [11], Efforts were also made to determine whether this population diversity are associated with genetic shuffling. Intragenic linkage disequilibrium (LD) among our study population was evaluated at ITS1 locus to identify potential recombination events within them. Moreover, any significant association of *Ascaris* genotypes with patient's age and sex was also evaluated.

2. Material and methods

2.1. Sample collection and detection of Ascaris sp.:

A total of 35 *Ascaris* isolates from human were included in our study. Fecal samples were collected from people of "low socio-economic community of Kolkata" through an on-going field project, studying the parasite burden of those communities. Poor hygiene, sanitation and malnutrition were common in those communities [12]. The ethical clearance for this study has been provided by NICED IEC (i.e. National Institute of Cholera and Enteric Diseases Institutional Ethical Committee). Informed consents have been obtained from the patients (in case of children consents have been obtained from their parents). The parasite's eggs within the fecal were primarily detected by microscopy [13]. DNA was isolated directly from microscopy positive fecal samples using STOOL DNA Minikit (QIAGEN, USA) as per manufacturer's protocol.

2.2. Polymerase chain reaction (PCR) amplification and DNA sequencing

Partial amplification of target gene (ITS1) was performed using gene specific primer pairs (Table 1). In all cases the PCR reaction was performed in 50 μl reaction volume containing approximately 0.4 μg and 0.1 μg of template DNA for primary and nested PCR respectively, 10 pM of each primer, 2.5 mM $MgCl_2$, 1 μg of Bovine Serum Albumin (SIGMA, USA), 200 μM dNTP and 2.5 U of *Taq* DNA polymerase (Bioline, USA) with the reaction parameters as initial denaturation for 5 min or 4 min (Primary and Nested respectively) at 94 °C. This was followed by 35 cycles of denaturation at 94 °C for 30 s, annealing at 65 °C or 60 °C (Primary and Nested respectively) for 30 s, extension at 72 °C for 30 s. This was again followed by the final extension

for 10 min at 72 °C. The amplified PCR products were then separated by electrophoresis on 1.5% agarose gels (SIGMA, USA) according to their size. PCR products of expected sizes were extracted from gels and purified (ROCHE, Germany). The Purified PCR products were then sequenced directly with specific primers (marked with [a] in Table 1) using the 'DigDye Terminator V3.1 cycle sequencing kit' (APPLIED BIOSYSTEMS, USA) as per the manufacturer's protocol. The labeled DNA fragments were further purified by sodium acetate and ethanol precipitation. The sequencing was carried out in an ABI 310 PRISM Automated Genetic Analyzer. Accuracy of DNA sequencing data has been confirmed by sequencing in both directions and also by repetition of DNA sequencing with a new PCR product for all study isolates.

2.3. Analysis of sequence polymorphisms

ITS1 sequences of our study isolates were aligned with all previously published sequences of corresponding loci (downloaded from NCBI GenBank database, accession numbers have been provided in Table 2) using ClustalW multiple alignment program of MEGA version 4 software [14]. Nucleotide position of each single nucleotide polymorphism (SNP) within the target loci was identified from the aligned sequences. The nucleotide positions of SNPs within the target loci were relative to the reference sequence of G1 genotype (GenBank accession number AJ554036) (Table 2). Variable sequences of our target loci (in respect to the reference sequence) were submitted to NCBI GenBank database with accession numbers JN176638 – JN176674.

Intragenic LD and number of recombination events at ITS1 locus among our study population were also assessed by using DnaSP version 5.10.01 (www.ub.es/dnasp/) software.

2.4. Statistical analysis

Associations of *Ascaris* genotypes with patient's age and sex have been evaluated by Epi-Info version 3.5.4 software [15].

2.5. Phylogenetic analysis

Phylogenetic trees were constructed from the previously aligned ITS1 sequences by MEGA version 4 software [14]. Two individual methods [i.e. Neighbor-Joining (NJ) and Maximum parsimony (MP)] were used to confirm the topology of the tree. In both cases, widely

---

Notes to Table 2
[a] Numbers correspond to nucleotide position on reference sequence AJ554036 [5].
[b] Previously reported *Ascaris* genotypes (G1– G6)/haplotypes (H1– H13).
[c] Ba: Bangladesh.
[d] Br: Brazil.
[e] Ja: Japan.
[f] Ch: China.
[g] Hu: human.
[h] P: pig.
[i] –: nucleotide deletion.
[j] S: nucleotide G or C.
[k] R: nucleotide A or G.
[l] W: nucleotide A or T.
[m] Al: *A. lumbricoides* (without nomenclature).
[n] Au: Australia.
[o] As: *A. suum* (without nomenclature).
[p] UK: United Kingdom.
[q] De: Denmark.
[r] IND: India.
[s] I: similar to reference sequence G1.
[t] [*]: identified in this present study.
[u] New sequence variants of ITS1, identified in our study.

distributed and most prevalent *Ascaris* genotype G1 (Genbank ID AJ554036) was considered as an out-group. The bootstrap values were also analyzed to estimate confidence intervals. Genetic distance analysis among our study isolates was also performed using MEGA version 4 software [14].

3. Results

Among our 35 study isolates, majority (27) corresponded to previously reported and widely distributed genotype G1 [5,10]. Along with genotype G1, 8 new sequence variants of ITS1 have also been identified. These new sequence variants have assigned alphanumerical codes beginning with letter 'IND' to indicate their Indian origin (i.e. IND1-IND8) (Table 2).

Sequence comparison of our study isolates with the global *Ascaris* population has revealed 17 new SNPs, which were present in 8 newly identified ITS1 sequences from our Indian *Ascaris* population (i.e. IND1-IND8) (Table 2). Phylogenetic comparison of our study isolates with the global *Ascaris* population using two individual methods (i.e. NJ and MP) generates trees with similar topology. In both cases, few of our Indian isolates (with newly identified ITS1 sequences) formed distinct cluster with high bootstrap value (marked with green color in both trees), which may indicates their distinct phylogenetic position in respect to global *Ascaris* population (Fig. 1). We have also performed genetic distance analysis among our study isolates. The result has been provided in data 1. Intragenic LD between pairs of polymorphic sites at ITS1 locus of our study population was also evaluated to identify potential recombination events within them. Among 171 pairwise comparisons, 92 were significant by Chi-square test and 89 were significant after Bonferroni correction (Table 3). An incomplete LD value ($|D'|$ Y = 0.9818 + 0.1974X, where Y is the LD value and X is the nucleotide distance in kilobases) was also detected (Table 3). Moreover, intragenic recombination analysis at ITS1 locus of our study isolates has identified 2 potential recombination events within our study population (Table 3). Moreover, any significant association of *Ascaris* genotypes with patient's age and sex was also studied, which revealed that G1 genotype was significantly present among female patients (co-efficient value = 0.815, p value = 0.000002) aged between 10 to 15 years (co-efficient value = 0.690, p value = 0.000105). The age and sex information of the patients, included in our study has been provided in Table 4.

4. Discussion

A. lumbricoides and *A. suum* are two of the world's most common soil transmitted nematode and together cause serious health and socio-economic problems. Ascariasis has been considered as Neglected Tropical Diseases (NTD) by WHO, since it is highly prevalent in poor urban and rural areas and has a certain impact on patient's health, physical fitness and productivity [1,2]. Morphological similarity of these two nematodes entails ongoing uncertainty concerning their taxonomic status and argues for the need to explore deeper into their molecular epidemiology [4]. A recent surveillance study among school children from south India revealed a highest prevalence of *Ascaris* species among all STHs infections. Co-infection with other STHs has also been reported [16]. Even though few surveillance studies on *Ascaris* infection have been conducted in India, diagnosis of this parasite was solely based on microscopy. Differentiation between *Ascaris* species certainly cannot be confirmed by microscopy but require detailed molecular epidemiological study based on genetic markers. In the present study, *Ascaris* population from human has been genetically characterized based on widely used genetic marker ITS1.

Sequence analysis of our study isolates has identified G1 as a dominant genotype. As much as 27 among 35 study isolates were corresponding to this widely distributed genotype. This result corroborates with previous report from China, where genotype G1 was dominant among human and G3 among pig [5]. Our study has also identified 8 new sequence variants of ITS1 (IND1-IND8) within our Indian *Ascaris* population. Similar finding was previously reported by Leles *et.al* from Brazil. They have also identified 13 new *Ascaris* haplotypes (H1–H13) from human [17]. Sequence comparison of our study isolates with previously reported *Ascaris* sequences has identified 17 new SNPs within our study isolates. Moreover, all of these SNPs were present within 8 newly identified sequence variants of ITS1 (IND1-IND8), which indicates their distinct genetic organization. This finding was further well supported by the observation of phylogenetic analysis. All the previously reported *Ascaris* sequences were retrieved from NCBI database and *Ascaris* sequences from our study isolates were phylogenetically compared with them. Phylogenetic analysis revealed an interesting scenario. Few of our Indian isolates (with new variations of ITS1 sequences) formed a separate cluster with high bootstrap value, indicating their distinct phylogenetic position in respect to the global *Ascaris* population. Moreover, Intragenic LD analysis between pairs of polymorphic sites at ITS1 locus has identified an incomplete LD value with two potential recombination events within our study population. This finding was quite compatible with a previous report by Li *et.al*. They have identified a similar type of observation (presence of intragenic LD value and recombination events) in *gp60* locus of another enteric parasite, *Cryptosporidium homonis* [18]. Since, *Ascaris sp.* multiply through sexual reproduction [11], genetic recombination during meiosis could be a natural phenomenon. Furthermore, a recent study has identified the molecular evidence of polyandry in *A.suum*. Single female of *A. suum* can mate with multiple males, which can also increase the chance of genetic variations [19]. Such high possibilities of genetic shuffling could be associated with increasing population diversity in a restricted geographic region [5,7,10,17] and frequent emergence of new sequence variants, identified in present as well as in previous studies [17]. Attempts were also made to determine whether any statistically significant association exists between the identified *Ascaris* genotypes and patient's age and sex. Genotype G1 was found to be significantly present among female patients (co-efficient value = 0.815, p value = 0.000002) aged between 10 to 15 years (co-efficient value = 0.690, p value = 0.000105). This finding was quite congruous with a previous report by Anuar *et.al* [20]. They have reported that Ascariasis was significantly related to patients aged <15 years and earning low household income.

Since, Ascariasis is one of the major Soil Transmitted Helminthiases (STHs) and has been declared as Neglected Tropical Diseases (NTD) by WHO, genome information of its infecting strains from different parts of the world is certainly crucial to investigate the disease epidemiology. This study explores the genetic organization of Indian *Ascaris* population for the first time; it will certainly include some fundamental information on the molecular epidemiology of Ascariasis.

Acknowledgments

This study was jointly supported by the grants from National Institute of Infectious Diseases, Japan, Okayama University Program of Founding Research Centre for Emerging and Re-emerging Infectious Disease (OUP 2-5), and Ministry of Education, Culture, Sports, Science and Technology of Japan. The authors also acknowledge to the patients, provide their fecal samples for the study. Authors would like to thank Mrs. Debarati Ganguly for her immense help regarding proof-reading of this manuscript.

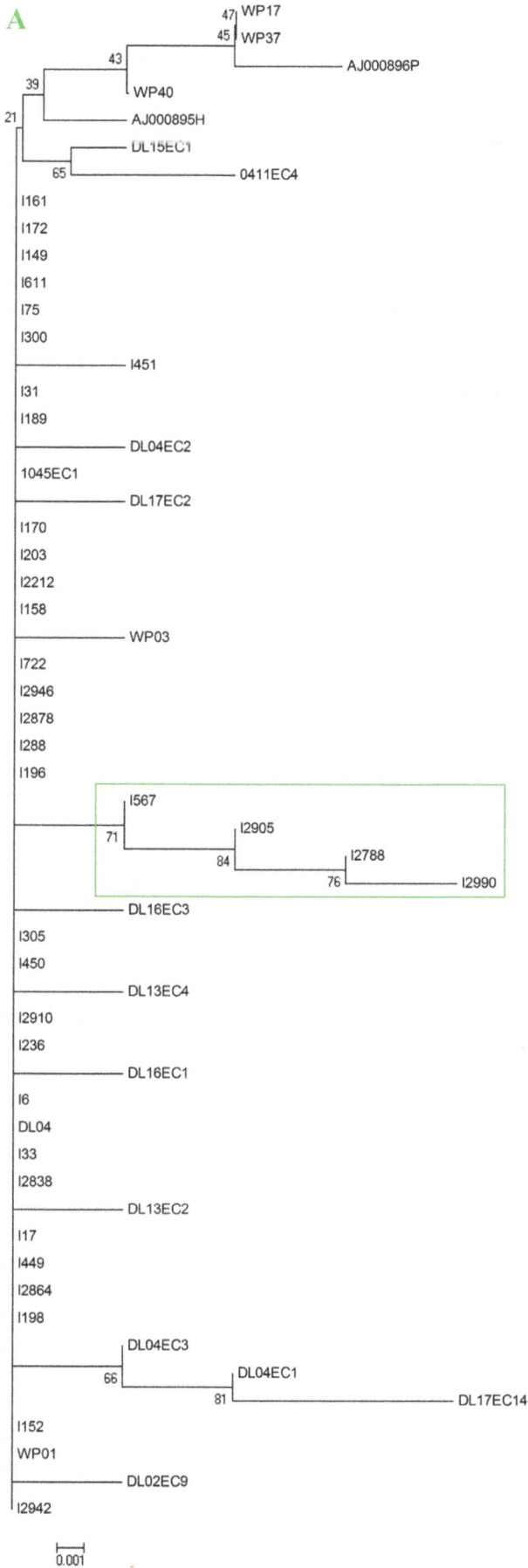

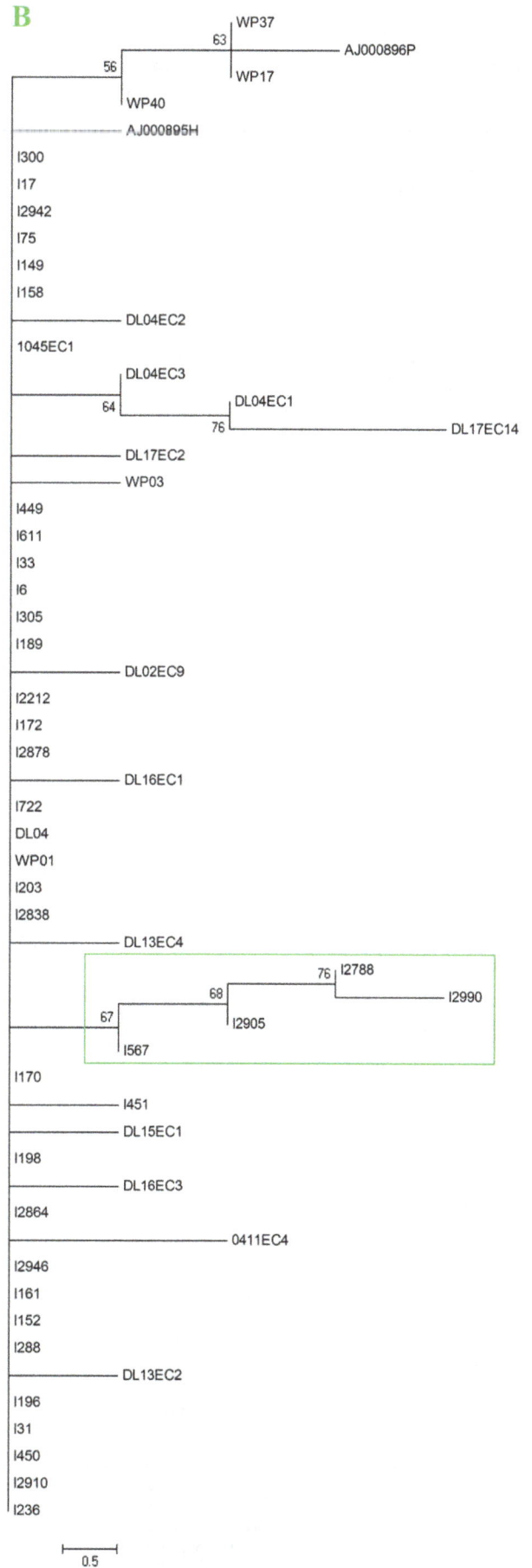

Table 3
Intragenic linkage disequilibrium (LD) and recombination analysis at internal transcribed spacer 1 (ITS1) region of our study isolates.

Population	Number of samples analyzed	Number of polymorphic sites analyzed	Number of pairwise comparisons	Number of significant pairwise comparisons [a]	Intragenic LD($|D'|$) [b]	Rm [c]
Indian	35	21	171	92(89)	$Y = 0.9818 + 0.1974X$	2

[a] Number of significant pairwise comparisons by Chi square test (after Bonferroni correction).
[b] Intragenic linkage disequilibrium (LD), where Y is the LD value and X is the nucleotide distance in kilobases.
[c] Minimum number of intragenic recombination events.

Table 4
Age and sex information of patients, included in the study.

Sample ID	Patient information		GenBank accession number	Genotypes/ haplotypes
	Age (Y:M)	Sex (M/F)		
I158	10.2	M	JN176638	G1[a]
I305	12.6	F	JN176639	G1[a]
I300	10.7	F	JN176640	G1[a]
I152	14.11	F	JN176641	G1[a]
I6	10.7	F	JN176642	G1[a]
I172	12.5	F	JN176643	G1[a]
I149	12.1	F	JN176644	G1[a]
I31	15.0	F	JN176645	G1[a]
I203	12.6	F	JN176646	G1[a]
I170	11.8	M	JN176647	G1[a]
I449	11.7	F	JN176648	G1[a]
I450	13.9	F	JN176649	G1[a]
I2878	15.10	F	JN176655	G1[a]
I198	27.1	F	JN176658	G1[a]
I189	14.5	M	JN176659	G1[a]
I33	12.6	F	JN176660	G1[a]
I2212	26.9	M	JN176663	G1[a]
I2946	9.7	M	JN176664	G1[a]
I236	14.4	F	JN176665	G1[a]
I2910	10.9	F	JN176666	G1[a]
I17	10.1	F	JN176667	G1[a]
I2864	11.3	F	JN176668	G1[a]
I2838	13.2	F	JN176669	G1[a]
I2942	12.8	F	JN176670	G1[a]
I611	10.8	F	JN176672	G1[a]
I75	15.4	F	JN176673	G1[a]
I161	14.3	F	JN176674	G1[a]
I451	8.0	F	JN176651	IND1[b]
I2990	6.5	M	JN176653	IND2[b]
I2788	9.7	M	JN176654	IND3[b]
I288	12.7	F	JN176656	IND4[b]
I196	9.2	M	JN176657	IND5[b]
I2905	7.6	M	JN176661	IND6[b]
I567	37.5	F	JN176662	IND7[b]
I722	23.5	F	JN176671	IND8[b]

[a] Previously reported *Ascaris* genoype G1, identified in our study.
[b] New sequence variants of target locus (i.e. ITS1), identified in our study.

References

[1] Dold C, Holland CV. Ascaris and ascariasis. Microbes Infect 2011;13:632–7.
[2] De Silva NR, Brooker S, Hotez PJ, Montresor A, Engels D, et al. Soil-transmitted helminth infections: updating the global picture. Trends Parasitol 2003;19:547–51.
[3] Nejsum P, Jr Parker ED, Frydenberg J, Roepstorff A, Boes J, et al. Ascariasis is a zoonosis in Denmark. J Clin Microbiol 2005;43:1142–8.
[4] Cavallero S, Snabel. V, Pacella F, Perrone V, D'Amelio S. Phylogeographical studies of Ascaris spp. based on ribosomal and mitochondrial DNA sequences. PLoS Negl Trop Dis 2013;7(4):2170.
[5] Peng W, Yuan K, Zhou X, Hu M, EL-Osta YG, et al. Molecular epidemiological investigation of Ascaris genotypes in China based on single-strand conformation polymorphism analysis of ribosomal DNA. Electrophoresis 2003;24:2308–15.
[6] Peng W, Yuan K, Hu M, Zhou X, Gasser RB. Mutation scanning-coupled analysis of haplotypic variability in mitochondrial DNA regions reveals low gene flow between human and porcine Ascaris in endemic regions of China. Electrophoresis 2005;26:4317–26.
[7] Zhu X, Chilton NB, Jacobs E, Boes J, Gasser RB. Characterization of Ascaris from human and pig hosts by nuclear ribosomal DNA sequences. Int J Parasitol 1999;29:469–78.
[8] Criscione CD, Anderson JD, Sudimack D, Peng W, Jha B, et al. Disentangling hybridization and host colonization in parasitic roundworms of humans and pigs. Proc Biol Sci 2007;274:2669–77.
[9] Bhattacharya D, Haque R, Singh U. Coding and noncoding genomic regions of *Entamoeba histolytica* have significantly different rates of sequence polymorphisms: implications for epidemiological studies. J Clin Microbiol 2005;43:4815–9.
[10] Leles D, Araújo A, Vicente ACP, Iñiguez AM. Molecular diagnosis of ascariasis from human feces and description of a new Ascaris sp. genotype in Brazil. Vet Parasitol 2009;163:167–70.
[11] Goldstein P. Spermatogenesis and spermiogenesis in *Ascaris lumbricoides Var. suum*. J Morphol 1977;154(3):317–37.
[12] Mukherjee AK, Chowdhury P, Bhattacharya MK, Ghosh M, Rajendran K, et al. Hospital-based surveillance of enteric parasites in Kolkata. BMC research notes, 2; 2009 110.
[13] Mukherjee AK, Chowdhury P, Das K, Raj D, Karmakar S, et al. Helminth burden among school going children of southern Bengal. India: a survey report, 2 (3); 2013 189–91(G.J.B.A.H.S).
[14] Tamura K, Dudley J, Nei M, Kumar S. MEGA4: Molecular Evolutionary Genetics Analysis (MEGA) software version 4.0. Mol Biol Evol 2007;24(8):1596–9.
[15] Dean AG, Arner TG, Sunki GG, Friedman R, Lantiga M, et al. Epi Info™, a database and statistic program for public health professional. Atlanta, Georgia, USA: Centers for Disease Control and Prevention; 2007.
[16] Ragunathan L, Kalivaradhan SK, Ramadass S, Nagaraj M, Ramesh K. Helminthic infections in school children in Puducherry, South India. J Microbiol Immunol Infect 2010;43(3):228–32.
[17] Leles D, Araújo A, Vicente ACP, Iñiguez AM. ITS1 intra-individual variability of Ascaris isolates from Brazil. Parasitol Int 2010;59:93–6.
[18] Li N, Xiao L, Cama VA, Ortega Y, Gilman RH, et al. Genetic recombination and *Cryptosporidium hominis* virulent subtype IbA10G2. Emerg Infect Dis 2013;19:1573–82.
[19] Zhou C, Yuan K, Tang X, Hu N, Peng W. Molecular genetic evidence for polyandry in *Ascaris suum*. Parasitol Res 2011;108(3):703–8.
[20] Anuar TS, Md Salleh F, Moktar N. Soil-transmitted Helminth infections and associated risk factors in three Orang Asli tribes in Peninsular Malaysia. Sci Rep 2014;4:4101.
[21] Ishiwata K, Shinohara A, Yagi K, Horii Y, Tsuchiya K, et al. Identification of tissue-embedded ascarid larvae by ribosomal DNA sequencing. Parasitol Res 2004;92:50–2.

Fig. 1. Phylogenetic comparison of our study isolates with the global *Ascaris* population: Internal transcribed spacer 1 (ITS1) sequences of our study isolates were aligned with ITS1 sequences from the global *Ascaris* population using ClustalW multiple alignment program of MEGA version 4 software. Two separate phylogenetic trees were generated from this alignment using two individual methods- [A] Neighbor-Joining (NJ) and [B] Maximum parsimony (MP). In both cases, widely distributed and most prevalent *Ascaris* genotype G1 (Genbank ID AJ554036) was considered as out-group. The bootstrap values were also analyzed to estimate confidence intervals. Phylogenetic comparison using these two individual methods (i.e. NJ and MP) generates trees with similar topology. In both cases, few of our Indian isolates (with newly identified ITS1 sequences) formed distinct cluster with high bootstrap value (marked with green color in both trees), which may indicates their distinct phylogenetic position in respect to global *Ascaris* population.

13

SNP2Structure: A Public and Versatile Resource for Mapping and Three-Dimensional Modeling of Missense SNPs on Human Protein Structures

Difei Wang [a,b,c,*], Lei Song [b], Varun Singh [b], Shruti Rao [b], Lin An [b], Subha Madhavan [a,b,**]

[a] Department of Oncology, Lombardi Comprehensive Cancer Center, Georgetown University Medical Center, Washington, DC 20007, USA
[b] Innovation Center for Biomedical Informatics, Georgetown University Medical Center, Washington, DC 20007, USA
[c] Department of Biochemistry and Molecular & Cellular Biology, Georgetown University, Washington, DC 20007, USA

ARTICLE INFO

ABSTRACT

One of the long-standing challenges in biology is to understand how non-synonymous single nucleotide polymorphisms (nsSNPs) change protein structure and further affect their function. While it is impractical to solve all the mutated protein structures experimentally, it is quite feasible to model the mutated structures in silico. Toward this goal, we built a publicly available structure database resource (SNP2Structure, https://apps.icbi. georgetown.edu/snp2structure) focusing on missense mutations, msSNP. Compared with web portals with similar aims, SNP2Structure has the following major advantages. First, our portal offers direct comparison of two related 3D structures. Second, the protein models include all interacting molecules in the original PDB structures, so users are able to determine regions of potential interaction changes when a protein mutation occurs. Third, the mutated structures are available to download locally for further structural and functional analysis. Fourth, we used Jsmol package to display the protein structure that has no system compatibility issue. SNP2Structure provides reliable, high quality mapping of nsSNPs to 3D protein structures enabling researchers to explore the likely functional impact of human disease-causing mutations.

Keywords:
Active site mutations
Protein structure
Molecular modeling
Disease causing SNPs
SNP database

1. Introduction

Next-generation sequencing (NGS) has enabled the rapid discovery of single nucleotide polymorphisms (SNPs) in humans [1]. SNPs account for about 90% of human genetic variation. Most of this genetic variation does not affect protein structure and function. However, non-synonymous SNPs (nsSNPs), which change the amino acid sequence of a protein, usually have a detrimental effect on protein structure and/or function and are frequently associated with human diseases.

In the last decade, considerable effort has been devoted to using nsSNP information to mapping the mutations onto the protein sequence and to predict potential functional impact and the association with human diseases [2–13]. Although it is possible to predict disease causing effects for nsSNP changes with about 70 to 90% accuracy using various

annotated databases (2, 5 –7, 11a), it is still challenging to understand such deleterious effects at the protein structure level. Several web portals have been built to predict the effects of mutations on protein function and association with disease [7,11,13]. These resources are limited in their visualization capabilities and accessibility; there is a need to build a more user-friendly resource to provide direct mapping of nsSNPs onto 3D protein structural models. Protein structure visualization provides critical visual information about how mutations impact biological protein function and helps to guide drug design and discovery.

The challenge of understanding how nsSNPs are associated with disease motivated us to build a database for all known human protein structures with modeled nsSNP mutation sites as a freely available resource for the scientific community. As our first step toward understanding the deleterious effects of protein mutations, we developed a novel pipeline to map all nsSNPs mutations in dbSNP (build 137) [14] onto the available human X-ray and solution NMR structures in the Protein Data Bank (PDB) [15]. In particular, we are focusing on missense SNPs (msSNP, a common type of nsSNPs). Our mapping pipeline addressed shortcomings in some of the existing web portals with a similar aim. We found that some of the existing web portals do not provide three-dimensional structures but instead are using static images of the structures [13]. The ability

* Correspondence to: D. Wang, Department of Biochemistry and Molecular & Cellular Biology, Georgetown University Medical Center, Washington, DC 20007, USA

** Correspondence to: S. Madhavan, Innovation Center for Biomedical Informatics, Georgetown University Medical Center, Washington, DC 20007, USA

E-mail addresses: Difei.Wang@georgetown.edu (D. Wang), Subha.Madhavan@georgetown.edu (S. Madhavan).

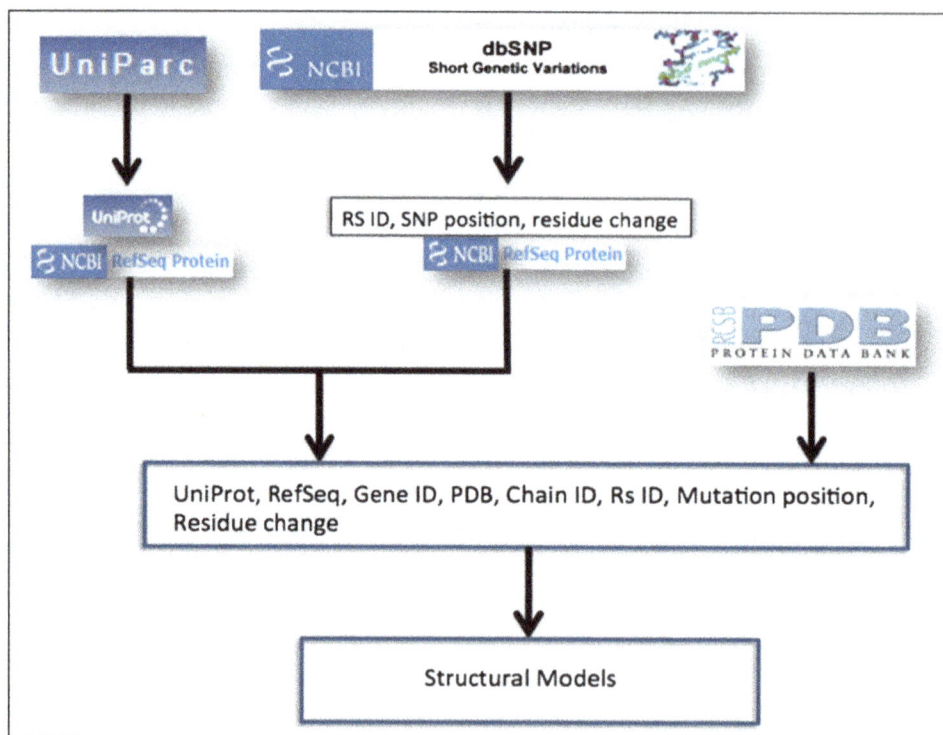

Fig. 1. Scheme of structural model building using publicly available resources.

to visualize residue mutations in 3D structures enables users to see if the mutation alters catalytic activity due to its proximity to the active site; to determine if the mutation alters key interactions inside the protein itself or between macromolecules; or if the mutation changes the stability of the folded structure. We found inconsistent residue numbering between PDB structures [15] and UniProt sequences [16], which can result in a position shift in mapping the mutated residue to the wild type structure [13]. The mapping discrepancies within these two resources have resulted in errors being propagated into derivative databases and could potentially lead to incorrect assignment of mutation site in the structure of interest based on these databases. It is not a trivial task to correct the discrepancies [17].

A significant novel feature in SNP2Structure is the direct comparison of two protein structures (wild type vs. mutated or mutated vs. mutated) with a user-friendly interface. SNP2Structure also provides information on all associated and interacting molecules (protein, DNA, RNA and small ligands) packed in the original crystal structure for displaying protein models, in contrast to other resources that focus on the protein structure only [7,11,13]. Our tool allows users to inspect protein mutations that may affect the interaction with other macromolecules. The implantation of Jsmol package (http://chemapps.stolaf.edu/jmol/jsmol/jsmol.htm) avoids the system compatibility issue for displaying protein structure. Finally, all mutated structures are downloadable for further structural analysis. We believe SNP2Structure will be a valuable resource for exploring potential structural and functional impact of missense mutations in various human diseases.

2. Methods

SNP2Structure is based on the integration of data from publically available bioinformatics resources to annotate mutated positions both in protein sequences and their associated X-ray and solution NMR structures. We took advantage of the existing mutation information in dbSNP and the annotation of protein sequences in UniProt and UniParc [16], (Fig. 1) to obtain the genomic location of each msSNP.

To obtain protein sequence information, we parsed the UniParc XML file to retrieve the human sequences and their UniProt and RefSeq IDs with active status [18]. No isoforms were considered. We then parsed the dbSNP gene records to get the RefSeq IDs and the corresponding dbSNP Reference SNP IDs (rs IDs), mutation positions, and residue name. Only mutations having a single amino acid change were considered. Indels, frame-shift, and synonymous mutations were discarded. These two lists were merged together and only the matched RefSeq IDs were stored for further analysis. The residue positional information in our dataset refers to the 'canonical' sequence in UniProt. The reasons we selected UniProt as our reference database are: 1) UniProt is a highly curated and manually reviewed database, and 2) UniProt has a strict rule to select the canonical sequence for all the protein products encoded by one gene. To obtain protein structure information, we retrieved all existing human protein structures in PDB and only considered the X-ray and solution NMR structures. We then merged these two lists (sequence and structure) by filtering out unmatched UniProt IDs.

Based on our findings, approximately 5 –10% of protein structures in PDB have incorrect amino acid residue numbering compared to the annotations in UniProt. This was a critical problem to address since the discrepancies can lead to errors when mapping mutation sites to protein structures. For example, the sequence of Alcohol Dehydrogenase 5 (ADH5) in the UniProt entry (P11766) contains residues Gly60, Cys61 and Phe62. However, in one of the ADH5 X-ray structures (PDB 1TEH), it has Gly60 followed by Cys60A and Phe61. The unconventional residue number Cys60A in PDB 1TEH is the cause of the residue numbers after Gly60 being changed by − 1, which is a problem if a single amino acid mutation occurs at any residues after Gly60. Two other examples include structures of Alcohol Dehydrogenase class 4 mu/sigma chain (ADH7), PDB 1AGN and 1D1S. In these structures, Thr118 directly follows Ile116. It looks like residue 117 is missing in the structures. When we aligned the actual PDB sequences to the corresponding UniProt sequence (P40394), we found that they correspond to Ile129 and Thr130, respectively. There is no missing residue between Ile129 and Thr130. It is clear that the difference between PDB numbering and

Fig. 2. Key components of SNP2Structure database.

UniProt numbering is not consistent from the start to the end residue. These discrepancies cause errors when assigning the mutation positions in the structures. For instance, rs59534319 corresponds to the Lys238Glu mutation of the ADH7 protein in UniProt. In one web portal instead of Lys226 (there is a 12 difference between UniProt and PDB residue numbering), it was assigned to residue 225 in PDB 1D1S due to the inconsistencies between UniProt and PDB [13].

To correct the problem of residue numbering in PDB structures, we aligned the actual PDB sequences to the corresponding UniProt sequences and reassigned the residue numbers in PDB structures according to residue numbers in UniProt sequences. The actual PDB sequences were generated by parsing the ATOM records in the PDB files. The needle program in the EMBOSS package [19] was used for sequence alignment between actual PDB sequence and the corresponding UniProt sequence. Only the structures without mutations and/or missing residues compared to its UniProt canonical sequence and more than 20 amino acids long in length in the mapped range were considered in the current release of SNP2Structure. Next, we renumbered all considered PDB structures with the correct residue numbers based on sequence alignment results. We then used the final table with all the information as input to build the structural models using Modeller 9.11 [20] (~280,000 models). The model building procedure is straightforward. Only the coordinates of atoms in the side chain of the mutated residues were changed. The rest of the coordinates were kept intact. We have built one model per mutation per PDB entry. For evaluating the mutation effect of msSNPs, we either calculated the PolyPhen2 score

or parsed the SIFT results table. PolyPhen2 and SIFT are the two popular tools to predict potential impact of an amino acid substitution on the protein structure and function of a human protein [5,6]. PolyPhen2 uses straightforward physical and comparative considerations while SIFT calculates the probability of the amino acid change at certain position in a queried sequence after comparing it with related protein sequences. These scores indicate if the mutation is deleterious or not. SNP2Structure also includes DrugBank ID for small molecules, active site, and metal binding site information. The active site/metal binding site information was extracted from the UniProt database. For comparison, we also included resolution for X-ray structure, chain information, and other key information in the database. In addition, the relative solvent accessible area for both wild-type and mutated structures were calculated using the NACCESS 2.1.1 program [21].

3. Results

3.1. Data Portal Features

SNP2Structure was designed to be a user-friendly and web-based portal. It has four key components: 1) search, 2) structural visualization, 3) structural comparison, and 4) download. The protein mutation information can be queried through an Oracle database (Fig. 2). The corresponding structures/models can be downloaded in PDB format for further analysis.

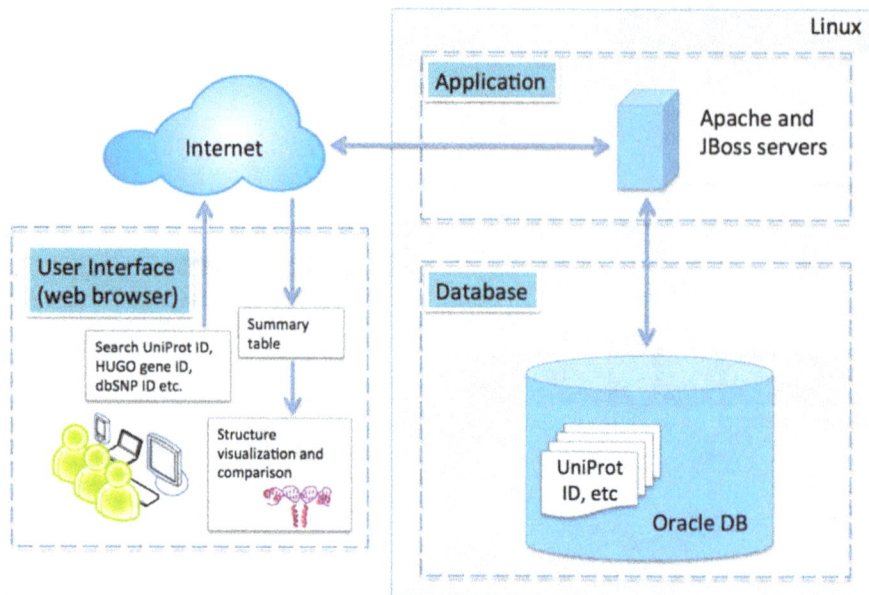

Fig. 3. SNP2Structure web portal architecture.

3.2. Search

Users are able to retrieve protein mutation information of interest by inputting the HUGO gene symbol, UniProt ID, dbSNP rs ID, or PDB ID in the search box on the homepage. For example, enter HUGO gene symbol RUNX1 [Runt-related transcription factor 1, also known as acute myeloid leukemia 1 protein (AML1) or core-binding factor subunit alpha-2 (CBFA2)] in the Search box. If it exists in the database, any record that contains RUNX1 as a string/substring will appear while the user is typing. Select the correct gene and click the button with "Find SNPs." For RUNX1, it outputs 26 msSNP mutation records. The output table is downloadable as a text file. If your query does not exist in the database, it will report a page with a "no records existing" message.

3.3. Structural Visualization and Comparison

The structural visualization component allows users to visualize protein structures of interest. The default display setting for protein is ribbons colored by secondary structures. Unlike existing portals, our portal provides an interactive display of the 3D protein structures in JSmol. By simply right clicking on the structure, a user can access various features in JSmol, such as, changing the style and color of the structure, highlighting hydrogen molecules and water molecules, highlighting the van der Waals surface on the protein structure, zoom in and out of a mutation site, spin the structure, etc. A user can also utilize the command line box to execute these functions. Moreover, hovering over the secondary protein structures will display the amino acid residue name and number.

Fig. 4. Comparison of two single amino acid mutations of p53 (Left: Arg280Ile and Right: Lys292Ile). DNA is shown as purple strands while p53 is displayed as colored ribbons. The mutated residues are displayed as green balls.

Since we used the structure in the asymmetric unit to generate the models, users may explore the likely interactions between mutated proteins and their associated macromolecules and ligands. Interactions, however, may be artificial due to the well-known crystal packaging effect. The use of biological assembly structures would be ideal, but they need extensive manual curation since more than one biological assembly model (author suggested, software calculated or both) may exist for a particular protein. The predicted results of PolyPhen2 and SIFT for the effect of mutations are displayed on the same page for reference. We also provide the relative surface area of the structures, which allows users to explore 1) the effect of a mutated structure compared to the wild type and 2) compare the effect of two different mutations on same structure.

In order to compare the structures of different mutations within the same protein, we designed two visualization windows. Side-by-side comparison can be done for either wild type structure vs. mutated structure or mutated vs. mutated structures for the same protein. For proteins with more than one X-ray structure, it is also possible to compare the mutated structures among those X-ray structures. This feature is valuable since it shows which parts of the protein structures may be conformationally variable.

3.4. Download Option and Portal Statistics

Users can download structures to their local computer or do analysis online. The search results can be exported as a table with all necessary information for the corresponding model retrieval. For example, the model with rs183443805 from PDB 1PML can be downloaded using the following link, https://apps.icbi.georgetown.edu/molecule3D/snp2str/DEC18/models/P00750_1PML_A_183443805_ARG_224_CYS.pdb. Since we have corrected the discrepancies between residue numbering between PDB and UniProt, the downloaded structures will have the corrected residue numbering. If users need the structures with original numbering, they shall visit the PDB website (http://www.rcsb.org) to download the original PDB files.

Our portal includes 1810 unique UniProt IDs, 7021 wild type protein structures (6135 X-ray and 886 NMR), and 26,097 unique dbSNP rs IDs. The total number of models is 289,709.

3.5. System Implementation

The SNP2Structure database was developed under the Linux system using common software packages for the web server including Apache web server and Oracle database management, Fig. 3. The web interface of SNP2Structure was written in Groovy using the Grails framework. Data is stored in an Oracle database. We also used the JSmol package that is a JavaScript version of Jmol for protein structure visualization (http://chemapps.stolaf.edu/jmol/jsmol/jsmol.htm). The advantage of using JSmol instead of Jmol is that any HTML5 compatible web browser can open the JSmol application. Jmol also needs the Java Virtual Machine (JVM) installed first and often causes compatibility issues [7,9]. We designed the system as simple as possible and made the database flexible, interactive, and intuitive for users. We recommend using Google Chrome, Firefox, or Safari to open the portal. The portal is not compatible well with Internet Explorer on a PC running Windows XP or Windows 7. SNP2Structure runs on m1.large instance (4 cores of CPUs with 7.5 Gb memory) on the Amazon cloud.

3.6. Application Example

Here, we give an example to illustrate structural comparison, one of the unique applications of SNP2Structure. P53 is the most studied tumor suppressor protein in cancer biology. Hundreds of missense mutations in p53 have been identified in the last two decades. One of the most frequently mutated residues is Arg280, which makes hydrogen-bond interactions with Gua in the p53 canonical DNA response element CATG.

Arg280Ile (rs121912660) (green balls, left, Fig. 4) mutation abolishes such important interactions and has deleterious effects on p53 tumor suppression function. Another less frequent p53 mutation Lys292Ile (rs121912663) seems to have a less deleterious effect since Lys292 (green balls, right, Fig. 4) is far from the DNA major groove. The PolyPhen prediction for Arg280Ile is deleterious and for Lys292Ile is 'neutral'. Moreover, the relative surface area for the protein structure with mutation at Arg280Ile is 42096.4 compared to 42032.1 for the structure with a mutation at Lys292Ile.

4. Conclusion

We have built a web portal to share, visualize, and analyze protein structures constructed in-house that are associated with msSNP mutations. Our web application features direct comparison of two related structures: either wild type vs. mutated, or mutated vs. mutated. In addition, we carefully corrected the numbering discrepancy of mutation positions between structures and sequences using public resources. It is user-friendly and the structural models are downloadable for further structural and functional analysis. We believe this resource is valuable to the basic research community for understanding and exploring the likely functional impact of human disease-causing msSNPs as well as to translational researchers exploring structure-based drug design.

Funding

This work was supported by the Food and Drug Administration (FDA) Centers of Excellence in Regulatory Science and Innovation (CERSI) program: [grant number FDA U01FD00413]; and the National Institutes of Health/National Cancer Institute [grant number U54-CA149147]. We also thank Amazon Corporation for providing the computing resources for this study.

Acknowledgments

We thank Dr. Baris Suzek, Dr. Peter McGarvey, Mr. Michael Harris and Mr. Shailendra Singh for valuable discussions. We also thank Mr. Kevin Rosso for assisting with the early stages of web portal development; and Dr. Laura Sheahan for text editing.

References

[1] Collins FS, Brooks LD, Chakravarti A. A DNA polymorphism discovery resource for research on human genetic variation. Genome Res 1998;8:1229–31.
[2] Adzhubei IA, et al. A method and server for predicting damaging missense mutations. Nat Methods 2010;7:248–9.
[3] Chasman D, Adams RM. Predicting the functional consequences of non-synonymous single nucleotide polymorphisms: structure-based assessment of amino acid variation. J Mol Biol 2001;307:683–706.
[4] Ferrer-Costa C, et al. PMUT: a web-based tool for the annotation of pathological mutations on proteins. Bioinformatics 2005;21:3176–8.
[5] Ng PC, Henikoff S. SIFT: Predicting amino acid changes that affect protein function. Nucleic Acids Res 2003;31:3812–4.
[6] Ramensky V, Bork P, Sunyaev S. Human non-synonymous SNPs: server and survey. Nucleic Acids Res 2002;30:3894–900.
[7] Reva B, Antipin Y, Sander C. Predicting the functional impact of protein mutations: applications to cancer genomics. Nucleic Acids Res 2011;39, e118.
[8] Espinosa O, Mitsopoulos K, Hakas J, Pearl F, Zvelebil M. Deriving a mutation index of carcinogenicity using protein structure and protein interfaces. PLoS One 2014;9, e84598.
[9] Douville C, Carter H, Kim R, Niknafs N, Diekhans M, Stenson PD, et al. CRAVAT: cancer-related analysis of variants toolkit. Bioinformatics 2013;29:647–8.
[10] Schwarz JM, et al. MutationTaster evaluates disease-causing potential of sequence alterations. Nat Methods 2010;7:575–6.
[11] a) Yue P, Melamud E, Moult J. SNPs3D: candidate gene and SNP selection for association studies. BMC Bioinf 2006;7:166.
      b) Wang Z, Moult J. SNPs, protein structure, and disease. Hum Mutat 2006;17: 263–70.

[12] Mottaz A, David FP, Veuthey A-L, Yip YL. Easy retrieval of single amino-acid poly-
     morphisms and phenotype information using SwissVar. Bioinformatics 2010;26:
     851–2.

[13] Ryan M, et al. LS-SNP/PDB: annotated non-synonymous SNPs mapped to Protein
     Data Bank structures. Bioinformatics 2009;25:1431–2.

[14] Sherry ST, et al. dbSNP: the NCBI database of genetic variation. Nucleic Acids Res
     2001;29:308–11.

[15] Berman HM, et al. The Protein Data Bank. Nucleic Acids Res 2000;28:235–42.

[16] Wu CH, et al. The Universal Protein Resource (UniProt): an expanding universe of
     protein information. Nucleic Acids Res 2006;34:D187–191.

[17] David FP, Yip YL. SSMap: a new UniProt-PDB mapping resource for the curation of
     structuralrelated information in the UniProt/Swiss-Prot Knowledgebase. BMC Bioinf
     2008;9:391–402.

[18] Pruitt KD, Tatusova T, Maglott DR. NCBI reference sequences (RefSeq): a curated
     nonredundant sequence database of genomes, transcripts and proteins. Nucleic
     Acids Res 2007;35:D61–65.

[19] Rice P, Longden I, Bleasby A. EMBOSS: the European Molecular Biology Open Soft-
     ware Suite. Trends Genet 2000;16:276–7.

[20] Sali A, Blundell TL. Comparative protein modelling by satisfaction of spatial re-
     straints. J Mol Biol 1993;234:779–815.

[21] Hubbard SJ, Thornton JM. NACCESS, Computer Program. Department of Biochemis-
     try and Molecular Biology, University College London; 1993.

14

Chimeric mitochondrial peptides from contiguous regular and swinger RNA

Hervé Seligmann

Unité de Recherche sur les Maladies Infectieuses et Tropicales Émergentes, Faculté de Médecine, URMITE CNRS-IRD 198 UMER 6236, Université de la Méditerranée, Marseille, France

ARTICLE INFO

ABSTRACT

Keywords:
RNA–DNA differences
Bijective transformation
Nucleotide substitution
Proteome
Systematic deletions
delRNA

Previous mass spectrometry analyses described human mitochondrial peptides entirely translated from swinger RNAs, RNAs where polymerization systematically exchanged nucleotides. Exchanges follow one among 23 bijective transformation rules, nine symmetric exchanges ($X \leftrightarrow Y$, e.g. $A \leftrightarrow C$) and fourteen asymmetric exchanges ($X \rightarrow Y \rightarrow Z \rightarrow X$, e.g. $A \rightarrow C \rightarrow G \rightarrow A$), multiplying by 24 DNA's protein coding potential. Abrupt switches from regular to swinger polymerization produce chimeric RNAs. Here, human mitochondrial proteomic analyses assuming abrupt switches between regular and swinger transcriptions, detect chimeric peptides, encoded by part regular, part swinger RNA. Contiguous regular- and swinger-encoded residues within single peptides are stronger evidence for translation of swinger RNA than previously detected, entirely swinger-encoded peptides: regular parts are positive controls matched with contiguous swinger parts, increasing confidence in results. Chimeric peptides are 200× rarer than swinger peptides (3/100,000 versus 6/1000). Among 186 peptides with >8 residues for each regular and swinger parts, regular parts of eleven chimeric peptides correspond to six among the thirteen recognized, mitochondrial protein-coding genes. Chimeric peptides matching partly regular proteins are rarer and less expressed than chimeric peptides matching non-coding sequences, suggesting targeted degradation of misfolded proteins. Present results strengthen hypotheses that the short mitogenome encodes far more proteins than hitherto assumed. Entirely swinger-encoded proteins could exist.

1. Introduction

Mitochondrial genomes apparently compensate for their reduced size by cumulating multiple functions for single sequences [12,14]. For example, tDNA, DNA templating for tRNAs, probably functions also occasionally as light strand replication origin [79,80,83,106–108,110]. The complementary strand of tDNA has similar secondary structure formation capacities and might template for additional functional tRNAs with anticodons usually corresponding to the inverse complement of the tRNA's regular anticodon [7,81,82,84]. Mitochondrial tRNA sidearm loops might also function as anticodons, potentially increasing further mitochondrial anticodon repertoires [90,95]. Translation of stop codons also increases protein coding repertoires [81], reassigning stop codons to amino acids [26,84–87,91,7,11,98].

These various mechanisms expand protein-coding potentials of DNA/RNA sequences. Multifunctional sequences, as suggested for tRNA synthetase genes [53,69,70] are presumably relics of ancient, short protogenomes, plausibly consisting of ancestors of ribosomal RNAs [72,73,111], where sequence multifunctionality was probably essential. Presumably, alternative codings are relics of mechanisms that increase sequence multifunctionality.

1.1. Swinger polymerization

A further little known phenomenon increases DNA's protein coding repertoire: nucleotide polymerization that systematically exchanges nucleotides. This alters gene and mRNA coding properties. Assuming this phenomenon enables to detect homology relationships of otherwise 'orphan' DNA and RNA sequences. The homology of these orphan sequences had not been determined because these apparently orphan sequences are so much transformed as compared to their 'parent' homologue that homology is undetectable without assuming a systematic exchange between nucleotides, but becomes obvious after taking the systematic exchange(s) into account. These transformations consist of systematic exchanges between nucleotides during DNA or RNA polymerization, producing so-called swinger sequences.

The first described swinger RNAs were from vertebrate mitogenomes, and correspond to a 3'-to -5' inversion, without complementing, of the homologous, template sequence [88,92], also called 'reversing' transformation [27,28]. When considering a specific sequence, this transformation follows the swinger rule $A \leftrightarrow T + C \leftrightarrow G$ (bijective transformation rule π9 according to the annotation system in [58]) of the negative

E-mail address: varanuseremius@gmail.com.

strand of the specific sequence, one among nine systematic symmetric exchanges, of type X ↔ Y, e.g. A ↔ C [93]. Fourteen asymmetric exchanges exist, of type X → Y → Z → X, e.g. A → C → G > A [94]. About hundred mitochondrial transcripts corresponding to one of these 23 swinger types have been detected within the human EST database of GenBank, with about twice as many from the nine symmetric exchanges than from the fourteen asymmetric exchanges.

Swinger RNAs matching eleven exchange types were detected within GenBank's EST database, six symmetric, and four asymmetric transformations. Most of these swinger RNAs (obtained by classical Sanger sequencing) are longer than 100 nucleotides and have >90% similarity with the mitogenome if the swinger transformation is assumed over their complete length [90,91]. All 23 swinger types exist in the human mitochondrial transcriptome among short reads produced by RNA seq (Illumina) ([103], data from [30]). Abundances of different swinger types as estimated from GenBank's ESTs (sequenced by classical methods) and next generation massive sequencing (RNA seq) are overall congruent (i.e. Figure 2 in [103]). This congruence between swinger RNA abundances is remarkable for two reasons: first because comparable results are obtained by two independent methods (Sanger versus next generation sequencing); and second because biological samples differ (not the same cells/mitochondrial lines were analyzed). This suggests that mitochondrial swinger transcription is general to mitochondria, not tissue- or line-specific.

Hence sequences potentially template for 23 swinger transformed versions, increasing considerably the potential coding density of any sequence. Swinger DNA was also detected for nuclear and mitochondrial genes [96,97], especially ribosomal RNAs [99], but for now only according to swinger rule A ↔ T + C ↔ G. Swinger sequences detected in Genbank originate from numerous independent research projects and laboratories, only this author describes them as swinger-transformed.

1.2. Swinger versus chimeric RNAs and peptides

Some detected sequences are not entirely swinger-transformed, sequences contiguous to the swinger sequence match the untransformed, contiguous DNA template, and hence are regular RNA [100]. These RNAs transcribed partly by regular, and partly by swinger transcriptions, are termed chimeric RNAs [100]. The transition from one to the other part is frequently abrupt, suggesting sudden switches in the polymerization mode of the same polymerase.

Analyzes here search for peptides matching translation of such chimeric RNAs, where contiguous parts of the peptide are translated from regular and swinger parts of the sequence. These peptides are also considered chimeric, and differ from previously described swinger peptides [103] because the latter are only translated from swinger-

transformed RNA, while chimeric peptides would be transcribed from RNA that is in part regular, and in part swinger-transformed.

The principle according to which chimeric sequences are produced is shown for a specific 120 nucleotides long sequence of the human mitogenome (Fig. 1). The mid-forty nucleotides are swinger transformed according to swinger rule A ↔ C + G ↔ T (second sequence in Fig. 1). Swinger RNAs consist solely of swinger-transformed regions such as the underlined transformed regions. Chimeric RNAs have at least one of the contiguous, untransformed (5′ and/or 3′) parts. Swinger peptides are solely translated from swinger-transformed sequences (such as the underlined sequence in Fig. 1). Chimeric peptides are translated from a sequence that stretches over a regular and a swinger-transformed RNA region. Only a minority of detected RNA sequences bearing swinger transformations are chimeric, most follow in their entirety a given nucleotide exchange rule [90,91,100]. Detection of chimeric peptides would be evidence independent of previous descriptions of swinger peptides (entirely encoded by swinger-transformed RNA [103]) for translation of swinger-transformed RNA. Much fewer chimeric human mitochondrial RNAs [100] than entirely swinger-transformed human mitochondrial RNAs [90,91] have been detected. Hence I expect to detect fewer chimeric peptides than for previous analyzes searching for entirely swinger-encoded peptides.

1.3. Swinger polymerization by regular polymerases?

Swinger polymerizations could result from unusual polymerization modes by regular polymerases because the principle of swinger polymerization does not differ from that of point nucleotide misinsertions. The difference is in the systematic change in templating rules, from f(A,C,G,T) = (A,C,G,T/U) (regular DNA replication/transcription), to a different rule, e.g. A ↔ C, which can also be annotated as f(A,C,G,T/U) = (C,A,G,T/U), stressing its systematic, rather than punctual nature. It seems plausible that point nucleotide misinsertions are due to switches to unstable, unusual conformations of polymerases, lasting the time of a misinsertion. Hypothetically, these unstable, misinsertion-inducing conformations are occasionally stabilized, so that the nucleotide exchange corresponding to that misinsertion occurs systematically along the sequence stretch polymerized while the polymerase is in that unusual conformation, producing a swinger DNA/RNA. Swinger RNAs are for now the only evidence indicating the existence of such unusual, stabilized polymerase states.

This hypothesis on polymerase conformations yields two testable predictions. The first prediction is that biochemical parameters experimentally estimated for point misinsertions by polymerases predict properties of swinger sequences. In this respect, the affinity (Km) and Vmax of each of the twelve misinsertions, and the four regular

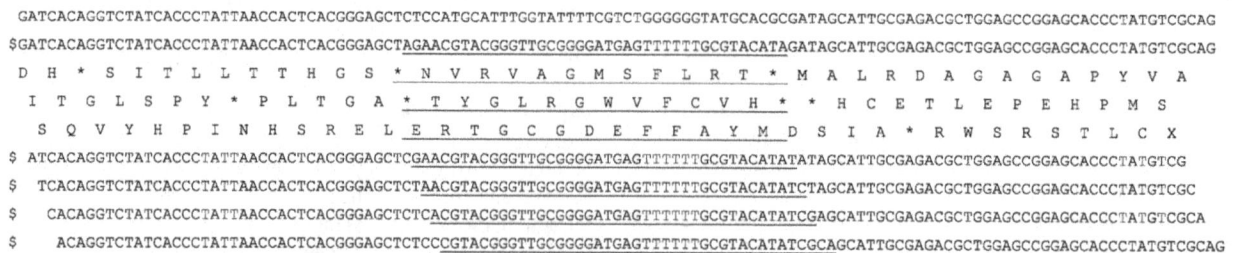

```
    GATCACAGGTCTATCACCCTATTAACCACTCACGGGAGCTCTCCATGCATTTGGTATTTTCGTCTGGGGGGTATGCACGCGATAGCATTGCGAGACGCTGGAGCCGGAGCACCCTATGTCGCAG

$GATCACAGGTCTATCACCCTATTAACCACTCACGGGAGCTAGAACGTACGGGTTGCGGGGATGAGTTTTTTGCGTACATAGATAGCATTGCGAGACGCTGGAGCCGGAGCACCCTATGTCGCAG

   D  H  *  S  I  T  L  L  T  T  H  G  S  *  N  V  R  V  A  G  M  S  F  L  R  T  *  M  A  L  R  D  A  G  A  G  A  P  Y  V  A
    I  T  G  L  S  P  Y  *  P  L  T  G  A  *  T  Y  G  L  R  G  W  V  F  C  V  H  *  *  H  C  E  T  L  E  P  E  H  P  M  S
     S  Q  V  Y  H  P  I  N  H  S  R  E  L  E  R  T  G  C  G  D  E  F  F  A  Y  M  D  S  I  A  *  R  W  S  R  S  T  L  C  X

$  ATCACAGGTCTATCACCCTATTAACCACTCACGGGAGCTCGAACGTACGGGTTGCGGGGATGAGTTTTTTGCGTACATATATAGCATTGCGAGACGCTGGAGCCGGAGCACCCTATGTCG
$  TCACAGGTCTATCACCCTATTAACCACTCACGGGAGCTCTAACGTACGGGTTGCGGGGATGAGTTTTTTGCGTACATATCTAGCATTGCGAGACGCTGGAGCCGGAGCACCCTATGTCGC
$  CACAGGTCTATCACCCTATTAACCACTCACGGGAGCTCTCACGTACGGGTTGCGGGGATGAGTTTTTTGCGTACATATCGACGCATTGCGAGACGCTGGAGCCGGAGCACCCTATGTCGCA
$  ACAGGTCTATCACCCTATTAACCACTCACGGGAGCTCTCCCGTACGGGTTGCGGGGATGAGTTTTTTGCGTACATATCGCAGCATTGCGAGACGCTGGAGCCGGAGCACCCTATGTCGCAG
```

Fig. 1. Example of running windows reduced to 120 nucleotides for illustration purposes, for the human mitochondrial genome, and swinger transformation of the mid-third 40 nucleotides. First row: regular genomic sequence; rows starting by $ are the first 5 running windows, with the mid third swinger transformed according to swinger rule A ↔ C + G ↔ T (as an example). Running windows used for actual analyses are 270 nucleotides long, and transformed according to each of the 23 swinger transformations, along the same principles as shown above for 120 nucleotides. The three peptides translated from the first running window sequence are also indicated, stops codons are translated here as '*'. Analyses searching for mass spectrometry data matching these predicted peptides consider the possibility that any amino acid is integrated at stops (19 possibilities, merging leucine and isoleucine, which are undistinguishable by the sequencing technique used here, because their molecular weights are identical). This means that for a single running window sequence after a single swinger transformation, $3 \times 19 = 57$ hypothetical peptides are considered. This number is doubled to 114 considering the inverse complement sequence, and multiplied by 23 considering all 23 potential swinger transformations. Hence 2622 peptides are translated from the 23 swinger transformations of each running window sequence indicated by $. This running window structure enables detection of chimeric peptides where the regular part is translated from the 5′, as well as from the 3′ side of the swinger part.

Table 1

Human mitochondrial peptides detected assuming abrupt switches between regular and swinger parts of RNA, for peptides where each regular and swinger parts have >8 amino acids (mass spectra from [33]). Columns are: 1. Peptide number; 2. swinger type; 3. amino acid inserted at stop(s) ('no' indicates lack of stops); 4. strand and frame; 5. peptide sequence; 6. PSMs; 7. Xcorr; 8. trypsin miscleavage; 9. PEP; 10–13. Positive strand positions of 5' and 3' extreme amino acids of regular and swinger parts of detected peptide; 14. Peptide extremity matching regular transcription. Underlined: peptide swinger part; *, $ marks swinger peptide parts covering previously described swinger reads, respectively previously described swinger peptides [103]. Peptides 8 and 9 differ in posttranslational amino acid modifications (not indicated). Highlighted peptide parts match both translations according to vertebrate mitochondrial, and nuclear (standard) genetic codes. Peptide parts not highlighted match only translation according to the vertebrate mitochondrial genetic code, and are incompatible with translation according to the nuclear genetic code. For example, peptide 3 could be translated in the cytosol on the base of RNA transcribed from mitochondrial inserts in the nuclear chromosome (numts), peptides 5 could not, as peptides 1 and 2 because at least one part of the peptide is not compatible with translation according to the nuclear genetic code. Further analyses (see text) show that fewer detected peptides are compatible with the nuclear genetic code than expected by chance, and that more peptides than expected by chance are compatible only with translation according to the mitochondrial genetic code.

1	2	3	4	5	6	7	8	9	10	11	12	13	14		
1	ac	a	r0	YGVSEGLAAPVGAYNVGAFAALYMAANFSFLNQDAVQVADK	3	2.77	0	0.374	3246	3156	3153	3123	5'	29	12
2	ac	v	r1	SCLLAFLMGSLMLTLIVGLSK	20	3.04	0	0.169	3465	3432	3492	3468	3'	12	9
3	ac	r	r0	AWGGGFDVDWWGSDDIVAMR	43	3.06	0	0.745	16188	16161	16158	16131	5'	10	10
4	ag	k	r2	AIGKVAFSTSVMLEVMFLVNK	185	3.06	1	0.082	2289	2262	2319	2292	3'	11	10
5	ag	f	r1	SYTFPPGSSSVACWLGCSPSPTLTLIFGLSK	49	3.38	0	1.000	3456	3432	3522	3459	3'	9	22
6	ag	s	r0	GGSPSDSTTSSQQLLSSILWSK	149	3.98	0	0.755	10125	10101	10164	10128	3'	10	12
7	ag	no	f0	LLGAVPLASASLTIGSLALAGMPFLTGFYSKDHIIETANMS	1	2.37	1	0.931	13704	13767	13644	13701	3'	29	12
8	at	no	f1	FIAYHSPGKVNFVPATAVTR	399	2.77	1	0.196	882	924	855	879	3'	11	9
9	at	no	f1	FIAYHSPGKVNFVPATAVTR	604	2.77	1	0.633	882	924	855	879	3'	11	9
10	at	p	f1	MPNSFNWDVGGNSSKLPVECLVEQGPEAR	1	2.31	1	0.605	1467	1500	1407	1464	3'	17	12
11	at	y	r2	TMSYALTLLLLQTCRGFSR	25	4.34	1	0.836	5601	5574	5631	5604	3'	9	10
12	at	no	r2	DMGDASVMGLSVNEASYDGK	19	2.17	0	0.121	6933	6906	6963	6936	3'	10	10
13	cg	no	f0	TLGQGVAHDLRTNPVDFVGDK	6	2.69	1	0.094	1344	1365	1368	1404	5'	9	12
14	cg	k	r2	LLASLPQPTVVPSTMPTISVRSGVLAGCLIGWWKPK	2	2.00	1	0.592	11103	11058	11163	11106	3'	16	20
15	cg	t	r2	TDNTNHHLTGSAIMTMTAPVK	540	4.89	0	0.653	11625	11601	11661	11628	3'	11	10
16	cg	no	r1	TSQTDLLTDPPITYEFLWAFSVNK$	8	2.03	1	0.592	12150	12126	12195	12153	3'	9	14
17	ct	m	f1	MWEDLMVEAMNSLSSATVGR	880	4.49	0	0.837	1995	2019	2022	2052	5'	9	11
18	ct	e	r0	GMGPMAYLASLALKENMVNNAEGFK	642	2.47	1	0.182	4233	4209	4206	4161	5'	9	16
19	ct	t	f0	NPSLSISVPSTRHVSMPITISSIPPQTTEMCLMK	3	2.54	1	0.866	4290	4350	4251	4287	3'	21	13
20	ct	a	r2	GVNWAKMNIAGYESSYNEQR	5	2.76	1	0.352	10512	10485	10542	10515	3'	10	10
21	ct	m	r0	AMMGDCAVCGTEMMSMCIK	13	2.03	0	0.42	13914	13890	13887	13860	5'	10	9
22	ct	v	r0	NVVWSVAVAAMMKGGVGVGMGGHMEMK	31	3.31	1	0.393	15321	15279	15276	15240	5'	14	13
23	ct	y	f1	DVSGPSSPSSSLMTLTLFSPDLLGDPDNYTLANPLNTPPY	2	2.27	0	0.724	15714	15801	15684	15711	3'	30	10
24	gt	q	r2	NNLFSLYCYLFQLWMMDPEHMNSMALK	1	2.55	0	0.464	13329	13287	13365	13332	3'	16	11
25	ac gt	s	r2	MVGSFMGSGDKPTEPGDSFGEPRSEAGPGPGSTLQSAR	38	2.43	1	0.459	1995	2049	2052	2109	5'	19	19
26	ac gt	h	r2	WSSSLAAPSAFVLVGMSSRHSLLVCGTHVYFFGHNWNK	1	2.71	1	0.218	4428	4389	4500	4431	3'	14	24
27	ac gt	s	f2	SGWVVEWSRHSVLLLLSLPVLAAGITMLLTDR	8	3.85	1	1.000	6591	6648	6558	6588	3'	21	10
28	ac gt	y	f0	EATASSAGNDASYDGQSGKDSQATPYTKPTPK	9	2.77	1	0.425	7221	7254	7161	7218	3'	12	20
29	ac gt	x	r1	GDKLFYDXGLLWGAQAGMVR	435	4.06	1	0.901	7503	7479	7476	7446	5'	9	11
30	ac gt	x	r1	RPLSPXGASLWSSVLXTYLR	129	5.31	0	0.516	7476	7452	7509	7479	3'	9	11
31	ac gt	n	f2	VMVTDLLQKSWSPHSYNNYITNR	6	2.39	1	0.971	8157	8193	8127	8154	3'	13	10
32	ac gt	n	f2	SNALNNAGKNAEGHYSSSPNNK	1	2.52	1	1.000	8559	8583	8520	8556	3'	9	13
33	ac gt	c	f1	TPGVVPEPAPAANVHSSCPPCPWLPCFPPSLPPSLTLTK	15	2.75	0	0.296	12564	12624	12510	12561	3'	22	17
34	ac gt	no	f0	SLKQNWDFSFNSSTMVVAGIFLLIR	1	2.11	1	1.000	13299	13338	13266	13296	3'	14	11
35	ac gt	x	f1	EMHLCSXEDSRAHNTWGXLK	13	2.02	1	0.989	16728	16752	16698	16725	3'	10	10
36	ag ct	k	r1	SLAPSGWSLLNLTNPLFSSMNLPTILLHKR	16	4.31	1	0.208	1728	1704	1788	1731	3'	11	19
37	ag ct	d	r1	LGDDWLEDMGNSNQNQLK	3	2.08	1	0.899	3426	3399	3396	3375	5'	9	9
38	ag ct	no	f2	WALFLSGTDSSSVSLAPLAATGSWGGLNQTQLR	7	2.96	0	0.143	5043	5076	4980	5040	3'	12	21
39	ag ct	y	f0	NPPYTWSDYMSIFCFVVCLGGLR	15	2.08	0	0.875	7485	7515	7518	7554	5'	10	13
40	ag ct	e	r1	YVGVEDESAVTNTSTNLTLPTIGQPSNGKK	2	2.14	1	0.472	7809	7779	7776	7719	5'	10	20
41	ag ct	q	r2	GDACWGPVPSQLGGQGQAGVVKGLQGLHQQGGPQNGGR	1	2.24	1	0.804	9456	9387	9384	9342	5'	23	15
42	ag ct	no	f1	AHVEAPIAGSMVLAVTSPGSNNR	37	3.74	0	1.000	11604	11643	11646	11670	5'	14	9
43	ag ct	v	r1	RSPLPGDQVDYVVVHGGMSVQFLWAFSVNK	34	3.08	1	1.000	12180	12126	12213	12183	3'	19	11
44	ag ct	f	f1	FNPFFGFVGPITKPTLNFNK	914	3.51	0	0.423	14874	14904	14847	14871	3'	11	9
45	ag ct	v	r2	HVHPEPSDEVAAYGANSIRCVGVGVVVMLVR	1	3.23	1	0.721	15015	14979	15069	15018	3'	13	18
46	at cg	no	f2	SSLRPYTKCVVFLASEEVK	4	2.41	1	1.000	3135	3129	3126	3096	5'	9	10
47	at cg	e	r2	QAEVFLSLQSSSQNHCFMQHISSGESASYVVPEK	215	3.23	0	0.224	3858	3822	3921	3861	3'	13	21
48	at cg	a	r0	FEDNKWDSFIDFYQTYFLGLAGNAGDCNGYGDMSYK	1	2.82	1	1.000	4074	4002	4107	4077	3'	24	12
49	at cg	e	f2	GISWPKLDEEGGGPFEAGEAPAGLK	59	3.82	1	0.169	5847	5874	5799	5844	3'	9	16
50	at cg	k	f1	SIAGVDVAMAVSGTKTLYLLHSNTHHNR	1	3.31	1	0.886	6201	6231	6147	6198	3'	10	18
51	at cg	r	f2	WLPWLGCSCGWCRLITSTPTYFPHYSR	1	2.49	1	0.941	8070	8097	8022	8067	3'	11	16
52	at cg	c	r2	CYLVGAFHCNLHNQENCK	27	3.92	0	0.690	8247	8223	8220	8196	5'	9	9
53	at cg	e	r1	AGEGLLEVWKASEPNSAVAK	8	3.17	1	0.171	9072	9042	9039	9012	5'	10	10
54	at cg	e	f1	EIFLSLLPQVGSGMGGESSR	11	3.57	0	0.249	9648	9669	9672	9705	5'	9	11
55	at cg	h	f2	GFLCIKLSCVGGCPHLLASSLYYFLTK$	8	2.08	1	0.935	11037	11070	10992	11034	3'	13	17
56	at cg	x	r0	DGGNXGSQGXGAMSSHVPMMKMNLNVVLK$	38	2.25	1	0.766	12321	12291	12288	12237	5'	12	17
57	at cg	r	f0	RPRLTSLPSLLNDINTILWSGGSAGSVNMGSVGEFVGR	1	2.94	1	0.124	15687	15735	15738	15798	5'	18	20
58	acg	n	r2	NTTTLSRTLNVGAVMNNVMVDVAGFNGSLVK	60	2.68	1	1.000	3315	3255	3345	3318	3'	21	10
59	acg	e	f0	SEHTPQLPTETTSSALSDRR	159	3.52	1	0.141	5085	5109	5112	5142	5'	10	10
60	acg	g	r0	WGGSTTNGGEPTGGSTLVGGEYKLQGDR	173	2.52	1	0.351	6483	6450	6531	6486	3'	12	16
61	acg	q	r2	QLVEQPKDTVEWQDMVEVGYNVVR	3	5.28	1	0.401	6891	6852	6924	6894	3'	13	11
62	acg	t	f2	TSKPHPTTTPPPSSSTPLQGLQCGGVRGVQAHQGAGHVQGR	73	3.03	1	0.876	13167	13215	13218	13287	5'	17	24
63	acg	t	f2	TSKPHPTTTPPPSSSTQISPITCGGVRGVQAHQGAGHVQGR	29	2.99	1	0.866	13167	13230	13233	13290	5'	22	19
64	acg	no	f2	GKQEPGLEQLCASSAALGEIPLPNNNPPLPK	17	2.36	1	0.236	13989	14022	13932	13986	3'	12	19
65	acg	e	f0	GLPQHQVHHKPHKPHYETHTQQK	2	2.05	0	1.000	14862	14901	14835	14859	5'	14	9

(continued on next page)

Table 1 (*continued*)

1	2	3	4	5	6	7	8	9	10	11	12	13	14		
66	acg	e	f0	GLPQHQVHHQPHKPHYETHTQQK	2	2.05	0	1.000	14865	14901	14835	14862	3'	13	10
67	acg	a	f1	GGFSGPSQILIATILCSFMGK	4	2.24	0	0.319	16266	16305	16242	16263	3'	12	9
68	act	e	r2	GENAGELEGWTATLSCSSPEGPTTLGPLR	6	2.59	0	0.156	3228	3186	3273	3231	5'	14	15
69	act	t	r1	KPAGASPAFFPGGGTSTLKPVDTGATLLTMEGETVSPGSVVK$	2	2.04	0	1.000	5598	5511	5508	5472	5'	30	12
70	act	k	f2	NLPILLLTNVEPQPFPPTPK	27	4.28	0	0.948	11097	11118	11064	11094	3'	9	11
71	act	k	f2	NLPILLLTNVDPQPFPPTPK	197	4.97	0	1.000	11097	11118	11064	11094	3'	9	11
72	act	n	f2	WPLQCRQMSTTSSNFCHFNPNNNPPLPK	3	2.93	0	0.970	13998	14022	13941	13995	3'	9	19
73	agc	r	f2	DQPNPLRPCPAPLTDAMAIR	45	2.19	0	0.272	3945	3969	3972	4002	5'	10	10
74	agc	k	r1	NNSSPIPVPVSSMVMPAAKTGK	365	3.86	1	0.382	6369	6336	6396	6372	3'	13	9
75	agc	v	r2	FQLQSMPVLGSVGGVGGAMGVMR	125	4.64	0	0.958	8157	8115	8184	8160	3'	14	9
76	agc	no	r2	TSYLLLQQPLLWWLCWFKLCVFWK	5	2.01	1	0.163	10749	10716	10785	10752	3'	12	12
77	agc	q	r1	MQQAAFALQWMLLWGQDWQQQWMR	8	2.16	0	0.264	12432	12390	12459	12435	3'	15	9
78	agc	w	r2	MSSTCGNDISMSRISGLFSAWGWK	4	2.04	1	0.739	13098	13068	13137	13101	3'	11	13
79	agc	q	f0	LQQMMQQADVLVAGIFLLIR	13	3.77	0	0.353	13314	13338	13281	13311	3'	9	11
80	agc	v	f0	IVAFSTSSQLGLMVLEVPVGVK	260	4.16	0	0.296	13458	13488	13491	13515	5'	13	9
81	agc	d	r1	FLVDLGLGGVGAGFGSDEGLDDGLCGVWCDFMTVSLLMNK	2	2.80	0	0.029	13971	13887	13884	13854	3'	30	10
82	agt	a	f0	TDAQSGGASAYKAHENILLR	1	2.25	1	1.000	2343	2370	2313	2340	3'	10	10
83	agt	n	f2	LIYSTSITLLPMTGGNGEGMR	194	4.00	0	0.586	5442	5475	5478	5502	5'	12	9
84	agt	d	f2	LIYSTSITLLPMTGGDGEGMR	61	3.96	0	0.460	5442	5475	5478	5502	5'	11	10
85	agt	r	r2	NGVSSSGGVEEGGVEVAVCLLLCVEWWLVRVCLVLLVR	1	2.41	1	0.180	6423	6369	6366	6312	5'	20	18
86	agt	g	r2	AASCGPPSCLPEVGINGGGNGMISTAAGGPGIVGAMNEANG*	1	2.29	0	0.275	8493	8409	8526	8496	3'	30	11
87	agt	a	r2	AAEGNSYAEEFYGEADAGGGYAVEAATAWGGPPLAEAVPR	1	2.43	0	0.518	10143	10068	10065	10026	5'	25	15
88	agt	d	r0	NYSSAMGACQGGSDESNDDGSGVCVWFARGPVVAAPGAVDR	6	2.55	0	0.953	13566	13491	13488	13446	5'	26	15
89	agt	y	f2	RIRPVEVVGAIPHYFLLQYPHHR*	1	2.81	0	0.173	13713	13749	13680	13710	3'	12	11
90	atc	h	f1	HPLWNLHVEGGFSSNTCAVRPAVMGNVESYMHKHK	12	2.55	1	0.124	1413	1452	1455	1515	5'	15	20
91	atc	q	f1	GTLTVQQQHNMPPTFLGNAR	61	3.19	0	0.659	2616	2643	2646	2673	5'	10	10
92	atc	v	r2	GLVIVVVKALGSVGGVGGAMGVMR	429	4.41	1	0.299	8157	8115	8187	8160	3'	14	10
93	atc	v	r2	GLVIVVVKAVGSVGGVGGAMGVMR	1057	4.48	1	0.140	8160	8115	8187	8163	3'	14	10
94	atc	x	r1	MXLMLMIALVAXEXIQMVVLFSVMAGMLGVVGWCR$	1	2.16	0	0.818	13263	13233	13335	13266	3'	11	24
95	atc	k	r0	FNYAFLGWGDWLLLWNYHMGMKVK	17	3.81	1	1.000	14049	14022	14019	13980	5'	10	14
96	atc	q	r2	LWLCSKGGQWLQLGFVMNQFLMNDPK	35	2.60	1	0.630	15408	15381	15378	15333	5'	11	15
97	atg	n	f0	TLGQGVAHEVANNGLHLRPLFSSCTR	1	2.13	0	0.076	1344	1389	1392	1419	5'	17	9
98	atg	d	r1	EPMMHQVSMGKKPVSGGDPPDEDDVDGIK	14	2.57	1	0.129	5061	5007	5088	5064	5'	19	10
99	atg	no	f0	TMASSSPPSVPPAPAGSASASVTVASPPLALVAPVR$	13	2.42	0	1.000	5376	5409	5412	5481	5'	9	27
100	atg	no	f2	GASFLFIWNSLYLLFGAWAGVLGTALSLLIRAELGQPGNL	1	2.24	1	1.000	6054	6141	6024	6051	3'	30	10
101	atg	m	r0	IMRMGAFGIGNMSGENTSTK	65	2.86	1	1.000	6291	6264	6261	6234	5'	10	10
102	atg	p	f0	APPTALVGTDSPHSTEAMWNDLLQCSEPPDSSFFSPPVA	10	2.64	0	0.564	6987	7074	6960	6984	3'	30	9
103	atg	k	r1	KPYTLPMESMNPFCSHSKAK	51	2.17	1	1.000	10284	10263	10314	10287	3'	10	10
104	atg	n	r1	GGAYQGNQSNNLLGGGLVVGWGLDNRLEGLFVVGLMSWSVG	3	3.02	1	0.924	12936	12846	12969	12939	3'	30	11
105	atg	x	r1	EWAEVSSCGEEGXGGADAASEEPTKTTGER	11	2.51	1	0.991	14826	14784	14781	14739	5'	10	19
106	cgt	q	r1	GLYWWDQQYGSGQGGWSLASPLDLGAQWTQGVGFR	1	2.15	1	1.000	228	189	186	126	5'	14	21
107	cgt	x	f0	HPKPKPWEMXLXIPLXIXLLVQXGTALWTLGK	1	2.72	0	0.632	2103	2166	2073	2100	3'	22	10
108	cgt	m	r0	LLQSDHTAFGSAPMSPQPK	2	2.13	0	0.217	2604	2577	2631	2607	5'	10	9
109	cgt	a	f2	AGASEGMTSYMMCLHTHYNLQHSPSNLANMSDK	6	2.06	0	0.034	4278	4347	4251	4275	3'	23	10
110	cgt	t	r2	RPPSGWPSCTSTLVSGATTTTGRGAGVTVETTECGSSTDNM	6	2.51	0	0.584	4989	4902	5028	4992	3'	30	11
111	cgt	no	r0	LFLGMCLKQENPVMMSGLK	393	4.09	1	0.162	12066	12033	12093	12069	3'	10	9
112	cgt	t	f0	VFLLTMTFNQNITLWIWQHIASTGHPGMNATMSCK	2	2.14	0	0.703	12330	12384	12387	12435	5'	19	17
113	cgt	f	f2	MNGTEGHVVHVPDGVASMIYPTLQLMFPTTNSPSK	8	2.57	0	0.942	13107	13161	13059	13104	3'	19	16
114	ctg	v	r0	VFVSLSLVSPFWLNPASTVAVNVNGYNEEVEVGHGYVVK	1	2.98	0	1.000	3126	3078	3192	3129	3'	17	22
115	ctg	e	f1	QSHMKSPEPVGDEEEDEER	26	4.14	1	0.032	3783	3807	3810	3837	5'	9	10
116	ctg	k	r2	DQVRPLVLCMVMLYFTIHLLHAYK	8	2.55	1	0.480	8601	8571	8568	8532	5'	12	12
117	ctg	k	f0	MNNKVFLVFVQTTIPLYLK	791	3.61	1	0.130	14001	14025	13971	13998	3'	9	10
118	ctg	no	f2	SPSSMYPNNKLEEGLYELK	70	4.56	1	0.048	15915	15945	15948	15972	5'	10	9
119	acgt	t	f1	GGAGGTPAATGTRTPSAGSDSVSTDFTQPTTSTTTPTNLK*	5	2.24	1	0.825	2520	2466	2586	2523	5'	18	22
120	acgt	a	r1	FVKAALFLLAGTYPSLGAR	79	4.05	1	0.342	2937	2913	2910	2883	5'	10	9
121	acgt	d	r1	MVEDMTGWADGLISTGDVDPTFSGVPKLSGGSAK	6	2.88	0	0.287	4305	4236	4233	4206	5'	23	11
122	acgt	f	r2	EEMLDGSFCGTFVFGGPVLALFSVMR	1	2.20	0	0.163	4452	4416	4413	4371	5'	14	12
123	acgt	no	r2	VPRQALVPFEVNEASYDGK	141	3.13	1	0.592	6930	6906	6960	6933	3'	9	10
124	acgt	no	r2	VPRQALVPFDVNEASYDGK	388	3.16	1	0.496	6930	6906	6960	6933	3'	9	10
125	acgt	no	f1	SNFLPTTLSRPIRNAPTLLGLGVMHAAHPAGQQMLEQAK	29	2.97	1	1.000	7287	7350	7353	7404	5'	21	18
126	acgt	q	r1	HGQAMLALPVLDPSVVVLGGCQGVGGK	1	2.74	0	0.031	10857	10833	10911	10860	3'	9	18
127	acgt	no	r2	VVGGVGWVPLAWLSLDMLQR	167	3.64	0	0.555	14448	14424	14481	14451	3'	9	11
128	acgt	a	r2	WMSGALILLGGAFCVLGSFRGGVGFVLLPGDVMADAGVER*	4	2.37	1	1.000	14772	14718	14715	14652	5'	18	22
129	actg	d	r1	SSYKDPFAEDADLDNDIALLSLGDLVPLVK	149	3.81	1	0.469	2733	2682	2769	2736	5'	17	13
130	actg	q	r2	GMGQGVHSQQAMVQAKVGAVMQQVMVDVAGGQNVGQPQGR	2	2.36	1	0.409	3363	3276	3273	3246	5'	30	10
131	actg	s	f0	NSVCSDGSARAVSPLAPGLSHK	25	2.75	1	0.560	3273	3303	3306	3315	5'	10	12
132	actg	s	r0	NSVCSDGSARSSSPLAPGLSHK	23	3.21	1	1.000	3273	3309	3312	3339	5'	12	10
133	actg	s	r0	MIMSAWSWKVMSSSMMETSMVEHLLDMIEIRPR	6	2.41	1	0.948	7527	7482	7479	7431	5'	17	16
134	actg	no	f2	SLSPFMITPSSVGVGVMVAVER	59	3.28	0	0.884	7767	7794	7797	7830	5'	11	11
135	actg	t	r0	TPKEQEIGEATVGTGIFNLTAK	2	2.42	1	1.000	8778	8748	8811	8781	3'	11	11
136	actg	no	f2	NQMIQALLITILLGLYFTLLSIVTAGTVFGLR	1	2.29	0	0.690	9834	9894	9897	9930	5'	20	12
137	actg	f	r1	SNAFGESKFTPTETMADTMAAFFFEYCGK	2	2.42	1	0.403	11400	11367	11451	11403	3'	12	17
138	actg	r	r2	WGSSNHEHGGAGCLMGMVQGR	571	4.35	0	0.165	11481	11460	11517	11484	3'	9	12
139	actg	k	r2	WGSSNHEHGGAGGLMGMVQGKGK	67	4.67	0	0.973	11478	11454	11517	11481	3'	10	13
140	actg	d	r0	AMLLDMGAWVSKVETWVDAR	9	4.42	1	1.000	12186	12153	12150	12126	5'	11	9
141	actg	d	r0	AMLLDMGAWVSQVETWVDAR	9	4.42	0	1.000	12186	12153	12150	12123	5'	10	10

Table 1 (*continued*)

1	2	3	4	5	6	7	8	9	10	11	12	13	14	
142	actg	no	f2	VSAASSGFYRPLPNNNPPLPK	81	3.67	0	1.000	13992	14022	13962	13989	3'	11	10
143	agct	k	r2	SKTLLLMWTLILIQSTSGK	110	3.76	1	0.377	2757	2730	2727	2703	5'	10	9
144	agct	s	f0	ASLSVALDPSGSTNTSLTAKHPNQLASIYFSR	2	2.24	1	0.987	5775	5832	5730	5772	3'	20	12
145	agct	no	f0	KAPNPCLAICALDACMLEK	1222	3.24	1	0.448	5931	5955	5958	5985	5'	9	10
146	agct	s	f0	GGPAGSVFWATIQANPMASMTFSKSYSK	4	2.67	1	0.429	7608	7653	7572	7605	3'	16	12
147	agct	h	r0	TLLNKTSPTFTYLSGEAHLTCHGEK*	4	2.19	1	0.342	13599	13569	13641	13602	3'	11	14
148	agtc	h	f0	SLLPSLSTQHHRYSGWVPGGSGLDIHAPK	1	2.81	1	0.420	2445	2472	2475	2529	5'	12	17
149	agtc	r	f2	AATDDERPTTTGLNSSTTTLLLSR	5	2.66	0	0.303	5214	5247	5163	5211	3'	11	13
150	agtc	e	f1	LTEPLTNGESSWEKASGSMVLAAVLLK$	155	2.13	1	0.932	11628	11658	11580	11625	3'	11	16
151	agtc	m	f0	IVAFSTSSQLGLMASEALAGMK	168	3.53	0	0.790	13458	13494	13497	13521	5'	13	9
152	agtc	w	f0	IVAFSTSSQLGLMASEALAGWK	434	4.72	0	0.715	13458	13494	13497	13521	5'	13	9
153	agtc	d	f0	IVAFSTSSQLGLMASEALAGDK	448	4.33	0	0.817	13458	13494	13497	13521	5'	12	10
154	agtc	r	r1	SLLPFVLFTFTPWSTLGLSMFLWVLGRGGLLFGR*	1	3.40	1	1.000	13758	13725	13821	13761	3'	13	21
155	agtc	no	r2	VAPVMSLIWFVLLPVSGPILEWCGRLMK	16	3.06	1	0.437	14934	14907	14991	14937	3'	9	19
156	atcg	k	r2	ATKTVGGVFGQTNQSPDPK*	1	2.68	1	0.251	321	294	291	267	5'	10	9
157	atcg	y	f1	SVGGGSAGYCVCGAAWGLGEPTKPQYHPPQFMYLTSSK	8	2.37	0	0.517	555	609	498	552	3'	19	19
158	atcg	e	r0	VISSEFIMQSQSPKHELEK	42	3.08	1	0.103	1629	1605	1602	1575	5'	10	9
159	atcg	a	r2	LYSQAFNSSSAQHTHGVGGCGVHWVRFEFK	1	2.25	1	0.669	3324	3369	3372	3411	5'	15	25
160	atcg	no	f2	LMPPLCKIHHESVALLVR	20	2.55	1	1	3936	3912	3909	3885	5'	9	9
161	atcg	g	r1	EKNQAVPEGPSMFISGPTQVK	3	2.47	1	0.573	4383	4353	4413	4386	3'	11	10
162	atcg	g	r1	EKNQAVPEGLAMFISGPTQVK	15	2.81	1	0.57	4380	4353	4413	4383	3'	11	10
163	atcg	g	r1	EKNQAVPEGLSMFISGPTQVK	20	2.80	1	0.57	4383	4353	4413	4386	3'	11	10
164	atcg	no	f0	AHTPKMLVMGPGLLPSGQGLGR	27	2.13	1	0.472	4503	4527	4530	4566	5'	10	12
165	atcg	x	r1	SEASASGSAKAAHDHLDDHPMMXLLFFVNSSMMAHLGK	1	2.79	1	0.34	5115	5064	5175	5118	3'	18	20
166	atcg	q	r2	QDCCDQDGSDEDPSNQNPQPAPKQER	1	2.42	1	0.309	6315	6282	6279	6240	5'	12	14
167	atcg	v	r2	GPVTVQAKVVGSVGGVGGAMGVMR	203	4.83	1	0.074	8160	8115	8187	8163	3'	15	9
168	atcg	v	r2	GPVTVQAKVLGSVGGVGGAMGVMR	181	4.59	1	0.082	8157	8115	8187	8160	3'	14	10
169	atcg	a	f2	AASHPVPVPMTLLMLGLLTNTLTMYQWWR	24	2.66	0	0.053	9477	9537	9450	9474	3'	19	10
170	atcg	a	r0	AMTLHAHAGAMFSEPAVLWVAISAMSAGAEPTAVANAK	2	2.44	1	0.693	11196	11115	11226	11199	3'	28	10
171	atcg	q	r2	GDAGEMLLVNAGLLGAQFLLASK	28	3.47	1	0.068	13719	13695	13692	13653	5'	10	13
172	atcg	e	r2	GDAGEMLLVIAGLLGAEFLLASK	29	3.54	0	0.147	13719	13695	13692	13653	5'	10	13
173	atcg	t	r0	SAEHSLGAGYHSGLMWGGVFKGLATVTLSGSPTTSGENT	1	2.93	1	0.885	15546	15456	15573	15549	3'	30	9
174	atgc	c	f2	NIPFLLFGVNSCCVIPSCNMPSACWINCKCLCK	1	2.36	1	0.239	1872	1911	1815	1869	3'	13	19
175	atgc	e	f0	SVESMLLGEENNFAEEAKAK	641	3.52	1	0.148	1902	1929	1872	1899	3'	11	9
176	atgc	w	f0	WDISQGKTFAVILNLVLYPHPPK	23	3.62	1	0.851	3240	3270	3204	3237	3'	11	12
177	atgc	k	f1	HYLYDMSPLNGIENHGKK	154	3.19	1	0.419	4272	4293	4296	4323	5'	9	9
178	atgc	h	f1	LMHHHYKSSAHHVHSPMIVHHNNYQYK*	35	2.40	1	0.591	6501	6522	6444	6498	5'	9	18
179	atgc	e	r1	IINITAVEENPSGRSSLHK	32	3.62	1	0.040	7155	7131	7128	7101	5'	10	9
180	atgc	e	r1	IINITAVEEIPSGRSSLHK	42	4.38	1	0.156	7155	7131	7128	7101	5'	10	9
181	atgc	no	f2	LLKECLSLASVPATPPYHTFEEPVYMK	12	2.48	1	0.681	7524	7560	7482	7521	3'	13	14
182	atgc	f	f1	AQWLFAFALFLKMPFPFEVMFHMSMK	5	2.10	1	1.000	9006	9051	9054	9081	5'	16	10
183	atgc	n	r1	NYLYYKSYCVSYSTTNNLSFNITK	14	3.54	1	0.304	11298	11268	11265	11229	5'	12	12
184	atgc	k	r0	SKNKPDTNASSNPVMMSGLK	232	3.49	1	0.793	12054	12033	12087	12057	3'	9	11
185	atgc	c	f1	VIFCQMVEFCVMVQVHSDNCADIIEAPLHKMTSK	1	2.13	1	0.623	13428	13452	13353	13425	3'	9	24
186	atgc	y	f1	VVYNGLQAMPEAYSQDFSLLTTFPPHPPSK	12	2.47	0	0.484	13941	14001	13914	13938	3'	21	9

insertions, as determined by Lee and Johnson [46]) (therein Table 1), were used to predict abundances of swinger RNAs. Indeed, these experimental kinetic parameters predict several properties of swinger RNAs [58,93,94], strengthening the hypothesis that regular polymerases are responsible for swinger polymerizations, by switching to unusual, stabilized 'swinger' conformations, similar (or even identical) to conformations causing point misinsertions, but lasting longer.

The second testable prediction of the hypothesis is that the same polymerase produces regular and swinger-transformed sequences. Hence occasionally, polymerases switch in the midst of replication/transcription, so that part of the sequence follows regular templating rules, and the other, contiguous part, is swinger transformed according to one among the 23 swinger rules. The fact that there are far more RNAs that are entirely swinger transformed than chimeric RNAs [100] suggests that such switches during polymerization are rare, and usually occur before or at the onset of polymerization.

The 16S rRNA gene in the complete mitogenome of *Kamimuria wangi* is a swinger transformed $A \leftrightarrow T + C \leftrightarrow G$ DNA sequence, embedded within an otherwise regular insect mitogenome [99]. The reasons why until now only DNA matching this $A \leftrightarrow T + C \leftrightarrow G$ swinger transformation has been detected, remain unknown. This $A \leftrightarrow T + C \leftrightarrow G$ exchange rule is also common among chimeric (part regular, part swinger) RNAs [100].

Peptides encoded by chimeric transcripts are detected for the first time here. An example of three peptides and the corresponding DNA, chimeric RNA is described in Fig. 1. Underlined parts are translated after swinger-transformation of the transcribed RNA sequence. When a detected peptide corresponds only to all or part of the underlined amino acid sequence, this peptide is considered a swinger peptide, as the peptides described in an earlier publication [103]. When the detected peptide encompasses part of the underlined, and part of the contiguous amino acid sequence(s) that are not underlined in Fig. 1, the peptide is considered chimeric, because translated from untransformed RNA (the part that is not underlined in Fig. 1), and from RNA that is swinger translated (underlined part in Fig. 1).

1.4. Previously detected swinger RNA

An anonymous reviewer suggested to add, for reader convenience, explanations on how swinger sequences described in previous publications had been detected in GenBank. These methods are not described in the Materials and methods section, as this would be inadequate and confusing: neither results nor analyses beyond those described in earlier publications on RNA were done in the context of the presently described proteomic analyses. The aim here is (chimeric) peptide detection. Following descriptions are only for the convenience of potential readers.

In a first step, the 23 swinger versions of the human mitogenome were transformed *in silico*. This means that for swinger transformation A ↔ C (as an example), all As in the human mitogenome are replaced by the 'replace' function in the software Word by 'X'. Then all Cs are replaced by A. The last step is to replace all Xs by C, producing a hypothetical, A ↔ C swinger transformed human mitogenome. Similar procedures produce all 23 possible swinger versions of the mitogenome. Each of these is then analyzed by BLASTn [2]. Two types of analyses have been done. The first analyzes (publications by Seligmann on swinger sequences prior to 2016, from 2012 on) search for alignments between the swinger transformed mitogenome and various sequence databases in GenBank, using standard default megablast parameters. This resulted in detecting long, highly similar sequences, as described in [88,92–94], for example. Such searches do not yield alignments with nuclear chromosome sequences, but detect about 100 ESTs (expressed sequence tags). The length of the alignments (>100 nucleotides) and the similarity with the hypothetical swinger-transformed mitogenome versions (>90%), as these were previously presented (Table 1 in [88,92], and Table 2 in [93,94]), are not compatible with randomly obtained results (as tested by simulations based on randomly shuffled swinger mitogenomes in [96] (therein Section 2.2.3.)).

These long EST swinger sequences were then confirmed by sequences detected within sequence read archives (SRA) of the human transcriptome published by Garzon et al. [30]), GenBank SRA entries SRX768406–SRX768440. For these analyses pertaining to short RNA reads (50 nucleotides, RNA seq, Illumina), BLASTn searched for 'somewhat similar sequences', using default search parameters, and detected swinger reads as these are described in [103] (therein supplementary data). These results on short swinger reads converge with those obtained by the first, EST-focused search ([103], therein Figure 2). Using the same search tool and criteria as used for the RNA seq reads, the 23 swinger-transformed mitogenome versions also align with nuclear chromosome sequences (they did not align with human nuclear chromosomes when using megablast as for the EST search).

1.5. Nuclear origins of swinger sequences (numts)

Alignments detected with relaxed criteria (BLASTn, see previous section) between swinger mitogenome sequences and each short reads and nuclear chromosome sequences suggest the possibility that swinger reads (but not swinger ESTs) could originate from the cytosol. To some extent, this is not relevant to the main issue at stake, the very existence of swinger polymerizations, but could be relevant because the very large human nuclear genome could by chance be the origin of these alignments, due to its size.

In addition, mitogenome copies (called numts, [51]) are inserted within nuclear chromosomes [36]. Because nuclear copies of regular mitogenomes exist, their occurrence for swinger-transformed mitogenomes is plausible, and would consist in itself a possible independent confirmation of the existence of swinger sequences, as previously discussed [103]. In addition, the possibility of swinger transcription of regular numts in the nucleus can't be ruled out.

Previous analyses [103] showed that the majority of detected swinger reads have mitochondrial origins. On average, alignments between swinger-transformed mitogenomes and RNA seq reads have higher identity percentages than between the same swinger mitogenome sequences and nuclear chromosome sequences. This is the case for a statistically significant majority of comparisons of identity percentages obtained between RNA reads and the swinger-transformed mitogenome, versus that between the same swinger-transformed mitogenome region and nuclear chromosome sequences. This suggests that most potential swinger numts diverge from their ancestral mitogenomic sequence, and that most RNA reads aligning with swinger-transformed mitogenomes have mitochondrial origins, because the swinger-transformed mitogenomes resemble on average more RNA reads than putative swinger numt(s) [103]. The 'Discussion' below develops these points in relation to potential nuclear origins of chimeric peptides.

1.6. Chimeric RNAs due to fusion between different RNAs

The term 'chimeric' transcripts has been used in the literature for a different type of RNA than the contiguous regular- and swinger-transcribed RNAs [101]. These other types of chimeric RNAs refer to two or more different transcripts produced each by regular polymerization, on the template of disjunct DNA regions. These RNAs are then fused by natural [60,132] or artificial reverse-transcription-associated phenomena [131]. These chimeric RNAs differ from the regular-swinger RNAs in the sense that for the latter the transcription process is chimeric (part regular-, part swinger transcriptions), but not in terms of their templating DNA regions, which are contiguous, not disjunct. It is possible that some unknown sequencing artifacts produce some of the detected swinger reads, but the non-random mapping of detected swinger peptides on detected swinger RNA reads, as previously described [103] shows that most swinger reads exist while translation occurs in the cell, and hence are not artifacts.

1.7. Swinger polymerization creates new genomic sequences

Another type of analyses detected swinger repeats within the regular mitogenome. Swinger repeats are usually short repeats that can only be detected when taking into account swinger transformations. These short sequences are inserted within the regular mitogenome, suggesting that natural retrotransposition of swinger RNAs produces novel DNA sequences [101]. They are more frequent and longer than expected by chance, and their length is proportional to the probability that the specific swinger transformation conserves circular code signals that presumably maintain the ribosomal translation frame in the gene. The natural circular code is a punctuation code within the genetic code consisting of 20 codons that as a group, have properties that enable protein coding frame retrieval [4,22,55–57]. This indicates that insertion of swinger sequences in the human mitogenome depends on their capacity to integrate protein coding genes without disrupting punctuation that presumably enables ribosomal detection of the coding frame.

1.8. Chimeric peptides

Recent analyses show convergent frequencies between swinger RNAs sequenced by classical and next generation (RNAseq) sequencing methods [103]. Hence swinger RNA occurrence is relatively well confirmed by data from independent methods and research teams. Here analyses complement at peptide levels results on chimeric transcripts. The existence of transcripts that are part regular (untransformed), part swinger RNAs, with an abrupt switch between these parts, predicts the existence of 'chimeric' peptides matching translation of such chimeric transcripts. Hence MS/MS mass spectra of peptide data (from [33]) previously used to detect swinger peptides [103] are reanalyzed here, using the same methods as by Seligmann [103]), searching for peptides matching in part the translation of the untransformed human mitogenome, and in part the translation of the swinger-transformed, contiguous mitogenome sequence. Chimeric peptides are peptides where swinger-encoded parts of a peptide are contiguous with parts translated from regular RNA. These would be stronger evidence for translation of swinger RNA than previous detections of entirely swinger-encoded peptides because the regular encoded parts function as matched positive controls, directly associated with swinger-encoded parts. In addition, chimeric peptides could suggest that swinger peptides are integrated within otherwise regular proteins, a further small step to understand functions associated with swinger phenomena.

2. Materials and methods

The revised Cambridge reference sequence for the human mitochondrial genome (NC_012920, [3]) was cut according to a running window of 270 nucleotides. Analyzes do not account for known mitochondrial polymorphisms, as this would expand analyzes beyond computing powers. The six frames of each of these nucleotide sequences of 270 bases were translated into the corresponding hypothetical peptides according to the vertebrate mitochondrial genetic code, after the 90 nucleotide-long mid-third of that sequence was swinger transformed according to each of the 23 swinger transformations. Hence each running window (around 16,300 in total) is represented by the six peptides translated from each of its 23 partly swinger-transformed versions (6 frames × 23 swinger versions = 138 hypothetical chimeric peptides for each of the 16,300 running windows). The window length of 90 codons/amino acids is designed to match the length of the longest (non-chimeric) peptides (up to 40 amino acids, [103]) previously detected in this dataset [33]. All translated hypothetical peptides are used by Thermo Protein Discoverer to predict a theoretical mass spectrometry distribution, which is matched with observed MS/MS mass spectrometry data from Gueugneau et al. ([33]).

Stops are translated as 'X', which Thermo Proteome Discoverer considers by default as leucine/isoleucine (these have equal masses, and are indistinguishable by mass spectrometry). Peptides including stops are duplicated 18 times, replacing 'X' by one of the 18 remaining amino acid species, excluding leucine and isoleucine. Hence predicted peptides include the possibility that any amino acid could be inserted at stops. Analyses assume that all stops in a single predicted peptide are translated by the same amino acid. Hence the 138 peptides for a single window of 270 nucleotides, if it includes at least one stop (the majority of cases), are represented 19 times, inserting X and each of the remaining amino acids at stops (19 x 138 = 2622 chimeric peptides). In total, approximately 42.7 million hypothetical chimeric peptides were tested.

Consensus searches were handled with the Sequest (Thermo Fisher Scientific, Illkirch) algorithm with the following mass tolerances: Parent = 1 Da and Fragment = 0.5 Da (monoisotopic masses). Fixed carbamidomethyl (C) and variable Oxydation (M) modifications were activated, as well as the lysine → pyrrolysine modification, and only one missed trypsin cleavage was allowed. False discovery rate was estimated against a reverse decoy database using the Percolator algorithm. No protein grouping was allowed since the database only contained non redundant entries. Peptides with false discovery rate q < 0.05 and score Xcorr > 1.99 were considered identified. The score Xcorr is a likelihood of match between expected and observed MS/MS data that is unaffected by peptide length. Further explanations on peptide detection and characterization by the software are given in the Discussion. Observed mass spectra were compared separately to predicted peptides 19 times, each time inserting a different amino acid at stops. Here analyzes test the existence of a specific group of peptides, namely chimeric peptides. The false discovery rate q is adapted to such populations of detected items [41]. Results also indicate the posterior error probability PEP, an estimate of detection error specific to each individual peptide, which might be useful in the future, when analyzes focus on specific peptides, rather than on a population of peptides. Results are not analyzed according to this criterion more adapted to studies focusing on specific individual peptides.

3. Results

3.1. Chimeric peptides

Analyses detect according to the filtering criteria 1301 chimeric peptides, among approximately 42.7 million chimeric peptides produced by combinations of stop codon-amino acid insertions, swinger transformations and frames for the running window of 270 nucleotides (illustration in Fig. 1). Hence chimeric peptides are detected for approximately 3 among 100,000 hypothetical chimeric peptides. This is 200 times less than the rate of detection for 'regular' non-chimeric swinger peptides, using the same criteria and the same data, approximately 6 per 1000 predicted swinger peptides [103].

Previously detected chimeric human mitochondrial RNAs are about 3% of all RNAs detected with at least some swinger part. Part of the discrepancy between chimeric RNA versus peptide detections probably results from the fact that proteomic analyses only considered abrupt switches between regular and swinger parts of peptides. Blast analyses detecting RNAs are not limited by this consideration, and can detect RNAs where the switch is not abrupt: in a transition sequence between regular and swinger transformed sequences, nucleotides seem random. Hence for practical reasons, detection of chimeric RNAs encompasses more possibilities than chimeric peptide detection, explaining lower rates of chimeric peptide detection (relative to rates of swinger peptide detection); these are lower than rates of chimeric RNA detection (relative to swinger RNA detection rates).

Here we focus specifically on chimeric peptides for which each regular and swinger parts have more than 8 amino acids. This is because considering 19 different amino acid species (merging leucine and isoleucine), the e value for 42.7 million potential chimeric peptides is about 0.0001 for amino acid sequences of 9 residues $(42,000,000 \times 1/19^{-9})$. Hence the match of each regular and swinger part of the detected peptide with the predicted chimeric peptide is unlikely to be due to chance, as estimated by this approximate e value.

This restricts the sample of 1301 detected chimeric peptides to 186 chimeric peptides of at least 18 residues, from various swinger transformations and stop-amino acid insertions (Table 1). Among these 186 chimeric peptides, the regular-encoded part of the peptide corresponds to the 5′ part of the peptide for 41% of the 186 chimeric peptides. This means that a statistically significant majority of chimeric peptides (two tailed sign test, P = 0.0061) correspond to the 5′ translation of swinger RNA and 3′ translation of regular RNA. Note that this statistically significant bias could not occur if detected chimeric peptides were due to random detection artifacts, strengthening the suspicion that results reflect a biological reality. Hence 41% of chimeric peptides reflect translation of regular transcripts that switch at a given point to swinger transcription. Frequencies and mean lengths of chimeric peptides, for each swinger type (Table 2)

Table 2

Frequencies and lengths of chimeric swinger peptides detected in Table 1: Columns indicate: swinger type; peptide number; mean PSMs; mean amino acid number in peptide non-swinger part; mean amino acid number in peptide swinger part.

Swinger type	N	PSMs	Reg	Swinger
A ↔ C	3	22.0	17.0	10.3
A ↔ G	4	96.0	14.8	14.0
A ↔ T	5	209.6	11.6	10.0
C ↔ G	4	139.0	11.3	14.0
C ↔ T	7	225.1	14.7	11.7
G ↔ T	1	1.0	16.0	11.0
A ↔ C + G ↔ T	11	59.6	13.8	14.2
A ↔ G + C ↔ T	10	103.0	13.2	14.4
A ↔ T + C ↔ G	12	31.2	12.3	14.8
A → C → G → A	10	52.2	14.6	13.7
A → C → T → A	5	47.0	14.2	13.6
A → G → C → A	9	91.9	14.1	10.2
A → G → T → A	8	33.3	18.3	12.4
A → T → C → A	7	230.3	12.1	14.7
A → T → G → A	9	18.8	18.3	12.8
C → G → T → C	8	52.4	18.4	12.9
C → T → G → C	5	179.2	11.4	12.6
A → C → G → T → A	10	82.1	14.0	14.3
A → C → T → G → A	14	71.8	13.6	11.9
A → G → C → T → A	5	268.4	13.2	11.4
A → G → T → C → A	7	173.1	12.1	13.6
A → T → C → G → A	18	33.7	14.6	12.3
A → T → G → C → A	13	92.6	11.8	12.7

show that the regular (non-swinger) part of the chimeric peptides is on average slightly longer than the swinger part, though this difference is not statistically significant.

3.2. Swinger peptides and chimeric peptides

Note that chimeric peptides, due to their part that matches translation of regular transcripts, differ in mass spectrometry properties from peptides entirely translated from swinger RNA, even if these have the exact same swinger sequence. Hence detection of chimeric peptides with swinger parts overlapping previously detected 'regular' swinger peptides (as the swinger peptides described by [103]) would be strong, independent methodological confirmation that positive results are not artifacts. Indeed, the swinger parts of eight chimeric peptides in Table 1 overlap with one of the 263 previously described swinger peptides [103]. These previously described swinger peptides cover on average 1.1% of the swinger-transformed mitogenome, expecting approximately 2 overlaps with chimeric peptides in Table 1 if no association exists between the two independent analyses. This means that chimeric peptides map on previously described swinger peptides 4 times more frequently than expected. This association between two independent searches confirms that results are not false positive matches between the mass spectrometry data and some predicted hypothetical chimeric peptides among a very large number of predicted hypothetical chimeric peptides. In addition, note that even if detected swinger and chimeric peptides correspond to the same swinger region, the corresponding MS/MS mass spectra differ, because for chimeric peptides mass spectra include also the adjacent residues translated from regular, untransformed RNA, while for swinger peptides, mass spectra do not include the latter residues. This non-random correspondence between swinger peptides and swinger parts of chimeric peptides suggests that translation of swinger RNAs is not random, and probably specific to some mitogenome regions.

3.3. Swinger RNA and chimeric peptides

Previously detected swinger peptides preferentially map on human mitogenome regions covered by independently detected swinger RNAs [103]. Their numbers increase with numbers of detected swinger transcripts. These positive associations between swinger RNA and swinger peptides can also be expected for chimeric peptides described in Tables 1–2. Such associations would confirm that the detected chimeric peptides actually exist, because they would match two independent material evidences, peptides, and RNA fragments.

The mean number of PSMs (peptide spectrum matches) for chimeric peptides increases as a function of the number of human mitogenome regions covered by swinger RNA (also called contigs), for the swinger type corresponding to the swinger part of the chimeric peptides (Fig. 2). Swinger transcriptomic data are from Seligmann [103]. Chimeric peptides presumably reflect translation of chimeric RNAs, along part regular, and part swinger transcription rules. Hence amounts of chimeric peptides should reflect numbers of possible transitions between regular and swinger RNAs, estimated by the number of swinger contigs previously described by Seligmann [103]. Indeed, a positive association between PSMs of chimeric peptides and swinger RNA contigs exists ($r = 0.64$, one tailed $P = 0.0006$), strengthening confidence in the validity of results, and corresponding with previous results for swinger peptides [103]. Note that similar correlation analyses for numbers (not PSMs) of detected chimeric peptides do not yield statistically significant associations with contig numbers.

The swinger part of 8 chimeric peptides (marked by * in Table 1) maps on human mitogenome regions also covered by the adequate type of swinger RNA (six swinger types, two matches for A → G → T → A and A → C → G → T → A, and one match for each A → G → C → T → A, A → G → T → C → A, A → T → C → G → A, and A → T → G → C → A swinger transformations). Considering the overall

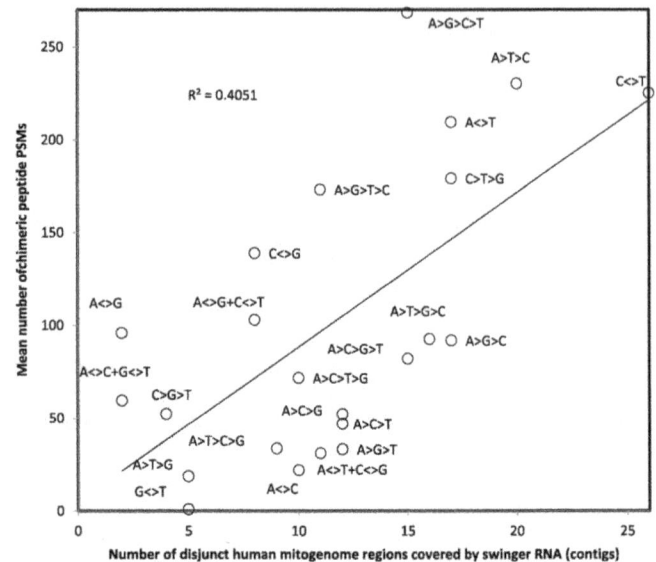

Fig. 2. Mean number of PSMs detected for chimeric peptides, as a function of the number of disjunct human mitogenome regions covered by swinger RNA (RNA data from [102]). The positive association indicates the expected causal link between swinger RNA and chimeric peptides.

mitogenome coverage by swinger RNAs (on average 2.6% of the genome), lack of association between swinger RNAs and the swinger part of chimeric peptides would expect 4.76 matches across all 23 swinger transformations, with 0.21 peptides for the average swinger transformation. This predicted number for specific swinger transformations was always <0.5 peptides. Detecting at least one match for six among 23 swinger types, when less than 0.5 are expected for all 23 swinger transformations has $P = 0.022$ according to a two-tailed Fisher exact test. This indicates that chimeric peptides associate with detected swinger RNA, though this association is weaker than the previously described association between swinger RNA and swinger peptides [103].

3.4. Chimeric peptides: strong validation of swinger sequences

Chimeric peptides are in terms of confirmation of swinger polymerization only secondary evidence, because peptides are translated from RNA, as compared to previous descriptions of swinger RNAs and chimeric swinger RNAs [100], which directly result from swinger polymerization. This point is also valid for swinger peptides. However, detection of (numerous) peptides matching translation of contiguous parts of the mitogenome, where one part reflects regular transcription, and the other swinger transcription, is a strong methodological confirmation for swinger phenomena and associated translation into peptides, which is not implied by the detection of 'pure' swinger peptides. This is because the non-swinger part of the peptide is a positive control paired to its contiguous swinger part. Hence in addition of describing a further aspect of the biological phenomenon of swinger polymerizations, chimeric peptides are also a further validation of the phenomenon's existence.

3.5. Chimeric peptides integrated in regular proteins?

An important question associated to swinger sequences is their function: among others, do they code for functional proteins, and are swinger peptides integrated into regular, perhaps functional proteins? A reanalysis of Table 1 yields a first insight into these important questions. The regular (non-swinger) part of eleven peptides matches the sequence of six among thirteen known, regular, mitogenome-encoded proteins. Their swinger parts correspond to the translation

Table 3

Chimeric peptides from Table 1 with regular part matching proteins translated from known mitogenome-encoded genes. Swinger parts are underlined, gene identity is followed by the position of the 'normal' part of the peptide matching the regular translation of the gene in the regular protein. The swinger transformation and the amino acid inserted at stop(s) are also indicated. Peptide parts matching translation according to both nuclear and mitochondrial genetic codes are highlighted: peptide 100 could be translated in the cytosol on the base of RNA transcribed from mitochondrial inserts in the nuclear chromosome (numts), all remaining peptides could not, as at least on part of the peptide is incompatible with translation according to the nuclear genetic code. Analyses (see text) show that there are fewer detected peptides compatible with the nuclear genetic code than expected, and more than expected peptides compatible only with the mitochondrial genetic code.

Table 1 #	Peptide	Gene	Position	Swinger rule	Stop
100	GASFLFIWNSLYLLFGAWAGVLGTALSLLIRAELGQPGNL	COX1	18–47	A > T > G	r
27	SGWVEWSRHSVLLLLSLPVLAAGITMLLTDR	COX1	205–213	A ↔ C + G ↔ T	s
181	LLKECLSLASVPATPPYHTFEEPVYMK	COX1	500–512	A > T > G > C	x
136	NQMIQALLITILLGLYFTLLSIVTAGTVFGLR	COX3	157–168	A > C > T > G	e
169	AASHPVPVPMTLLMLGLLTNTLTMYQWWR	COX3	41–59	A > T > C > G	a
23	DVSGPSSPSSSLMTLTLFSPDLLGDPDNYTLANPLNTPPY	Cyt B	238–267	C ↔ T	y
19	NPSLSISVPSTRHVSMPITISSIPPQTTEMCLMK	ND1	305–318	C ↔ T	t
38	WALFLSGTDSSSVSLAPLAATGSWGGLNQTQLR	ND2	165–176	A ↔ G + C ↔ T	n
34	SLKQNWDFSFNSSTMVVAGIFLLIR	ND5	249–262	A ↔ C + G ↔ T	f
80	IVAFSTSSQLGLMVLEVPVGVK	ND5	301–313	A > G > T > C	d
7	LLGAVPLASASLTGSLALAGMPFLTGFYSKDHIIETANMS	ND5	374–402	A ↔ G	

of the contiguous swinger transformation of these genes, along nine (four symmetric, and five asymmetric) systematic nucleotide exchange rules (Table 3). Note that up to three chimeric peptides are detected for two large mitochondrial proteins (cytochrome c oxidase I and NADH:ubiquinone oxidoreductase subunit 5). It is plausible that such peptides are integrated within complete proteins. These sequence alterations could modulate (or not) the regular function of the protein, and not necessarily impair function.

These 11 chimeric peptides integrated in regular proteins represent 5.9% of all 186 detected chimeric peptides. Considering that regular mitochondrion-encoded proteins have a total length of 3789 amino acids, the regular proteins represent 11.43% of the total number of amino acids that could be translated from the positive and negative strands of the human mitogenome. This means that chimeric peptides embedded within regular coding sequences are half as frequent as expected (5.9 versus 11.43%). This principle is further strengthened when examining the number of PSMs (number of identified peptide spectra matching a hypothetical peptide) for these 11 regular-protein-integrated chimeric peptides, as compared to the mean number of PSMs for all chimeric peptides detected for that swinger transformation: their PSMs is in all but one case (peptide 80 in Table 1) lower than the mean PSMs of other chimeric peptides for that swinger transformation. Hence chimeric peptides within regular proteins are rarer, and less expressed (as far as PSMs numbers can be trusted to reflect peptide abundances), than chimeric peptides translated from non-coding sequences, and non-coding frames of regular protein coding genes.

3.6. The natural circular code and swinger RNA, peptides and chimeric peptides

An anonymous reviewer suggested examining whether properties of chimeric peptides can be predicted from frameshift error-correcting properties of the natural circular code. Indeed, abundances of detected swinger RNAs in GenBank's EST database are proportional to reading frame retrieval (RFR) after swinger transformation of the natural circular code [58]. In this context, RFR, which estimates the capacity of the natural circular code to retrieve the protein coding frame, is calculated for the 20 codons that form the natural circular code, after each of the 23 swinger transformations: some codons belonging to the natural circular code are transformed into another codon included in the natural circular code, meaning that this property is invariant in relation to that codon and swinger transformation. RFR estimates this across all 20 codons of the natural circular code, for each swinger transformation. The length of swinger repeats in the human mitogenome is proportional to the RFR of the swinger transformation [101], which suggests that RFR

affects insertion rates of swinger repeats in protein coding regions, and hence could also affect chimeric peptide production.

The association between RFR and swinger RNA abundances for EST sequences occurs also for mitogenome coverage by swinger RNA reads sequenced by RNAseq in the transcriptome by Garzon et al. [30]) (Pearson correlation coefficient $r = 0.528$, one tailed $P = 0.005$). For swinger peptides as described by Seligmann [103]), the mean number of PSMs also increases with RFR ($r = 0.364$, one tailed $P = 0.044$). This positive association between mean PSMs numbers and RFR is also detected for chimeric peptides from Table 1 ($r = 0.367$, one tailed $P = 0.043$). These two results are independent, also because mean PSMs of swinger and chimeric peptides are only weakly correlated ($r = 0.24$, $P > 0.05$). Hence detections of chimeric and swinger peptides are proportional to extents by which swinger transformations conserve natural circular code 'frame' punctuations. Note that RFR, as mitogenome contig numbers in a previous section, associate with mean PSMs, rather than numbers of detected peptides, suggesting that in the context of these specific data, PSMs are better quantitative estimates than other variables.

4. Discussion

4.1. Statistical validity of peptide detections by mass spectrometry

An anonymous reviewer of a previous version indicated that detection of peptides with masses approximately matching the numerous possibilities produced by translation of all potential chimeric RNAs could be due to chance, due mainly to the large number of hypothetical chimeric peptides. Indeed, considering all 19 possible amino acids inserted at stops introduces a 'fudge' factor that enables adapting many hypothetical peptides to an actual fragment with a similar mass. Note that 28 among 186 (15%) detected peptides lack stops, invalidating this argument for several detected chimeric peptides. Independently of this, there are three reasons why this important point does not invalid the remaining results on chimeric peptides presented here. This is first because mean chimeric peptide PSMs converge with corresponding swinger RNA contig numbers, an independent type of data unrelated to the problems of proteomic analyses, already discussed above.

The other two points relate to the nature of the MS/MS mass spectrometry analyses themselves. The factor 'detection by chance' is integrated into the detection software used by Thermo Proteome Discoverer. The software compares the match between the mass spectrum of the actual fragment and the predicted mass spectrum of the hypothetical peptide, and its match with a dataset of decoy (false, negative controls) predicted peptides. The q value estimates the false detection rate (FDR, see explanations by [41]) of a peptide based on comparing matches by the actual predicted peptides and the decoy peptide database. This q is a probability of detection corrected for the false detection rates within the population of positive results (classical P values consider the whole population of statistical decisions, not only the subpopulation of positives). Hence the reported detections account for matches due to chance, considering the various parameters of the samples analyzed/compared, among them in particular sample sizes.

The third point relates to the nature of the statistic whose distribution is used to evaluate the above mentioned q (FDR). It is Xcorr, the cross correlation of the goodness of fit between the experimental peptide fragments and theoretical mass spectra. This integrates fits with each b and y ions, which correspond to asymmetry in the physical fragmentation of peptide bonds within the detected peptide, resulting into shorter peptide subparts: b ions occur when the residue's N-terminal is charged, y ions when the C-terminal is charged. Hence the match between the observed and the predicted peptide is not based solely on the similarity between their total masses, but also on fit between distributions of masses of sub-fragments of the (expected and observed) peptides, and this separately for b and y ions. The Xcorr statistic accounts, in addition to peptide size, for the number of matching masses of such sub-fragments. This allows inferring more precisely the residue sequences in the peptide, and means that peptide detection is not based only on a single measure, its total mass, but also on the mass of several subfragments.

In this context, the peptide ACD can function as a simplified example. Its mass corresponds to six possible peptides, ACD, ADC, CAD, CDA, DAC and DCA. Hence if ACD results from translation of swinger RNA, one can't assert that the observed mass is due to this peptide rather than any of the other five possibilities. However, Xcorr also considers the masses of subfragments of this peptide. Detection of a subfragment matching the mass of AC excludes four among the six possible peptides. A fragment matching the mass of CD matches only two peptides. If both subfragments AC and CD are detected, the characterization of the peptide ACD can be considered as assessed.

In addition, this process is done separately for b and y ions, because mass spectrometry analyses are in principle sufficiently precise to distinguish between these ions (remember that the precision of 0.5 Da of the analyzed data means a precision of half the mass of a hydrogen atom, which is also far less than the difference between amino acids with similar molecular masses). Hence Xcorr integrates information from both b and y ions, evaluating whether that information is congruent with the observed data. This procedure, coupled to q values based on comparisons of the Xcorr distribution obtained for negative controls (decoy peptides), renders detections relatively robust, despite fuzzy factors. In fact, large numbers of predicted peptides are necessary to estimate properly the distribution of random Xcorrs. The last point stresses that q (as P) values account for numbers of predicted peptides.

4.2. Confirmation of chimeric peptides by Waters technology

An anonymous reviewer suggested to confirm the existence of chimeric peptides by additional, independent mitochondrial proteomic data. In this context, I focused on another analysis of trypsinized human mitochondrial peptides [1], extracted by a more up to date MS/MS technology (Waters, Milford, MA, http://www.waters.com). This technique yields more accurate mass estimates than the method used by [33] (0.5 Da for the latter versus 5 ppm for the Waters method, hence about 10× more accurate estimates).

Analyses of the twelve samples from Alberio et al. [1]) by the software PLGS yield relatively few hits matching chimeric peptides considering only peptides where each regular and swinger-encoded parts are each at least nine amino acids long. One peptide matches significantly according to PLGS a chimeric peptide whose swinger part (underlined) matches swinger transformation A → T → G → A, LVSASVEMNQQQVPGSAGR (the regular part are residues 4228–4237 translated from the third frame of the negative strand of the

human mitogenome). The other peptide detected in these data has a swinger part that matches transformation C → G → T → C, SAAAARAGSACCLTSTAVTDRNLNTTF, the regular-encoded part corresponds to COX1, residues 211–219 in that regular mitochondrion-encoded protein.

Hence a different technology detects within independent mitoproteomic data peptides matching translations of chimeric RNAs, with one part regular, the other swinger transformed RNA. Hence, at least qualitatively, these independent data and technology confirm the existence of chimeric peptides and their integration in regular mitochondrial proteins. A more detailed description of 'regular' swinger peptides (meaning peptides entirely coded by swinger transformations of the mitogenome (unlike chimeric peptides that are in part regular-, in part swinger-encoded)) detected in the data from Alberio et al. [1]) will be presented elsewhere.

These results from data by Alberio et al. [1]) are too scarce to indicate whether chimeric peptides are produced according to a non-random profile. However, the non-random convergence between chimeric and entirely swinger peptides (detected in the same dataset from [33]) noted in a previous section in Results is in itself an indication that swinger-encoded peptides or parts of peptides are non-randomly produced.

4.3. Nuclear mitogenome copies

Previous transcriptomic analyses that detected non-canonical RNAs transformed according to systematic rules, such as deletions of mono- and dinucleotides after each transcribed trinucleotide (producing delRNAs, [102]), and swinger transformations [103], included controls that account whether the transformed mitogenome versions match nuclear chromosome sequences: mitogenome analyses are frequently contaminated by such chromosomic pseudogenes [9,10,48–51,62,66, 67,123–125,133,134].

These previous analyses blasted the swinger-transformed mitogenome versions versus the (regular) human nuclear chromosomes. For transformed mitogenome regions aligning with both transcriptomic reads and chromosomes, similarities between the transformed mitogenome and the RNA contigs were compared with the corresponding similarities between the same transformed mitogenome region and the chromosomes. For each del- and swinger RNAs (non-canonical RNAs), similarities with RNA contigs were greater than those with chromosome sequences in significant majorities of cases [102,103], as already discussed above for swinger RNA reads.

These results indicate two major issues. First, overall, RNA contigs result from non-canonical transcriptions of the mitogenome, the point that was being tested. Second, the observation that chromosome sequences match transformed versions of the mitogenome suggests that chromosomes include inserts of mitogenomic origins that were transformed according to systematic rules. The observation that these are on average less similar to the transformed mitogenome than RNA contigs suggests that these transformed mitochondrial sequences inserted in nuclear chromosomes mutated apart from the original sequence, as expected for inserts lacking function in the cell's nucleus [16,24,29,34–36,38,40,52,54,63,65,71,77,113,116–119,126,128,129].

4.4. Peptides translated according to nuclear or vertebrate mitochondrial genetic codes

Similar-minded analyses at the peptide level can test whether chimeric peptides in Table 1 were translated according to the human mitochondrial or the nuclear genetic codes. For that purpose, the regular and swinger transformed versions of the human mitogenome were translated according to the standard genetic code, which differs from the vertebrate mitochondrial genetic code by the reassignment of codon ATA from Met to Ile, of TGA from Trp to stop, and AGR from stop to Arg [23]. These four codons are 6.25% of all 64 codons.

Each swinger- and regular-encoded part of detected chimeric peptides has at least 9 amino acids. Hence the probability of detecting chimeric peptides that would have identical sequence according to both genetic codes is $(1 - 0.0625)^{-k}$, where k is the total length of the peptide. This principle is applied to the chimeric peptides in Table 1 so as to calculate the predicted number of peptides, for each size category, that is expected to match translation according to both nuclear and mitochondrial genetic codes. Lengths of chimeric peptides in Table 1 range from 18 to 42 residues, a total of 24 length categories. The observed number of chimeric peptides compatible with translation according to both genetic codes (in total 30 among the 186 chimeric peptides) is lower than expected in 16 among 24 size categories. Obtaining this result has P = 0.038 according to a one-tailed sign test. This means that, considering the length of chimeric peptides, there are statistically significantly fewer than expected peptides with sequences compatible with translation according to the nuclear genetic code.

The same principle can be applied to chimeric peptides in Table 1 whose sequences are only compatible with translation according to the mitochondrial genetic code, separately for each the regular- and the swinger-encoded parts. Here, the observed number (54) should be larger than the predicted number, if the sample is biased towards mitochondrion-encoded/translated peptides. Considering that 6.25% of codons differ in codon-amino acid assignments between the two genetic codes, the total expected number of chimeric peptides, considering their size, containing at least one of the 4 codons with coding assignment differing between nuclear and mitochondrial genetic codes is 35.97. This number is far lower than the observed 54 according to a chi-square test (P = 0.0027). Hence chimeric peptides with sequences compatible only with translation according to the mitochondrial genetic code are significantly more frequent than expected. This bias confirms the mitochondrial origin of chimeric peptides in Table 1. The number of peptide length categories where more observed peptides than expected are only compatible with mitochondrial translation is again 16 among 24 length categories, which has P = 0.038 according to a one tailed sign test.

These analyses show that detected peptides are more likely translated according to the mitochondrial genetic code than according to the nuclear genetic code. Note that translation, within the mitochondrion, according to the nuclear code is possible: it potentially depends for some codons upon the presence of cytosolic tRNAs, which could be occasionally imported in mitochondria [21,32,39,44,75,76,78,112,114]. However, this rationale is not symmetric: cytosolic translation according to the mitochondrial genetic code is much less probable than the opposite, so that nuclear origins are not compatible with the results obtained.

In fact, whether peptides have cytosolic or mitochondrial origins does not actually affect the main point that is addressed here, which is that these peptides were translated in part from swinger RNA. The same point applies to the potential nuclear (numt) origin of swinger-transformed mitochondrial DNA: independently of the location of the process, detection of chimeric peptides implies that swinger transformations occurred, whether during transcription of the regular mitogenome or nuclear inserts, or during numt insertion, possibly by natural swinger retrotranscription. This does not exclude the possibility that some detected chimeric peptides originate from the cytosol, but stresses the fact that most are mitochondrial, and that this issue is not directly relevant to the fact that swinger RNAs, chimeric RNAs, and corresponding peptides, exist, independently of the question of which cellular compartments produce them.

4.5. Few chimeric peptides in regular proteins translated from mitochondrion-encoded genes

Chimeric peptides in Table 3 have regular parts that match sequences of regular mitochondrial proteins encoded by mitochondrial genes. These are about 5% of all 186 detected chimeric peptides.

Peptides translated from regular mitochondrial genes represent about 11% of the total length potentially translated from the complete mitogenome, considering all six frames. Hence these 11 chimeric peptides potentially integrated in regular mitochondrial proteins are half as frequent as one could expect. Their PSMs is lower than for other chimeric peptides. These are hence rarer and less expressed than one could expect. Possibly, chimeric peptides integrated in regular proteins perturb proper protein folding. Incorrect folding induces various degradation mechanisms associated with mitophagy [5,42,122], which could explain that only few chimeric peptides are detected within regular proteins. These findings are not incompatible with the possibility that at least some swinger transcripts and peptides are functional.

4.6. Secondary structure formation by swinger transformed RNA and swinger RNA detection

Secondary structure formation by self-hybridization of DNA/RNA groups bijective transformations into three classes of each eight transformations. These share self-hybridization properties within each class [27,28]. This means that seven bijective transformations (including $A \leftrightarrow T + C \leftrightarrow G$) conserve self-hybridization properties of the original, untransformed sequence. Secondary structure formation by swinger RNA associates with swinger RNA detection [104], but these groupings/properties do not correlate with differences in chimeric or swinger peptide abundances/PSMs (not shown). The issue of regulation of alternative mitochondrial transcriptions, respectively post-transcriptional splicing, in relation to secondary structure formation by transformed RNA [61] remains unclear: a positive association exists between RNA occurrence and secondary structure formation for regular and swinger RNAs, but for transcripts resulting from systematic deletions (delRNAs), a negative association exists between secondary structure formation after deletions and delRNAs [105].

4.7. Swinger transformations, RNA–DNA differences (RDDs) and heteroplasmy

Specific non-random point differences occur between DNA and RNA sequences, either due to nucleotide substitutions [47] or inserts/deletions [15], including for human mitochondrial transcripts [8,37,59]. These RDDs appear shortly after transcripts exit polymerases [130], suggesting RDDs are due to post-transcriptional edition. The systematic repetition of transformations over long sequence stretches that characterize swinger RNA seem less likely produced by post-transcriptional edition than some unusual stabilized polymerase state, however, at this point, no possibility can be excluded, and potential connections of del- and swinger RNAs with RDDs should be kept in mind.

For the same reason, and by definition, punctual mitochondrial heteroplasmies [45,74,115,121] could not account for swinger parts of chimeric peptides, because these have to be translated from sequences differing from standard mitogenome sequences by far more than punctual nucleotide substitutions. Mitochondrial length heteroplasmies are common (49% of individuals, [64]), and in principle could, by chance correspond to swinger-like inserts in the mitogenome. Considering the seven regions containing length heteroplasmies described by Ramos et al. [64]) (therein table 3), only three among 186 chimeric peptides in Table 1 (peptide numbers 3, 156 and 157) potentially overlap (and this only in part) with these length heteroplasmies. Hence length heteroplasmies map non-randomly on chimeric peptides (3 among seven). Hence some presumed chimeric peptides might be translated from regions presenting length heteroplasmies, but this explanation is compatible with, at most, a small minority of chimeric peptides. Hence heteroplasmy could not explain chimeric peptides.

4.8. Translation increasing codon size or transcription systematically deleting nucleotides

It is important to note in the broader context of the discussion of results that further little known mechanisms increase the coding potential of sequences. A different, sometimes tRNA-based mechanism produces an alternative decoding of sequences, that of systematic frameshifting, which expands the codon from three to four (or five) nucleotides [6], called here tetracodons or pentacodons. This could result from systematic ribosomal slippages, a phenomenon that would correspond to programmed frameshifts (e.g. [25,43]), but occurring systematically, and serially; and/or from translational activities of tRNAs with expanded anticodons [68,120,127]. These cases relate to previously described isolated frameshift mutations, interpreted as isolated tetra-, pentacodons.

The hypothesis of an early genetic code based on quadruplets was suggested by Baranov et al. [6] to solve the problem that the weak triplet codon-anticodon interactions could not occur from a thermodynamic point of view in the absence of ribosomes, especially if these occurred at high temperatures [17–20]. Molecules as complex as ribosomes probably were absent at proto-life stages. Codon-anticodon interactions between four (or more) base pairs are more stable than those between three base pairs. Symmetry considerations also enable the deduction that the primeval genetic code was based on a subset of 64 quadruplets, called the tesserae, specifically for the vertebrate mitochondrial genetic code [31].

The expanded codon hypothesis is that modern genes include overlapping coding regions that consist of series of tetra- or pentacodons. This hypothesis is compatible with bioinformatic analyses where all eight frames of mitochondrial genes were translated assuming tetracodons. Blast analyses detected alignments between parts of these hypothetical tetracoded peptides and regular proteins in GenBank. Several other analyses based on codon usages in these tetracoding sequences confirm their special coding status, including higher GC contents than in non-tetracoding neighboring mitochondrial sequences. This corresponds to the prediction that tetracoding is an adaptation to translation at high temperature [89]. This point was further confirmed by a positive correlation between predicted tetracoding in lizard mitogenomes and mean body temperature in these lizard species [109]. Accordingly, overlap coding by tetracodons increases with temperature.

At this point, and besides the proven existence of decoding mechanisms for isolated tetracodons, the strongest further evidence for the existence of protein coding regions based on tetracodons is the coevolution between predicted tetracoding regions and the predicted antisense mitochondrial tRNAs with expanded anticodons, which is observed in mammal and *Drosophila* mitochondria [88,90,95]. In addition, mitochondrial peptides matching translation of regular and swinger RNAs according to tetra- and pentacodons have been detected [103], as well as translation of delRNAs (or dRNAs), RNAs transcribed while systematically deleting every fourth, or every fourth and fifth nucleotide. Peptide translation of such transcripts uses regular tRNAs but produces peptides identical to those resulting from decoding by tRNAs with expanded anticodons of regular transcripts [102]. These delRNAs are produced by systematic deletions, every third nucleotide, and correspond at deletion level, to systematic nucleotide substitutions/exchanges. This predicts that chimeric peptides consisting in part of regular-translated, and in part tetra- or pentacoded peptides, might exist.

The strongest evidence for swinger-encoding is the association between detected swinger RNAs and detected swinger peptides. Analyses detecting mass spectra matching predictions according to translations of tetra- and pentacodons suffer the caveat that evidence is based solely on mass spectra, with the above discussed difficulties in asserting the robustness of results based only on proteomics. However, further analyses detected peptides matching translations, according to expanded codons, of swinger-transformed sequences, and showed their association with detected swinger RNA [103]. Hence from a methodological point of view, translation according to expanded codons of swinger RNAs is stronger evidence for tetra- and pentacoding than such translation of regular RNA because it is confirmed by the independent detections of two 'unusual' types of molecules, swinger RNA and corresponding peptides matching expanded codons.

4.9. Robustness of experimental design

An anonymous reviewer indicated that analyzes comparing transcriptome and proteome make sense only if data originate from individuals with the same phenotypes, and if possible the same tissues and even the same individual(s), however analyzes compare tumor transcriptome [30] with normal proteome [33]. This setup is indeed suboptimal. However, considering this point, RNA and peptide data converge (also in previous analyzes, [102,103]) despite that RNA and peptide data originate from different tissues/individuals/phenotypes. This indicates that the phenomenon is general, and robust. This should not be surprising, because analyzes consider only RNA and peptides corresponding to the mitogenome. Most tissue-specific differences in mitochondrial RNA and protein profiles relate to molecules imported from the cytosol [13].

Methods used to detect the various types of unusual peptides take into account the large numbers of possibilities in matching observed and hypothetical mass spectra, so that positive detections are robust, and could not be due to chance.

Beyond methodological issues, occurrence of peptides coded by combinations of presumably unusual coding systems (translation of stops, together with translation according to expanded codons, and this for swinger RNAs), suggests that these basically ignored mechanisms expand more frequently than presumed the coding potential of genes, at least of the short mitogenomes. Detections of chimeric peptides, consisting of peptide parts corresponding to regular translation, adjacent to peptide parts matching translation of contiguous swinger RNA, strengthen confidence in the validity of results as positive controls, and expand our understanding of the phenomenon: swinger peptides are occasionally integrated in regular mitochondrion-encoded proteins, but their occurrence is downregulated.

5. Conclusions

1. Analyses of MS/MS mass spectrometry data detect peptides matching the translation of chimeric transcripts, RNA following in part regular, and in part swinger-transformed transcription, assuming abrupt switches between regular and swinger transformed parts of the RNA.

2. The 186 detected chimeric peptides (peptides consisting of a part encoded by regular RNA and a contiguous part encoded by swinger RNA) represent 3/100,000 among potential chimeric peptides, about 200 times fewer (6/1000) than detected swinger peptides (peptides entirely encoded by swinger RNA) in the same data. Eleven among these 186 chimeric peptides have a regular-encoded part that corresponds to proteins translated from classical mitochondrion-encoded genes.

3. Chimeric peptides map on previously detected swinger RNA. This association is weaker than a previously described association between 'regular' swinger peptides and swinger RNAs [103].

4. The vertebrate mitochondrial genetic code differs from the nuclear genetic code for four codons. Numbers of detected chimeric peptides that could be translated from human mitogenome sequences according to the nuclear genetic code are significantly fewer than expected considering the differences between the two genetic codes. This means that the majority of detected chimeric peptides are not cytosolic contaminations and were translated in the mitochondrion.

5. Previous detections of swinger peptides (predicted products of translation of swinger RNA) suggested that swinger transformed RNA is translation-competent [103]. Chimeric peptides where the regular part corresponds to known mitochondrion-encoded proteins might be incorporated into the respiratory chain complexes. Chimeric and swinger peptides might affect known mitochondrial functions despite low abundances if they have regulatory functions. Results are compatible with the possibility that some proteins are encoded by swinger transformations, with yet unknown functions.

Acknowledgments

This work has been carried out thanks to the support of the A*MIDEX project (no ANR-11-IDEX-0001-02) funded by the « Investissements d'Avenir » French Government program, managed by the French National Research Agency (ANR). I thank Nicolas Armstrong for coaching the mass spectrometry analyses, and anonymous reviewers for constructive comments.

References

[1] Alberio T, Bondi H, Colombo F, Alloggio I, Pieroni L, Urbani A, et al. Mitochondrial proteomics investigation of a cellular model of impaired dopamine homeostasis, an early step in Parkinson's disease pathogenesis. Mol BioSyst 2014;10:1332–44.

[2] Altschul SF, Madden TL, Schaeffer AA, Zhang J, Zhang Z, Miller W, et al. Gapped BLAST and PSI-BLAST: a new generation of protein database search programs. Nucleic Acids Res 1997;25:3389–402.

[3] Andrews RM, Kubacka I, Chinnery PF, Lightowlers RN, Turnbull DM, Howell N. Reanalaysis and revision of the Cambridge reference sequence for human mitochondrial DNA. Nat Genet 1999;23:147.

[4] Arquès DG, Michel CJ. A complementary circular code in the protein coding genes. J Theor Biol 1996;182:45–58.

[5] Ashrafi G, Schwarz TL. The pathways of mitophagy for quality control and clearance of mitochondria. Cell Death Differ 2013;20:31–42.

[6] Baranov PV, Venin M, Provan G. Codon size reduction as the origin of the triplet genetic code. PLoS One 2009;4:e5708.

[7] Barthélémy R-X, Seligmann H. Cryptic tRNAs in chaetognath mitochondrial genomes. Comput Biol Chem 2016;62:119–32.

[8] Bar-Yaacov D, Levin AG, Richards AL, Hachen N, Rebolledo Jaramillo B, Nekrutenko A, et al. RNA–DNA differences in human mitochondria restore ancestral form of 16S ribosomal RNA. Genome Res 2013;23:1789–96.

[9] Bensasson D, Zhang D, Hartl DL, Hewitt GM. Mitochondrial pseudogenes: evolution's misplaces witnesses. Trends Ecol Evol 2001;16:314–21.

[10] Bensasson D, Feldman MW, Petrov DA. Rates of DNA duplication and mitochondrial DNA insertion in the human genome. J Mol Evol 2003;57:343–54.

[11] Beznosková P, Gunišová S, Valášek LS. Rules of UGA-N decoding by near-cognate tRNAs and analysis of readthrough on short uORFs in yeast. RNA 2016; 22:456–66.

[12] Breton S, Milani L, Ghiselli F, Guerra D, Stewart D, Passamonti M. A resourceful genome: updating the functional repertoire and evolutionary role of animal mitochondrial DNAs. Trends Genet 2014;30:555–64.

[13] Calvo SE, Clauser KR, Mootha VK. MitoCarta2.0 : an updated inventory of mammalian mitochondrial proteins. Nucleic Acids Res 2016;44:D1251–7.

[14] Capt C, Passamonti M, Breton S. The human mitochondrial genome may encode for more than 13 proteins. Mitochondrial DNA 2015. http://dx.doi.org/10.3109/19401736.2014.1003924.

[15] Chen C, Bundschuh R. Systematic investigation of insertional and deletional RNA–DNA differences in the human transcriptome. BMC Genomics 2012;13:616.

[16] Dayama G, Emery SB, Kidd JM, Mills RE. The genomic landscape of polymorphic human nuclear mitochondrial insertions. Nucleic Acids Res 2014;42:12640–9.

[17] Di Giulio M. The universal ancestor lived in a thermophilic or hyperthermophilic environment. J Theor Biol 2000;203:203–13.

[18] Di Giulio M. The universal ancestor was a thermophile or a hyperthermophile. Gene 2001;221:425–36.

[19] Di Giulio M. The universal ancestor and the ancestor of bacteria were hyperthermophiles. J Mol Evol 2003;57:721–30.

[20] Di Giulio M. The universal ancestor was a thermophile or a hyperthermophile: tests and further evidence. J Theor Biol 2003;221:425–36.

[21] Duchêne AM, Pujol C, Maréchal-Drouard L. Import of tRNAs and aminoacyl-tRNA synthetases into mitochondria. Curr Genet 2009;55:1–18.

[22] El Soufi K, Michel CJ. Circular code motifs near the ribosome decoding center. Comput Biol Chem 2015;59A:158–76.

[23] Elzanowski A, Ostell J. The genetic codes. NCBI. 2013 accessed 10V2016 http://www.ncbi.nlm.nih.gov/Taxonomy/Utils/wprintgc.cgi?mode=t.

[24] Ermakov OA, Simonov E, Surin VL, Titov SV, Brandler OV, Ivanova NV, et al. Implications of hybridization, NUMTs, and overlooked diversity for DNA barcoding of Eurasian ground squirrels. PLoS One 2015;10:e0117201.

[25] Farabaugh PJ. Programmed translational frameshifting. Microbiol Rev 1996;60:103–34.

[26] Faure E, Delaye L, Tribolo S, Levasseur A, Seligmann H, Barthélémy RM. Probable presence of an ubiquitous cryptic mitochondrial gene on the antisense strand of the cytochrome oxidase I gene. Biol Direct 2011;6:56.

[27] Fimmel E, Danielli A, Struengmann L. On dichotomic classes and bijections of the genetic code. J Theor Biol 2013;336:221–30.

[28] Fimmel E, Giannerini S, Gonzalez DL, Struengmann L. Circular codes, symmetries and transformations. J Math Biol 2015;70:1623-164.

[29] François E, Gomes F, Arias MC. A protocol for isolating insect mitochondrial genomes: a case study of NUMT in Melipona flavolineata (Hymenoptera: Apidae). Mitochondrial DNA A DNA MappSeq Anal 2016;27:2401–4.

[30] Garzon R, Volinia S, Papaioannou D, Nicolet D, Kohlschmidt J, Yan PS, et al. Expression and prognostic impact of lncRNAs in acute myeloid leukemia. Proc Natl Acad Sci U S A 2014;111:18679-1868.

[31] Gonzalez DL, Giannerini S, Rosa R. On the origin of the mitochondrial genetic code: towards a unified mathematical framework for the management of genetic information. Nat Precedings 2012. http://dx.doi.org/10.1038/npre.2012.7136.1.

[32] Gowher A, Smirnov A, Tarassov I, Entelis N. Induced tRNA import into human mitochondria: implication of a host aminoacyl-tRNA-synthetase. PLoS One 2013; 8:e66228.

[33] Gueugneau M, Coudy-Gandilhon C, Gourbeyre O, Chambon C, Combaret L, Polge C, et al. Proteomics of muscle chronological ageing in post-menopausal women. BMC Genomics 2014;15:1165.

[34] Hazkani-Covo E, Graur D. A comparative analysis of numt evolution in human and chimpanzee. Mol Biol Evol 2007;24:13–8.

[35] Hazkani-Covo E. Mitochondrial insertions into primate nuclear genomes suggest the use of numts as a tool for phylogeny. Mol Biol Evol 2009;26:2175–9.

[36] Hazkani-Covo E, Zeller RM, Martin W. Molecular poltergeists: mitochondrial DNA copies (numts) in sequenced nuclear genomes. PLoS Genet 2010;6:e1000834.

[37] Hodgkinson A, Idaghdour Y, Gbeha E, Grenier JC, Hip-Ki E, Bruat V, et al. High resolution genomic analysis of human mitochondrial RNA sequence variation. Science 2014;344:413–5.

[38] Hu QX, Fan Y, Xu L, Pang W, Wang S, Zheng YT, et al. Analysis of the complete mitochondrial genome and characterization of diverse NUMTs of Macaca leonina. Gene 2015;571:279–85.

[39] Igloi GL, Aldinger CA. Where have all the inosines gone? Conflicting evidence for A-to-I editing of the anticodon of higher eukaryotic tRNAACGArg questions the dogma of a universal wobble-mediated decoding of CGN codons. IUBMB Life 2016;68:419–22.

[40] Jensen-Seaman MI, Wildschutte JH, Soto-Calderón ID, Anthony NM. A comparative approach shows differences in patterns of numt insertion during hominoid evolution. J Mol Evol 2009;68:688–99.

[41] Käll L, Storey JD, MacCoss MJ, Noble WS. Posterior error probabilities and false discovery rates: two sides of the same coin. J Proteome Res 2008;7:40–4.

[42] Kim S-H, Park Y-Y, Yoo Y-S, Cho H. Self-clearance mechanism of mitochondrial E3 ligase MARCH5 contributes to mitochondria quality control. FEBS J 2015. http://dx.doi.org/10.1111/febs.13568.

[43] Klobutcher LA, Farabaugh PJ. Shifty ciliates: frequent programmed translational frameshifting in euplotids. Cell 2002;111:763–6.

[44] Koley S, Adhya S. A voltage-gated pore for translocation of tRNA. Biochem Biophys Res Commun 2013;439:23–9.

[45] Korpelainen H. The evolutionary processes of mitochondrial and cholorplast genomes differ from those of nuclear genomes. Naturwissenschaften 2004;91: 505–11.

[46] Lee HR, Johnson KA. Fidelity of the human mitochondrial DNA polymerase. J Biol Chem 2006;281:36236–40.

[47] Li M, Wang IX, Li Y, Bruzel A, Richards AL, Toung JM, et al. Widespread RNA and DNA sequence differences in the human transcriptome. Science 2011;333: 53–8.

[48] Lopez JV, Cevario S, O'Brien SJ. Complete nucleotide sequences of the domestic cat (Felis catus) mitochondrial genome and a transposed mtDNA tandem repeat (Numt) in the nuclear genome. Genomics 1996;33:229–46.

[49] Lopez JV, Cumver M, Stephens JC, Johnoson WE, O'Brien SJ. Rates of nuclear and cytoplasmic mitochondrial DNA sequence divergence in mammals. Mol Biol Evol 1997;14:277–86.

[50] Lopez JV, Stephens JC, O'Brien SJ. The long and short nuclear mitochondrial DNA (Numt) lineages. Trends Ecol Evol 1997;12:114.

[51] Lopez JV, Yukhi N, Masuda R, Modi W, O'Brien SJ. Numt, a recent transfer and tandem amplification of mitochondrial DNA to the nuclear genome of the domestic cat. J Mol Evol 1994;39:174–90.

[52] Mao X, Dong J, Hua P, He G, Zhang S, Rossiter SJ. Heteroplasmy and ancient translocation of mitochondrial DNA to the nucleus in the Chinese Horseshoe Bat (Rhinolophus sinicus) complex. PLoS One 2014;9:e98035.

[53] Martinez-Rodriguez L, Erdogan O, Jimenez-Rodriguez M, Gonzalez-Rivera K, Williams T, Li L, et al. Functional class I and II amino acid-activating enzymes can be coded by opposite strands of the same gene. J Biol Chem 2015;290: 19710–25.

[54] Michalovova M, Vyskot B, Kejnovsky E. Analysis of plastid and mitochondrial DNA insertions in the nucleus (NUPTs and NUMTs) of six plant species: size, relative age and chromosomal localization. Heredity 2013;111:314–20.

[55] Michel CJ. Circular code motifs in transfer and 16S ribosomal RNAs: a possible translation code in genes. Comput Biol Chem 2012;45:17–29.

[56] Michel CJ. An extended genetic scale of reading frame coding. J Theor Biol 2015; 365:164–74.

[57] Michel CJ. The maximal C(3) self-complementary trinucleotide circular code X in genes of bacteria, eukaryotes, plasmids and viruses. J Theor Biol 2015;380: 156–77.

[58] Michel CJ, Seligmann H. Bijective transformation circular codes and nucleotide exchanging RNA transcription. Biosystems 2014;118:39–50.

[59] Moreira S, Valach M, Aoulad-Aissa M, Otto C, Burger G. Novel modes of RNA editing in mitochondria. Nucleic Acids Res 2016;44:4907–19.

[60] Frenkel-Morgenstern M, Gorohovski A, Lacroix V, Rogers M, Ibanez K, Boullosa C, et al. ChiTaRS: a database of human, mouse and fruit fly chimeric transcripts and RNA-sequencing data. Nucleic Acids Res 2013;41:D142–51.

[61] Ojala D, Montoya J, Attardi G. tRNA punctuation model of RNA processiong in human mitochondria. Nature 1981;290:470–4.

[62] Olson LE, Yoder AD. Using secondary structure to identify ribosomal numts: cautionary examples from the human genome. Mol Biol Evol 2002;19:93–100.

[63] Ramos A, Barbena E, Mateiu L, del Mar González M, Mairal Q, Lima M, et al. Nuclear insertions of mitochondrial origin: database updating and usefulness in cancer studies. Mitochondrion 2011;11:946–53.

[64] Ramos A, Santos M, Mateiu L, del Mar Gonzalez M, Alvarez L, Azevedo L, et al. Frequency and pattern of heteroplasmy in the complete human mitochondrial genome. PLoS One 2013;8:e74636.

[65] Ren T, Liang S, Zhao A, He K. Analysis of the complete mitochondrial genome of the Zhedong White goose and characterization of NUMTs: reveal domestication history of goose in China and Euro. Gene 2016;577:75–81.

[66] Ricchetti M, Tekala F, Dujon B. Continued colonization of the human genome by mitochondrial DNA. PLoS Biol 2004;2:E273.

[67] Richly E, Leister D. NUMTs in sequenced eukaryotic genomes. Mol Biol Evol 2004; 21:1081–4.

[68] Riddle DL, Carbon J. Frameshift suppression: a nucleotide addition in the anticodon of a glycine transfer RNA. Nat New Biol 1973;242:230–4.

[69] Rodin SN, Ohno S. 2 Types of aminoacyl-transfer-RNA synthetases could be originally encoded by complementary strands of the same nucleic-acid. Orig Life Evol Biosph 1995;23:393–418.

[70] Rodin S, Rodin A, Ohno S. The presence of codon–anticodon pairs in the acceptor stem of tRNAs. Proc Natl Acad Sci U S A 1996;93:4537–42.

[71] Rogers HH, Griffiths-Jones S. Mitochondrial pseudogenes in the nuclear genomes of *Drosophila*. PLoS One 2012;7:e32593.

[72] Root-Bernstein M, Root-Bernstein R. The ribosome as a missing link in the evolution of life. J Theor Biol 2015;367:130–58.

[73] Root-Bernstein M, Root-Bernstein R. The ribosome as a missing link in prebiotic evolution II: ribosomes encode ribosomal proteins that bind to common regions of their own mRNAs and rRNAs. J Theor Biol 2016;397:115–27.

[74] Rose G, Passarino G, Scornaienchi V, Romeo G, Dato S, Bellizzi D, et al. The mitochondrial DNA control region shows genetically correlated levels of heteroplasmy in leukocytes of centenarians and their offspring. BMC Genomics 2007;8:293.

[75] Rubio MA, Hopper AK. Transfer RNA travels from the cytoplasm to organelles. Wiley Interdiscip Rev RNA 2011;2:802–17.

[76] Salinas T, Duby F, Larosa V, Coosemans N, Bonnefoy N, Motte P, et al. Co-evolution of mitochondrial tRNA import and codon usage determines translational efficiency in the green alga *Chlamydomonas*. PLoS Genet 2012;8:e1002946.

[77] Schmitz J, Piskurek O, Zischler H. Forty million years of independent evolution: a mitochondrial gene and its corresponding nuclear pseudogene in primates. J Mol Evol 2005;61:1–11.

[78] Schneider A. Mitochondrial tRNA import and its consequences for mitochondrial translation. Annu Rev Biochem 2011;80:1033–53.

[79] Seligmann H. Hybridization between mitochondrial heavy strand tDNA and expressed light strand tRNA modulates the function of heavy strand tDNA as light strand replication origin. J Mol Biol 2008;379:188–99.

[80] Seligmann H. Mitochondrial tRNAs as light strand replication origins: similarity between anticodon loops and the loop of the light strand replication origin predicts initiation of DNA replication. Biosystems 2010;99:85–93.

[81] Seligmann H. Avoidance of antisense, antiterminator tRNA anticodons in vertebrate mitochondria. Biosystems 2010;101:42–50.

[82] Seligmann H. Undetected antisense tRNAs in mitochondrial genomes? Biol Direct 2010;5:39.

[83] Seligmann H. Pathogenic mutations in antisense mitochondrial tRNAs. J Theor Biol 2011;269:287–96.

[84] Seligmann H. Two genetic codes, one genome: frameshifted primate mitochondrial genes code for additional proteins in presence of antisense tRNAs. Biosystems 2011;105:271–85.

[85] Seligmann H. An overlapping genetic code for frameshifted overlapping genes in *Drosophila* mitochondria: antisense antitermination tRNAs UAR insert serine. J Theor Biol 2012;298:51–76.

[86] Seligmann H. Overlapping genetic codes for overlapping frameshifted genes in Testudines, and *Lepidochelys olivacea* as special case. Comput Biol Chem 2012;41: 18–34.

[87] Seligmann H. Coding constraints modulate chemically spontaneous mutational replication gradients in mitochondrial genomes. Curr Genomics 2012;13:37–54.

[88] Seligmann H. Overlapping genes coded in the 3′-to-5′-direction in mitochondrial genes and 3′-to-5′ polymerization of non-complementary RNA by an 'invertase'. J Theor Biol 2012;315:38–52.

[89] Seligmann H. Putative mitochondrial polypeptides coded by expanded quadruplet codons decoded by antisense tRNAs with unusual anticodons. Biosystems 2012; 110:84–106.

[90] Seligmann H. Pocketknife tRNA hypothesis: anticodons in mammal mitochondrial tRNA side-arm loops translate proteins? Biosystems 2013;113:165–76.

[91] Seligmann H. Putative protein-encoding genes within mitochondrial rDNA and the D-Loop region. In: Lin Z, Liu W, editors. Ribosomes: molecular structure, role in biological functions and implications for genetic diseases; 2013. p. 67–86 [chapter 4].

[92] Seligmann H. Triplex DNA:RNA, 3′-to-5′ inverted RNA and protein coding in mitochondrial genomes. J Comput Biol 2013;20:660–71.

[93] Seligmann H. Polymerization of non-complementary RNA: systematic symmetric nucleotide exchanges mainly involving uracil produce mitochondrial transcripts coding for cryptic overlapping genes. Biosystems 2013;111:156–74.

[94] Seligmann H. Systematic asymmetric nucleotide exchanges produce human mitochondrial RNAs cryptically encoding for overlapping protein coding genes. J Theor Biol 2013;324:1–20.

[95] Seligmann H. Putative anticodons in mitochondrial sidearm loops: Pocketknife tRNAs? J Theor Biol 2014;340:155–63.

[96] Seligmann H. Mitochondrial swinger replication: DNA replication systematically exchanging nucleotides and short 16S ribosomal DNA swinger inserts. Biosystems 2014;25:22–31.

[97] Seligmann H. Species radiation by DNA replication that systematically exchanges nucleotides? J Theor Biol 2014;363:216–22.

[98] Seligmann H. Phylogeny of genetic codes and punctuation codes within genetic codes. Biosystems 2015;129:36–43.

[99] Seligmann H. Sharp switches between regular and swinger mitochondrial replication: 16S rDNA systematically exchanging nucleotides A ↔ T + C ↔ G in the mitogenome of *Kamimuria wangi*. Mitochondrial DNA 2016;27:2440–6.

[100] Seligmann H. Swinger RNAs with sharp switches between regular transcription and transcription systematically exchanging ribonucleotides: case studies. Biosystems 2015;135:1–8.

[101] Seligmann H. Systematic exchanges between nucleotides: genomic swinger repeats and swinger transcription in human mitochondria. J Theor Biol 2015; 384:70–7.

[102] Seligmann H. Codon expansion and systematic transcriptional deletions produce tetra-, pentacoded mitochondrial peptides. J Theor Biol 2015;387:154–65.

[103] Seligmann H. Translation of mitochondrial swinger RNAs according to tri-, tetra- and pentacodons. Biosystems 2016;140:38–48.

[104] Seligmann H. Swinger RNA self-hybridization and mitochondrial non-canonical swinger transcription, transcription systematically exchanging nucleotides. J Theor Biol 2016;399:84–91.

[105] Seligmann H. Systematically frameshifting by deletion of every 4th or 4th and 5th nucleotides during mitochondrial transcription: RNA self-hybridization regulates delRNA expression. Biosystems 2016;142:43–51.

[106] Seligmann H, Krishnan NM. Mitochondrial replication origin stability and propensity of adjacent tRNA genes to form putative replication origins increase developmental stability in lizards. J Exp Zool B Mol Dev Evol 2006;306:433–49.

[107] Seligmann H, Krishnan NM, Rao BJ. Possible multiple origins of replication in primate mitochondria: alternative role of tRNA sequences. J Theor Biol 2006;241: 321–32.

[108] Seligmann H, Krishnan NM, Rao BJ. Mitochondrial tRNA sequences as unusual replication origins: pathogenic implications for *Homo sapiens*. J Theor Biol 2006; 243:375–85.

[109] Seligmann H, Labra A. Tetracoding increases with body temperature in Lepidosauria. Biosystems 2013;447:155–63.

[110] Seligmann H, Labra A. The relation between hairpin formation by mitochondrial WANCY tRNAs and the occurrence of the light strand replication origin in Lepidosauria. Gene 2014;542:248–57.

[111] Seligmann H, Raoult D. Unifying view of stem-loop hairpin RNA as origin of current and ancient parasitic and non-parasitic RNAs, including in giant viruses. Curr Opin Microbiol 2016;31:1–8.

[112] Sharma A, Sharma A. *Plasmodium falciparum* mitochondria import tRNAs along with an active phenylalanyl-tRNA synthetase. Biochem J 2015;465:459–69.

[113] Shi H, Dong J, Irwin DM, Zhang S, Mao X. Repetitive transpositions of mitochondrial DNA sequences to the nucleus during the radiation of horseshoe bats (*Rhinolophus*, Chiroptera). Gene 2016;581:161–9.

[114] Sieber F, Duchêne AM, Maréchal-Drouard L. Mitochondrial RNA import: from diversity of natural mechanisms to potential applications. Int Rev Cell Mol Biol 2011;287:145–90.

[115] Smigrodzki RM, Khan SM. Mitochondrial microheteroplasmy and a theory of aging and age-related disease. Rejuvenation Res 2005;8:172–98.

[116] Song S, Jiang F, Yuan J, Guo W, Miao Y. Exceptionally high cumulative percentage of NUMTs originating from linear mitochondrial DNA molecules in the *Hydra magnipapillata* genome. BMC Genomics 2013;14:447.

[117] Song H, Moulton MJ, Whiting MF. Rampant nuclear insertion of mtDNA across diverse lineages within Orthoptera (Insecta). PLoS One 2014;9:e110508.

[118] Soto-Calderón ID, Clark NJ, Wildschutte JV, DiMattio K, Jensen-Seaman MI, Anthony NM. Identification of species-specific nuclear insertions of mitochondrial DNA (numts) in gorillas and their potential as population genetic markers. Mol Phylogenet Evol 2014;81:61–70.

[119] Soto-Calderón ID, Lee EJ, Jensen-Seaman MI, Anthony NM. Factors affecting the relative abundance of nuclear copies of mitochondrial DNA (numts) in hominoids. J Mol Evol 2012;75:102–11.

[120] Sroga GE, Nemoto F, Kuchino Y, Bjork GR. Insertion (sufB) in the anticodon loop or base substitution (sufC) in the anticodon stem of tRNA(Pro)2 from *Salmonella typhimurium* induces suppression of frameshift mutations. Nucleic Acids Res 1992;20:3463–9.

[121] Stefano GB, Kream RM. Mitochondrial DNA heteroplasmy in human health and disease. Biomed Rep 2016;4:259–62.

[122] Taylor EB, Rutter J. Mitochondrial quality control by the ubiquitin-prooteasome system. Biochem Soc Trans 2011;39:1509–13.

[123] Thalman O, Hebler J, Poinar HN, Pääbo S, Vigilant L. Unreliable mtDNA data due to nuclear insertions: a cautionary tale from analysis of humans and other great apes. Mol Ecol 2004;13:321–35.

[124] Thalman O, Serre D, Hofreiter M, Lukas D, Eriksson J, Vigilant L. Nuclear insertions help and hinder inference of the evolutionary history of gorilla mtDNA. Mol Ecol 2005;14:179–88.

[125] Tourmen Y, Baris O, Dessen P, Jacques C, Malthièry Y, Reynier P. Structure and chromosomal distribution of human mitochondrial pseudogenes. Genomics 2002;80: 71–7.

[126] Tsuji J, Frith MC, Tomii K, Horton P. Mammalian NUMT insertion is non-random. Nucleic Acids Res 2012;40:9073–88.

[127] Tuohy TM, Thompson S, Gesteland RF, Atkins JF. Seven, eight and nine-membered anticodon loop mutants of tRNA(2Arg) which cause + 1 frameshifting. Tolerance of DHU arm and other secondary mutations. J Mol Biol 1992;228:1042–54.

[128] Verscheure S, Backeljau T, Desmyter S. In silico discovery of a nearly complete mitochondrial genome Numt in the dog (Canis lupus familiaris) nuclear genome. Genetica 2015;143:453–8.

[129] Wang B, Zhou X, Shi F, Liu Z, Roos C, Garber PA, et al. Full-length Numt analysis provides evidence for hybridization between the Asian colobine genera Trachypithecus and Semnopithecus. Am J Primatol 2015;77:901–10.

[130] Wang IX, Core LJ, Kwak H, Brady L, Bruzel A, McDaniel L, et al. RNA-DNA differences are generated in human cells within seconds after RNA exits polymerase II. Cell Rep 2014;6:906–15.

[131] Xie B, Yang W, Chen L, Jiang H, Liao Y, Liao DJ. Two RNAs or DNAs may artificially fuse together at a short homologous sequence (SHS) during reverse transcription or polymerase chain reactions, and thus reporting an SHS-containing chimeric RNA requires extra caution. PLoS One 2016;11:e0154855.

[132] Yang W, Wu J-M, Bi A-D, Ou-yang Y, Shen H-H, Chim G-W, et al. Possible formation of mitochondrial-RNA containing chimeric or trimeric RNA implies a post-transcriptional and post-splicing mechanism for RNA fusion. PLoS One 2013;8: e77016.

[133] Yao YG, Kong QP, Salas A, Bandelt HJ. Pseudomitochondrial genome haunts disease studies. J Med Genet 2008;45:769–72.

[134] Zhang DX, Hewitt GM. The long and short of nuclear mitochondrial DNA (Numt) lineages Reply from D-X. Zhang and G.M. Hewitt. Trends Ecol Evol 1997;12:114.

15

Predicting a double mutant in the twilight zone of low homology modeling for the skeletal muscle voltage-gated sodium channel subunit beta-1 (Na$_v$1.4 β1)

Thomas Scior [a], Bertin Paiz-Candia [a], Ángel A. Islas [b], Alfredo Sánchez-Solano [b], Lourdes Millan-Perez Peña [c], Claudia Mancilla-Simbro [b], Eduardo M. Salinas-Stefanon [b]

[a] Facultad de Ciencias Químicas, Universidad Autónoma de Puebla, Puebla, Mexico
[b] Laboratorio de Biofísica, Instituto de Fisiología, Universidad Autónoma de Puebla, Puebla, Mexico
[c] Centro de Química, Instituto de Ciencias, Universidad Autónoma de Puebla, Puebla, Mexico

ARTICLE INFO

Keywords:
Ig-like
CDR1
MD-2
Patch-clamp
Site-directed mutagenesis
Analogy modeling

ABSTRACT

The molecular structure modeling of the β1 subunit of the skeletal muscle voltage-gated sodium channel (Na$_v$1.4) was carried out in the twilight zone of very low homology. Structural significance can *per se* be confounded with random sequence similarities. Hence, we combined (i) not automated computational modeling of weakly homologous 3D templates, some with interfaces to analogous structures to the pore-bearing Na$_v$1.4 α subunit with (ii) site-directed mutagenesis (SDM), as well as (iii) electrophysiological experiments to study the structure and function of the β1 subunit. Despite the distant phylogenic relationships, we found a 3D-template to identify two adjacent amino acids leading to the long-awaited loss of function (inactivation) of Na$_v$1.4 channels. This mutant type (T109A, N110A, herein called TANA) was expressed and tested on cells of hamster ovary (CHO). The present electrophysiological results showed that the double alanine substitution TANA disrupted channel inactivation as if the β1 subunit would not be in complex with the α subunit. Exhaustive and unbiased sampling of "all β proteins" (Ig-like, Ig) resulted in a plethora of 3D templates which were compared to the target secondary structure prediction. The location of TANA was made possible thanks to another "all β protein" structure in complex with an irreversible bound protein as well as a reversible protein–protein interface (our "Rosetta Stone" effect). This finding coincides with our electrophysiological data (disrupted β1-like voltage dependence) and it is safe to utter that the Na$_v$1.4 α/β1 interface is likely to be of reversible nature.

1. Introduction

1.1. The function and structure of Na$^+$ channels

Ion channels are a ubiquitous class of membrane-spanning proteins. They accomplish electrochemical functions and specifically regulate ion movements (Na$^+$, K$^+$, Ca^{++}cations or Cl$^-$ anions) through their gating mechanism, understood as the transition between open active, inactive and closed states. A typical channel is a multimeric protein complex. It is assembled from a pore-forming α subunit that is often assisted by other subunits labeled β, γ, δ, etc. [1] Mammalian Na$^+$ channels are heterotrimers, composed of one central α subunit of four variable repeat units or domains (DI to DIV) and two or more auxiliary β subunits.

Nine α isoforms and 4 β isoforms have been described for this class [2]. For many ion channels (Na$^+$, Ca^{++}, GABA, and NMDA) subunit cooperativity is paralleled by small molecule modulation through interaction sites other than the pore region with its outer and inner vestibules. Such ligand binding sites are often referred to as allosteric, modulatory or regulatory [3].

1.2. The Na$^+$ channel β1 subunit (Na$_v$β1)

Na$^+$ channel β subunits were functionally characterized as channel gating modulators and channel protein expression regulators at the plasma membrane level and were structurally identified as "cell adhesion molecules" [4,5]. The β subunit modulation confers differential activity depending on the channel isoform and tissue type where the protein complex is expressed. The primary sequence of the sodium channel β subunit (Na$_v$β1) is the same for all α subunit isoforms [6]. The presence of Na$_v$β1 is a necessary but not sufficient prerequisite to modulate channel activity. The extracellular domain of β1 is necessary

Abbreviations: MD-2, myeloid differentiation factor 2 (MD-2); SDM, site-directed mutagenesis; Na$_v$1.4, skeletal muscle voltage-gated sodium channel; TLR4, Toll-like receptor type 4; Trk, tyrosine receptor.

E-mail addresses: tscior@gmail.com, thomas.scior@correo.bupa.mx (T. Scior).

and sufficient to modulate the channel gating of α subunit isoforms $Na_v1.2$ and $Na_v1.4$; this subunit accelerates channel inactivation and recovery from inactivation [7–9]. In more explicit terms the Na^+ channel gating, in presence of the β1 subunit, changes from slow to fast mode at different extents in practically every isoform except in $Na_v1.5$, which predominates in cardiac myocytes and exhibits fast gating on its own [10]. In stark contrast, the skeletal muscle isoform $Na_v1.4$ requires the co-expression of β1 to reconstitute the native fast Na^+ currents [4,11].

1.3. Na^+ channel α and β subunit models

At present no crystal structure of a full α subunit eukaryote Na^+ channel has been published. Currently the best template to model a mammalian α subunit constitutes the bacterial channel Na_vAb (PDB codes: 4EKW [12] and 3RVY [13]) which has a 33% identity (E value of $2\,e^{-13}$) with respect to $rNa_v1.4$ isoform (100%). Homology is a prerequisite for reliable 3D template modeling of target proteins with unknown structure. An intriguing question for ion channel researchers over the recent years has been how to gain insight into the cooperativity between α and β1 subunits of the Na^+ channels despite the absence of crystallographic data.

Heterotetrameric voltage-gated Ca^{++} and Na^+ channels α subunits are thought to be homologous, sharing a common ancestral K^+ channel and being originated by gene duplication separately, or Na^+ channels having evolved from Ca^{++} channels. This reasoning comes from the interesting fact that the four domains DI to DIV of the Na^+ channels are more similar to the corresponding four repeats of Ca^{++} channels than resemblance between each other [14,15]. Each domain in both ion channels possesses six transmembrane segments (S1 to S6) and the central pore region is constituted by a S5-p-S6 fold unit, while transmembrane helix S4 is considered the voltage sensor giving response to the electrical depolarization stimuli and thusly initiating the channel opening for ion flux [2,16].

The naming convention of accessory subunits among these channels, however, is inconsistent, for instance the β subunit of the voltage-gated Ca^{++} channel (Ca_v), which has been crystallized in complex with its Ca_vα interface (PDB code: 1T0J [17]) is located on the intracellular side and it has a 13.4% identity to Na_vβ1 but unlike the latter, the former belongs to the P-loop containing nucleotide triphosphate hydrolase superfamily. Conversely, the Ca_vα2δ subunit resembles more our target Na_vβ1. Although named "delta" it embraces a domain with the same fold unit as target Na_vβ1. Moreover it shows a two-peptide complex linked by two disulfide bridges [18]. Thus we dismissed the Ca_v templates for Na_vβ1 modeling. No wonder, former homology models of target Na_v channels have been based on another oligomeric channel type, namely the voltage-gated potassium channel (K_v). Although it possesses a β subunit (K_vβ, KCNAB family) which modulates channel gating, K_v differs due to its function as an oxidoreductase enzyme as well as its location in complex with α cytoplasmic segments (PDB code: 1QRQ [19]).

Surprisingly, Na_vβ1 subunit also modulates members of a K^+ channel subfamily. Mutational studies of Shaker K^+ channels have assisted in the generation of a computational model of the $K_v1.2$–Na_vβ1 interface [20]. Today, much better resolved crystal structures of voltage-gated K^+ channels have lend detailed insight into their topologies (PDB codes: 3LUT [21]). Unlike eukaryotic α K_v channels (composed of 4 subunit chains), the Na_v channels share the same topology of four variable transmembrane domains, loops and voltage sensor on a single subunit chain with Ca_v channels [22].

The advent of crystal structures of full bacterial voltage-gated Na^+ channels such as NaChBac and Na_vAb (PDB code: 4DXW [23] and 3RVY [13]) has allowed the modeling of full heteromeric α eukaryote subunits by homology [24,25]. Prior to the advent of bacterial Na^+ templates extant models had been generated from K^+ channel templates such as the bacterial MthK channel. They included only pore-forming domains. At that time pore width and other geometrical data

for modeling were inferred from channel blocker ligand studies [25–28]. Up to now, more structural insight is in need, e.g. the loop lengths and overall geometries of the highly variable segments or the outer and inner vestibules. At the end of the present study about predicting the interacting residues, the crystal structures of Na_vβ3 and Na_vβ4 were published (PDB codes: 4L1D [29]; 4MZ2 [30]). Yet, due to insufficient data the molecular mechanism of interaction between Na_vα and β1 subunits remains to be elucidated at an atomic level.

Homology models of the α subunit for the wild type isoforms $Na_v1.4$ and β1 subunit had already been used in our laboratory to assist the experimental work [11,28,31]. Here we combined computed protein structure prediction, site-directed mutagenesis (SDM) and electrophysiological studies to investigate the possible function of relevant amino acids, involved in the inactivation process. Our results showed that two adjacent residues (threonine T109 and asparagine, N110) had a critical role in the inactivation process. When we mutated threonine 109 and asparagine 110 to alanine (T → A; N → A, called TANA), the kinetic process of inactivation was affected generating a general loss of function.

2. Materials and methods

2.1. Searching 3D template for the generation of the target subunit β1

Since the crystal structure of the mammalian Na^+ channel subunit β1 (Na_vβ1) has not been elucidated the 3D target model was generated from its primary sequence data from UniProt [4,32] and a related 3D template (PDB database [33]). To this end we searched for homologous crystal templates to generate the target 3D model among the known PDB database entries by FASTA and BLAST [34,35]. All templates belonged to the general class of "all β protein" structures, in particular the immunoglobulin superfamily with the so-called immunoglobulin-like fold motif (Ig-like domain). As a data subset we collected all 3D templates of an Ig-like protein in complex with any other protein, regardless of its relatedness (homology) to the Na_vα subunit. In this way, the study was based on a combined homology and analogy approach. On the one hand, homology was used for the Na_vβ1 model generation, while on the other hand, analogous interfaces were studied which were formed between cell adhesion proteins of the "all β protein" class and any other protein type whether or not it was found homologous or not (analogous) to the Na_vα subunit. Of note, the denomination "beta" for the channel's subunit β protein coincides with the "all β protein" class, but without intention to label a "beta fold" motif as such. This becomes evident in the case of calcium channel β subunits which do not belong to the "all β protein" class. Here Greek letters merely label the subunits (proteins in complex).

2.2. Alignments of sequences and secondary structure determination

It is also noteworthy to state, Chimera [36] or Swiss PDB Viewer [37] could not resolve automated structure alignments in all cases. On occasion, it became necessary to help out by manual superposition (Vega ZZ [38]). To this end, some atoms were selected in the N-terminal and C-terminal regions, others in the loop regions, namely the turns between strands A to B and E to F.

Automated multiple sequence alignments (MSA) and sequence identity determinations were carried out with web-based programs and software package tools (Clustal W, Chimera) [39,36]. The secondary structure for the target sequence was estimated as a consensus (overlay) of results by prediction tools NPSA and JPRED [40,41]. In addition, hair pin loops were assessed, i.e. type II turns with the general pattern "XG", where X is any (one) amino acid [42]. The result was compared to the secondary structures of known crystal structures with Ig-like domains [43].

A

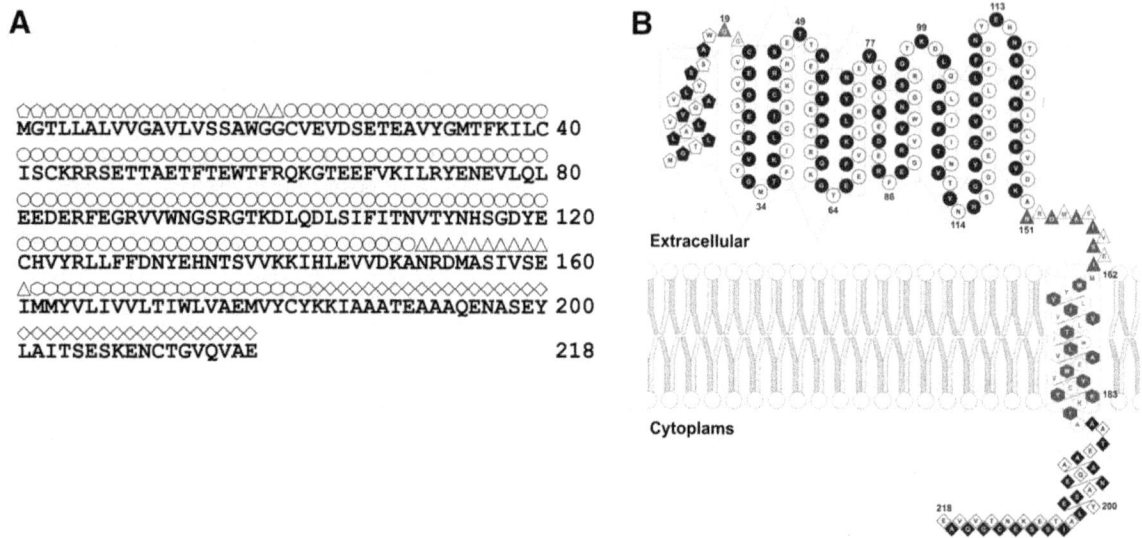

```
⌇⌇⌇⌇⌇⌇⌇⌇⌇⌇⌇⌇⌇⌇⌇⌇⌇△⌇⌇⌇⌇⌇⌇⌇⌇⌇⌇⌇⌇⌇⌇⌇⌇⌇⌇⌇⌇⌇
MGTLLALVVGAVLVSSAWGGCVEVDSETEAVYGMTFKILC 40
⌇⌇⌇⌇⌇⌇⌇⌇⌇⌇⌇⌇⌇⌇⌇⌇⌇⌇⌇⌇⌇⌇⌇⌇⌇⌇⌇⌇⌇⌇⌇⌇⌇⌇⌇⌇⌇⌇⌇⌇
ISCKRRSETTAETFTEWTFRQKGTEEFVKILRYENEVLQL 80
EEDERFEGRVVWNGSRGTKDLQDLSIFITNVTYNHSGDYE 120
⌇⌇⌇⌇⌇⌇⌇⌇⌇⌇⌇⌇⌇⌇⌇⌇⌇⌇⌇⌇⌇⌇⌇⌇⌇⌇⌇⌇⌇⌇△△△△△△
CHVYRLLFFDNYEHNTSVVKKIHLEVVDKANRDMASIVSE 160
△△△△
IMMYVLIVVLTIWLVAEMVYCYKKIAAATEAAAQENASEY 200
⌇⌇⌇⌇⌇⌇⌇⌇⌇⌇⌇⌇⌇⌇
LAITSESKENCTGVQVAE                       218
```

B

Fig. 1. Schematic display of the main chain with its principal domains for the Na$_v$β1 subunit. It documents the location of the major structural domains of Scn1b sodium channel voltage-gated subunit β type 1 *Rattus norvegicus* (accession code: Q00954) [4]. Shown are the sequence (A) and topology (B) which embraces the signal peptide (pentagons, length: 1–18), extracellular immunoglobulin domain (circles), transmembrane domain (hexagons), and intracellular domain (diamonds). The triangles symbolize the linker domain. Every second symbol is filled black or left white to mark the alternative neighbors.

2.3. Manual subunit β1 modeling

The present model was generated using SCRWL [44] and Vega ZZ [38] for manual threading of the target sequence (UniProt [4,32]) across the chosen template structure – in contrast to our earlier multiple template models generated with I-TASSER [45,11] and Modeller [46]. In particular, I-TASSER – an acronym for *iterative threading assembly refinement server* – required the target sequence of the unknown structure in FASTA format as input data. The program automatically launches a 3D-template search (psi-Blast) and reports the homologous proteins from the protein data bank (PDB [33]), assisted by their sequence profiles (psi-pred), while the query sequence is threaded through a collection of possible 3D templates (multiple template construction) [47]. Our topological analyses were documented by web-based tool Topo 2D/TMRPres2D [48]. Moreover, Vega ZZ was served as a general purpose modeling tool [38].

A step-wise description of the combined homology/analogy modeling approach is given in the following Results section.

2.4. Chinese Hamster Ovary (CHO) cell co-transfection

CHO-K1 cells were transiently transfected with rat Na$_v$1.4 cDNA (UniProt accession number P15390) which was cloned into the pGW1H (1 µg) and with cDNA of either native or mutated rNa$_v$β1 (2.5 µg each). Then cDNA was mixed with Lipofect AMINE Plus reagent (Gibco, Invitrogen). CHO-K1 cells were maintained in Dulbecco's modified Eagle's medium (Invitrogen) supplemented with 6% fetal bovine serum (Gibco, Invitrogen), 0.1 mM hypoxanthine, and 0.01 mM thymidine at 37 °C in a 5% CO_2 humidified incubator. The transfected cells were given fresh Dulbecco's modified Eagle's medium containing 1000 U penicillin, 0.1 mg streptomycin + 0.25 µg of amphotericin B per ml, and were passaged at 2- to 3-day intervals with a brief trypsin–EDTA treatment. The cells were dissociated and seeded onto glass coverslips (12-mm diameter; Fisher Scientific, Pittsburgh, PA, USA) in a 35-mm dish 1 day before use. For electrophysiological experiments, coverslips with attached cells were transferred to a recording chamber (RC-13; Warner Instruments, Hamden, CT, USA). The chamber was superfused at a rate of 0.5 ml min^{-1} with normal external solution at 36 ± 1 °C.

2.5. Site-directed mutagenesis and electrophysiology

Briefly, alanine substitutions in positions 109 and 110 were introduced in the rNav β1 construct (Scnb1: Q00954) and cloned into a pGEMHE new vector with a single pair of mutagenic primers. Standard procedures and electrophysiology protocols were performed and

Table 1

Listing of inspected ion channel structures in search of suited 3D templates.

PDB code	Observations	Ref.
1TOJ	Chains A and B: voltage-gated calcium channel subunit beta2a which is an intracellular domain and does not correspond to the ectodomain Na$_v$β1. Chain C: voltage-dependent l-type calcium channel alpha-1c. Comparable to 4DEY.	[17]
4DEY	Chain A: voltage-dependent l-type calcium channel subunit beta. Chain B: voltage-dependent l-type calcium channel subunit alpha.	[50]
1ZSX	Chain A: voltage-gated potassium channel beta-2 subunit. It forms a heteromeric complex, without channel interface, albeit an analogous salt bridge exists (R74---E1114) between two regions with similarities to Na$_v$β1.	[51]
3LUT	Chain A: voltage-gated potassium channel subunit beta-2. Chain B: potassium voltage-gated channel alpha subunit. The intracellular beta-2 segment has not an Ig-like fold. It has nothing in common with target Na$_v$β1. The 499 residue-long alpha1 structure also presents the extra and intracellular loops. The latter have no homology with target loop regions.	[21]
2A79	Chain A: voltage-gated potassium channel beta-2 subunit. Chain B: potassium voltage-gated channel alpha subunit. Compare to newer 3LUT.	[52]
4EKW	Voltage-gated sodium channel in potentially inactivated state, from *Arcobacter butzleri* at 3.2 Å resolution, related to 3RVY, 3RVZ and 3RW0.	[12]
3RVY	Crystal structure of the voltage-gated sodium channel mutant Ile217Cys, at 2.7 resolution.	[13]
1QRQ	Mammalian beta subunit of K$^+$ channels forming a structurally differing four-fold symmetric structure at 2.8 Å resolution, not related to target.	[19]
4L1D	Human sodium channel β3 subunit folds into an Ig domain, showing a homotrimeric complex in its crystal asymmetric unit.	[29]

applied as previously described [49]. Values are reported as the mean $\pm$ SEM. Statistical comparisons between two mean values were conducted by the unpaired Student's t-test. Graphs were built and fitted using Sigmaplot 11.0 (SPSS, Inc., Chicago, IL, USA) and Origin 8.02 (Origin Lab Corp., Northampton, MA, USA).

2.6. Electrophysiological recordings and data analysis

The cells were allowed to stand for 5 min to facilitate precipitation and adhesion. They were then perfused with external solution from

Table 2
Listing of Ig-like templates with analogous interfaces. The percent identities of primary sequence against target amino acid sequence (NCBI Reference Sequence: NM_017288.1) are given for those templates which are discussed in more details (cf. asterisk).

PDB code	Observations	Ref.
1FHG	Telokin, C-terminus of smooth muscle myosin light chain kinase. It lacks the disulphide bridge between β-strands B and F.	[57]
1WWA	Tyrosine kinase receptor A, High affinity nerve growth factor receptor. Belongs to the I-set domain family of Immunoglobulins.	[53]
1WWB	Neurotrophic tyrosine kinase receptor type 2. Ligand binding domain of TrkB. Belongs to the I-set domain family of Immunoglobulins.	[53]
1WWC	Neurotrophic tyrosine kinase receptor type 3. NT3 binding domain of TrkC receptor. Belongs to the I-set domain family of Immunoglobulins.	[53]
1HCF	* (Brain-derived) Neurotrophic tyrosine kinase receptor type 2 in complex with neurotrophin-4 (NTF-4). Upon binding to NTF-4 TrkB undergoes homodimerization, autophosphorylation and activates. The receptor possesses an Ig-like β-sandwich fold; it belongs to the I-set domain family of Immunoglobulins, and the Ligand binding domain of TrkB. The neurotrophin has a cysteine-knot cytokines fold. (id score: 13%).	[59]
1JPS	* Chain i: light chain of immunoglobulin fab d3h44. Chain h: heavy chain of Immunoglobulin fab d3h44. The fab fragments embrace an Ig-like fold. The antibody shows an interface with an analogous flap of the tissue factor as its ligand (antigen). (id score: 18%)	[60]
3GRW	* Chain A: The ligand is the fibroblast growth factor receptor 3. Chain L: Fab light chain. Chain H: Fab heavy chain. The analogous interface is between the Ig-like fold domain of the antibody and FGFR3 comparable to 1JPS. (id score: 15%)	[61]
3KLD	* Chain A: contactin 4. Fragment: Ig-like domains 1-4. Chain B: tyrosine-protein phosphatase gamma. Fragment: carbonic anhydrase-like domain. The bound proteins form an analogous interface. (id score: 16%)	[62]
1HE7	Tyrosine kinase receptor A, High affinity Nerve Growth factor receptor. Belongs to the I-set domain family of immunoglobulins.	[69]
1HXM	T-cell receptor δ chain C region in complex with T-cell receptor γ-2 chain C region. They are heavy chains of immunoglobulins possessing C1-set (constant) and V-set (variable) domains. They have an Ig-like β-sandwich fold.	[70]
1BD2	T cell receptor complex, formed by an HLA class I histocompatibility antigen, A-2 α chain, a leukemia viral peptide and an alpha-beta T cell receptor (TCR), B7. Two human T cell receptors bind in a similar diagonal mode to the HLA-A2/Tax peptide complex using different TCR amino acids.	[71]
1KXQ	* Camelid heavy chain variable domain (Vhh) antibody in complex with porcine pancreatic α-amylase. (id score: 19%)	[72]
3D2F	Chains A and C: Heat shock proteins complex (Hsp). Nucleotide exchange factor (NEF) Sse1p of Hsp110 bound to nucleotide-binding domain (NBD) of Hsp70. The yeast Sse1p Hsp110 possess a 2 layer (bred buns) sandwich architecture.	[58]
3FXI	* Toll-like receptor (TLR4) in complex with myeloid differentiation factor 2 (MD-2) and a bacterial lipopolysaccharide (LPS). The Toll-like receptor has a leucine-rich repeats (LRR), flanked by cysteine-rich domains common in cell adhesion molecules among other proteins. (id score: 11%)	[73]
1NEU	* Chain A: extracellular domain from the major structural protein of peripheral nerve myelin with a typical Ig-like fold; five residues at the C-terminus are disordered, suggesting a flexible linkage to the membrane. (id score: 23%)	[74]

electronic valves remote operated by a programmable controller (Val-505®; CIDES Technology, Puebla, Mexico) to allow the exchange and perfusion of the different solutions used. All experiments were performed at $36 \pm 1\ °C$, which was regulated by a bipolar temperature controller (Medical Systems Corporation, Boston, MA, USA).

3. Results

In the following we lay out the technical procedure in seven steps – all of which were paramount to identify relevant amino acids on $Na_v\beta1$ for the possible protein–protein interface between $Na_v1.4\ \alpha$ and $\beta1$ subunits. After the seven modeling steps we show the experimental results (Step 8).

3.1. Step 1: the target subunit β1

The primary sequence of the rat Na^+ channel subunit $\beta1$ ($rNa_v\beta1$) was retrieved from the UniProt web service (accession code: Q00954

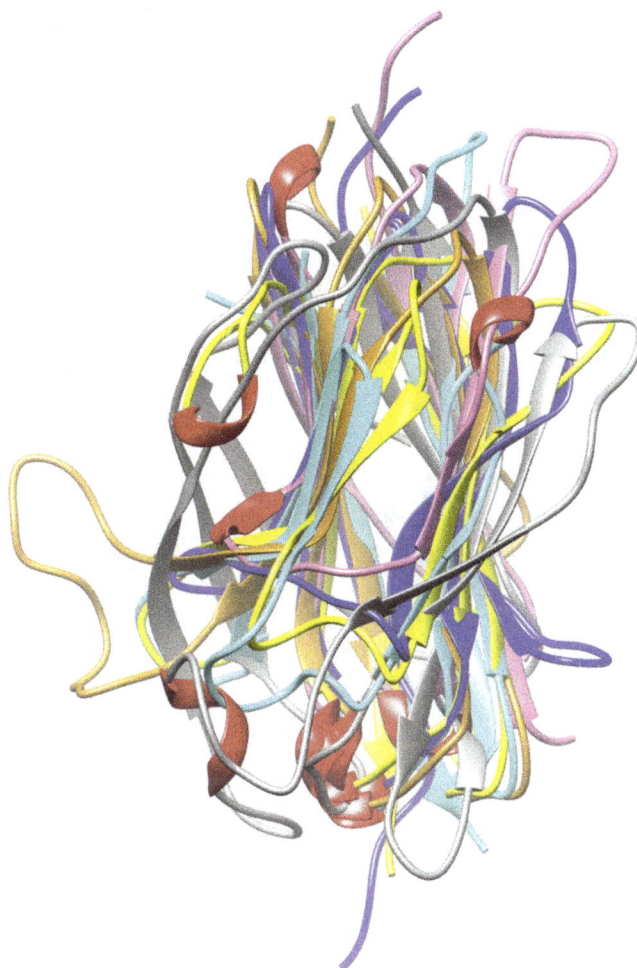

Fig. 2. Display of Ig-like 3D templates in superposition. While the central strands follow a more regular pattern [42] the variable loops wrap up the common Ig-like fold at the surface. Protruding backbone lines (e.g. orange or gray tubes on either side) represent longer loops. Small 3_{10} helical segments (red) and beta strands of sheets (ribbons as arrows) are visible. In representation of all others, six superposed samples (cf. * in Table 2) were displayed and colored individually (dark blue 1HCF, pink 1JPS, light blue 1NEU, gray 1FXI, orange 3GRW, yellow 1KXQ) [59–61,72–74]. The N-terminal segments start on top (e.g. pink line), the C-terminal parts end toward the bottom (e.g. dark blue, light blue, yellow tubes).

Table 3
Multiple sequence alignment by Clustal W [39]. Homology between template (rBeta1) and 3D templates was very low. Na$_v$β3 (4L1D) and Na$_v$β4 (4MZ2) were not available at the time of modeling. Homology was expressed as percentage identity score between the aligned sequences. The threshold below which the twilight zone (uncertainty) for homology modeling exists was estimated according to literature (Fig. 14.4, in [75]).

PDB code	Residue length	Overall id score	Threshold (approx.)	Twilight Zone	Ref [4]
rBeta1	218	100%			
4L1D_A	127	45%	30%	No	[29]
4MZ2_A	129	20%	30%	Yes	[30]
1HCF_X	101	19%	30%	Yes	[59]
3KLD_A	384	18%	20%	borderline	[62]
1NEU_A	124	27%	28%	borderline	[74]
3FXI_C	140	11%	25%	Yes	[73]
3GRW_A	241	15%	22%	Yes	[61]
1KXQ_E	120	19%	28%	Yes	[72]
1JPS_H	225	16%	23%	Yes	[60]

[4,32]). The extracellular domain was limited to 142 residues excluding the signal peptide (Fig. 1).

3.2. Step 2: inspection of known sodium channel structures

The initial search of suited 3D models of the voltage-gated ion channels left us with more open questions than reliable answers

(Table 1). Although collecting structures of ion channels is a straight-forward task, some implications fairly limit their practical use as 3D templates: (1) the types and (2) numbers of subunits (chains) of extant crystal structures (homo- or heterotetrameric repeat units), (3) the sequence similarities or (4) the specific residue variations responsible for ion selectivity in the repeat units, (5) the specific residues of the α/β1 interface situated in the structurally unknown loops or elsewhere, (6) in addition to residue changes due to phylogenetic distances among the published data for different species. None of the primary sequences of the ion channels (Table 1) showed homology to the heterotetrameric Na$_v$α subunit [Clustal W [39]]. With no reliable crystallographic data for the entire multimeric channel at hand we continued searching for suited 3D templates of the subunit Na$_v$β1 alone.

3.3. Step 3: phylogeny of the target Na$_v$β1 protein and its homology to 3D templates

According to the SCOP classification and annotation system, from all PDB entries (101,046 as of June 2014) over 48,700 structures fell into the top-level phylogenetic class of "all β proteins". "All beta" means that the proteins are composed of β strands building up beta sheets. Within this lineage class, over 6500 structures belong to the folding motif "immunoglobulin-like β-sandwich". The domain fairly resembles

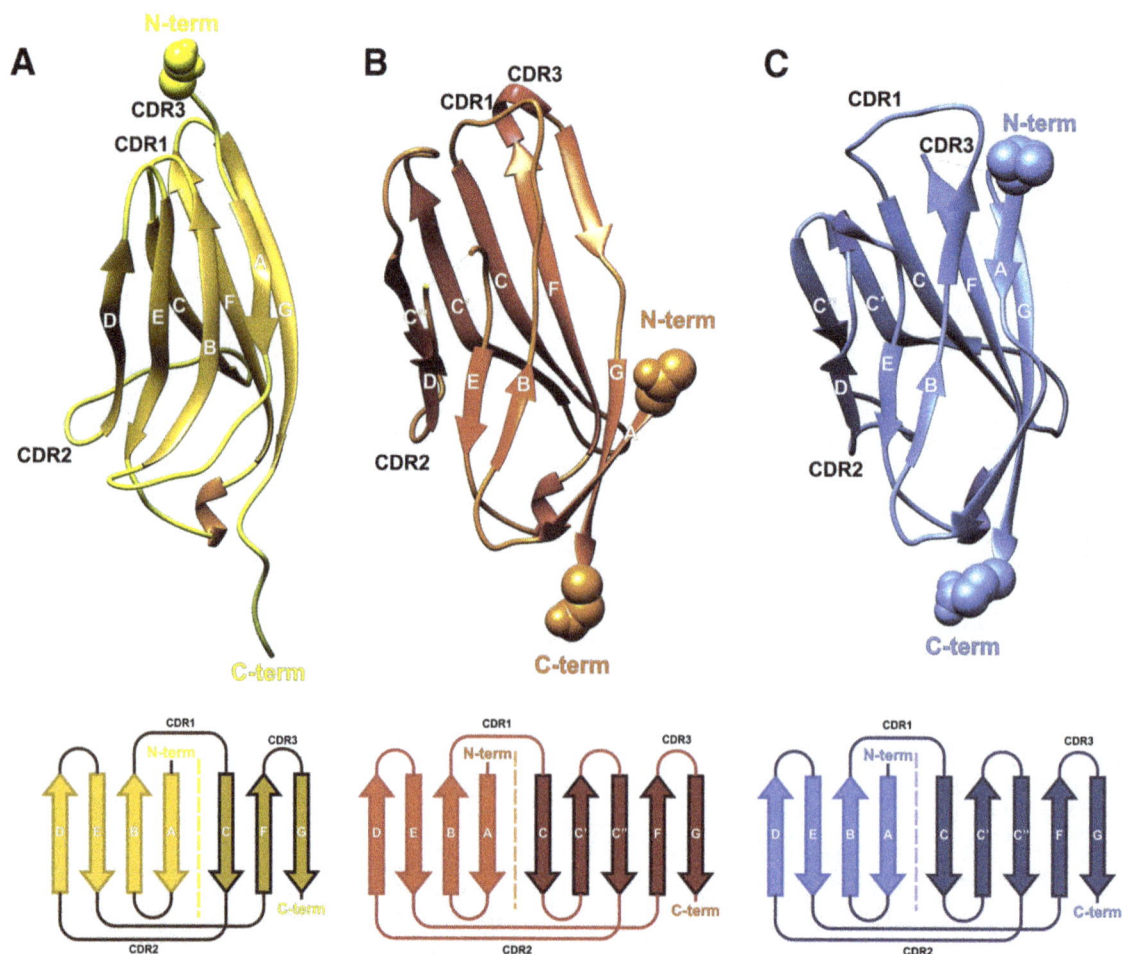

Fig. 3. Display of protein models (first row) and topology schemes (bottom row) of the used template (panel A, 1HCF [59]) in comparison to Na$_v$β4 (panel B, 4MZ2 [30]) and myelin ectodomain (panel C, 1NEU [74]). The β strands (arrow symbols) are labeled. Between β strands C and D (leftmost strand) a larger loop segment allows the formation of two extra β strands, labeled C' and C" (panels B and C) [42]. They were analyzed in the following step.

```
NPSA      CC?HHHHHH?EEEEEECCCCEEEECCCC?EEHCC??EEEEEECCCCCCCC?HEHHHHHHC
Jpred     --HHHHHHHHHHHH----EEEEEE---EEEEE---EEEEEEE--------EEEEEEEEE
NavBeta1  mgtllalvvgavlvssawGGCVEVDSETEAVYGMTFKILCISCKRRSETTAETFTEWTFR


NPSA      CCCCHHHEEEEE??HHCCCCCCCCCCCCCCEEEEECCCCCCCCCCCCCEEEEEEEEEECCCCCEE
Jpred     ------EEEEEEE---------------EEEEE----------EEEEEE------EEEE
NavBeta1  QKGTEEFVKILRYENEVLQLEEDERFEGRVVWNGSRGTKDLQDLSIFITNVTYNHSGDYE


NPSA      EEEEEEEEEECCCCCCCCHHHHEEEEHHHHCCCHHHHHHHHHHHHHHHHHHHHHHHHHH
Jpred     EEEEE----------E---EEEEEEEEE------HHHHHHHHHHHHHHHHHHHHHHHHH
NavBeta1  CHVYRLLFFDNYEHNTSVVKKIHLEVVDKANRDMASIVSEIMMYVLIVVLTIWLVAEMVY


NPSA      HHHHHHHHHHHHHHHHHHHHHHHEEEEECCCCCCCCCEEEECC
Jpred     HHHHHH----------HHHHH----------------
NavBeta1  CYKKIAAATEAAAQENASEYLAITSESKENCTGVQVAE
```

Fig. 4. Consensus estimation for the secondary structure of the target Na$_v$β1 amino acid sequence (third line) by NPSA (first line) and JPRED (second line) web tools [40,41]. Symbols: one letter coded amino acids, in red (helical, H), in brown (beta strands/sheet, E), in blue (loop/coiled C). The signal peptide (in positions1 to 18) was typed in lower case letters.

two bred buns like a sandwich (with nothing put in-between) for eating. Commonly, the domain possesses a Greek key architecture and seven or more β strands to form the two β sheets.

The PDB data base [32] was searched by FASTA and BLAST [34,35] for potential 3D templates of the target primary sequence of Na$_v$β1. A great plethora of crystal structures with an Ig-like β sandwich fold exist. At this stage the study concluded with a trade-off between the sheer numbers of sequences versus a reduced sample set of 3D templates (Table 2) which were amenable to inspection and yet covering a wider range of structural variations (Fig. 2). Many Ig-like motifs are seen in extracellular parts of transmembrane proteins where they are involved in protein–protein interactions. Historically they were labeled by a collective name as "cell adhesion molecules" – although the name "cell protein adhesion molecule" would be more appropriate in our case [53–56]. Despite their different functions they all share a common fold unit, the Ig-like β sandwich structure. For instance, telokin (PDB code: 1FHG) [57] or the chaperone family (or heat shock proteins, HSP; PDB code: 3D2F [58]), the receptor tyrosine kinases (TrkA, B, C; PDB codes: 1WWA [53], 1WWB [53], 1WWC [53], 1HCF [59]); immunoglobulins (antibodies, PDB codes: 1JPS [60], 3GRW [61], 3KLD [62], 2GKI_A (rat) [63], 1KAC_B [64], and 3BKJ_H [65]); and antimicrobial protein tachycitin 1XT5_A [66], 1EAJ_A [67] or myelin protein zero 3OAI [68].

As can be judged by eyesight the sampled crystal structures show a wide range of loop variations (Fig. 2). In consequence, those crystal structures with a common fold unit were inspected which embraced a protein-liganded complexes regardless of the degree of overall sequence conservation (Table 3). Prior to the appearance of Na$_v$β3 and β4 subunits (PDB codes: 4L1D [29]; 4MZ2 [30]) we used the hitherto known PDB entries as 3D templates (Table 2). At the time of modeling – during 2012 to 2013 – the homology between target primary sequence and potential 3D templates was found to be extremely weak (cf. * in Table 2). Below the threshold of around 30% for a 100 to 150 residue-sized domain, sequence alignments of template structures against the target sequence fell into the twilight zone of very low homology (Table 3).

3.4. Step 4: topology, sequence alignments of templates and threading of target Na$_v$β1

At the time of modeling the sequence identities ranged between 11% and 23% for a residue length ranging from 101 to 384 – risking randomly aligned sequences as a direct result of confounding relevant with irrelevant residue positions by chance conservation (Table 3). Hence, the level of complexity was lowered to safer grounds of structural knowledge. To this end, two-dimensional topology diagrams of representative Ig-like β sandwich proteins were compared (cf. * in Table 2): BDNF/NT-

3 growth factors receptor (PDB code: 1HCF_X [59]), antigen binding fragment Fab (PDB code: 1JPS_H [60]), Fibroblast growth factor receptor (PDB code: 3GRW_A [61]), Contactin-4 (PDB code: 3KLD_A [62]), Camelid Vhh antibody (PDB code: 1KXQ [72], Myelin protein P0 (PDB code: 1NEU_A [74]) and TLR4/MD-2 (PDB code: 3FXI_C [73]). In particular, their common structural elements – adjacent strands and loops, bonded or nonbonded neighbor residues, cysteine bridges, hydrogen-bond network, gaps in loops, β-turn-β and hair pin motifs – were analyzed (Fig. 2).

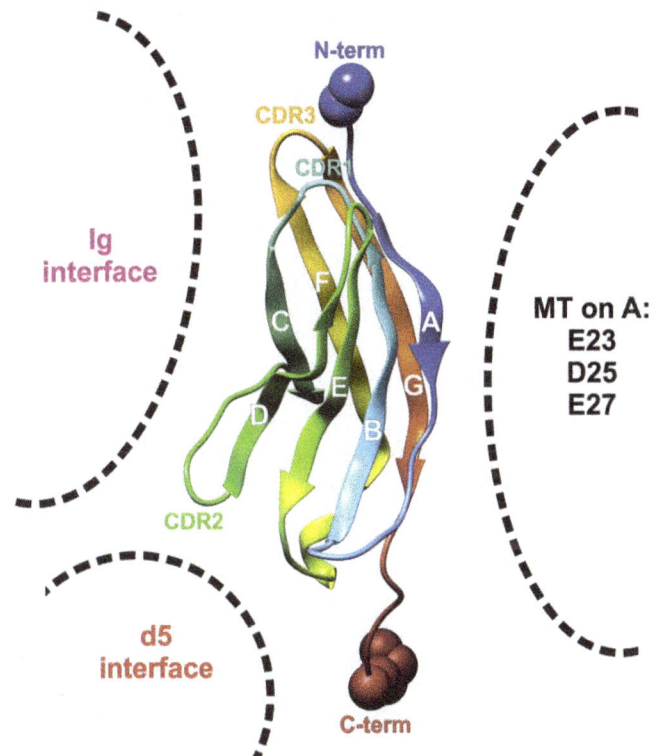

Fig. 5. Once selected the 3D template, the known reversible and irreversible contact surface areas where localized by projecting the corresponding areas from superposed templates (Table 2, Fig. 3). According to the working hypothesis, only areas of reversible interfaces (d5) should be inspected as possible Na$_v$1.4 α/β1 interface to determine residues for SDM studies. The literature [5,11] attests minor activities for mutant types (MT), e.g. glutamate 23 and 27 or aspartate 25 sitting on the C-term end (blue arrow point) of β strand A. CDR1, 2, 3 are the segments for irreversible association to antigens by antibodies (immunoglobulins, Ig).

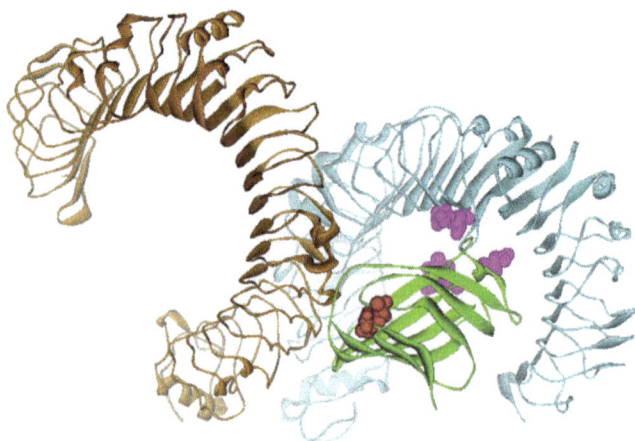

Fig. 6. Main chains of TLR4 (blue), counter TLR4 (orange or brownish), all β protein MD-2 (green, equivalent Na$_v$β1). All atoms are not displayed except for the irreversible Ig-like interface CDR1, 2, 3 regions (magenta) of an invisible, superimposed immunoglobulin and the reversible interface of 3D template 1HCF chain X [59]. Note: the template and Ig backbones of the 3D are omitted, as well as the counter MD-2 was redundant and not depicted. While the (blue) TLR4 is permanently bound to the MD-2, the counter TLR4 may leave upon antagonist binding into the MD-2 pocket [76,77]. This location coincides with the projected location of the reversible interface of neurotrophin-binding domain d5 of TrkB (red space-filling atoms) [59]. Note: CDR1, 2, 3 consist of a variable range of amino acids, but were formally represented by 1, 2 or 3 central residues, respectively (magenta colored space-filling atoms).

At the end of the present modeling study the crystal structures of Na$_v$β3 and β4 were published (PDB codes: 4L1D [29]; 4MZ2 [30]). Now – during Spring 2015 – we carried out fully automated homology modeling of the target protein and compared the results to our manually generated model (Fig. 3) [59]. It was found that aiming at a higher overall id score was not necessary. A lower id score could include a better structural conservation in just the local hot spot(s). Despite its

poor id score (19% in Table 3) our 3D template (PDB code: 1HCF) [59]) yet outperformed the higher scoring templates (PDB codes: 4L1D [29]; 4MZ2 [30]) for four good reasons: (i) allowing to pinpoint unexplored target surface areas in search of residues to be mutated (hot spots). (ii) Even templates with lower id score still conserve the Ig-like fold unit (β-sandwich core). (iii) Even with a higher id score above the twilight zone of homology, templates like 1NEU or Na$_v$β3 and β4 do not present reliable loop coordinates. (iv) Albeit the overall structure of our 3D template had large variation to show in the loop parts – and therein it was not worse than any other template – the lengths, distances or twists of its fold geometry closely resembled that of the two crystal structures of β3 and β4 subunits: strand A–turn–strand B, strand C–(initial part of longer) loop–strand D, strand E–turn–strand F, or strand F–turn–strand G.

After secondary structure prediction (Fig. 4) the primary sequence of Na$_v$β1 was threaded through the aforementioned PDB templates, resulting in the empirical selection of our 3D template where the strands, loops and turns matched the predicted secondary structure of the target subunit.

With the secondary structure prediction at hand (Fig. 4) the appropriate 3D template was found in the chain X of 1HCF (PDB code: 1HCF_X [59]) which constitutes an "all β domain", called d5 of the neurotrophic tyrosine kinase receptor type 2, known as cell surface receptor TrkB. As a most valuable asset d5 of TrkB showed a reversible interface with neurotrophin proteins (cf. literature for further details [59]) whereas higher scoring 1NEU [74] did not (Tables 2 and 3).

Furthermore, the immunoglobulins (antibodies, cf. literature for further details [60,61,72]) bind to antigens in a practically irreversible fashion (cf. antibody–antigen clumping in diagnostics).

According to our ongoing electrophysiological study at that point in time it was hypothesized that Na$_v$1.4 α and β1 subunits interact reversibly, all of which would be reflected by a Na$_v$1.4 α/β1 interface with a reversible contact zone [11]. Subsequently, all target surface areas which correspond to the antigen binding sites of antibodies (CDR1, 2 and 3) could not interact with the channel's Na$_v$1.4 α subunit.

```
             10         20         30         40         50         60 template ids
     mgtllalvvg avlvssawGG CVEVDSETEA VYG_MTFKILC ISCKRRSETT AETFTEWTFR
                    AWGG CVEVDSETEA VYGgMTFKILC ISC_____ _ETFTEWTFR Phe318Arg319
     Ser283His284   SHMA PTITFLESPT SDHHWCIPFTV KGN          PKPALQWFY Phe318Tyr319
         n-Term     StrandAAAAAAAAA ATT_StrandB BBLLLLLLLL LLLStrandC
     Signal peptide (1-18)                        CDR1   of  IG


             70         80         90        100        110        120 template ids
     QKGTEEFVKI LRYENEVLQL EEDERFEGRV VWNGSRGTKD LQDLSIFITN VTYNHSGDYE
     QK     FVKI LRYE_____ ____RFEGRV VWN____ KD LQDLSIFITN VTYNHSGDYE Tyr359Glu360
     NG    AILN ESKY_____ ____ICTKIH VTN      HT EYHGCLQLDN PTHMNNGDYT Tyr359Thr360
     CLLLLLLLLL LLLLLLLLLL LLLLLStrandDDDTTLLLLLL StrandEEEE LLLLLLTTStrandF
              RYE_____ ___R ??                            DN  (=Asp349Asn350)
                                                     Nav α adhesion motif
          CDR2 variable domain of Ig                 TANA double mutant
          additional Loop of Ig-like                 IAA  V mutant type of Navβ1


            130        140        150        160        170        180 template ids
     CHVYRLLFFD NYEHNTSVVK KIHLEVVDKA NRDMASIVSE IMMYVLIVVL TIWLVAEMVY
     CHVYR_____ _____SVVK KIHLEVVDKA NRDM                           Asp382Met383
     LIAKN           EYGK DEKQISAHFM GWPG                          Pro382Gly383
     FFFFF           LLLS trandGGGG         c-Term
                  CDR3 variable domain of Ig
```

Fig. 7. Manual construction of the target model after manual threading of the target sequence through the primary sequence of the 3D template (PDB code: 1HCF [59]). The final positions were achieved by accommodating beta strand and loop lengths and the *AnyGly* motif of type II turns [42]. The manual (not unattended) construction of the target 3D model of Na$_v$β1 was achieved by SCWRL [44]. Line 1: aa (amino acid) count. The aligned blocks are flanked by residues (three-letter codes) with their respective id numbers as given in the 3D template, e.g. aligned E120 = template Glu360 [59]. Line 2: aa seq. of Na$_v$β1 from Q00954 [4]. Line 3: aa seq. of Na$_v$β1 manually threaded onto 1HCF chain X (Loops ___). The small capital "g" shows a glycine residue cut out. Then the local geometry was healed under the built-in GROMACs force field using SPDBV [37]. Line 4: aa seq. of TrkB d5 (3D template 1HCF_X [59]). Line 5: secondary structure: LLLL etc symbolizes; strandAAAA etc is the beta strands A; TT is a type 2 hair pin loop with a XG motif; "??" marks a doubtful hair pin XG motif. Line 6: hints about structures and functions of target and templates.

Table 4

Listing of the observed nonbonded interactions between the ectodomain (d5 or Ig2) of TrkB-d5 and its neurotrophin-4/5 ligand (Fig. 8).

Protein interface	Ligand residues	Receptor residues
Observed nonbonded interactions for columns 2 and 3	Two observed complexes with NGF/NT4/5 [59,69]	NT-binding domain d5 of NTR: TrkA/B/C [59]
No/No	No/No	T325/S327/S345
No/No	No/No	S326/K328/K346
No/npHb/Weak pHb	No/No	F327/Y329/I347
	E35/E37 + R114	
+−/+−	R103/R114	N349/**D349**/N366
Hb/Hb	H84/Q94	Q350/**N350**/K367
wHb/Hb	H2O/E13	H297/H299/R316*
		(* not adjacent L315 or H317)
np/Hb	I6/R10 (bb)	L333/H335/Y353*
		(* not adjacent Y352, H349)

The amino acids of the analogous protein–protein interface to the Na$_v$ α/β1 interaction site are represented with their one-letter codes. The analogy data were retrieved from crystal structures (PDB codes: 1WWW [53,86] and 1HCF [59]). Legend of symbols: (w or p)Hb = (water-mediated or polar hydrogen bonds; (+-) = salt bridge; (no) = not observed; (np) = nonpolar or hydrophobic; (bb) = protein backbone or main chain. The two residues in bold face (D349, N350) correspond to T109, N110 of TANA.

On the contrary, the β1 subunit in contact with the α subunit would rather correspond to a reversible interface like that seen in the TrkB complex [59]. In order to create "research exclusion zones" in the Na$_v$1.4 α/β1 contact area under investigation, the CDR1, 2 and 3 regions of immunoglobulins were projected (by superposition) onto the 3D template [59] in addition to the mutated residues that only showed minor electrophysiological effects (Fig. 5) [5,11].

3.5. Step 5: the proof of concept: a multimeric protein complex with reversible and irreversible interfaces to a central "all β protein" (TLR4/MD-2 as the Rosetta Stone)

In the innate immune system, the Toll-like receptor (TLR) complex is situated on the cell surface to signal the presence (invasion) of smallest amounts of bacterial lipopolysaccharide (LPS) [76,77]. We used the crystal structure of the LPS-liganded human TLR4/MD-2 complex (PDB code: 3FXI [73]). The central myeloid differentiation factor 2 (MD-2) binds LPS as well as to two TLR4 proteins. It is a cell adhesion molecule.

MD-2 folds into seven strands with a Greek-key motif building up two β sheets. Its shape resembles convex lenses but is open on one side to accommodate lipids. Moreover, MD-2 belongs to the Ig-like β-sandwich; E-set domain (early Ig-like fold family) is possibly related to the immunoglobulin family and implicated in lipid (LPS) recognition. It is

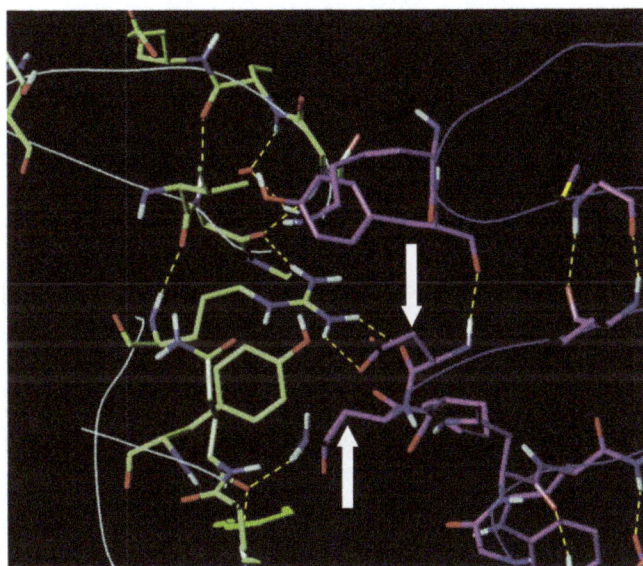

Fig. 8. The three-dimensional model displays the amino acids on both sides of the ligand–receptor interface between neurotrophin (green) and receptor domain d5 of TrkB (purple) [59]. The hydrogen-bonding network is displayed (dashed yellow lines). Aspartic acid and adjacent asparagine (white arrows) mark the central part sitting on a loop turn (D349, N350). They were proposed to become the double mutant TANA (Fig. 7). Color code: carbon atoms of neurotrophin in green, O atoms in red, N atoms in blue, polar H atoms in light blue. All C atoms of binding domain d5 are held in purple color. The backbones of both proteins are displayed as lines. Nonpolar hydrogen atoms omitted for better viewing.

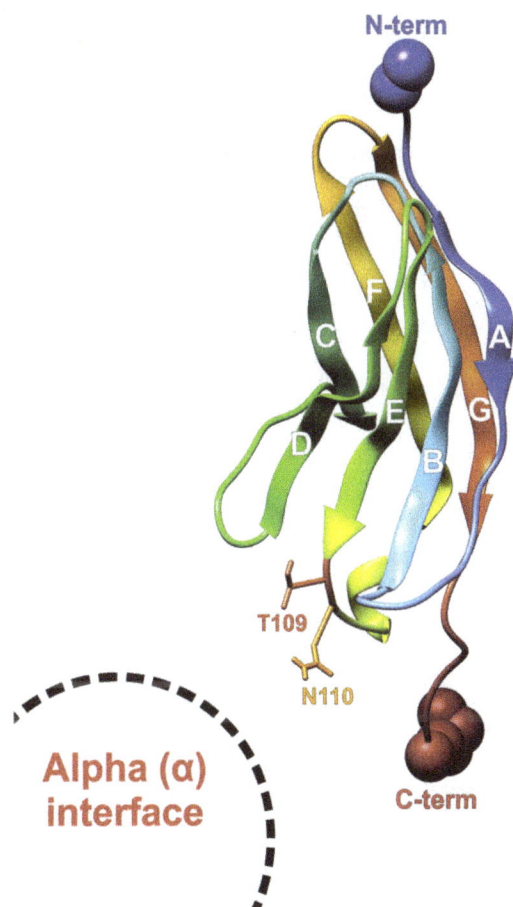

Fig. 9. Display of the final Na$_v$β1 3D model with the postulated interface with the α subunit. The location of the successful double mutant TANA becomes evident when comparing it to the template with the projected reversible and irreversible contact surface areas (Fig. 5). Each beta strand is labeled by its letter (from A to G). TANA (T109 → A N110 → A) lies on a prominent loop (flap) between strands E and F. The amino terminal side shows toward the extracellular space, while the carboxy terminal end of the ectodomain of Na$_v$β1 is followed by the transmembrane and intracellular parts (Fig. 1). The protein backbone is displayed in rainbow colors from blue (N-term) over green, yellow and orange to red (C-term). Space-filling atoms mark the model endings.

attached to TLR4 and counter TLR4. The MD-2/TLR4 interface is very pronounced and the area of interaction enlarged during evolution. The MD-2/counter TLR4 interface adopts an enlarged area, though less pronounced. It is assumed that MD-2 associates to TLR4 permanently in contrast to the reversible association to the counter (second) TLR4 [73,76,77].

Once in superposition onto MD-2 our $Na_v\beta1$ target 3D model was inspected for potential protein–protein interaction areas on its surface in order to propose residues for mutation (Fig. 6). According to our prior studies [76,77] it has been cryptographically known that the TLR4/MD-2 complex binds reversibly a counter TLR4 (orange)/counter MD-2 (PDB code: 3FXI [73]). The evolutionary adaptation of enlarged interface areas (flaps) can also be observed in the case of MD-2 which permanently binds to TLR4 (cf. green and blue flaps flanked by the magenta space-filling atoms in Fig. 6) and to a lesser extent to the rightmost part of MD-2. In consequence, the only remaining area to look for potential zones of subunit–subunit interactions on $Na_v\beta1$ is located towards the counter TLR4 (orange in Fig. 6) which becomes a "leaving group" when LPS antagonists bind into the MD-2 pocket(s) [73,78]. Two arguments prove our working hypothesis: (i) the irreversible interface between MD-2 (green) and TLR4 (blue) matches exactly the irreversible interface between an antigen and its antibody (magenta). (ii) The reversible interface between MD-2 (green) and counter TLR4 (orange) matches exactly the reversible interface between template d5 of Trk and its neurotrophin ligand (red). All told, the TLR4/MD-2 complex had the same relevance for us as the "Rosseta Stone" for French Egyptologist Champollion to decipher the hieroglyphs.

3.6. Step 6: generating the 3D model of the rat $Na_v\beta1$ target subunit

The 3D model of rat $Na_v\beta1$ was generated using SCWRL 4 and Vega ZZ 3.0 [44,38]. At this stage – after the manual threading of the target sequence through the selected 3D template – the formation of type II hair pin loop turns (XG motif) and the Cys–Cys bridge was considered

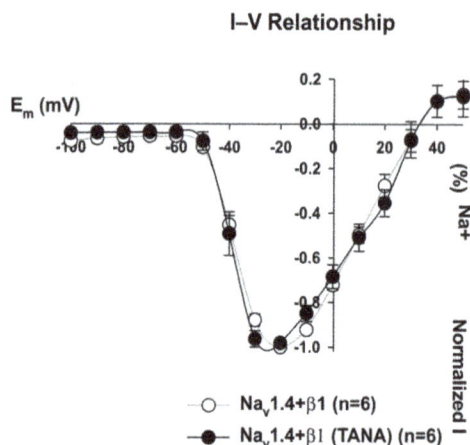

Fig. 11. Recovery graph to compare wild type (WT: $Na_v1.4$ and $\beta1$ subunit) and mutant type (MT: $Na_v1.4$ and $\beta1$-TANA). Note, at time = 2 ms around (on the x axis) 50% of WT channels (0.5 on y axis) have recovered from inactivation, while only 10% of MT have recovered. After normalizing to percentage basis (100), WT = 100%, and MT is about 20%, yielding a difference of around 80%.

a key aspect to verify the residue positions (Fig. **7**). Since the template complex showed a protein–protein interface between d5 of TrkB and a neurotrophin ligand (PDB Code: 1HCF [59] the target residues in the analogous $\alpha/\beta1$ subunits interface should also bear typical side chains (interface forming Asp, Glu, Asn, Gln, Arg, Lys, His, Thr, Tyr; but deprecating Val, Leu, Iso, Ala, Pro, Met, Phe).

3.7. Step 7: identifying $Na_v\beta1$ residues for mutational studies at the $Na_v \alpha/\beta1$ interface

The reversible protein–protein contact zone between TrkB d5 and its neurotrophin ligand in the template complex indicated that surface area of $Na_v\beta1$ is flanked by the irreversible CDR1, 2, 3 sites and the fruitless mutation points (Figs. 5 and 6). The template's physiological role is to

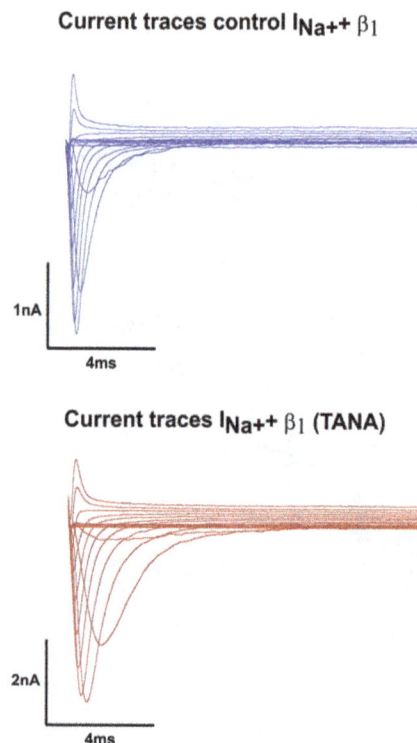

Fig. 10. Representative current traces of the $Na_v1.4$ sodium channel. Its α subunit was co-transfected with wild type (WT) $\beta1$ subunit (blue) or the double mutation TANA (red). Leftmost panel (black), the electrophysiological effects of a ß1 subunit mutation on the I-V relationship of Na_v 1.4. Sodium currents were generated by step depolarizations from a holding potential of − 100 mV in 10 mV increments from − 100 mV to + 50 mV (30 ms duration) in several mammalian CHO cells transfected with either WT α together with WT ß1 subunits as control or WT α subunit with TANA (red traces). Calibrations as shown, $n = 6$.

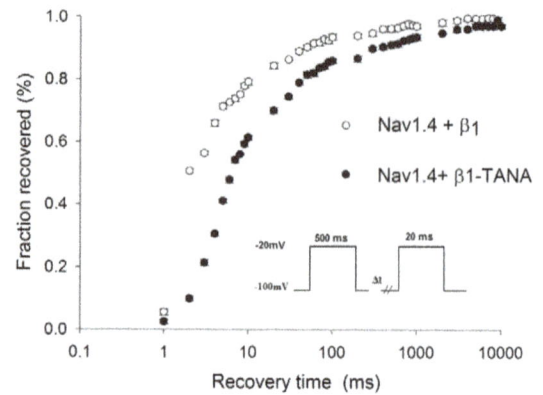

trigger cell growth, differentiation and protection upon binding to tyrosine kinase receptors (Trk proteins) on the neural cell surfaces. Although the potential contact zone between both molecules (receptor: domain d5 of TrkB; ligand: neurotrophin) is far more extended, only a few side chains associate with noncovalent bonds which is in keeping with literature reports (Table 4) [80,81].

Next, the spatial and chemical features at the template's reversible protein–protein interface were studied in details [79,81]. Historically, attempts to correlate mutation analysis to predict the 3D structure of proteins from the correlations in their aligned sequences fell short of expectations [80,82]. The reasons thereof are manifold. The explanations deviate from the scope and we refer to the literature instead [83,84, 82]. Ligand binding and enzymatic activities are frequently located at protein and domain interfaces [85]. The analogue protein–protein complexes (Table 2) taught us nature's lesson that a typical noncovalent association of two distinct polypeptide chains does not exceed ten to twenty amino acids at all (Fig. 8) [79]. In addition, the template topology parallels that of the target $\alpha/\beta1$ interface: of all five extracellular domains, the d5 domain is adjacent to the transmembrane helical segment in the primary sequence [86].

In good keeping with the literature attesting a protein binding role to prominent functional loop residues, the template's amino acids aspartate 349 and asparagine 350 were identified (Table 4, Fig. 8) and the corresponding target segment documented (cf. label "Asp349Asn350" in Fig. 7). Then both adjacent residues were mutated into alanine (T109A, NA) which gave the double mutant its name: TANA (Fig. 9). The combined SDM/electrophysiological studies with wild and mutant type TANA led to the long-awaited general loss of function of $Na_v1.4$ channels in biological tests. Hence, the 3D model of $Na_v\beta1$ successfully predicted two residues to disturb the interaction between the $\alpha/\beta1$ subunit of the $Na_v1.4$ channel.

3.8. Step 8: modulation of the inactivation of the voltage-gated sodium channel $Na_v1.4$ by β_1 subunit

Voltage-gated Na^+ channels formed by α and β subunits, characteristically display gating kinetics on millisecond time scales, ensuring rapid electrical communication between cells [2]. $Na_v\beta1$ subunits interact non-covalently with pore-forming α subunits in the extracellular space, which accelerates gating kinetics, and modifies voltage dependence [2,4,8,87,88]. In this part of our study we observed that TANA disrupted inactivation, delayed recovery from inactivation and disrupted $\beta1$-like voltage-dependence (Fig. 10). TANA had neither an effect on the current–voltage (I–V) curve nor in the total amplitude of the current demonstrating that TANA did not change the peak voltage of activation (cf. leftmost inlay chart in Fig. 10). The observed loss of function can be classified as a general loss of function type because TANA had a double effect on: (i) kinetics of recovery from inactivation and (ii) high frequency stimulation. The general loss was about 80% by looking at the maximum difference between wild type (WT) and mutant type (MT) TANA (Fig. 11).

4. Discussion

The identification of the two aforementioned residues in the $\beta1$ subunit to interact with the α subunit was made possible based on a mixed homology and analogy approach exploiting hitherto unrelated topological and structural data of liganded proteins regardless of the degree of phylogenetic relatedness. In more explicit terms, the overall similarity percentage or identity score of MSA studies was not the driving force for decision taking. On the other hand the proposed mixed low-homology/analogy concept was not fool-proof either, on the contrary it required more personal expertise and user-attended modeling. The protein modeling was based on the secondary structure prediction and the proper identification of turns and loop segments. Myeloid differentiation factor 2 from our prior work with an "all β protein" complex

helped us to distinguish between reversible and irreversible protein association interfaces. Supported by our electrophysiological data we postulated that the $\beta1$ subunit contacts α subunit in a reversible fashion. Then we identified two adjacent amino acids (T109, N110) because the corresponding residues (D349, N350) were also key binders in the reversible template complex (1HCF). After mutating both residues a general loss of sodium channel function was detected in our electrophysiology experiments.

5. Conclusions

The computed structure–function studies have resulted in the correct prediction of two adjacent functional residues which led to a loss of function in subsequent electrophysiological studies. Only a single attempt to identify two residues was necessary because the 3D model correctly pinpointed the subunit interface location. Our experimental results surpass previous electrophysiological attempts that have partially elucidated residues at the $\alpha/\beta1$ interface. Our results contribute to the understanding of channel modulation and suggest a sequential interplay of both, α and $\beta1$ subunits, while the association of $\beta1$, through a distinct extracellular domain, accelerates gating.

Acknowledgment

This work was supported by a graduate students grant from the National Council of Science and Technology of Mexico (CONACyT): grant numbers 239388 and 232486 awarded to and 237103 awarded to Bertin Paiz-Candia. We are grateful for financial support from the Mexican PROMEP funds for "Cuerpo Academico" BUAP CA-120 in 2013. Thanks to VIEP-BUAP for financial support for ESS in 2013/4 (SCJT-NAT-13-G and SCJT-NAT-14-G).

References

[1] Hille B. Ion channels of excitable membranes. 3rd ed. Sunderland MA: Sinauer Associates Inc. Press; 2001[ISBN-13: 978-0878933211].
[2] Catterall WA. Voltage-gated sodium channels at 60: structure, function and pathophysiology. J Physiol 2012;590(11):2577–89.
[3] Changeux JP. Allostery and the Monod–Wyman–Changeux model after 50 years. Annu Rev Biophys 2012;41:103–33.
[4] Isom LL, De Jongh KS, Patton DE, Reber BF, Offord J, Charbonneau H, et al. Primary structure and functional expression of the beta 1 subunit of the rat brain sodium channel. Science 1992;256(5058):839–42.
[5] Isom LL. Sodium channel beta subunits: anything but auxiliary. Neuroscientist 2001; 7(1):42–54.
[6] Brackenbury WJ, Isom LL. Voltage-gated Na+ channels: potential for beta subunits as therapeutic targets. Expert Opin Ther Targets 2008;12(9):1191–203.
[7] Wallner M, Weigl L, Meera P, Lotan I. Modulation of the skeletal muscle sodium channel alpha-subunit by the beta 1-subunit. FEBS Lett 1993;336(3):535–9.
[8] McCormick KA, Srinivasan J, White K, Scheuer T, Catterall WA. The extracellular domain of the beta1 subunit is both necessary and sufficient for beta1-like modulation of sodium channel gating. J Biol Chem 1999;274(46):32638–46.
[9] Chen C, Cannon SC. Modulation of Na+ channel inactivation by the beta 1 subunit: a deletion analysis. Pflugers Arch 1995;431(2):186–95.
[10] Makita N, Bennett Jr PB, George Jr AL. Voltage-gated Na+ channel beta 1 subunit mRNA expressed in adult human skeletal muscle, heart, and brain is encoded by a single gene. J Biol Chem 1994;269(10):7571–8.
[11] Islas AA, Sánchez-Solano A, Scior T, Millan-PerezPeña L, Salinas-Stefanon EM. Identification of Navβ1 residues involved in the modulation of the sodium channel Nav1.4. PLoS One 2013;8(12):e81995.
[12] Payandeh J, Gamal El-Din TM, Scheuer T, Zheng N, Catterall WA. Crystal structure of a voltage-gated sodium channel in two potentially inactivated states. Nature 2012; 486(7401):135–9.
[13] Payandeh J, Scheuer T, Zheng N, Catterall WA. The crystal structure of a voltage-gated sodium channel. Nature 2011;475(7356):353–8.
[14] Marban E, Yamagishi T, Tomaselli GF. Structure and function of voltage-gated sodium channels. J Physiol 1998;508(3):647–57.
[15] Goldin AL. Evolution of voltage-gated Na(+) channels. J Exp Biol 2002;205(5): 575–84.
[16] Hofmann F, Flockerzi V, Kahl S, Wegener JW. L-type CaV1.2 calcium channels: from in vitro findings to in vivo function. Physiol Rev 2014;94(1):303–26.
[17] Van Petegem F, Clark KA, Chatelain FC, Minor Jr DL. Structure of a complex between a voltage-gated calcium channel beta-subunit and an alpha-subunit domain. Nature 2004;429(6992):671–5.
[18] Walker D, De Waard M. Subunit interaction sites in voltage-dependent Ca^{2+} channels: role in channel function. Trends Neurosci 1998;21(4):148–54.

[19] Gulbis JM, Mann S, MacKinnon R. Structure of a voltage-dependent K$^+$ channel beta subunit. Cell 1999;97(7):943–52.

[20] Nguyen HM, Miyazaki H, Hoshi N, Smith BJ, Nukina N, Goldin AL, et al. Modulation of voltage-gated K$^+$ channels by the sodium channel β1 subunit. Proc Natl Acad Sci U S A 2012;109(45):18577–82.

[21] Chen X, Wang Q, Ni F, Ma J. Structure of the full-length Shaker potassium channel Kv1.2 by normal-mode-based X-ray crystallographic refinement. Proc Natl Acad Sci U S A 2010;107(25):11352–7.

[22] Nerbonne JM. Molecular basis of functional voltage-gated K$^+$ channel diversity in the mammalian myocardium. J Physiol 2000;525(2):285–98.

[23] Zhang X, Ren W, DeCaen P, Yan C, Tao X, Tang L, et al. Crystal structure of an orthologue of the NaChBac voltage-gated sodium channel. Nature 2012; 486(7401):130–4.

[24] O'Reilly AO, Eberhardt E, Weidner C, Alzheimer C, Wallace BA, Lamper A. Bisphenol A binds to the local anesthetic receptor site to block the human cardiac sodium channel. PLoS One 2012;7(7):e41667.

[25] Tikhonov DB, Zhorov BS. Architecture and pore block of eukaryotic voltage-gated sodium channels in view of NavAb bacterial sodium channel structure. Mol Pharmacol 2012;82(1):97–104.

[26] Lipkind GM, Fozzard HA. Voltage-gated Na channel selectivity: the role of the conserved domain III lysine residue. J Gen Physiol 2008;131(6):523–9.

[27] Tikhonov DB, Zhorov BS. Modeling P-loops domain of sodium channel: homology with potassium channels and interaction with ligands. Biophys J 2005;88(1): 184–97.

[28] Scior TRF, Martínez-Morales E, Cruz SL, Salinas-Stefanon EM. In silico modeling of toluene binding site in the pore of voltage-gate sodium channel. J Recept Ligand Channel Res 2009;2009(2):1–2.

[29] Namadurai S, Balasuriya D, Rajappa R, Wiemhöfer M, Stott K, Klingauf J, et al. Crystal structure and molecular imaging of the Nav channel β3 subunit indicates a trimeric assembly. J Biol Chem 2014;289(15):10797–811.

[30] Gilchrist J, Das S, Van Petegem F, Bosmans F. Crystallographic insights into sodium-channel modulation by the β4 subunit. Proc Natl Acad Sci U S A 2013;110(51): E5016–24.

[31] Scior TF, Islas AA, Martínez-Morales E, Cuanalo-Contreras K, Millan-Perezpeña L, Salinas-Stefanón EM. An in silico approach to primaquine binding to Trp756 in the external vestibule of sodium channel Nav1.4. J Recept Ligand Channel Res 2011; 2011(4):41–8.

[32] UniProt Consortium. Activities at the Universal Protein Resource (UniProt). Nucleic Acids Res 2014;42(Database issue):D191–8.

[33] Bernstein FC, Koetzle TF, Williams GJ, Meyer Jr EE, Brice MD, Rodgers JR, et al. The Protein Data Bank: a computer-based archival file for macromolecular structures. J Mol Biol 1977;112(3):535–42.

[34] Goldstein T, Studer C, Baraniuk R. A field guide to forward-backward splitting with a FASTA implementation. Numer Anal 2014 [arXiv:1411.3406v3, cs.NA] http://arxiv.org/abs/1411.3406.

[35] Altschul SF, Gish W, Miller W, Myers EW, Lipman DJ. Basic local alignment search tool. J Mol Biol 1990;215(3):403–10.

[36] Meng EC, Pettersen EF, Couch GS, Huang CC, Ferrin TE. Tools for integrated sequence-structure analysis with UCSF Chimera. BMC Bioinformatics 2006;12(7):339.

[37] Johansson MU, Zoete V, Michielin O, Guex N. Defining and searching for structural motifs using DeepView/Swiss-PdbViewer. BMC Bioinformatics 2012;13: 173.

[38] Pedretti A, Villa L, Vistoli G. VEGA—an open platform to develop chemo-bioinformatics applications, using plug-in architecture and script programming. J Comput Aided Mol Des 2004;18(3):167–73.

[39] Larkin MA, Blackshields G, Brown NP, Chenna R, McGettigan PA, McWilliam H, et al. Clustal W and Clustal X version 2.0. Bioinformatics 2007;23(21):2947–8.

[40] Combet C, Blanchet C, Geourjon C, Deléage G. NPS@: network protein sequence analysis. Trends Biochem Sci 2000;25(3):147–50.

[41] Cole C, Barber JD, Barton GJ. The Jpred 3 secondary structure prediction server. Nucleic Acids Res 2008;36(Suppl. 2):W197–201.

[42] Branden C, Tooze J. Introduction to protein structure. 2nd ed. New York: Garland Publishing Inc.; 1999[ISBN-13: 978-0815323051].

[43] Joosten RP, te Beek TA, Krieger E, Hekkelman ML, Hooft RW, Schneider R, et al. A series of PDB related databases for everyday needs. Nucleic Acids Res 2011;39(Database issue):D411–9.

[44] Krivov GG, Shapovalov MV, Dunbrack Jr RL. Improved prediction of protein side-chain conformations with SCWRL4. Proteins 2009;77(4):778–95.

[45] Zhang Y. I-TASSER: fully automated protein structure prediction in CASP8. Proteins 2009;77(9):100–13.

[46] Webb B, Sali A. Comparative protein structure modeling using MODELLER. Curr Protoc Bioinformatics 2014;47:5.6.1–5.6.32.

[47] Roy A, Kucukural A, Zhang Y. I-TASSER: a unified platform for automated protein structure and function prediction. Nat Protoc 2010;5(4):725–38.

[48] Spyropoulos IC, Liakopoulos TD, Bagos PG, Hamodrakas SJ. TMRPres2D: high quality visual representation of transmembrane protein models. Bioinformatics 2004; 20(17):3258–60.

[49] Halperin I, Wolfson H, Nussinov R. Correlated mutations: advances and limitations. A study on fusion proteins and on the Cohesin-Dockerin families. Proteins 2006; 63(4):832–45.

[50] Almagor L, Chomsky-Hecht O, Ben-Mocha A, Hendin-Barak D, Dascal N, Hirsch JA. The role of a voltage-dependent Ca^{2+} channel intracellular linker: a structure-function analysis. J Neurosci 2012;32(22):7602–11.

[51] Wennerstrand P, Johannsson I, Ugochukwu E, Kavanagh K, Edwards A, Arrowsmith C, et al. Crystal structure of human potassium channel kv beta-subunit (kcnab2); 2005. http://dx.doi.org/10.2210/pdb1zsx/pdb. [to be published].

[52] Long SB, Campbell EB, Mackinnon R. Crystal structure of a mammalian voltage-dependent Shaker family K$^+$ channel. Science 2005;309(5736):897–903.

[53] Ultsch MH, Wiesmann C, Simmons LC, Henrich J, Yang M, Reilly D, et al. Crystal structures of the neurotrophin-binding domain of TrkA, TrkB and TrkC. J Mol Biol 1999;290(1):149–59.

[54] Chothia C, Jones EY. The molecular structure of cell adhesion molecules. Annu Rev Biochem 1997;66:823–62.

[55] Harpaz Y, Chothia C. Many of the immunoglobulin superfamily domains in cell adhesion molecules and surface receptors belong to a new structural set which is close to that containing variable domains. J Mol Biol 1994;238(4):528–39.

[56] Williams AF, Barclay AN. The immunoglobulin superfamily-domains for cell surface recognition. Annu Rev Immunol 1988;6:381–405.

[57] Holden HM, Ito M, Hartshorne DJ, Rayment I. X-ray structure determination of telokin, the C-terminal domain of myosin light chain kinase, at 2.8 A resolution. J Mol Biol 1992;227(3):840–51.

[58] Polier S, Dragovic Z, Hartl FU, Bracher A. Structural basis for the cooperation of Hsp70 and Hsp110 chaperones in protein folding. Cell 2008;133(6):1068–79.

[59] Banfield MJ, Naylor RL, Robertson AG, Allen SJ, Dawbarn D, Brady RL, et al. Specificity in Trk receptor:neurotrophin interactions: the crystal structure of TrkB-d5 in complex with neurotrophin-4/5. Structure 2001;9(12):1191–9.

[60] Faelber K, Kirchhofer D, Presta L, Kelley RF, Muller YA. The 1.85 A resolution crystal structures of tissue factor in complex with humanized Fab D3h44 and of free humanized Fab D3h44: revisiting the solvation of antigen combining sites. J Mol Biol 2001;313(1):83–97.

[61] Qing J, Du X, Chen Y, Chan P, Li H, Wu P, et al. Antibody-based targeting of FGFR3 in bladder carcinoma and t(4;14)-positive multiple myeloma in mice. J Clin Invest 2009;119(5):1216–29.

[62] Bouyain S, Watkins DJ. The protein tyrosine phosphatases PTPRZ and PTPRG bind to distinct members of the contactin family of neural recognition molecules. Proc Natl Acad Sci U S A 2010;107(6):2443–8.

[63] Kim YR, Kim JS, Lee SH, Lee WR, Sohn JN, Chung YC, et al. Heavy and light chain variable single domains of an anti-DNA binding antibody hydrolyze both double- and single-stranded DNAs without sequence specificity. J Biol Chem 2006;281(22): 15287–95.

[64] Bewley MC, Springer K, Zhang YB, Freimuth P, Flanagan JM. Structural analysis of the mechanism of adenovirus binding to its human cellular receptor, CAR. Science 1999; 286(5444):1579–83.

[65] Miles LA, Wun KS, Crespi GA, Fodero-Tavoletti MT, Galatis D, Bagley CJ, et al. Amyloid-beta–anti-amyloid-beta complex structure reveals an extended conformation in the immunodominant B-cell epitope. J Mol Biol 2008;377(1):181–92.

[66] Hernández Prada JA, Haire RN, Allaire M, Jakoncic J, Stojanoff V, Cannon JP, et al. Ancient evolutionary origin of diversified variable regions demonstrated by crystal structures of an immune-type receptor in amphioxus. Nat Immunol 2006;875–82.

[67] van Raaij MJ, Chouin E, van der Zandt H, Bergelson JM, Cusack S. Dimeric structure of the coxsackie virus and adenovirus receptor D1 domain at 1.7 A resolution. Structure 2000;8(11):1147–55.

[68] Liu Z, Wang Y, Yedidi RS, Brunzelle JS, Kovari IA, Sohi J, et al. Crystal structure of the extracellular domain of human myelin protein zero. Proteins 2012;80(1): 307–13.

[69] Robertson AG, Banfield MJ, Allen SJ, Dando JA, Mason GG, Tyler SJ, et al. Identification and structure of the nerve growth factor binding site on TrkA. Biochem Biophys Res Commun 2001;282(1):131–41.

[70] Allison TJ, Winter CC, Fournié JJ, Bonneville M, Garboczi DN. Structure of a human gamma delta T-cell antigen receptor. Nature 2001;411(6839):820–4.

[71] Ding YH, Smith KJ, Garboczi DN, Utz U, Biddison WE, Wiley DC. Two human T cell receptors bind in a similar diagonal mode to the HLA-A2/Tax peptide complex using different TCR amino acids. Immunity 1998;8(4):403–11.

[72] Desmyter A, Spinelli S, Payan F, Lauwereys M, Wyns L, et al. Three camelid VHH domains in complex with porcine pancreatic alpha-amylase. Inhibition and versatility of binding topology. J Biol Chem 2002;277(26):23645–50.

[73] Park BS, Song DH, Kim HM, Choi BS, Lee H, Lee JO. The structural basis of lipopolysaccharide recognition by the TLR4-MD-2 complex. Nature 2009;458(7242):1191–5.

[74] Shapiro L, Doyle JP, Hensley P, Colman DR, Hendrickson WA. Crystal structure of the extracellular domain from P0, the major structural protein of peripheral nerve myelin. Neuron 1996;17(3):435–49.

[75] Bourne PE, Weissig H. Structural bioinformatics (methods of biochemical analysis) paperback. First ed. Hoboken NJ: Wiley-Liss; 2003[ISBN-13: 978-0471201991].

[76] Scior T, Alexander C, Zaehringer U. Reviewing and identifying amino acids of human, murine, canine and equine TLR4/MD-2 receptor complexes conferring endotoxic innate immunity activation by LPS/lipid A, or antagonistic effects by Eritoran, in contrast to species-dependent modulation by lipid IVa. Comput Struct Biotechnol J 2013;5:e201302012.

[77] Scior T, Lozano-Aponte J, Figueroa-Vazquez V, Yunes-Rojas JA, Zähringer U, Alexander C. Three-dimensional mapping of differential amino acids of human, murine, canine and equine TLR4/MD-2 receptor complexes conferring endotoxic activation by lipid A, antagonism by Eritoran and species-dependent activities of Lipid IVA in the mammalian LPS sensor system. Comput Struct Biotechnol J 2013;7: e201305003.

[78] Anwar MA, Panneerselvam S, Shah M, Choi S. Insights into the species-specific TLR4 signaling mechanism in response to *Rhodobacter sphaeroides* lipid A detection. Sci Rep 2015;5:7657.

[79] Xu D, Tsai CJ, Nussinov R. Hydrogen bonds and salt bridges across protein–protein interfaces. Protein Eng 1997;10(9):999–1012.

[80] Moreira IS, Fernandes PA, Ramos MJ. Backbone importance for protein–protein binding. J Chem Theory Comput 2007;3(3):885–93.

[81] Aguiari G, Manzati E, Penolazzi L, Micheletti F, Augello G, Vitali ED, et al. Mutations in autosomal dominant polycystic kidney disease 2 gene: reduced expression of PKD2 protein in lymphoblastoid cells. Am J Kidney Dis 1999;33(5):880–5.

[82] Nilsson AM, Wijaywardene M, Gkoutos G, Wilson KM, Fernández N, Reynolds CA. Correlated mutations in the HLA class II molecule. Int J Quantum Chem 1999; 73(2):85–96.

[83] van Gunsteren WF, Mark AE. Prediction of the activity and stability effects of site-directed mutagenesis on a protein core. J Mol Biol 1992;227(2):389–95.

[84] Hubbard SJ, Gross KH, Argos P. Intramolecular cavities in globular proteins. Protein Eng 1994;7(5):613–26.

[85] Horn F, Bywater R, Krause G, Kuipers W, Oliveira L, Paiva ACM, et al. The interaction of class B G protein-coupled receptors with their hormones. Receptors Channels 1998;5(5):305–14.

[86] Wiesmann C, Ultsch MH, Bass SH, de Vos AM. Crystal structure of nerve growth factor in complex with the ligand-binding domain of the TrkA receptor. Nature 1999; 401(6749):184–8.

[87] Makita N, Bennett PB, George Jr AL. Molecular determinants of beta 1 subunit-induced gating modulation in voltage-dependent Na$^+$ channels. J Neurosci 1996; 16(22):7117–27.

[88] Qu Y, Rogers JC, Chen SF, McCormick KA, Scheuer T. Functional roles of the extracellular segments of the sodium channel alpha subunit in voltage-dependent gating and modulation by beta1 subunits. J Biol Chem 1999; 274(46):32647–54.

16

KENeV: A web-application for the automated reconstruction and visualization of the enriched metabolic and signaling super-pathways deriving from genomic experiments

Eleftherios Pilalis [a], Theodoros Koutsandreas [a], Ioannis Valavanis [a], Emmanouil Athanasiadis [b], George Spyrou [b], Aristotelis Chatziioannou [a,*]

[a] *Metabolic Engineering and Bioinformatics Programme, Institute of Medicinal Chemistry and Biotechnology, National Hellenic Research Foundation, Athens, Greece*
[b] *Biomedical Research Foundation, Academy of Athens, Athens, Greece*

ARTICLE INFO

ABSTRACT

Keywords:
KEGG
Enrichment analysis
Molecular pathways
Gene expression
Microarrays
Next generation sequencing

Gene expression analysis, using high throughput genomic technologies,has become an indispensable step for the meaningful interpretation of the underlying molecular complexity, which shapes the phenotypic manifestation of the investigated biological mechanism. The modularity of the cellular response to different experimental conditions can be comprehended through the exploitation of molecular pathway databases, which offer a controlled, curated background for statistical enrichment analysis. Existing tools enable pathway analysis, visualization, or pathway merging but none integrates a fully automated workflow, combining all above-mentioned modules and destined to non-programmer users.

We introduce an online web application, named *KEGG Enriched Network Visualizer* (*KENeV*), which enables a fully automated workflow starting from a list of differentially expressed genes and deriving the enriched KEGG metabolic and signaling pathways, merged into two respective, non-redundant super-networks. The final networks can be downloaded as SBML files, for further analysis, or instantly visualized through an interactive visualization module.

In conclusion, *KENeV* (available online at http://www.grissom.gr/kenev) provides an integrative tool, suitable for users with no programming experience, for the functional interpretation, at both the metabolic and signaling level, of differentially expressed gene subsets deriving from genomic experiments.

1. Introduction

Gene expression analysis using DNA microarrays or next-generation sequencing (NGS) technologies is commonly employed, in order to derive lists of genes that are differentially expressed among various experimental conditions. Obtaining a comprehensive list of genes is routinely performed today, given the high-throughput biological experiments and the plethora of methods for further statistical analysis of the data obtained. The subsequent association of a list of genes with particular cellular functionalities, through the identification of specific molecular pathways, highlights the modular character of the cellular response to the change of conditions, for instance the chemical environment, disease states, or drug-induced effects [1]. From a Systems Biology perspective, it is becoming obvious how critical is to identify entire biological processes and complex interactions, found in various metabolic and signaling pathways, rather than only highlight isolated biological factors

[2]. Moreover, the over- or under-expression of a single gene does not give significant information about the actual effect in cellular physiology, unless all known interactions are taken into account.

The most common approaches for pathway analysis are based on the statistical enrichment scores of various annotations [3–5], for instance Gene Ontology (GO) terms [6] and identifiers of pathway databases, for instance the *Kyoto Encyclopedia of Genes and Genomes* (KEGG) biological pathway database [7] and the *Molecular Signatures Database* (*MSigDB*) [3]. Such methods that exploit knowledge hosted in public repositories perform knowledge-based driven pathway analysis and are opposed to methods that use solely molecular measurements derived from an experiment [8]. Knowledge-based driven methods include over-representation analysis (ORA) methods, functional-scoring class (FCS) methods and pathway-topology (PT) based methods. ORA methods were developed firstly with GO terms emergence in order to cover the immediate need for functional analysis. Briefly, they use the statistical enrichment score in annotations found in various controlled vocabularies to statistically evaluate a fraction of genes in a particular pathway, which are found among the set of genes showing changes in

* Corresponding author.
  E-mail address: achatzi@eie.gr (A. Chatziioannou).

expression. Examples of ORA tools are *Onto-express* [9] and *GoMiner* [10]. In FCS approaches, *GeneTrail* for example [11], a gene-level statistic is firstly computed using differential gene or protein expression, gene-level statistics for all genes in a pathway are then aggregated into a single pathway-level statistic and finally all statistical significance for all pathways are evaluated. PT-based methods [12] (e.g. *Pathway-Express* [13] and *ScorePAGE* [14]) incorporate additional pathway topology information included in the public databases in order to compute gene-level statistics, e.g. gene products that interact with each other in a given pathway, how they interact (activation, inhibition, etc.), and where they interact (cytoplasm, nucleus, etc.).

In this work, we have developed an online web application based on the ORA approach, named *KEGG Enriched Network Visualizer* (*KENeV*), which exploits the widely used KEGG pathway database, in order to automatically derive and visualize the enriched molecular networks. KEGG is a library of molecular networks that has been widely used as a reference point for biological interpretation of large-scale datasets. The application takes as input a list of significant genes, from an expression analysis experiment, possibly accompanied by corresponding fold change measurements. The workflow [Fig. 1] starts with pathway enrichment analysis of the input gene list, using the *StRAnGER* algorithm [15,16], which prioritizes statistically significant enrichments of terms that belong to a controlled annotation vocabulary, as for instance, Gene Ontology terms or the identifiers of KEGG pathways. Subsequently, the application identifies and separates the promoted pathways in two levels, metabolic and signaling (protein–protein, and protein–small molecule interaction). These pathways are automatically downloaded, without any intervention by the user, converted to SMBL format [19], merged into two super-pathways, a metabolic and a signaling one, and finally delivered in two respective SBML files that can be downloaded or visualized on-the-fly in an embedded visualization module. In addition to the two super-pathways, the application provides a gene-pathway mapping visualization that depicts the pathways in which each selected gene participates. Finally, the application is demonstrated using a list of genes that has been associated to Type I Diabetes Mellitus and was found in the Autworks database [17].

2. Materials and methods

2.1. Pathway enrichment analysis

The pathway enrichment analysis is performed by the *StRAnGER* algorithm, described in detail in [15,16] which employs a combination of a parametric (Hypergeometric) and a non-parametric statistical test (bootstrap resampling) [18]. The algorithm initially uses the Hypergeometric test to assess the over-representation of KEGG pathways to the input gene list, and finally ranks the over-represented pathways. However, the final pathway ranking is based on a non-parametric, empiric algorithm which avoids assumptions about the distribution of term enrichments and thus can be adapted to any set of terms. With the bootstrap resampling, *StRAnGER* avoids the utilization of multiple test correction approaches (Bonferroni, FDR) that are very restrictive and problematic in regard to the finite nature of annotation vocabularies. The main problem of these methodologies is that they tend to promote enrichments yielding very low p-values (close to 0), but holding a very low biological content (e.g. enrichments such as 2/2 and 1/2). Bootstrapping is a non-parametric, empirical alternative to multiple test correction, which provides a corrected measure for the statistical significance of the enrichments based on their frequencies of observation. Instead of adjusting the p-values, the algorithm reorders the initial distribution and prioritizes the less frequently observed enrichments. The enrichments are derived as statistically significant if they satisfy

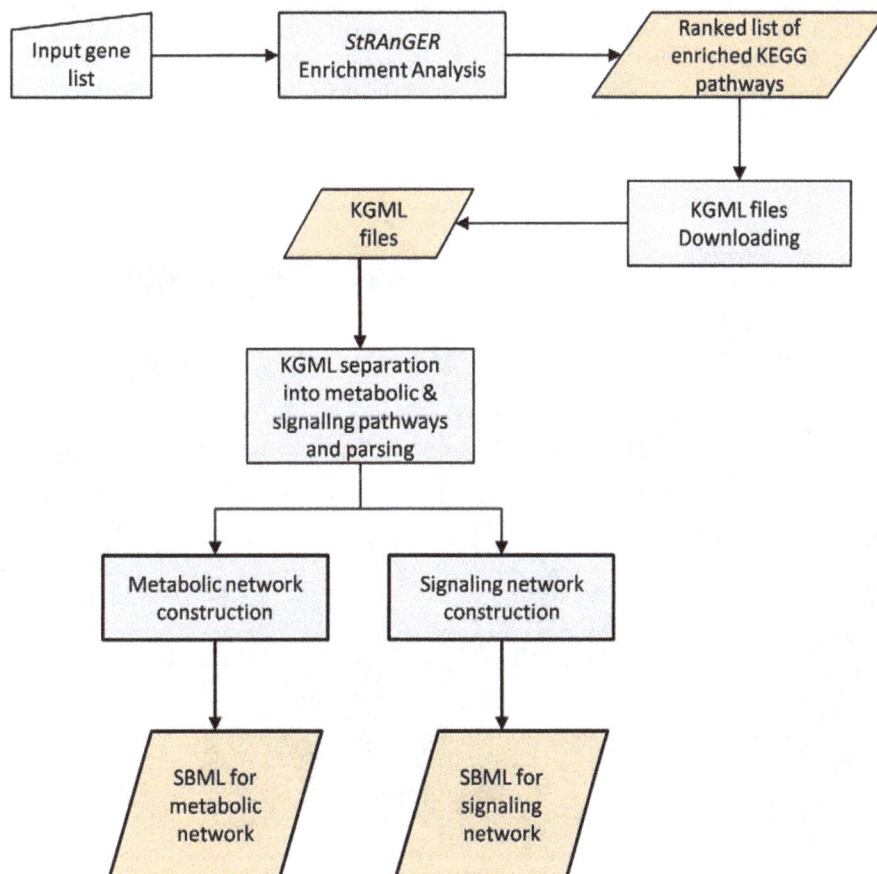

Fig. 1. Overall application workflow.

both the hypergeometric and the bootstrap thresholds, but they are ranked according to their bootstrap p-value. Hence, the algorithm prioritizes pathways with less frequent enrichments which tend to represent broader pathways or functions and, thus, are of stronger biological content. The final output of significant KEGG IDs is spanned by the KEGG pathways that are mapped to the significant elements.

2.2. XML parsing and merging

The statistically important pathways derived from the enrichment analysis are automatically downloaded from the REST interface of the KEGG database as KGML format files, which is the XML format used by KEGG. Subsequently, the KGML files are parsed (using the *ElementTree* Python XML parser) and their fields are stored as respective objects. Different types of compounds (simple molecules, glycans, drugs, proteins, enzymes) and relations (metabolic reactions, signaling cascades etc.) are detected using the KEGG annotation fields and are stored in the database. Each compound is then mapped to a unique node and to all reactions in which it participates. Finally, two merged networks are constructed using the *libsbml* library [19,20], one containing only the metabolic reactions and another containing all other interactions. In the case of the metabolic reactions, the corresponding genes are stored as reaction modifiers, whereas in the case of the signaling network, the type of interaction (e.g. phosphorylation, protein–protein interaction, gene expression relation) is stored as the corresponding reaction name. Finally, two distinct SBML files are constructed, which do not contain redundancies.

2.3. Implementation

The application was implemented in the *Web2Py* framework, in *Python* language. A *NoSQL* database (*MongoDB*), was used for the storage of KEGG data (organism–gene, organism–pathway and gene–pathway mappings). The application was programmed to automatically download the relevant files on a weekly basis, using the KEGG REST API (freely accessible for academic use). The Network Visualization Tool was implemented using a custom made *JavaScript* algorithm complemented by *jQuery* (http://jquery.com/) and "*sigma.js*" (http://sigmajs.org/) *JavaScript* libraries.

3. Results

3.1. Application

The online application is available under the url http://www.grissom.gr/kenev. The form accepts as input a list of differentially expressed genes, derived from a genomic experiment, optionally accompanied by their expression fold change values (natural scale) in tab-delimited format. The user is able to select the examined organism species from the complete list of KEGG organisms (currently including 3677 species). Additionally, the user is able to choose the cut-off values of the two aforementioned statistical tests. On form submission, the pathway enrichment analysis is performed, resulting to a final ranked list of KEGG pathways. The user is then able to click on a link that performs the pathway merging step and yields the signaling and metabolic super-pathways. Remark that it is possible that the pathway list does not include both metabolic and signaling pathways. In this case, only one type of SBML file is returned. The SBML files are available for downloading, in order to be processed or visualized by external software, for instance by the popular programs *Cytoscape* [21] and *CellDesigner* [22]. An embedded interactive visualization module described next is provided, as well.

3.2. Visualization module

For the case of the metabolic/signaling graphs using the generated SBML files (see Fig. 2), the network visualization can be performed using either a circular, a random or a force-directed (Fruchterman Reingold) layout. The node's size is proportional to the number of its first neighbors that interact with the specific node. Reactions, reactants, products, modifiers and both reactants and products nodes are colored and depicted in five corresponding concentric circles with red, green, blue, yellow and suntan, respectively. In addition, users are able to modify the node's colors based on the fold change of each gene if the respective values have been provided by the user. In the case of the latter color modification, green and red nodes correspond to up- and down-regulated

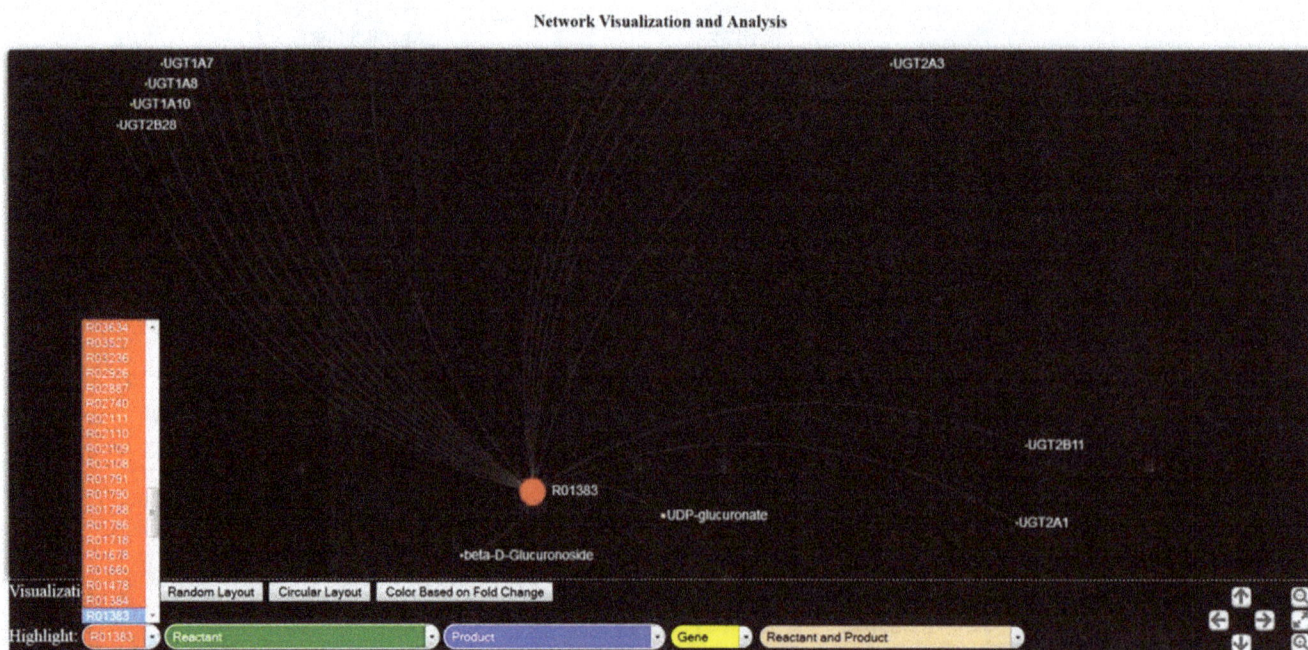

Fig. 2. In this screenshot, visualization of a sample SBML output file is presented. Reaction with ID "R01383" has been randomly selected from the "Reaction" drop-down menu and the corresponding genes, reactants and products that comprise the selected reaction are highlighted.

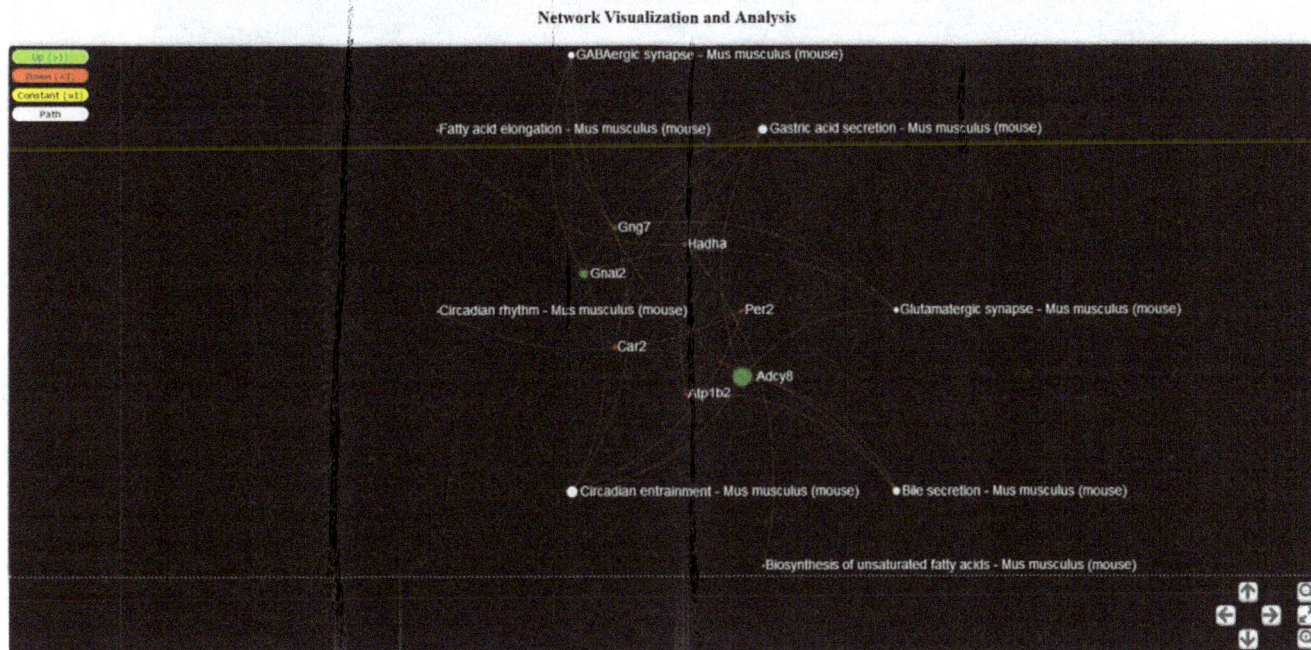

Fig. 3. In the present screenshot, a sample gene–pathway interaction network is illustrated. Genes (inner circle) Gng7, Gnai2 and Adcy8 are up-regulated (green color), while Car2, Atp1b2, Per2 and Hadha are down-regulated (red color). By clicking on a pathway (outer circle with white colors), genes that found on the selected pathway are highlighted and vice-versa.

genes, while nodes with no information are presented in gray color. Users are also able to locate and highlight a specific gene interactively, either by using the appropriate drop-down menu, or by clicking on the node. Finally, further visualization filters can be applied regarding node's labels and interactions, using the appropriate slider on the bottom of the web application.

For the case of genes–pathways mapping (see Fig. 3), visualization was implemented in a circular layout with two concentric circles, the inner one for the genes and the outer one for the KEGG pathways. Colors on the specific graph corresponded to up- (red)/down- (green) regulated genes, constant (yellow) and KEGG pathways (white). If a gene is found on a specific KEGG pathway, then an edge between the gene and the pathway is depicted, as shown in Fig. 3.

3.3. Case study

KENeV is showcased here using as input a well defined list of genes that has been related to Type I Diabetes Mellitus and was found deposited in the *Autworks* database (http://autworks.hms.harvard.edu/) [17]. *Autworks* began as a cross-disease biology web portal in order to store genes lists that have been found associated with autism or related disorders. Currently, it contains ~2300 additional disorders that can be compared with autism in terms of common gene networks beneath disorders.

The list of 407 genes associated to Type I Diabetes Mellitus was submitted to *KENeV* and the workflow was executed up to the visualization module (*p*-value threshold: 0.01, bootstrap distribution cutoff

Table 1
Enriched KEGG pathways as extracted by *KENeV* for the analysis of Type I Diabetes Mellitus gene list.

Rank	Term ID	Term description	Hypergeometric *p*-value	Enrichment	Bootstrap *p*-value
1	path:hsa04940	Type I diabetes mellitus — *Homo sapiens* (human)	5.86E-13	28/45	4.10E-05
2	path:hsa05320	Autoimmune thyroid disease — *Homo sapiens* (human)	2.78E-12	23/54	7.75E-05
3	path:hsa05321	Inflammatory bowel disease (IBD) — *Homo sapiens* (human)	3.36E-12	33/67	1.13E-04
4	path:hsa05164	Influenza A — *Homo sapiens* (human)	3.54E-12	40/177	1.47E-04
5	path:hsa04668	TNF signaling pathway — *Homo sapiens* (human)	4.84E-12	27/110	1.82E-04
6	path:hsa05332	Graft-versus-host disease — *Homo sapiens* (human)	5.57E-12	21/43	2.16E-04
7	path:hsa05145	Toxoplasmosis — *Homo sapiens* (human)	5.82E-12	31/120	2.50E-04
8	path:hsa05310	Asthma — *Homo sapiens* (human)	5.85E-12	18/32	2.84E-04
9	path:hsa05162	Measles — *Homo sapiens* (human)	6.27E-12	31/134	3.18E-04
10	path:hsa05152	Tuberculosis — *Homo sapiens* (human)	6.53E-12	40/179	3.53E-04
11	path:hsa05330	Allograft rejection — *Homo sapiens* (human)	7.27E-12	25/39	3.88E-04
12	path:hsa04612	Antigen processing and presentation — *Homo sapiens* (human)	7.33E-12	22/79	4.23E-04
13	path:hsa04672	Intestinal immune network for IgA production — *Homo sapiens* (human)	1.01E-11	21/49	4.57E-04
14	path:hsa05144	Malaria — *Homo sapiens* (human)	1.01E-11	22/49	4.91E-04
15	path:hsa05323	Rheumatoid arthritis — *Homo sapiens* (human)	1.13E-11	27/91	5.25E-04
16	path:hsa05168	Herpes simplex infection — *Homo sapiens* (human)	1.16E-11	46/186	5.59E-04
17	path:hsa05143	African trypanosomiasis — *Homo sapiens* (human)	1.39E-11	16/34	5.92E-04
18	path:hsa05140	Leishmaniasis — *Homo sapiens* (human)	1.87E-11	33/74	6.24E-04
19	path:hsa04060	Cytokine–cytokine receptor interaction — *Homo sapiens* (human)	2.09E-11	48/265	6.60E-04
20	path:hsa04620	Toll-like receptor signaling pathway — *Homo sapiens* (human)	5.27E-11	25/106	6.91E-04
21	path:hsa05166	HTLV-I infection — *Homo sapiens* (human)	7.18E-11	42/261	7.26E-04
22	path:hsa05416	Viral myocarditis — *Homo sapiens* (human)	2.25E-10	18/60	7.58E-04

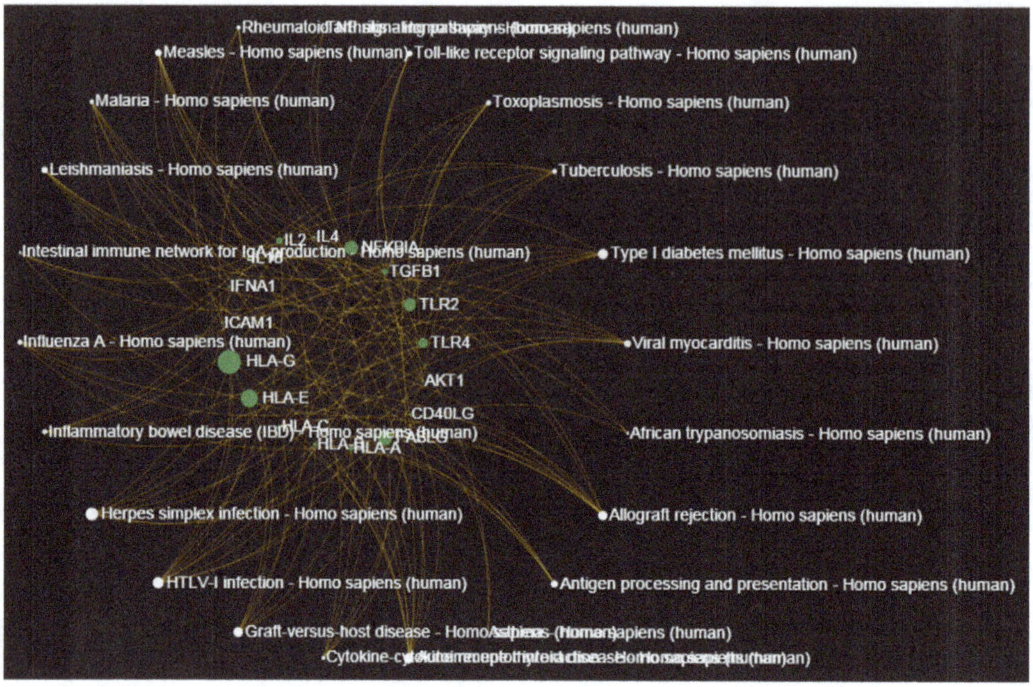

Fig. 4. Genes to pathways mapping as constructed by *KENeV* for the analysis of the Type I Diabetes Mellitus gene list.

percentage: 90%). Table 1 reports the enriched KEGG pathways extracted by *KENeV* inorder of statistical significance, based on the bootstrap extracted *p*-value (see last column). Some interesting pathways are related to autoimmune disorders and inflammation, while the pathway for the disease itself appears first in the output. Fig. 4 presents the genes to pathways mapping as constructed by the tool. An instance of the signaling network constructed by *KENeV*, focusing on the cross-talk between the NF-kB/RelA and PI3K pathways, is shown in Fig. 5. These pathways are known to control, among various cellular processes, inflammation and metabolism and they are implicated in metabolic and immunity-related diseases [23,24]. Finally, a screenshot of the merged

signaling network imported in SBML format in *CellDesigner* [22] is shown in Fig. 6.

4. Discussion

4.1. Comparison with similar tools

There are several software tools that allow the retrieval, visualization and/or the analysis of KEGG pathways. However they all have different features and orientations and none combines all key features of *KENeV* in a fully automated integrative online application, namely

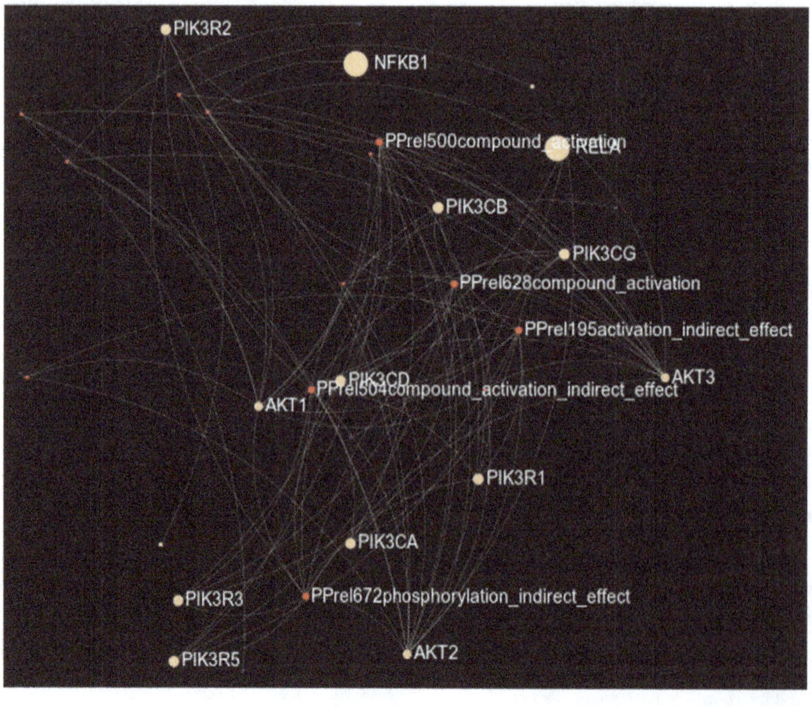

Fig. 5. Screenshot of a signaling network instance using the Type I Diabetes Mellitus gene list, showing the cross-talk between the NF-kB/RelA and PI3K pathways.

pathway enrichment analysis, automatic pathway retrieval, handling of metabolic and signaling pathways, handling of KGML inconsistencies, SBML conversion, SBML merging and visualization of the super-pathways, including highlighting of the differentially expressed genes.

For instance, two available KEGG pathway converters, *KEGG2SBML* and *KGML2BioPAX* [25] translate the KGML format to SBML and BioPAX formats respectively. However, they do not perform any handling of the KGML data correction of reactions and entries annotation enhancement, whereas they cannot handle signaling KEGG pathways. Also, they perform simple one-to-one KGML conversions and they cannot be used for complex networks construction.

Pathview [26] is an *R/Bioconductor* package, which provides KEGG pathway visualization, with very useful features, for instance automatic pathway retrieval, biochemical data modification and enhancement and integration of biological data provided by the user (gene expression, protein expression, metabolite level, genetic association, genomic variation, literature records etc.). However, *Pathview* does not include functions for merging multiple pathways into super-networks.

KEGGgraph [27] is another toolbox implemented in *R* and *Bioconductor*, which enables KGML data retrieval, parsing and visualization. Although this package supports merging of multiple graphs (metabolic and signaling), all functionalities are performed manually through specific R commands, and thus the toolbox is not intended to be used automatically by non programming experts.

KEGGParser [28] is a KEGG pathway semiautomatic parsing, editing and illustrating tool which is based on *MATLAB* functions, and on Bioinformatics and Image Processing toolboxes. *KEGGParser*'s advantage is the automatic correction of inconsistencies between KGML files and static pathway maps.

A more thoroughly developed KEGG pathway illustrator is *KEGGTranslator* [29], an application for KGML data visualization and conversion to multiple formats. *KEGGTranslator* performs multi layered correction/augmentation of KGMLs biological data, such as the removal of examined-organism unrelated nodes, the completion of reactions, the annotation of each entry with multiple identifiers and the addition of stoichiometric information. Additionally, it provides a functional graphical user interface and translates KGML files into various file formats (SBML, BioPAX). These output files can be used for simulation analysis due to the integrated stoichiometric information.

Another web application is *Pathway Projector* [30], which provides the illustration of metabolic pathways participating in KEGG Atlas, visualized in a Zoomable User Interface (ZUI). Additionally, it offers various functionalities, such as the simultaneous implementation of various "-omics" experimental data. However *Pathway Projector* is not able to visualize signaling and gene regulation networks. Furthermore, *Pathway Projector* does not extract pathway data in an exchangeable format such as SBML and BioPAX.

KEGG Converter [31], automatically retrieves, converts to SBML and merges KEGG pathways. However, as it is a simulation-oriented tool, it handles only metabolic pathways, it does not provide any automatic

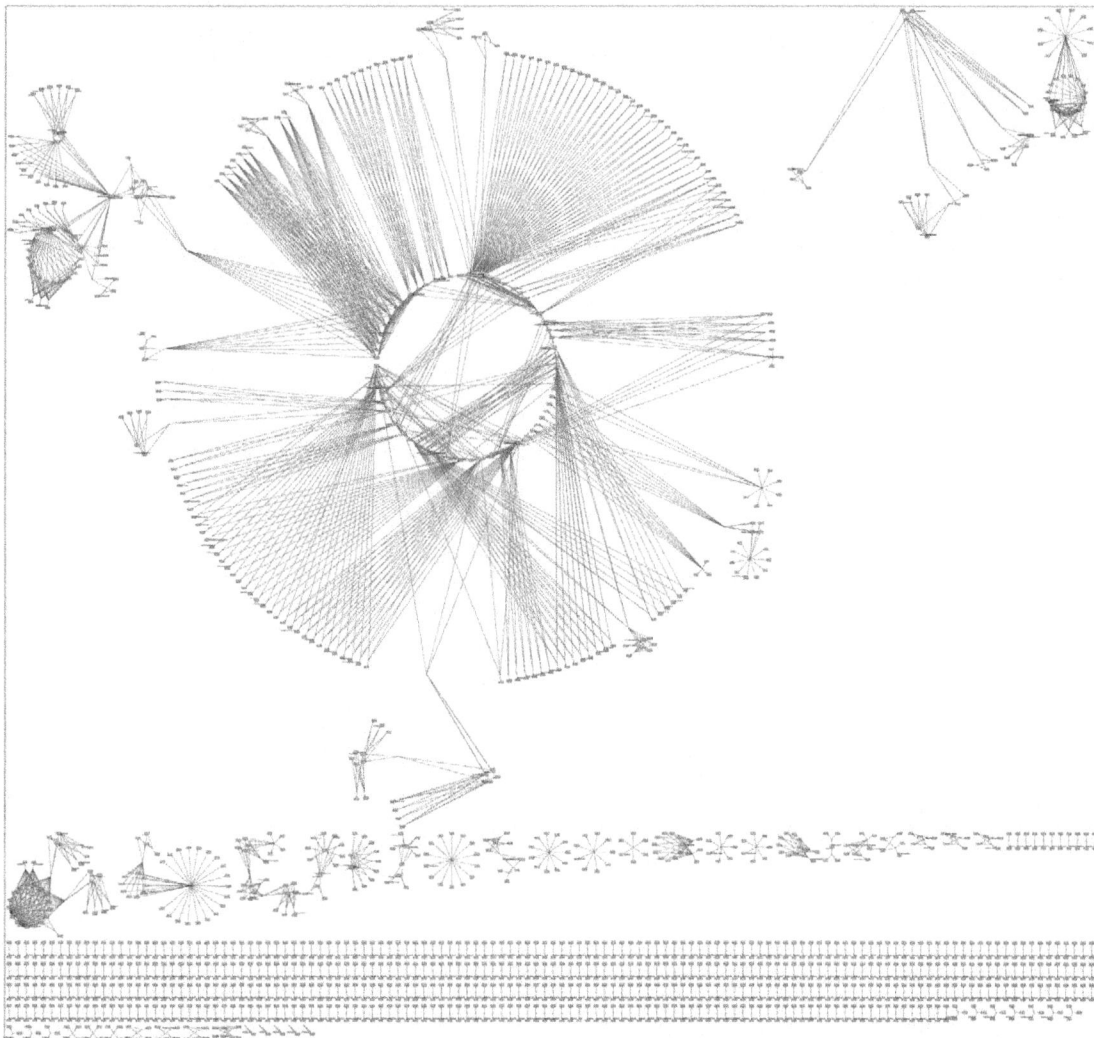

Fig. 6. Screenshot of the merged signaling network imported in *CellDesigner* in SBML format.

visualization and it is not integrated with any pathway enrichment analysis.

PathwayLinker [32] is an online tool that can provide estimates for possible signaling effects of modifying single proteins or selected groups of proteins (e.g., by microRNAs or as drug targets). It integrates protein–protein interaction and signaling pathway data from several sources, thus reducing any manual literature mining, and provides visualization features. It differs from *KENeV* in that it does not perform KEGG pathway enrichment calculations and pathways prioritization, and does not provide visualizations for the metabolic network.

With respect to the pathway over-representation analysis, there are tools that include the KEGG pathways as canonical pathway backgrounds, such as *DAVID* [33], *GeneTrail* [11] and *Webgedstalt* [34]. A direct comparison of *KENeV* to other gene-enrichment tools given the results/lists of KEGG pathways is out of the scope of this manuscript and could not be adequately informative for a thorough comparison. Different tools may provide different or re-arranged lists of pathways due to the different timing in updating the KEGG database (*KENeV* updates on a weekly basis the KEGG database). In terms of functionality, we note here that the following derived from comparing *KENeV* to *DAVID*, *Webgestalt* and *GeneTrail*: *KENEV* does not require a least number of genes per KEGG pathway since bootstrapping excludes such pathways from being reported in the final list of pathways. *KENeV* provides the bootstrapping method instead for providing the user the choice for a series of *p*-value correction methods (user could be unfamiliar to such methods). Importantly, *KENEV* merges pathways into super-pathways and provides all visualization features described. However, we provide in an .xls file (provided as Supplementary data) the lists of significantly enriched KEGG pathways derived when the list of 407 genes associated to Type I Diabetes Mellitus (submitted to *KENeV* as described in the Results section) was uploaded to *DAVID*, *Webgestalt* and *GeneTrail* tools. It can be observed that *KENeV* provides a shorter list of pathways, which is however similar in the firstly prioritized KEGG pathways to the longer ones obtained by the other tools.

Overall, the majority of the aforementioned tools, except from *KEGGgraph*, *KEGG Converter* and *PathwayLinker*, cannot perform multiple pathway integration into a merged network. Another drawback, specifically concerning users without any experience in programming, is that some tools are available as programming packages or simple scripts, instead of automated web applications, in contrast to *KENeV*, which enables the automated reconstruction of both signaling and metabolic enriched networks, using as only input a list of gene identifiers with their respective fold changes.

4.2. Future work

In the current version, the application is oriented towards visualization and therefore does not deliver stoichiometric-enabled SBML files for simulation purposes. Nevertheless, the KEGG maps are themselves designed for visualization purposes and hence they are not consistent for simulation. Notably, they contain gaps whereas they do not contain neither the necessary stoichiometry, nor the reactions' cofactors, which are essential for the oxidoreduction and energy balance. This information is however separately available in the *KEGG Reaction* database and it will be used for a future implementation of the application that will derive a simulation-enabled metabolic SMBL file, in addition to the current ones that are enabling visualization. Important steps towards this direction are the effective handling of inconsistencies in matter of pathway gaps, reaction mass balance and missing co-factor information.

Additionally, a merged metabolic/signaling hyper-network can be derived from the integration of metabolic and signaling sub-networks, provided the overcoming of a number of challenges, with respect to the expected high network complexity, especially the handling of complex protein–protein interactions presenting high connectivity.

Finally, more interaction databases, for instance *MSigDB* [2] and *Reactome* [35], will be exploited in order to enrich the pathway information.

5. Conclusion

Pathway enrichment analysis and visualization approaches provide valuable tools for the interpretation of genomics experimental results, typically characterized by increased dimensionality and complex connectivity. From this perspective, the *KENeV* web application presented here enables the automatic execution of a workflow that combines pathway enrichment analysis, reconstruction of the enriched metabolic and signaling networks, and finally instant interactive visualization of the obtained networks. In conclusion, the application provides an effective tool for the interpretation of the underlying biological mechanisms, which are detected by genomic experiments that derive a subset of genes with differential expression among various conditions.

Acknowledgments

The presented work in this article has been cofunded by European and Greek National Funds, through the Greek National Strategic Reference Framework Operational Program "Competitiveness and Entrepreneurship", Action COOPERATION (09ΣYN-11-675), entitled "PIK3CA Oncogenic Mutations in Breast and Colon Cancers: Development of Targeted Anticancer Drugs and Diagnostics" (POM), and (09ΣΥN-21-1078) entitled "Development of novel Angiogenesis-Modulating Pharmaceuticals by screening of natural compounds and synthetic analogues" (DAMP), and the National Operational Program "Education and Lifelong Learning" Action THALIS – University of Patras (MIS:377001),entitled "Development of Systems Biology and Bioinformatics Tools to Study the Dynamics of Cellular Aging" (MAESTRO).

References

[1] Lopez-Maury L, Marguerat S, Beahler J. Tuning gene expression to changing environments: from rapid responses to evolutionary adaptation. Nat Rev Genet 2002;9(8): 583–93.
[2] Kitano H. Computational systems biology. Nature 2002;420(6912):206–10.
[3] Subramanian A, et al. Gene set enrichment analysis: a knowledge-based approach for interpreting genome-wide expression profiles. Proc Natl Acad Sci U S A 2005; 102(43):15545–50.
[4] Zheng Q, Wang XJ. GOEAST: a web-based software toolkit for Gene Ontology enrichment analysis. Nucleic Acids Res 2008;36:W358–63 (Web Server issue).
[5] Huang da W, Sherman BT, Lempicki RA. Bioinformatics enrichment tools: paths toward the comprehensive functional analysis of large gene lists. Nucleic Acids Res 2009;37(1):1–13.
[6] Ashburner M, et al. Gene ontology: tool for the unification of biology. The gene ontology consortium. Nat Genet 2000;25(1):25–9.
[7] Kanehisa M, Goto S. KEGG: Kyoto encyclopedia of genes and genomes. Nucleic Acids Res 2000;28(1):27–30.
[8] Khatri P, Sirota M, Butte AJ. Ten years of pathway analysis: current approaches and outstanding challenges. PLoS Comput Biol 2012;8(2):e1002375. http://dx.doi.org/ 10.1371/journal.pcbi.1002375.
[9] Draghici S, Khatri P, Bhavsar P, Shah A, Krawetz SA, Tainsky MA. Onto-tools, the toolkit of the modern biologist: onto-express, onto-compare, onto-design and onto-translate. Nucleic Acids Res 2003;31(13):3775–81.
[10] Zeeberg BR, Feng W, Wang G, Wang MD, Fojo AT, Sunshine M, et al. GoMiner: a resource for biological interpretation of genomic and proteomic data. Genome Biol 2003;4(4):R28. http://dx.doi.org/10.1186/gb-2003-4-4-r28.
[11] Backes C, Keller A, Kuentzer J, Kneissl B, Comtesse N, Elnakady YA, et al. GeneTrail — advanced gene set enrichment analysis. Nucleic Acids Res 2007;35:W186–92. http://dx.doi.org/10.1093/nar/gkm323 [Web Server issue].
[12] Mitrea C, Taghavi Z, Bokanizad B, Hanoudi S, Tagett R, Donato M, et al. Methods and approaches in the topology-based analysis of biological pathways. Front Physiol 2013;4(278). http://dx.doi.org/10.3389/fphys.2013.00278.
[13] Draghici S, Khatri P, Tarca AL, Amin K, Done A, Voichita C, et al. A systems biology approach for pathway level analysis. Genome Res 2007;17(10):1537–45. http://dx. doi.org/10.1101/gr.6202607.
[14] Rahnenfuhrer J, Domingues FS, Maydt J, Lengauer T. Calculating the statistical significance of changes in pathway activity from gene expression data. Stat Appl Genet Mol Biol 2004;3.
[15] Chatziioannou AA, Moulos P. Exploiting statistical methodologies and controlled vocabularies for prioritized functional analysis of genomic experiments: the StRAnGER web application. Front Neurosci 2011;5:8.

[16] Pilalis E, Chatziioannou A. Prioritized functional analysis of biological experiments using resampling and noise control methodologies. 13th IEEE International Conference on BioInformatics and BioEngineering, IEEE BIBE 2013; 2013.

[17] Nelson TH, Jung J-Y, DeLuca TF, Hinebaugh BK, St Gabriel KC, Wall DP. Autworks: a cross-disease network biology application for autism and related disorders. BMC Med Genomics 2012;5(56). http://dx.doi.org/10.1186/1755-8794-5-56.

[18] Efron B. Bootstrap methods: another look at the jackknife. Ann Stat 1979;7(1):1–26.

[19] Hucka M, et al. The systems biology markup language (SBML): a medium for representation and exchange of biochemical network models. Bioinformatics 2003;19(4): 524–31.

[20] Bornstein BJ, et al. LibSBML: an API library for SBML. Bioinformatics 2008;24(6): 880–1.

[21] Smoot ME, et al. Cytoscape 2.8: new features for data integration and network visualization. Bioinformatics 2011;27(3):431–2.

[22] Matsuoka Y, et al. Modeling and simulation using Cell Designer. Methods Mol Biol 2014;1164:121–45.

[23] Tornatore L, Thotakura AK, Bennett J, Moretti M, Franzoso G. The nuclear factor kappa B signaling pathway: integrating metabolism with inflammation. Trends Cell Biol 2012;22:557–66.

[24] Vanhaesebroeck B, Stephens L, Hawkin P. PI3K signalling: the path to discovery and understanding. Nat Rev Mol Cell Biol 2012;13(3):195–203.

[25] Funahashi A, Jouraku A, Kitano H. Converting KEGG pathway database to SBML. 8th Annual International Conference on Research in Computational Molecular Biology (RECOMB 2004); 2004.

[26] Luo W, Brouwer C. Pathview: an R/Bioconductor package for pathway-based data integration and visualization. Bioinformatics 2013;29(14):1830–1.

[27] Zhang JD, Wiemann S. KEGGgraph: a graph approach to KEGG PATHWAY in R and bioconductor. Bioinformatics 2009;25(11):1470–1.

[28] Arakelyan A, Nersisyan L. KEGGParser: parsing and editing KEGG pathway maps in Matlab. Bioinformatics 2013;29(4):518–9.

[29] Wrzodek C, Drager A, Zell A. KEGGtranslator: visualizing and converting the KEGG PATHWAY database to various formats. Bioinformatics 2011;27(16):2314–5.

[30] Kono N, et al. Pathway projector: web-based zoomable pathway browser using KEGG atlas and Google Maps API. PLoS One 2009;4(11):e7710.

[31] Moutselos K, et al. KEGGconverter: a tool for the in-silico modelling of metabolic networks of the KEGG Pathways database. BMC Bioinf 2009;10:324.

[32] Farkas IJ, Szanto Varnagy A, Korcsmaros T. Linking Proteins to Signaling Pathways for Experiment Design and Evaluation. PLoS One 2012;7(4):e36202. http://dx.doi.org/10.1371/journal.pone.0036202.

[33] Huang DW, Sherman BT, Lempicki RA. Systematic and integrative analysis of large gene lists using DAVID bioinformatics resources. Nat Protoc 2009;4(1):44–57.

[34] Wang J, Duncan D, Shi Z, Zhang B. WEB-based GEne SeT AnaLysis Toolkit (WebGestalt): update. Nucleic Acids Res 2013;41:W77–83. http://dx.doi.org/10.1093/nar/gkt439 (Web Server issue).

[35] Matthews L, et al. Reactome knowledgebase of human biological pathways and processes. Nucleic Acids Res 2009;37(Database issue):D619–22.

17

Effect of mutations on the thermostability of *Aspergillus aculeatus* β-1,4-galactanase

Søs Torpenholt [a,1], Leonardo De Maria [b,2], Mats H.M. Olsson [a], Lars H. Christensen [b], Michael Skjøt [b,1], Peter Westh [c], Jan H. Jensen [a], Leila Lo Leggio [a,*]

[a] *Department of Chemistry, University of Copenhagen, Universitetsparken 5, 2100 Copenhagen, Denmark*
[b] *Novozymes A/S, Smørmosevej 25, 2880 Bagsværd, Denmark*
[c] *NSM, Research Unit for Functional Biomaterials, University of Roskilde, Universitetsvej 1, 4000 Roskilde, Denmark*

ARTICLE INFO

ABSTRACT

Keywords:
β-1,4-galactanase
Thermostability
Computational prediction
Protein design
GH53

New variants of β-1,4-galactanase from the mesophilic organism *Aspergillus aculeatus* were designed using the structure of β-1,4-galactanase from the thermophile organism *Myceliophthora thermophila* as a template. Some of the variants were generated using PROPKA 3.0, a validated pK_a prediction tool, to test its usefulness as an enzyme design tool. The PROPKA designed variants were D182N and S185D/Q188T, G104D/A156R. Variants Y295F and G306A were designed by a consensus approach, as a complementary and validated design method. D58N was a stabilizing mutation predicted by both methods. The predictions were experimentally validated by measurements of the melting temperature (T_m) by differential scanning calorimetry. We found that the T_m is elevated by 1.1 °C for G306A, slightly increased (in the range of 0.34 to 0.65 °C) for D182N, D58N, Y295F and unchanged or decreased for S185D/Q188T and G104D/A156R. The T_m changes were in the range predicted by PROPKA. Given the experimental errors, only the D58N and G306A show significant increase in thermodynamic stability.

Given the practical importance of kinetic stability, the kinetics of the irreversible enzyme inactivation process were also investigated for the wild-type and three variants and found to be biphasic. The half-lives of thermal inactivation were approximately doubled in G306A, unchanged for D182N and, disappointingly, a lot lower for D58N. In conclusion, this study tests a new method for estimating T_m changes for mutants, adds to the available data on the effect of substitutions on protein thermostability and identifies an interesting thermostabilizing mutation, which may be beneficial also in other galactanases.

1. Introduction

1.1. β-1,4-galactanases

Endo-β-1,4-galactanases (EC 3.2.1.89) are glycoside hydrolases which hydrolyse the β-1,4-O-glycosidic linkages in β-1,4-galactan and arabinogalactan type I, parts of pectin found in the non-woody plant cell wall [1]. In the carbohydrate active enzyme database (CAZY) [2] they are classified in GH53. β-1,4-Galactanases have received some

attention in industrial applications as demonstrated by a number of patents in the 1990s/2000s (eg [3,4]) and recent renewed interest in the academic literature for their potential in producing prebiotic oligosaccharides [5]. In several bacteria, for example the thermophile *Geobacillus stearothermophilus*, *Erwinia chrysanthemi*, *Bacillus subtilis* and *Bifidobacterium breve* [6–10], β-1,4-galactanase genes have been shown to be part of gene clusters involved in galactan utilization, consisting, in addition to the endo-galactanase, of a GH42 β-galactosidase, a galactooligosaccharide transport system, and a transcriptional regulator.

Both fungal and bacterial β-1,4-galactanases have been characterized biochemically and crystallographically. Fungal galactanases tend to release shorter products and are able to hydrolyse to some extent small oligosaccharides and small chromogenic substrates [11,12] while the investigated bacterial enzymes tend to release larger oligosaccharides, have a more extended substrate binding groove and cannot act on small substrates, presumably because of non-productive substrate

Abbreviations: AaGal, β-1,4-galactanase from *Aspergillus aculeatus*; AZCL-galactan, azurine-crosslinked galactan; CAZY, carbohydrate active enzyme database; DSC, differential scanning calorimetry; MtGal, β-1,4-galactanase from *Myceliophthora thermophila*; Tm, melting temperature; TsGal, *Talaromyces stipitatus* galactanase; WT, wild type.

 * Corresponding author
   E-mail address: leila@chem.ku.dk (L. Lo Leggio).
[1] Novozymes A/S, Smørmosevej 25, 2880 Bagsværd, Denmark.
[2] Novo Nordisk A/S, Novo Nordisk Park, 2760 Måløv Denmark.

binding [6,11,12]. Several 3D structures have been elucidated for GH53 β-1,4-galactanases, among which a bacterial galactanase [11], while for fungal galactanases these represent both mesophilic and thermophilic enzymes [13–15].

1.2. Engineering thermostability

Protein engineering and site-directed mutagenesis are used routinely to establish biological function and role of amino-acid residues in proteins, like the stability. The protein stability can be changed significantly by single or multiple mutations of specific amino acids, sometimes resulting in a beneficial effect. This type of work is however expensive and time-consuming so rapid prediction of how mutations might affect the stability of a protein is desirable. An important parameter in determining the protein stability, including thermostability, is the determination of the pK_a value of all the ionizable residues in a protein, which however, experimentally, is also challenging and time consuming. Computational pKa prediction programs have therefore been developed [16]. One of these programs is PROPKA [17], that utilize a very fast empirical approach of pKa prediction. PROPKA is today one of the most widely used programs due to its ease of use and speed, at the same time yielding accurate results as compared to other programs [18]. Details of the method and implementation in PROPKA 3.0, the version used in this study, can be found elsewhere [19–21].

1.3. Aim of this study

Within the scope of this article, we pursue two goals. First of all, we investigate if PROPKA can be applied to design thermostable proteins and predict the magnitude of the gain in thermostability upon mutation. We use as model system the mesophilic *Aspergillus aculeatus* β-1,4-galactanase (AaGal, GenBank ID AAA32692.1, PDB IDs: 1FHL, 1FOB; [13]) using as template for mutation *Myceliophthora thermophila* (also known as *Thielavia heterothallica*) β-1,4-galactanase (MtGal, GenBank ID AAE73520.1, PDB ID: 1HJS [14]). We pursue therefore also the secondary goal of obtaining a more thermostable AaGal. A consensus approach [22,23] was used as a complementary design tool to obtain more thermostable variants, which were also evaluated by PROPKA. The thermostability of these variants together with that of the wild type enzyme was experimentally determined by their melting temperature, T_m, and thermal irreversible inactivation rate.

2. Material and methods

2.1. Cloning, expression and purification of enzymes

Cloning, expression and purification of WT AaGal is described in detail in [24]. The enzyme was more than 95% pure as judged from the SDS-PAGE gel.

Variants of AaGal were made using the original expression vector [24] as template. Mutations were introduced by PCR using a method essentially as the Quick Change Site Directed Mutagenesis Kit (Roche Applied Science, Indianapolis, IN). The primers used for mutagenesis are shown in Table 1. G104D/A156R was generated by consecutive

rounds of PCR. All constructs were sequence verified prior to transformation into *Aspergillus oryzae*. The expression plasmids were transformed into *Aspergillus* as described in [25]. For each of the constructs 4–6 strains were isolated, purified and cultivated in microtiter plates. Expression was determined using the AZCL-galactan assay as described in Section 2.3 and the best producing strain was fermented in shake flasks.

The variants of AaGal were purified by hydrophobic interaction as essentially described in [14]. Ammonium sulphate (1 M), sodium chloride (1 M) and buffer (50 mM sodium acetate buffer, pH 4.5) were added to the culture filtrate and the pH adjusted to 4.5. The culture filtrate was then loaded on a Toyopearl Butyl 650 M from Tosoh. A gradient was applied from this loading buffer to 100% buffer B (50 mM sodium acetate buffer, pH 4.5) over ten column volumes, during which the enzymes eluted. The variants D58N, D182N, G104D/A156R were purified further by anion exchange using Q-sepharose 16/10 with a gradient of 0 to 1 M sodium chloride in 50 mM sodium acetate buffer, pH 4.5. 5 mL fractions were collected and the ones with galactanase activity (AZCL-galactan assay) were pooled and checked for purity by SDS-PAGE. Fractions with estimated 95% or more purity were used in further work.

2.2. Protein concentration and electrophoresis

The protein concentration was determined by absorbance measurements at 280 nm using a ND-1000 spectrophotometer from Fischer scientific. The purity was judged by SDS-PAGE on polyacrylamide gels (4–20% from Biorad or 4–12% from Nupage) after heat denaturation in the presence of DTT. Gels were stained/destained with Instant Blue from Expedeon.

2.3. Enzyme assays

2.3.1. AZCL galactan

This semiquantitative assay was used to identify expressing transformants and galactanase containing fractions. The assay was carried out in a micro titer plate. 20 µL sample (culture broth or elution fractions from purification, or MilliQ water as blank) and 200 µL 4 mg/mL AZCL-galactan (lupin) (from Megazyme International Ireland.) in 0.1 M sodium acetate buffer, pH 4.0, containing 0.01% Triton X-100, was added to each well and incubated for 10 min at room temperature with shaking and then 5 min at room temperature without shaking, to let the insoluble particles settle. The absorbance at 595 nm was then measured on 100 µL of the supernatant without stopping the reaction.

2.3.2. Azo-galactan

Azo-galactan is partly debranched potato galactan dyed with Remazol Brilliant Blue R (about one dye molecule per 20 sugar residues) and was purchased from Megazyme International Ireland. The assay was carried out essentially using the manufacturer's procedure (described in Megazyme booklet AGALP 11/99) with 10 µL pre-equilibrated enzyme solution (0.1 mg/mL in 50 mM sodium acetate buffer pH 4.5), 500 µL pre-equilibrated substrate solution (2% w/v azo-galactan in 50 mM sodium acetate buffer, pH 4.5) and 490 µL pre-heated buffer (50 mM sodium acetate buffer pH 4.5) in Eppendorf tubes. The reaction was incubated for 10 min at 40 °C and stopped by addition of 2.5 mL 95% ethanol with vigorous stirring, which will precipitate high molecular weight galactan. After incubation at room temperature for 10 min, the tubes were centrifuged for 10 min at 1800 g, and the A_{590} of the supernatant measured. A reaction mixture to which 10 µL buffer was added instead of enzyme solution was used as blank. The A_{590} is correlated to the extent of small molecular weight dyed galactan fragments released, which is linearly correlated to the amount of reducing sugars released. All azo-galactan assay measurements were made in triplicates.

Table 1
Primers used to construct the described variants.

D182N	atgatccatttggacAATggctggagctggg	Gtccaaatggatcatgatcttgggg
S185D, Q188T	ttggacgacggctggGATtgggatACCcaga actacttttacga	Ccagccgtcgtccaaatggatcatg
G104D	tccacgaccgatctcGATactttgaaatggc	gagatcggtcgtggaccagccgaa
A156R	tattcgaacatcggcAGActgctgcactcg	Gccgatgttcgaatagctgctcgtt
D58N	agcgacggcagctacAATctggactacaa	gtagctgccgtcgctggggttcacc
Y295F	gatggtctgggagtgTTCtattgggagcc	cactcccagaccatcagtagtggcc
G306A	tggatcggcaatgctGCCttgggttcgag	agcattgccgatccaggctggctcc

2.4. Irreversible thermal inactivation rate

The rate of irreversible thermal inactivation was determined at 55 °C by incubation of 0.1 mg/mL enzyme. Aliquots were taken out at appropriate intervals and immediately cooled on ice. Activity was measured using the azo-galactan assay. A is the activity remaining after heating for time t, while A_0 is the original activity for each variant without incubation at 55 °C. The inactivation rate was determined by non-linear regression of the A vs time (min) curves to the equation of a biphasic deactivation process (see Results for additional details).

2.5. Differential scanning calorimetry (DSC)

All protein samples were buffer changed into 50 mM sodium acetate buffer, pH 4.5 using a PD-10 (Sephadex™ G-25) column or a 10.000MWCO slide-A-lyser dialysis cassette from Pierce and diluted to 0.2 or 0.5 mg/mL with 50 mM sodium acetate buffer, pH 4.5, degassed for 10 min and equilibrated for 10 min, before each DSC experiment. The DSC experiments were performed using a nano DSC III from TA instruments with cell volumes of 300 μL. The heating scan rate was 60 °C/h. The calorimetric reversibility of the thermally induced transition was checked by reheating the protein solution several times in the calorimetric cell after cooling at 120 °C/h. The heating temperatures were chosen so they covered the endothermic peak (see below for details).

Measurements were carried out in triplicate for WT and in duplicate for the variants. Raw data was analysed with NanoAnalyze Data Analysis version 2.0.1 from TA Instruments by subtracting the buffer/buffer scan from the protein/buffer scan of each experiment and normalized regarding protein concentration. The data were fitted directly to the equation of an irreversible unfolding model enabling calculation of an apparent melting temperature (T_m) (see below for more details).

2.6. Identification of mutation sites by PROPKA

In order to screen possible mutations we used PROPKA to identify stabilizing motifs that were present in the thermophilic template but not in the mesophilic variant. PROPKA predicts protein pK_a values, and can therefore be used to examine the electrostatic part of the protein folding free energy, $\Delta G^{uf \to f}$. This is related to shifts in protein pK_a values because the interactions that stabilize the folded protein compared to the unfolded protein are the same interactions that shifts the pK_a values of the residues in the unfolded state to its folded state. The total electrostatic protein unfolding free energy can be made into a thermodynamic cycle where the unfolding free energy is calculated as

$$\Delta G^{f \to uf}(pH) = \Delta G^{f \to uf}_{neutral} + \Delta\Delta G^{f \to uf}_{elstat}\left(pH_{ref}\right) + \Delta\Delta G^{f \to uf}_{elstat}\left(pH_{ref}, pH\right). \quad (1)$$

The first term is the unfolding free energy when all ionizable residues are in their neutral form, the second term is the unfolding contribution when ionizable residues are charged to a reference pH value, pH_{ref}, and the third term is the unfolding contribution when changing the charge of the ionizable residues according to an actual pH value. In this particular case we use a pH_{ref} where all acids are uncharged and all bases are fully charged, which makes the contribution from the second term in the equation simple, and focus on the last term. The last term in the equation is the pH dependent term and can be calculated from the difference in charge between the folded and unfolded protein, $\Delta Q(pH)$ [26], as

$$\Delta\Delta G^{f \to uf}\left(pH_{ref}, pH\right) = 2.3RT \int_{pH_{ref}}^{pH} \Delta Q(pH)\, dpH. \quad (2)$$

The protein pK_a values determine the integrand charge difference, T is the temperature, and R is the gas constant. This charge difference between the folded and unfolded protein is calculated for all protein pK_a values by PROPKA.

PROPKA relates a protein pK_a value to its 3D protein structure. The pK_a values of all the ionizable residues in a protein are calculated by determining perturbations, ΔpK_a, to a reference water pK_a value, pK_a^{model}. Perturbations from desolvation (ΔpK_a^{DS}), protein dipole interactions ($\Delta pK_a^{Q\mu}$), and charge-charge interactions (ΔpK_a^{QQ}) based on the position and orientation of groups surrounding each ionizable residue are included. Thus, the protein pK_a value is expressed as

$$pK_a^{protein} = pK_a^{model} + \Delta pK_a^{model \to protein} = pK_a^{model} + \Delta pK_a^{DS} + \Delta pK_a^{Q\mu} + \Delta pK_a^{QQ}. \quad (3)$$

The empirical equations used in this expression are described in [17] and its later modifications in [19].

The calculated $\Delta G^{f \to uf}$, an indication of the change in thermostability, can then be converted to T_m using Gibbs' relationship.

$$\Delta G^{f \to uf} = \Delta H - T_m \cdot \Delta S. \quad (4)$$

We use the approximate approach of Robertson and Murphy [27] to relate the protein entropy, ΔS, with the number of protein residues. Once we have an estimate for ΔS, we can obtain ΔH from the calculated $\Delta G^{f \to uf}$ at ambient temperature (298 K), and thereby estimate T_m as the temperature where $\Delta G^{f \to uf} = 0$.

The $\Delta G^{f \to uf}$ of the template MtGal was obtained from protein pK_a values predicted with PROPKA using the published crystal structure (PDB code 1HJS and 1.87 Å resolution, [14]). All configurations in the crystallographic asymmetric unit and side-chain alternative conformations (4 × 2 for MtGal) were taken into account. These pK_a values were used to identify stabilizing groups of residue in MtGal. The groups (including their interaction partners) were then subsequently 'moved combinatorially' into AaGal by *in silico* modelling and the total $\Delta G^{f \to uf}$ of the mutated protein was re-evaluated. Mutants with significant elevated $\Delta G^{f \to uf}$ were constructed by site-directed mutagenesis.

2.7. Identification of mutation sites by a consensus approach

The multiple sequence homology alignment was performed by first collecting all the sequences of AaGal homologues of 40% or higher sequence identity. Subsequently, a multiple sequence alignment of all the so obtained sequences was created using ClustalW2 [28], the alignment columns were automatically scanned and the positions where the original AaGal amino acid was under represented identified. Mutants of AaGal were constructed to make a specific position more alike the majority of known homologues.

3. Results

3.1. Generation of mutants

The temperature optima for activity are 50 °C at pH 3.5 for AaGal and 65 °C at pH 6.5 for MtGal, despite a high sequence identity of 57% and correspondingly high structural similarity [14]. Residues that potentially contribute to a higher melting temperature, T_m, in MtGal were identified with PROPKA. Additional mutations were identified by a consensus method. The mutations in AaGal identified with PROPKA were D182N, S185D/Q188T, G104D/A156R, while Y295F and G306A were identified by consensus. D58N was predicted to be stabilizing by both methods. These variants were constructed by site-directed mutagenesis. All variants had at least half of wild type activity as measured by the azo-galactan assay, see Table 2.

3.2. Determination of thermodynamic stability

The thermal stability of a protein is typically defined by its melting temperature, T_m, which is the temperature where the transition between the folded and unfolded form of a protein is most pronounced, defined as the maximum point of the heat capacity curve. The PROPKA predicted change in T_m of each single variant is between 0.37 and 1.1 °C, listed in Table 2. Since the small changes in T_m constitute an experimental challenge, DSC was chosen as a technique to measure thermostability. The heat capacity is measured as a function of temperature. All enzymes showed one unfolding peak in the DSC experiments.

In many proteins an irreversible denaturation upon heating is observed. The reversibility of denaturation was therefore checked by reheated scan; a heated sample is cooled and then reheated with the same rate. The end-temperature of each repeated scan was gradually increased, to include the entire range of the endothermic peak. The original curves were reproduced until T_m was reached. This indicates a partially reversible denaturation of AaGal. The irreversible denaturation is most likely a result of aggregation during the heating process. Aggregates were observed in recovered samples.

The irreversible denaturation is believed to follow a reversible step described by the Lumry and Eyring model [29]:

$$N \underset{k_2}{\overset{k_1}{\rightleftharpoons}} U \overset{k_3}{\rightarrow} I.$$

The first step is a reversible unfolding of the native catalytically active protein, followed by an irreversible alteration of the unfolded protein to give the final irreversible state. If the rate constant k_3 of the second irreversible step U → I is much higher than the rate constant k_2 of the reversible step of U → N then most of the U molecules will be converted to I, reducing the Lumry and Eyring model to:

N→I.

The denaturation process can then be considered as a one-step process following first-order kinetics. The excess heat capacity determined by DSC can for this process be described as Eq. (5) [30,31]:

$$C_p^{ex} = \frac{\Delta H E_a}{RT_m^2} exp\left(\frac{\Delta H(T-T_m)}{RT_m^2}\right) \times exp\left(-exp\left(\frac{\Delta H(T-T_m)}{RT_m^2}\right)\right). \quad (5)$$

Thus by fitting the DSC curves to this equation it was possible to obtain the T_m value. The T_m is an apparent T_m-value, which can here be taken as an informative value because irreversible aggregation was shown to be essentially limited to after the melting point. In our study the pH, buffer, scan rate, equilibration time, heating range were kept constant, enabling a comparison of the apparent T_m measured for wild type AaGal and its variants.

The recorded T_m-values are listed in Table 2 along with the PROPKA calculated change in T_m upon mutation. The effect on thermostability of individual mutation was found to be rather limited. The variants Y295F

and G306A designed by consensus, the variant D182N designed by PROPKA, and the variant D58N identified by both methods are stabilized compared to WT AaGal. The gain in stability is in the range predicted by PROPKA.

A typical example of recorded denaturation curve is displayed in Fig. 1 along with its curve fit in Fig. 2. The asymmetric appearance of the endothermic peak is characteristic for an irreversible denaturation process.

3.3. Thermal denaturation kinetics

Although thermodynamic stability is most easily related to theoretical calculations, the kinetic stability of proteins is often of higher practical significance in applications. Therefore the kinetic stability of stabilized variants was further investigated. The rate of irreversible thermal inactivation was determined for WT, D58N, D182N and G306A β-1,4-galactanase, variants with an observed increase in T_m of 0.5 or above, by measuring the residual activity at 55 °C after incubation. The resulting A vs time (min) curves are shown in Fig. 3. Inspection of semi-logarithmic plots such as the one displayed as example in the inset of Fig. 3 for the WT enzyme, shows that both for WT and variants the data cannot be fit by a single first order process. Rather the data display a characteristic pattern of a biphasic denaturation process, which can be interpreted as an intermediate as in [33]. The biphasic denaturation process can be described as:

$$N \overset{k_1}{\rightarrow} X \overset{k_2}{\rightarrow} D.$$

N is the native enzyme form, X is an intermediate, D the denatured enzyme form, k_1 and k_2 the rate constants. The specific activity (A) at time (t) can for this process be described as in [33]:

$$A = \left(A_1 - \frac{A_2 k_1}{k_1 - k_2}\right) exp(-k_1 t) + \left(\frac{A_2 k_1}{k_1 - k_2}\right) exp(-k_2 t). \quad (6)$$

A_1 and A_2 are the specific activities of the native enzyme N and the intermediate X, respectively. The rate constants were obtained by

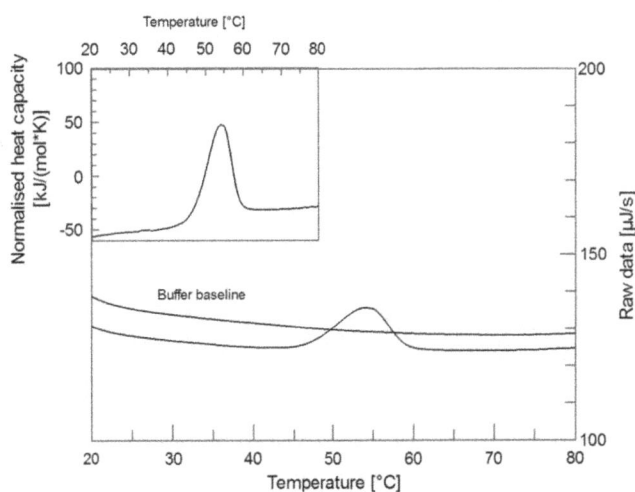

Fig. 1. Schematic DSC curves of 0.5 mg/ml WT AaGal in 50 mM sodium acetate buffer, pH 4.5 (sample) and buffer alone. Raw data [μ cal/s] was measured as a function of temperature [°C] with buffer/buffer and sample/buffer in the cells. The inset shows the appearance of a normalized DSC curve after subtraction of the buffer/buffer curve from the sample/buffer curve and protein concentration normalization. This figure was made with Grafit 7.0.0 [32].

Table 2
Predicted T_m, experimental T_m and relative activities of AaGal WT and variants. Activity was measured by the azo-galactan assay.

Variant	Activity (%)	DSC	PROPKA	DSC
		T_m (°C)	ΔT_m (°C)	ΔT_m (°C)
WT	100	60.87 ± 0.15	–	–
D58N	67	61.52 ± 0.08	0.75	0.65
Y295F	51	61.21 ± 0.34	0.41	0.34
D182N	88	61.40 ± 0.26	1.1	0.53
G306A	87	62.00 ± 0.01	0.37	1.1
S185D/Q188T	67	60.98 ± 0.27	0.57	0.11
G104D/A156R	82	60.69 ± 0.03	1.1	−0.18

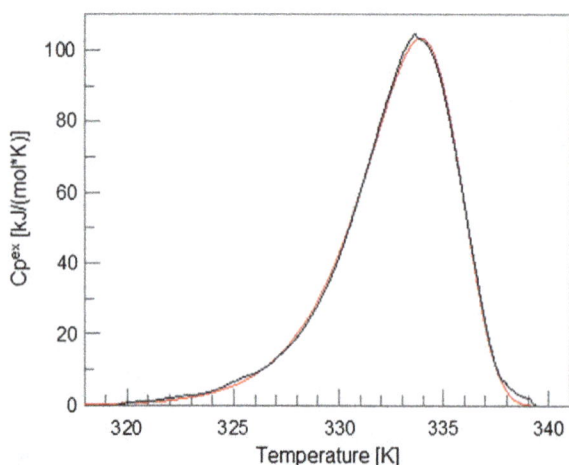

Fig. 2. Fit of a DSC thermogram to an irreversible model. The fit line is given in red and the experimental data in black. The experimental data given for 0.5 mg/mL WT AaGal in 50 mM sodium acetate buffer, pH 4.5 has been normalized according to the enzyme concentration, subtracted the buffer scan values and integrated baseline. The fit was made with Grafit 7.0.0 [32].

non-linear fitting of the data to Eq. (6) in Grafit 7.0.0 [32]. They are shown in Table 3, together with the half-lives calculated by:

$$t_{1/2} = \frac{\ln(2)}{k}. \tag{7}$$

The inactivation rate constants k_1 and k_2 were 0.0390 min^{-1}, 0.193 min^{-1}, 0.0405 min^{-1}, 0.0158 min^{-1} and 0.0514 min^{-1}, 0.231 min^{-1}, 0.0510 min^{-1}, 0.0291 min^{-1} corresponding to half-lives $t_{1/2,1}$ and $t_{1/2,2}$ of 17.8 min, 3.6 min, 17.1 min, 43.9 min and 13.5 min, 3.0 min, 13.6 min, 23.8 min for WT, D58N, D182N and G306A, respectively, also listed in Table 3.

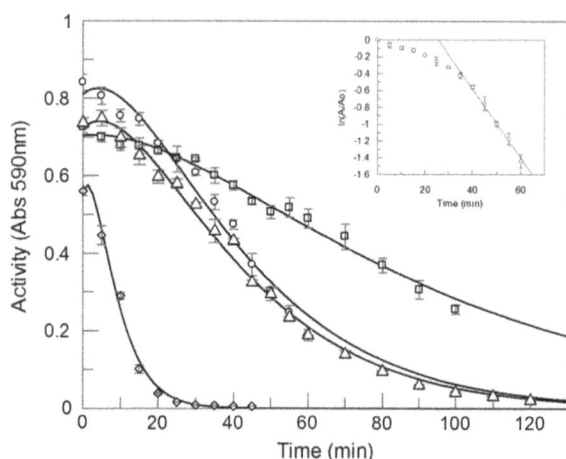

Fig. 3. Thermal inactivation of galactanases. WT AaGal is marked with circles, D182N with triangles, G306A with squares and D58N with diamonds. The enzymes were heated at 55 °C for variable times prior to activity determination by the azo-galactan method (see Materials and Methods section). Specific activity can be expressed as change in A_{590} since the same enzyme amount was used in each assay. In the inset, we show an example of semi-logarithmic plot for WT AaGal, clearly showing that the data cannot be reasonably modelled by a single first order rate constant (to visualize this, the latter part of the plot has been fit to a straight line, clearly showing that this is not a satisfactory model for all data). In the inset, A/A_0 is the residual activity (activity at time t divided by activity at time 0). The figure was made with Grafit 7.0.0 [32].

Table 3
Rate constants (k) and half-lives ($t_{1/2}$) of thermal inactivation of AaGal and variants.

Variant	k_1 (min^{-1})	k_2 (min^{-1})	$t_{1/2,1}$ (min)	$t_{1/2,2}$ (min)
WT	0.0390 ± 0.0050	0.0514 ± 0.0077	17.8	13.5
D58N	0.193 ± 0.029	0.231 ± 0.041	3.6	3.0
D182N	0.0405 ± 0.0028	0.0510 ± 0.0041	17.1	13.6
G306A	0.0158 ± 0.0012	0.0291 ± 0.0032	43.9	23.8

4. Discussion

4.1. Rational behind the designed variants

Elevated thermostability is often a desirable protein property, since many industrial applications require enzymes with activity at high temperatures. A great interest has therefore been in isolating proteins from thermophilic or hyperthermophilic organisms, to use directly for these applications or as template protein for protein engineering study. In this study we used the thermophilic Mtgal as template for PROPKA to design new variants of the mesophilic Aagal.

A consensus approach was chosen as complementary protein engineering tool. Mutations in AaGal generated with PROPKA were D182N, S185D/Q188T and G104D/A156R, while Y295F, G306A with the consensus approach, while D58N was identified by both. All mutated residues are shown in the alignment in Fig. 4. The target point residues of D182N, S185D/Q188T, G104D/A156R, D58N and Y295F are all represented in the MtGal. This is not the case for G306A. Gly306 in MtGal is an asparagine, also observed in β-1,4-galactanase from *Humicola insolens*, another thermophilic fungus, but is an Ala in the majority of sequences used for the consensus strategy which do not have a Gly, as shown in Fig. 4.

4.2. Predicted structural impact of the mutations

In order to illustrate the structural impact the mutations might have, the structures of AaGal and MtGal (used as a PROPKA design template) were superimposed, as displayed in Fig. 5. The superimposition reveals that the stabilizing effect of the mutation D182N and D58N might come from removal of the repulsive force between two aspartates (Fig. 5A) and (Fig. 5D), formation of a hydrogen bond in S185D/Q188T (Fig. 5B), formation of a coulombic interaction in G104D/A156R (Fig. 5C). Although suggested by a consensus strategy, the stabilizing effect for Y295F (Fig. 5E) is hard to rationalize in terms of structure, as the mutation is expected to remove a hydrogen bond between Oη in Tyr295 and Oγ in Ser43 in the core of the enzyme. However this mutation is also predicted to be slightly stabilizing by PROPKA. Also hard to fully rationalize in terms of structure is the stabilizing effect for G306A suggested by the consensus strategy. A general thermostabilizing effect of mutation of Gly, a flexibility inducing residue, to non-Gly, is often ascribed in the literature to a beneficial entropic effect (the entropy of the unfolded state is reduced) as first proposed by Matthews [35], so this could be an explanation. On the other hand there is no lack in the literature of studies showing mutation from non-Gly to Gly as strongly stabilizing, for example, a single Ser to Gly substitution increases the T_m of a *Clostridium thermocellum* endoglucanase by 7 °C [36]. An additional His to Gly thermostabilizing mutation was later identified for the same protein [37].

In AaGal, Gly306 is located in a loop region approximately 10 Å away from the catalytic active site residue Glu246 at the surface of the protein, see Fig. 5F. The loop also contains Cys310 involved in a disulphide bond to Cys253. One possible specific effect of mutation of Gly306 to an alanine could be to strengthen this disulphide bond by subtle structural alterations, thus improving the overall stability.

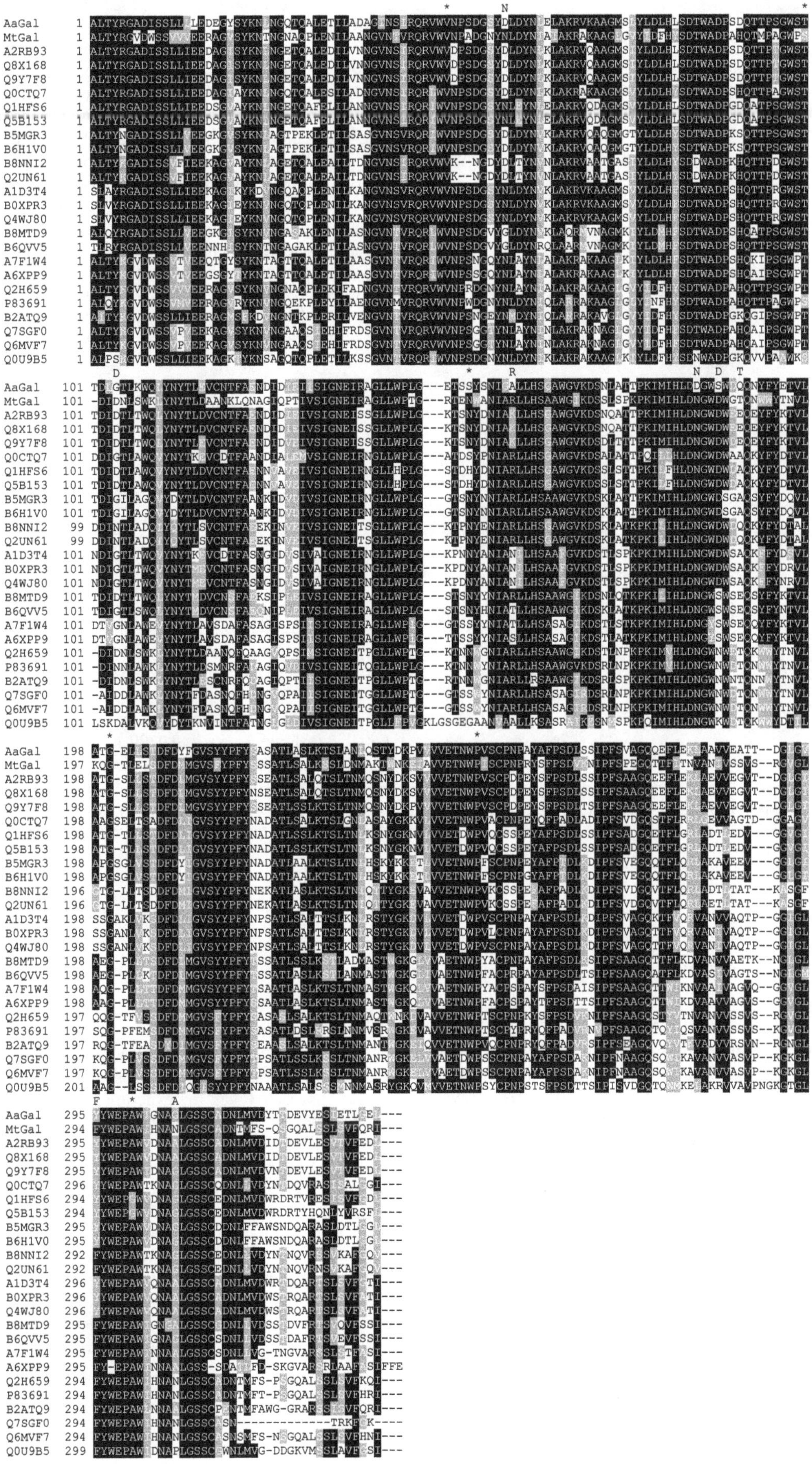

AaGal
MtGal
A2RB93
Q8X168
Q9Y7F8
Q0CTQ7
Q1HFS6
Q5B153
B5MGR3
B6H1V0
B8NNI2
Q2UN61
A1D3T4
B0XPR3
Q4WJ80
B8MTD9
B6QVV5
A7F1W4
A6XPP9
Q2H659
P83691
B2ATQ9
Q7SGF0
Q6MVF7
Q0U9B5

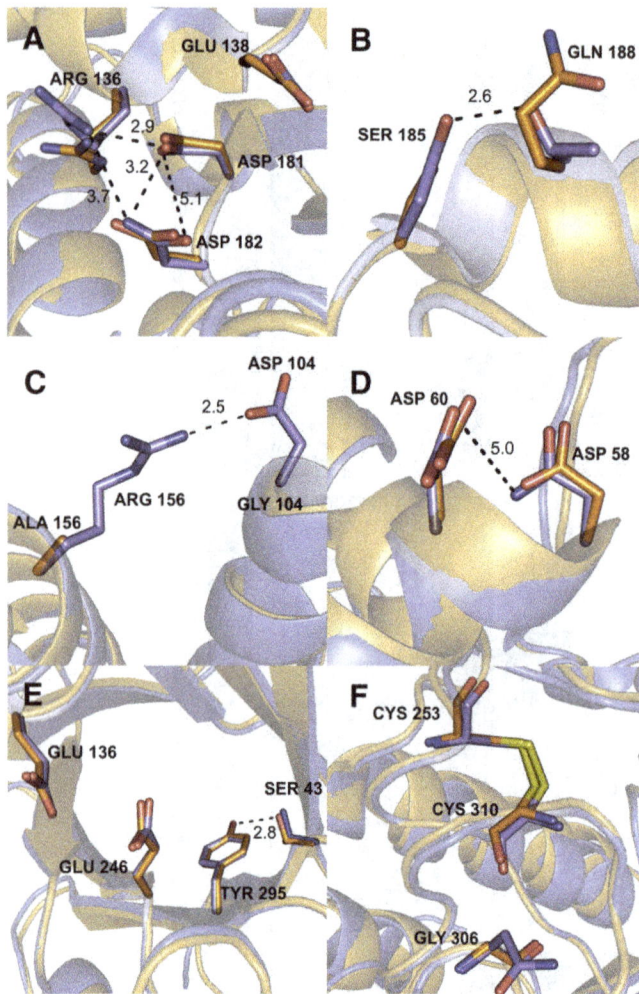

Fig. 5. Structural environment of the generated variants. A) D182N, B) S185D/Q188T, C) G104D/A156R, D) D58N, E) Y295F and F) G306A. AaGal (PDB code 1FOB) is coloured orange, MtGal (used as design and structural template, PDB code 1HJS) in blue). The black dashed lines show selected putative hydrogen bonds. The labelling is according to AaGal. The figure was made with Pymol [38].

4.3. Analysis of the change in stability upon mutation

DSC measurements on the wild-type enzyme and the four single mutants D182N, D58N, G306A and Y295F showed a slight increase in the T_m for the variants in correspondence with the PROPKA predicted values, see Table 2. The standard deviations were however in the range of 0.03 to 0.34 °C, emphasizing the difficulties in measuring very low ΔT_m in proteins. Taking into account these standard deviation we estimate that the increase in G306A is definitely significant, that the increase in T_m observed for D58N is probably significant, but that the increase observed for D182N and Y295F are within experimental error. These results do however show that PROPKA 3.0 might be a valuable tool in predicting the magnitude of the change in T_m upon mutation before experiments are taken on. Also, that the higher thermal stability of MtGal compared to AaGal obviously cannot be assigned to one specific residue, but is a combined effect of many small contributions.

The double mutants S185D/Q188T and G104D/A156R showed no change or decrease in T_m upon mutation, suggesting that the predicted hydrogen bond is not formed between Asp185 and Thr188 in S185D/Q188T, and neither is the predicted salt bridge between Asp104 and Arg156 in G104D/A156R, presumably due to unforeseen interactions with neighbouring residues. This emphasizes the importance of accurate predictions of mutant structures when redesigning enzymes, restricting the use of PROPKA 3.0 alone for this specific application.

There are ways to improve these predictions, such as relaxing the structure around the mutations with molecular dynamics, but the computational effort and general uncertainties increase as these techniques become more complex making it less of an attractive option for screening. A compromise can be reached by using scoring functions and only consider rotation around dihedral angles. This reduces the dimension of the problem, but it again requires accurate energy expressions for the protein structure and it is not obvious that it actually refines the geometry.

4.4. Analysis of the thermal inactivation of AaGal

WT AaGal and its variants G306A, D182N and D58N, with a ΔT_m of at least 0.5 °C were also compared with respect to their thermostabilities by recording time courses of the irreversible decay of enzyme activity at 55 °C. The inactivation temperature of 55 °C was chosen to obtain practically useful inactivation rates. T_m was recorded to be in the range of 60.87 to 62.00 °C. The inactivation time courses were obtained for the enzymes under identical conditions. The decrease of azogalactan activity (A) for each variant was plotted as a function of preincubation time. The so obtained activity vs time curves displayed a biphasic deactivation pattern, see Fig. 3, more clearly seen in semilogarithmic plots (shown in the inset of Fig. 3 for WT).

The rate constant k_1 at 55 °C is approximately 2.5 times lower and k_2 is approximately 1.8 times lower of G306A compared to that of WT AaGal, see Table 3, (k_1 is 0.0390 min^{-1} and 0.0158 min^{-1}, k_2 is 0.0514 min^{-1} and 0.0219 min^{-1} corresponding to half-lives of $t_{1/2,1}$ of 17.8 min and 43.9 min and $t_{1/2,2}$ of 13.5 min and 23.8 min, for WT and G306A, respectively), implying that the stability indeed has been increased upon mutation. The mutation of Asp182 to an asparagine did however not change the thermal irreversible inactivation rate, implying that this mutation do not alter the thermostability of the enzyme. The mutation of Asp58 to an asparagine, was observed to increase the rate constants k_1 and k_2 from 0.0390 min^{-1} and 0.00514 min^{-1} to 0.193 min^{-1} and 0.231 min^{-1} and thereby lowering the thermostability of the enzyme significantly.

5. Conclusions

In this study four single point mutations were identified giving a very small increase in T_m, one designed by PROPKA and three by a consensus approach. Given the experimental errors, the increases only seem significant for the D58N and G306A variants. However, only the G306A variant gives also rise to an increase in the lifetime of the enzyme at an elevated temperature. Interestingly one of the mutations which was mildly stabilizing in terms of T_m, D58N, shows a drastically reduced half-life, highlighting the importance of kinetic as well as thermodynamic stability for practical applications.

Some important lessons were learnt in the course of this study. Variants involving a double mutation, where the introduced residues were supposed to interact, failed to produce the desired effect, probably because the relatively simplistic approach used here does not correctly

Fig. 4. Multiple sequence alignment of fungal β-1,4-galactanases. The sequences are labelled using the UniProtKB database identifier (http://www.uniprot.org/) except for AaGal and MtGal (UniProtKB P48842 and P83692, respectively), which are indicated with their name. The signal peptide is not included in the sequences. Residues mutated in AaGal in this study (D58, G104, A156, D182, S185, Q188, Y295 and G306) are indicated by the new residue type above the mutated residue in the sequence of AaGal. Asterisks indicate residues 50, 100, 150, 200, 250 and 300 in the AaGal sequence. The alignment was generated with Clustal W2 [28] and visualized with Boxshade [34] with a 0.5 threshold.

predict the variant structures. Without applying more sophisticated approaches to structure prediction of variants, which are in themselves time consuming, we suggest that a combinatorial approach exploiting the often additive effect of independent amino-acid substitutions maybe effective, as PROPKA performed satisfactorily in predicting the magnitude of the effect on the thermostability of individual substitutions. Although there is a danger that a substitution with adverse effect will cancel out the effect of a favourable substitution, we would recommend that a good strategy would be to predict and experimentally produce combinations of spatially separated triple mutations, giving predicted increase in T_m of at least 2 °C. A sparse matrix approach could be used to reduce the number of variants to be produced and tested and still obtain informative results.

In this study we do identify a mutation, G306A, that gives a moderate but significant increase in stability. This single mutation can increase significantly the lifetime of AaGal, and allow a decrease of enzyme load in applications. Interestingly, a recent study published during processing of this manuscript, reports significant thermostabilization of a *Talaromyces stipitatus* galactanase (TsGal) by the G305A mutation [39]. TsGal shares high sequence identity (67%) with AaGal, and the mutation is in a structurally equivalent loop between the last strand and helix in the $(\beta\alpha)_8$ barrel. However the thermostabilizing mutation in TsGal is not at an equivalent amino acid, but its neighbour, since the WT sequences of the two enzymes are GNGALG for TsGal and GNAGLG for AaGal (with the mutated Gly residues underlined in the respective sequences). This loop seems therefore to be a stability 'hotspot' for GH53 galactanases, which may be related to a conserved disulphide bond in the vicinity of the loop, as discussed in the previous section.

It remains to be established if Ala is the most favourable residue at this hotspot or whether the requirement is simply for a non-Gly. Yet it would seem that there is a sequence bias towards Ala. Since the corresponding residue is an Asn in some already thermostable enzymes like MtGal, if this is a specific effect of a substitution to Ala, insertion in an already thermostable background like MtGal could further improve the stability, and conversely it would be interesting to study the effect of a G306N substitution in AaGal.

Much knowledge of protein thermostability is gained in systems which are fully reversible and with simple first order inactivation kinetics. Unfortunately, the majority of industrially interesting enzymes does not fit these simple models. This is also the case here, as we show that the AaGal unfolding is partly irreversible, and that the kinetics of unfolding are biphasic. There is a strong need of additional experimental data available in the public domain, including the apparently negative results, where predicted effects were not obtained, to build up a solid data basis against which to test theoretical approaches.

Acknowledgement

The authors acknowledge funding from the NABIIT program (from the Danish Council for Strategic Research, grant number 2106-07-0030 to JHJ) and the Program Commission on Sustainable Energy and Environment (grant number 2104-07-0028 to PW). Kristina Mortensen (NOVOZYMES), Johnny Christensen (NOVOZYMES) and Dorthe Boelskifte (University of Copenhagen) are thanked for their technical assistance, Pernille Galberg (University of Copenhagen) for scientific discussions and Tobias Tandrup (University of Copenhagen) for help with final formatting of Fig. 5.

References

[1] Carpita NC, Gibeaut DM. Structural models of primary-cell walls in flowering plants consistency of molecular with the physical-properties of the walls during growth. Plant J 1993;3:1–30.

[2] Lombard V, Golaconda Ramulu H, Drula E, Coutinho P, Henrissat B. The Carbohydrate-active enzymes database (CAZy) in 2013. Nucleic Acids Res 2014;42:D490–5.

[3] Bjørnvad M, Clausen I, Schülein M, Bech L, Østergaard P et al. (2000) Bacterial galactanases and uses thereof. WO0047711 patent.

[4] De Maria L, Svendsen A, Borchert TV, Christensen LLH, Larsen S et al. (2004) Galactanase variants. WO2004056988 patent.

[5] Michalak MT, Thomassen LV, Roytio H, Ouwehand AC, Meyer AS, et al. Expression and characterization of an endo-1,4-β-galactanase from *Emericella nidulans* in *Pichia pastoris* for enzymatic design of potentially prebiotic oligosaccharides from potato galactans. Enzyme Microb Technol 2013;50: 121–9.

[6] Tabachnikov O, Shoham Y. Functional characterization of the galactan utilization system of *Geobacillus stearothermophilus*. FEBS J 2013;280:950–64.

[7] Shipkowski S, Brenchley J. Bioinformatic, genetic, and biochemical evidence that some glycoside hydrolase family 42 β-galactosidases are arabinogalactan type I oligomer hydrolases. Appl Environ Microbiol 2006;72:7730–8.

[8] Delangle A, Prouvost A, Cogez V, Bohin J, Lacroix J, et al. Characterization of the *Erwinia chrysanthemi* Gan locus, involved in galactan catabolism. J Bacteriol 2007; 189:7053–61.

[9] Daniel R, Haiech J, Denizot F, Errington J. Isolation and characterization of the lacA gene encoding β-galactosidase in *Bacillus subtilis* and a regulator gene, lacR. J Bacteriol 1997;179:5636–8.

[10] O'Connell Motherway M, Fitzgerald GF, van Sinderen D. Metabolism of a plant derived galactose-containing polysaccharide by *Bifidobacterium breve* UCC2003. Microb Biotechnol 2011;4:403–16.

[11] Ryttersgaard C, Le Nours J, Lo Leggio L, Jørgensen CT, Christensen LLH, et al. The Structure of Endo-β-1,4-galactanase from *Bacillus licheniformis* in complex with two oligosaccharide products. J Mol Biol 2004;341:107–17.

[12] Torpenholt S, Le Nours J, Christensen U, Jahn M, Withers S, et al. Activity of three β-1,4-galactanases on small chromogenic substrates. Carbohydr Res 2011;346: 2028–33.

[13] Ryttersgaard C, Lo Leggio L, Coutinho PM, Henrissat B, Larsen S. *Aspergillus aculeatus* β-1,4-Galactanase: Substrate Recognition and Relations to Other Glycoside Hydrolases in Clan GH-A. Biochemistry 2002;41:15135–43.

[14] Le Nours J, Ryttersgaard C, Lo Leggio L, Østergaard PR, Borchert TV, et al. Structure of two fungal β-1,4-galactanases: Searching for the basis for temperature and pH optimum. Protein Sci 2003;12:1195–204.

[15] Otten H, Michalak M, Mikkelsen J, Larsen S. The binding of zinc ions to *Emericella nidulans* endo-β-1,4-galactanase is essential for crystal formation. Acta Crystallogr F 2013;69:850–4.

[16] Alexov E, Mehler EL, Baker N, Baptista AM, Huang Y, et al. Progress in the prediction of pK(a) values in proteins. Proteins Struct Funct Bioinf 2011;79: 3260–75.

[17] Li H, Robertson AD, Jensen JH. Very fast empirical prediction and rationalization of protein pKa values. Proteins Struct Funct Bioinf 2005;61:704–21.

[18] Davies M, Toseland C, Moss D, Flower DR. Benchmarking pKa prediction. BMC Biochem 2006;7:18.

[19] Olsson MHM, Sondergaard CR, Rostkowski M, Jensen JH. PROPKA3: consistent treatment of internal and surface residues in empirical pK(a) predictions. J Chem Theory Comput 2011;7:525–37.

[20] Rostkowski M, Olsson MHM, Sondergaard CR, Jensen JH. Graphical analysis of pH-dependent properties of proteins predicted using PROPKA. BMC Struct Biol 2011; 11:6.

[21] Olsson MHM. Improving the desolvation penalty in empirical protein pK(a) modeling. J Mol Model 2012;18:1097–106.

[22] Lehmann M, Pasamontes L, Lassen S, Wyss M. The consensus concept for thermostability engineering of proteins. BBA 2000;1543:408–15.

[23] Godoy-Ruiz R, Ariza F, Rodriguez-Larrea D, Perez-Jimenez R, Ibarra-Molero B, et al. Natural selection for kinetic stability is a likely origin of correlations between mutational effects on protein energetics and frequencies of amino acid occurrences in sequence alignments. J Mol Biol 2006;362:966–78.

[24] Christgau S, Sandal T, Kofod LV, Dalbøge H. Expression cloning, purification and characterization of a β-1,4-galactanase from *Aspergillus aculeatus*. Curr Genet 1995;27:135–41.

[25] Huge-Jensen B, Andreasen F, Christensen T, Christensen M, Thim L, et al. *Rhizomucor miehei* triglyceride lipase is processed ans secreted from transformed *Aspergillus oryzae*. Lipids 1989;24:781–5.

[26] Yang AS, Honig B. On the pH-dependence of protein stability. J Mol Biol 1993;231: 459–74.

[27] Robertson AD, Murphy KP. Protein structure and the energetics of protein stability. Chem Rev 1997;97:1251–67.

[28] Larkin MA, Blackshields G, Brown NP, Chenna R, McGettigan PA, et al. Clustal W and clustal X version 2.0. Bioinformatics 2007;23:2947–8.

[29] Lumry R, Eyring H. Conformation changes of proteins. J Phys Chem 1954;58: 110–29.

[30] Sanchez-Ruiz JM, Lopezlacomba JL, Cortijo M, Mateo PL. Differential scanning calorimetry of the irreversible thermal denaturation of thermolysin. Biochemistry 1988;27:1648–52.

[31] Sanchez-Ruiz JM. Theoretical analysis of Lumry–Eyring models in differential scanning calorimetry. Biophys J 1992;61:921–35.

[32] Leatherbarrow RJ. GraFit version 7. Horley, UK: Erithacus Software Ltd.; 2009.

[33] Violet M, Meunier JC. Kinetic study of the irreversible thermal denaturation of *Bacillus licheniformis* α-amylase. Biochem J 1989;263:665–70.

[34] Hofmann K, Baron M. B BOXSHADE version 3.21. http://sourceforge.net/projects/boxshade/.

[35] Matthews BW, Nicholson H, Becktel WJ. Enhanced protein thermostability from site-directed mutations that decrease the entropy of unfolding. Proc Natl Acad Sci U S A 1987;84:6663–7.

[36] Anbar M, Lamed R, Bayer EA. Thermostability enhancement of *Clostridium thermocellum* cellulosomal endoglucanase Cel8A by a single glycine substitution. ChemCatChem 2010;2:997–1003.

[37] Yi ZL, Pei XQ, Wu ZL. Introduction of glycine and proline residues onto protein surface increases the thermostability of endoglucanase CelA from *Clostridium thermocellum*. Bioresour Technol 2011;102:3636–8.

[38] DeLano WL. The PyMOL molecular graphics system, version 1.2r1. Schrödinger: LLC; 2002.

[39] Larsen D, Nyffenegger C, Swiniarska M, Thygesen A, Strube M, et al. Thermostability enhancement of an endo-1,4-β-galactanase from *Talaromyces stipitatus* by site-directed mutagenesis. Appl Microbiol Biotechnol 2014. http://dx.doi.org/10.1007/s00253-014-6244-z.

18

Comparison of Metabolic Pathways in *Escherichia coli* by Using Genetic Algorithms

Patricia Ortegon [a,1], Augusto C. Poot-Hernández [a,b,1], Ernesto Perez-Rueda [b,c,*], Katya Rodriguez-Vazquez [a,**]

[a] *Departamento de Ingeniería de Sistemas Computacionales y Automatización, IIMAS, Universidad Nacional Autónoma de México, Mexico*
[b] *Departamento de Ingeniería Celular y Biocatálisis, Instituto de Biotecnología, Universidad Nacional Autónoma de México, Cuernavaca, Morelos, Mexico*
[c] *Unidad Multidisciplinaria de Docencia e Investigación, Sisal Facultad de Ciencias, Sisal, Yucatán, UNAM, Mexico*

ARTICLE INFO

Keywords:
Metabolism
Genetic algorithms
KEGG database
k-medoids
Comparative genomics

ABSTRACT

In order to understand how cellular metabolism has taken its modern form, the conservation and variations between metabolic pathways were evaluated by using a genetic algorithm (GA). The GA approach considered information on the complete metabolism of the bacterium *Escherichia coli* K-12, as deposited in the KEGG database, and the enzymes belonging to a particular pathway were transformed into enzymatic step sequences by using the breadth-first search algorithm. These sequences represent contiguous enzymes linked to each other, based on their catalytic activities as they are encoded in the Enzyme Commission numbers. In a posterior step, these sequences were compared using a GA in an all-against-all (pairwise comparisons) approach. Individual reactions were chosen based on their measure of fitness to act as parents of offspring, which constitute the new generation. The sequences compared were used to construct a similarity matrix (of fitness values) that was then considered to be clustered by using a *k*-medoids algorithm. A total of 34 clusters of conserved reactions were obtained, and their sequences were finally aligned with a multiple-sequence alignment GA optimized to align all the reaction sequences included in each group or cluster. From these comparisons, maps associated with the metabolism of similar compounds also contained similar enzymatic step sequences, reinforcing the *Patchwork Model* for the evolution of metabolism in *E. coli* K-12, an observation that can be expanded to other organisms, for which there is metabolism information. Finally, our mapping of these reactions is discussed, with illustrations from a particular case.

1. Introduction

The study of the evolution of metabolism is fundamental to understanding the adaptive processes of cellular life, the emergence of high levels of organization (multicellularity), the diversity of cellular organization in the three major domains of life, *Archaea*, *Bacteria*, and *Eukarya*, and the complexity of the living world [4]. At present, the large scale of information derived from genomics and proteomics studies has allowed the construction of diverse databases devoted to organizing the metabolic processes, such as the KEGG [21], and MetaCyc [5]. Therefore, the information contained in these databases can be used to generate an integrative perspective on cellular functioning.

Metabolism can be considered one of the most ancient biological networks; in such a network, the nodes represent substrates or enzymes and the edges represent the relationships among them. From a global perspective, the comparative analysis of metabolic pathways aims to identify similarities and differences among them, providing insights for the identification of evolutionary events, such as enzyme recruitment and duplication events. In this regard, metabolic pathways exhibit high retention of duplicates within functional modules and a preferential biochemical coupling of reactions. This retention of duplicates may result from the biochemical rules governing substrate–enzyme–product relationships [1,8,14,19].

In this context, diverse studies have evaluated the variations among pathways, both intra- and interspecies [7,27], comparing the pathways based on their Enzyme Commission (EC) numbers and excluding information on the compounds. In addition, a method of path-and-graph matching has been proposed to query metabolic pathways based on a predefined graph, where a similarity measure based on EC numbers [30] and the distance between pathways as a combination of distances between compounds and between enzymes associated with amino acid biosynthesis networks are considered [11].

* Correspondence to: E. Perez-Rueda, Departamento de Ingeniería Celular y Biocatálisis, Instituto de Biotecnología, Universidad Nacional Autónoma de México, Cuernavaca, Morelos, Mexico.
** Correspondence to: K. Rodriguez-Vazquez, Departamento de Ingeniería de Sistemas Computacionales y Automatización, IIMAS, Universidad Nacional Autónoma de México, Mexico.
E-mail addresses: erueda@ibt.unam.mx (E. Perez-Rueda),
katya.rodriguez@iimas.unam.mx (K. Rodriguez-Vazquez).
[1] These authors contributed equally to this work.

In this work, we evaluated whether there are groups of similar reactions in different pathways, which might suggest a transfer of enzymatic activities, and whether these groups can be used to define common and variable regions of an organism's metabolism. This analysis was addressed using EC numbers, coded as a succession of reaction steps. To this end the metabolic maps of the bacterium *Escherichia coli* K-12, as deposited in the KEGG database, were transformed into linear Enzymatic Step Sequences (ESS), to be compared via a genetic algorithm (GA). The sequences compared were used to construct a similarity matrix to identify groups of conserved reactions based on a *k*-medoids clustering analysis, and then a multiple-sequence alignment (MSA) GA was optimized to align all the reaction sequences included in a group. Finally, we consider our comparisons in terms of the clues they provide in reinforcing the *Patchwork Model* in the evolution of metabolism for *E. coli* K-12 and probably for other organisms beyond this bacterium.

2. Methods

2.1. Construction of Enzymatic Step Sequences (ESS)

The KGML files (version 0.71) that describe the metabolic maps (pathways) of *E. coli* K-12 as of June 2011 were downloaded from the KEGG database (Fig. 1). Pathways were transformed into linear ESS by using the breadth-first search (BFS) algorithm [26], which infers the closer neighbor of each enzyme by considering a common compound, a substrate or product. In brief, a directed graphical representation of each metabolic map was created in which the nodes represented enzymes and the edges represented a shared substrate/product between two enzymes. This representation takes into account the reversibility of the reactions. Then, a group of BFS trees was generated for each metabolic map from a set of initialization nodes, which were used as roots. In this work, an initialization node was defined by two criteria: (i) a node whose substrate is not catalyzed by another enzyme in the metabolic map, and (ii) a node whose substrate comes from another metabolic map and has two or fewer neighbors in the graph. These criteria represent the metabolic input for each pathway; the first criterion considers the substrates not created in the same pathway, and the second one considers the connections with other pathways. Each initialization node was used as a root for the construction of a BFS tree. Thus, each tree was used as a guide for the construction of the corresponding ESS. In this way, a BFS tree creates as many ESS as the number of branches it contains. Finally, the first three levels of EC numbers are used to represent an enzyme as a string or sequence (Fig. 1). ESS constructed per metabolic map

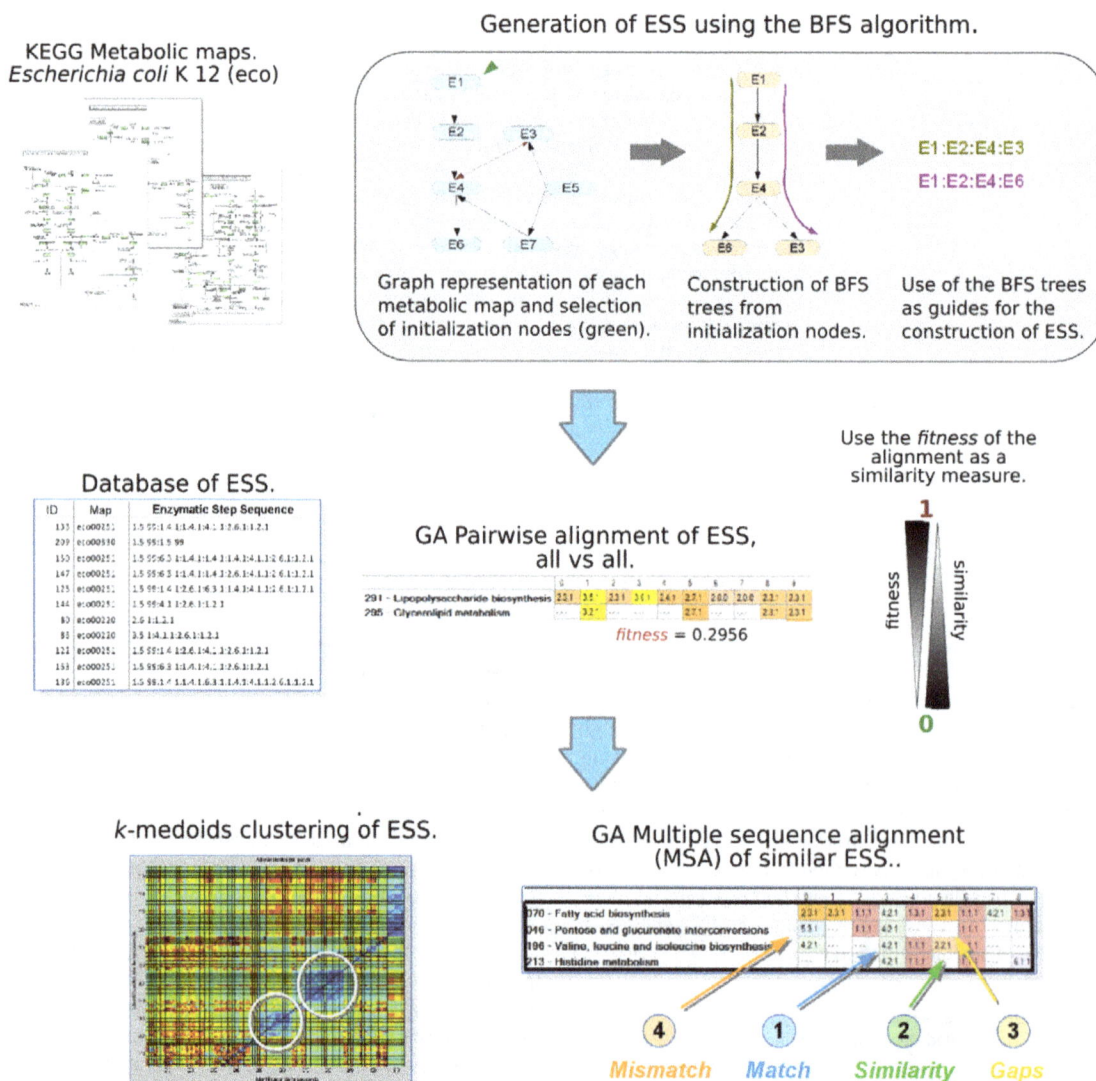

Fig. 1. General strategy for the comparative analysis of *E. coli* K-12 metabolism. The metabolic maps from KEGG were converted to ESS by using the breadth first search (BFS) algorithm. For each map a graphical representation was created, where nodes represent enzymes and edges are product-substrate relationships. Then, a set of initialization nodes was selected (green arrowhead) as roots for BFS trees. Those trees were used as guide for ESS construction. Afterwards all the ESS were compared against each other by GA pairwise alignments. The similarities among ESS were used to conduct a clustering analysis based on the *k*-medoids algorithm. Finally, clusters of similar sequences were aligned using an MSA approach.

is shown in Fig. 2 and Table S1, with mean lengths ± standard deviation (SD).

2.2. Sequence Alignments Obtained via Genetic Algorithms

The proposed GA starts by creating a random initial population of variable-length chromosomes, which represents the potential for alternative alignments. For each iteration or generation, the population evolves by means of selection, crossover, and mutation. Here, a tournament selection that randomly selected a subset of individuals and chose the best individual of each subset was used.

2.3. Crossover Operator

In this work, one-point crossovers were considered; however, it is important to note that for crossover two alternative metabolic pathway sequence alignments, they must be in general, of different sizes (i.e. have a different number of columns). Then, this operator is used to select a position in the first parent and the second parent is cut, keeping in mind the EC numbers conserved for the left side of the first parent. As shown in Fig. 3, offspring 1 is generated by combining the left of parent 1 and the right of parent 2, inserting a gap in the first column for the first three rows of the right side of parent 2. Offspring 2 is

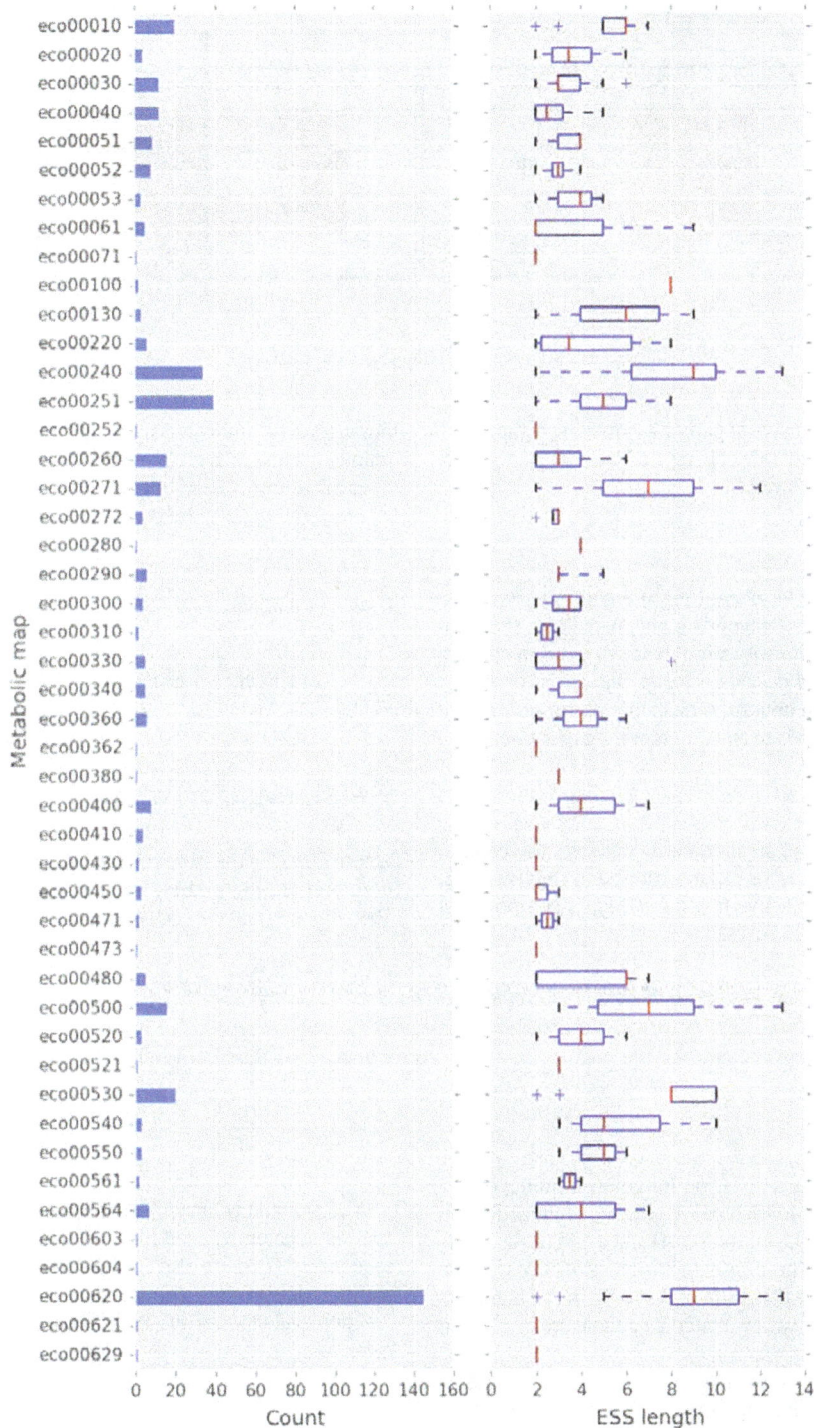

Fig. 2. Statistics for the ESS per metabolic map. Only the 45 metabolic maps that generate at least one sequence are shown. In the left panel, the number of ESS generated by the metabolic map are shown; in the right panel, the distribution of lengths of those sequences is shown.

Parent 1

Parent 2

Offspring 1

Offspring 2

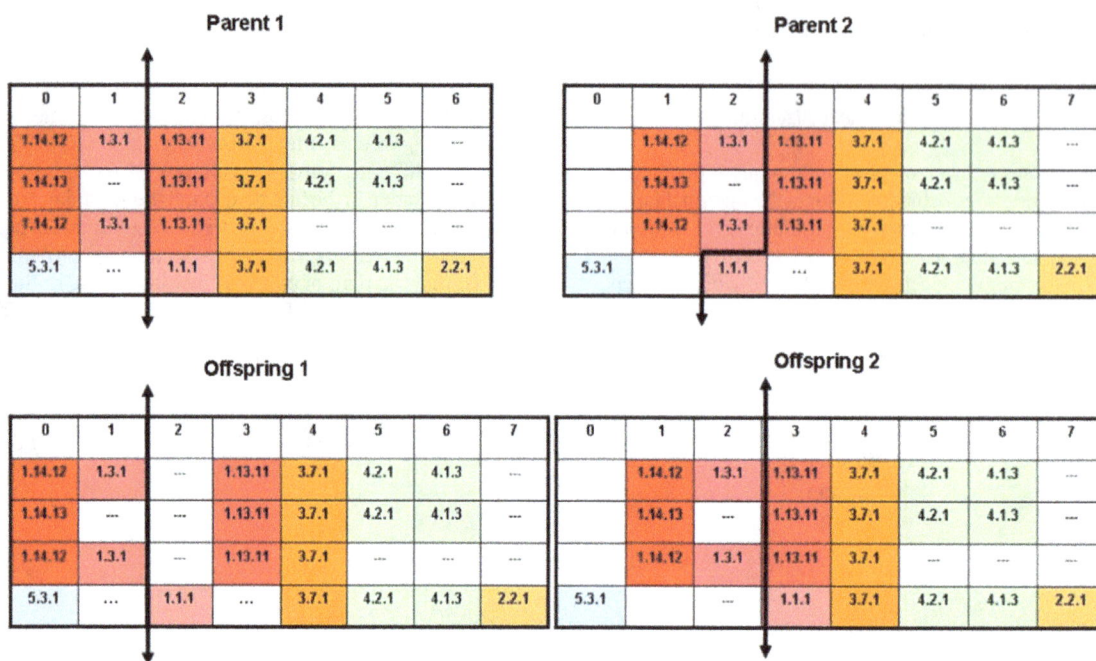

Fig. 3. Crossover for MSA of metabolic pathways.

produced by combining the left side of parent 2 (inserting a gap at the end of the last row) with the right side of parent 1. Thus, the EC numbers are kept constant, and only gaps vary.

2.4. Mutation

A binary codification was used, where 1 represented an enzyme and 0 represented a gap. The gap insertion was highly penalized. The algorithm was designed to find the best alignment with the maximum score. Therefore, if the selected position is a gap, this can be extended or reduced in one unit with an uniform probability; on the contrary, if the selected position corresponds to an EC number, a gap is inserted.

2.5. Objective Function

Assessment of the quality of an alignment considers the column homogeneity, with penalizations for gap insertions and column increments. Thus, the proposed objective function (O.F.) is composed of three weighted terms, defined in Eq. (1):

$$O.F. = 0.9 \ Homogeneity + 0.05 \ Gap \ Penalty + 0.05 \ Column \ Increment. \tag{1}$$

2.6. Homogeneity

The sum of pairs is a popular criterion for column homogeneity evaluation, as it assigns a cost to each pair of aligned codifications in each column of an alignment (substitution cost) and a cost to gap insertions (gap costs). The sum of these costs yields the global cost of the alignment. In this work, rating the grade of diversity in the elements of a given position (column) was evaluated as a measure of column homogeneity.

The EC numbers (columns in the alignment) are represented by three levels (subcolumns), evaluating the normalized entropy of each level based on Shannon's entropy. These results are weighted, giving a higher value to the first level. Gaps are considered one more symbol, and a penalty is applied to compensate for the possibility of "false

homogeneity" indicated by a high number of gaps in a column. The following equations give details for the evaluation of the homogeneity of each column. Given alignment M, where m_j is the jth column and m_{j1}, m_{j2}, and m_{j3} are the three levels of the EC number represented by m_j:

$$H(m_j) = 0.6E(m_j) + 0.4Gaps(m_j) \tag{2}$$

$$E(m_j) = \frac{\omega_1 E(m_{j1}) + \omega_2 E(m_{j2}) + \omega_3 E(m_{j3})}{\omega_1 + \omega_2 + \omega_3} \tag{3}$$

$$Gaps(m_j) = \frac{Number \ of \ Gaps}{Number \ of \ Sequences - 1} \tag{4}$$

where $E(m_{jk})$ is the normalized entropy for the kth level of the jth column and ω_1, ω_2, and ω_3 have values of 15, 10 and 5, respectively. Then, the entropy for each column is estimated as follows,

$$E(m_{jk}) = -\sum_a c_{jk}(a) \log_2 p_{jk}(a) \tag{5}$$

where $a \ Î \ \{all \ different \ symbols \ in \ column \ j\}$ and,

$$p_{jk}(a) = \frac{c_{jk}(a)}{\sum c_{jk}(a')}. \tag{6}$$

The probability of symbol a in column jk is the ratio of the symbol a counter in column jk and the sum of all symbol counters a' (number of sequences). Gaps are also considered as a symbol [9]. It is important to mention that a column of only gaps is removed previously to objective function evaluation. Thus, the homogeneity for the whole alignment is given as,

$$Homogeneity = \frac{\sum_{j=1} H(m_j)}{Number \ of \ Columns}. \tag{7}$$

2.7. Gap Penalty

The gaps concentration criterion is used to penalize the number of gap blocks in the whole alignment. This criterion is defined as:

$$GC = \frac{\bar{S}_{GB}}{GP} \tag{8}$$

where $\bar{S}_{GB}$ is the average length of gap blocks and GP corresponds to the total number of individual gaps in the alignment. It is important to mention that initial and final gaps are not taken into account for the gaps penalization. If there are no gaps in the alignment, the GC value is zero.

This criterion serves to reward alignments where gap codifications are more concentrated, that is, where there are few larger blocks of gaps rather than blocks of smaller lengths, in contrast to Equation 4, which penalizes the number of gaps in a column. The column increment term penalizes the addition of columns into the aligned matrix and is defined as follows:

$$CI = \frac{C_0}{C_1} \tag{9}$$

where C_0 is the length of the longest sequence in the unaligned matrix and C_1 corresponds to the number of columns of the aligned matrix. Thus, the defined objective function (O.F.) has to be minimized to attain better alignments.

2.8. Clustering

The k-medoids algorithm is used to cluster the sequences previously compared and included in the similarity matrix. The number of clusters is settled by using the elbow criterion and plotting the cluster quality from $k = 2$ to $k = 100$. A total of 20 replicates for each k value are evaluated, and the best result is selected. To define the cutoff for the clustering, the second derivative associated with this function is used, and when k is maximized, this value represents the greater slope, i.e., the greater difference between two k' values.

3. Results

3.1. Construction of Enzymatic Step Sequences (ESS)

In order to evaluate the conserved and variable catalytic steps in metabolic pathways of the bacterium *E. coli* K-12, a collection of ESS was generated and compared using a GA. In this work, an ESS was defined as a linear collection of consecutive enzymatic reactions from a given substrate to a given product, similar to the method used for a previously proposed definition of metabolic pathways [7]. Therefore, each ESS was reconstructed by following subsequent reactions in each metabolic map. The enzymes related to each reaction were represented by using the first three levels of the EC classification to describe their general type of chemical reaction, as previously suggested [17]. In total, 452 ESS associated with 47 metabolic maps of *E. coli* K-12 were obtained [21] (see Fig. 2 and also Table SI in the supplementary material). From these, the average of sequences generated per metabolic map was 9.6 with an average length of 3.91 ± 1.1 ESS, where the map associated with pyruvate metabolism showed the highest number of ESS (145, with an average length of 9 ± 2.27 steps), probably because it represents an intersection of key pathways of energy metabolism. The second most frequent pathway corresponds to the glutamate map (39 ESS, with an average length of 5 ± 1.53 steps). Indeed, almost all maps considered in this analysis generated ESS, and the pyruvate metabolism map was also the map with the highest diversity of ESS. Finally, we found 5 pathways with only two ESS and 11 pathways with only one ESS.

Therefore, it seems that the numbers and lengths of sequences of enzymatic steps depend on the size of the metabolic map and reflect the number of alternative pathways that can be traced, beginning from the start nodes, i.e., large metabolic maps generate more ESS than small metabolic maps as a consequence of the complexity of the map. In this regard, the pyruvate metabolism is represented by a complex map, as it is the endproduct of glycolysis and the starting point for gluconeogenesis, and it can be generated by transamination of alanine. It can be converted by the pyruvate dehydrogenase complex to acetyl-CoA [6,23,25,28], which can enter the TCA cycle or serve as the starting point for the synthesis of long-chain fatty acids, steroids, and ketone bodies.

3.2. Significance of ESS Alignments

To evaluate the significance of the alignments, 10 random sets of ESS were constructed by shuffling the EC numbers from the real sequences, where the length and the EC composition were conserved (Figure S1). Because we were interested in evaluating the most significant scores, a threshold of fitness of 0.4 that considers "$\bar{x} - 3\sigma$" was chosen, including 7.16% of the real ESS alignments. In contrast, an average of $0.34\% \pm 0.029$ of the total random sequence alignments were below the fitness threshold of 0.4. According to these data, we expect an average of 30 ESS alignments included in our dataset to be possible false positives, suggesting that GAs infer efficiently the significant and similar sequences.

Alternatively, we assessed the consistency of the GA approach in a different scenario, one which considered a Dynamic Programing (DP) Needleman–Wunsch algorithm for the pairwise alignment of ESS. This algorithm uses the same objective function as the GA, minimizing the score of the alignment. The DP algorithm exhibits a similar (Gumbel-like) distribution as the GA fitness values, suggesting that our approach is consistent with the previous results described (Figure S2).

3.3. ESS Comparisons Identify Groups of Similar Reactions in Different Pathways

In order to evaluate the similarity of the metabolic maps for the bacterium *E. coli* K-12, 452 ESS were used to carry out all-against-all pairwise alignments. Therefore, the GA previously described was applied based on the objective function (O.F.), with which we evaluated the entropy per column using a population of 100 chromosomes, a 1% mutation rate and 90% crossover rate. Mutation is applied per individual and it means that an individual will be mutated each generation. Therefore, an individual in the population is a particular possible solution to the ESS alignment. The algorithm was concluded when 20 generations were reached with no changes in the objective function. Ten replicates of the GA were performed, and the best result was chosen. The GA evaluates the alignment by using a normalized entropy-based function defined above in Eq. (1) (see the Methods section). Since the fitness value is a measure of the entropy for each column in the alignment, a value near 0 corresponds to an alignment with homogeneous columns, i.e., columns containing similar EC numbers. Conversely, values near 1 correspond to alignments with columns that are less homogeneous, i.e., columns containing dissimilar EC numbers. Therefore, the fitness value is a measure of how similar two sequences are. Based on similarity values from the all-against-all comparisons, a matrix that considered the similarity (fitness) values of 452 vs 452 sequences was constructed. This matrix was posteriorly analyzed by using the k-medoids algorithm and identified groups with similar ESS that may share similar catalytic properties. To determine the number of clusters (k) that could be generated, the elbow criterion was used. In addition, diverse k values were used to construct different clusters: $k = 21, 27,$ and 34, as these exhibited the higher peaks in the analysis plot. As a criterion to discriminate between the clustering at $k = 21, 27,$ and 34, the clusters were depurated, eliminating any sequence whose mean fitness with the rest of the sequences of its cluster was above 0.4. After depuration, 93, 71, and 37 sequences were eliminated for $k = 21, 27,$ and 34, respectively.

In this way, we selected a *k* value of 34 to minimize the number of sequences eliminated during depuration.

From this analysis, similar ESS associated with diverse pathways were clustered together. For instance, the ESS constructed from the pyruvate map were differentially clustered in nine groups with sequences belonging to different pathways, including those for the citrate cycle (TCA cycle), glycolysis/gluconeogenesis, and lysine, glycine, serine, and threonine metabolism (Fig. 4). This finding correlates with the fact that pyruvate is the endproduct at which diverse metabolic pathways converge, including glycolysis, gluconeogenesis, and alanine metabolism, among others. Based on these results, we suggest that similar catalytic processes are required to metabolize different compounds, to converge at a compound, probably by diverse recruitment events, where the catalytic activity is preferentially coupled, as previously noticed using as the comparative the first two digits associated with EC numbers [2,8,19]. In addition, some clusters are comprised almost exclusively of sequences belonging to pyruvate metabolism, such as clusters 7, 8, 13, and 15, which implies that some ESS have probably been duplicated to increase this metabolic pathway and its product.

Alternatively, cluster 4 contains sequences from different maps related to lipid metabolism, including lipopolysaccharide, glycerolipid, glycerophospholipid, and steroid metabolism, pyrimidine metabolism, selenium amino acids, the pentose phosphate pathway, and galactose metabolism. The similar catalytic activities in all these pathways suggest the phenomenon of recruitment events. Finally, cluster 23 contains exclusively ESS from the glycerolipid and glycerophospholipid metabolic pathways, suggesting a common origin shared by these two metabolic pathways. These results correlate with the origin of fatty acid metabolism by gene duplication [8]. Thus, an ancestral pathway catalyzed by both fatty acid degradation and biosynthesis could have originated in a first step. The direction of this ancestral pathway would be dependent

on the acyl carriers and fatty acids available, providing evidence of the existence of two different pathways [8,15,19].

Similar results were observed in the ESS included in clusters 18, 24, 22, 25, and 30, which included reactions from diverse amino acid metabolism pathways, whereas clusters 2, 11, 17, 21, 23, and 29 included ESS from carbohydrate metabolism. In particular, clusters 17 and 21 included ESS belonging to galactose, starch, and sucrose metabolic pathways; cluster 12 included ESS from glycosphingolipid biosynthesis; and cluster 21 included ESS from the galactose, pentose, sucrose, mannose, and glycolysis/gluconeogenesis pathways. All these findings suggest that the assimilation of these carbon sources involved identical enzymatic processes that may have arisen from recruitment events, according to the *Patchwork Model* of evolution.

3.4. Multiple Sequence Alignments (MSA) of Similar ESS

In order to maximize the identification of functional similarities among sequences included in a common cluster, MSAs were performed using a progressive GA alignment and an objective function based on entropy. This function considered three main factors, determined empirically, to evaluate the enzymatic number level (15, 10, or 5). The progressive GA works as follows: the ESS are sorted according to their similarity values, and this order is used to guide the sequence alignment, considering the more similar pair and adding a third one, and so on. This algorithm increases column homogeneity among the sequences aligned and penalizes the insertion of gaps; hence, the fitness function has to be minimized to obtain better alignments. From this analysis, in Fig. 5 we show the MSA of cluster 21. Cluster 21 contains 17 ESS from five different maps related to carbohydrate metabolism: glycolysis/gluconeogenesis, fructose and mannose metabolism, galactose metabolism, the pentose phosphate pathway, and pentose and glucoronate

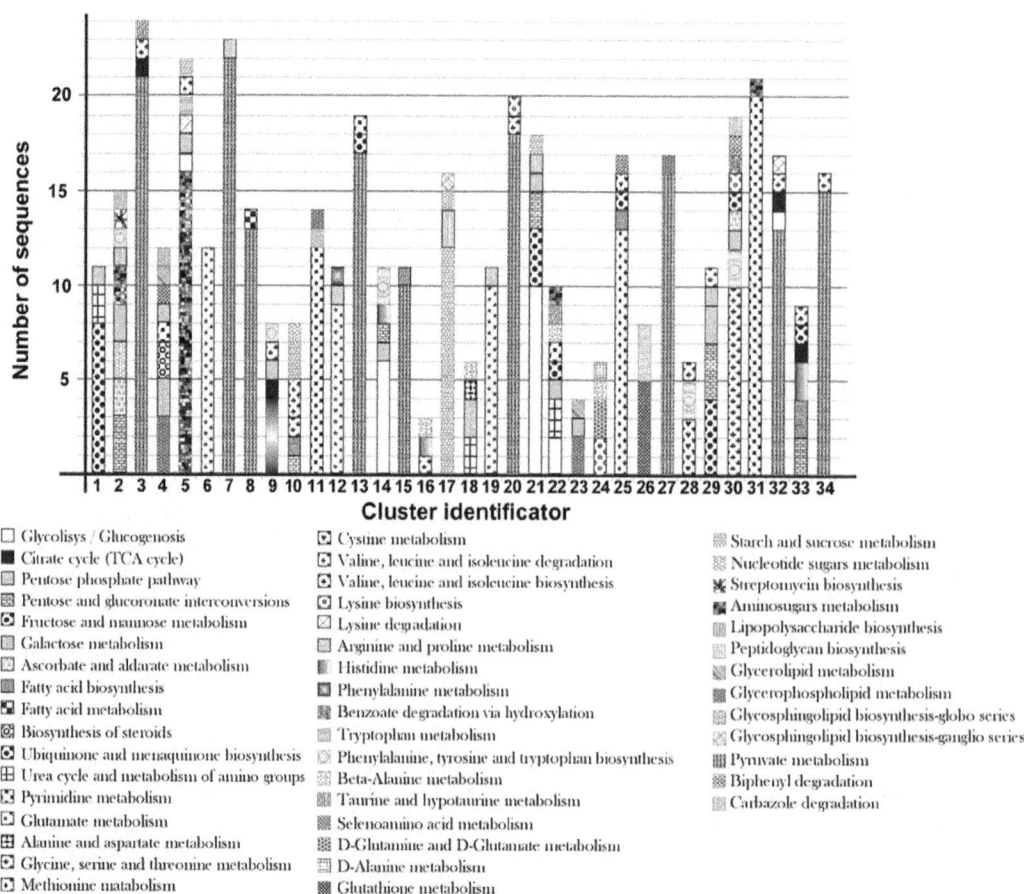

Fig. 4. Distribution of ESS among the 34 identified clusters. Bars represent the number of sequences per cluster.

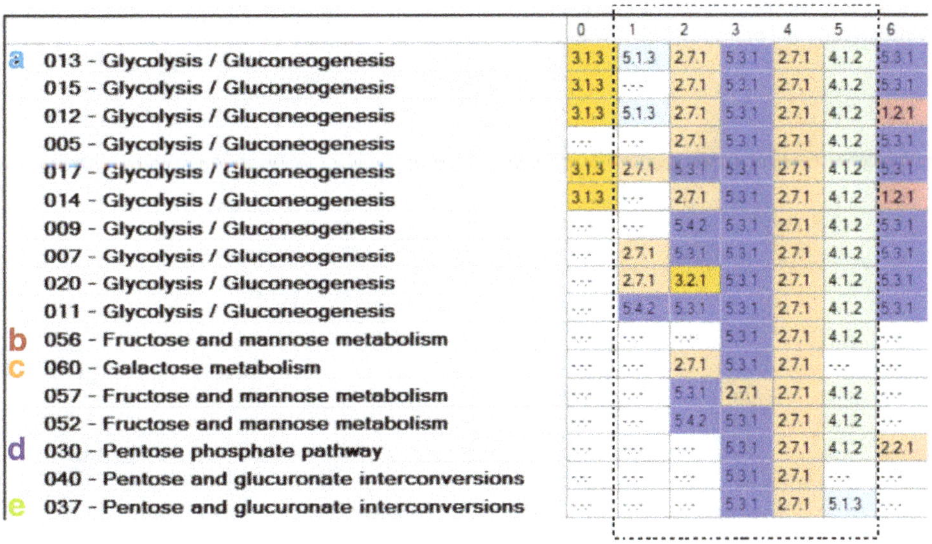

		0	1	2	3	4	5	6
a	013 - Glycolysis / Gluconeogenesis	3.1.3	5.1.3	2.7.1	5.3.1	2.7.1	4.1.2	5.3.1
	015 - Glycolysis / Gluconeogenesis	3.1.3	...	2.7.1	5.3.1	2.7.1	4.1.2	5.3.1
	012 - Glycolysis / Gluconeogenesis	3.1.3	5.1.3	2.7.1	5.3.1	2.7.1	4.1.2	1.2.1
	005 - Glycolysis / Gluconeogenesis	...	...	2.7.1	5.3.1	2.7.1	4.1.2	5.3.1
	017 - Glycolysis / Gluconeogenesis	3.1.3	2.7.1	5.3.1	5.3.1	2.7.1	4.1.2	5.3.1
	014 - Glycolysis / Gluconeogenesis	3.1.3	...	2.7.1	5.3.1	2.7.1	4.1.2	1.2.1
	009 - Glycolysis / Gluconeogenesis	...	...	5.4.2	5.3.1	2.7.1	4.1.2	5.3.1
	007 - Glycolysis / Gluconeogenesis	...	2.7.1	5.3.1	5.3.1	2.7.1	4.1.2	5.3.1
	020 - Glycolysis / Gluconeogenesis	...	2.7.1	3.2.1	5.3.1	2.7.1	4.1.2	5.3.1
	011 - Glycolysis / Gluconeogenesis	...	5.4.2	5.3.1	5.3.1	2.7.1	4.1.2	5.3.1
b	056 - Fructose and mannose metabolism	...	...	...	5.3.1	2.7.1	4.1.2	...
c	060 - Galactose metabolism	...	...	2.7.1	5.3.1	2.7.1	...	...
	057 - Fructose and mannose metabolism	...	...	5.3.1	2.7.1	2.7.1	4.1.2	...
	052 - Fructose and mannose metabolism	...	...	5.4.2	5.3.1	2.7.1	4.1.2	...
d	030 - Pentose phosphate pathway	...	...	...	5.3.1	2.7.1	4.1.2	2.2.1
	040 - Pentose and glucuronate interconversions	...	...	...	5.3.1	2.7.1	...	...
e	037 - Pentose and glucuronate interconversions	...	...	...	5.3.1	2.7.1	5.1.3	...

Fig. 5. MSA of the ESS included in cluster 21.

interconversions. In this cluster, columns 3 and 4 are identical in almost all the sequences, suggesting a core of enzymatic reactions conserved among all these pathways and a common metabolic mechanism used to metabolize these carbohydrates. This *core* of ESS includes an isomerization (EC 5.3.1) reaction followed by a phosphorylation (EC 2.7.1) reaction, suggesting that these reactions are ancestral to all these pathways or that they were recruited in small catalytic modules [8], whereas surrounding reactions could be added to increase the other metabolic pathways.

Nevertheless, it is expected that similar enzymatic steps found in different metabolic maps actually represent different proteins. Therefore, to examine this phenomenon, five ESS were selected from cluster 21 that represented each of the five metabolic maps, and the information for protein sequences associated with each EC number was obtained. Based on this approach, the ESS from the glycolysis/gluconeogenesis, fructose and mannose metabolism, galactose metabolism, and pentose and glucoronate metabolism pathways, representing different enzymes, as shown in Fig. 6, were compared. Sequences representing the pentose phosphate pathway are a subset of the glycolysis/gluconeogenesis pathway; this is a consequence of how KEGG organizes and represents graphically the metabolic maps. Therefore, the proposed method is able to identify similar ESS assigned to different metabolic contexts.

Based on these data, it was suggested that the similar enzymatic steps identified by this strategy may share a common origin, i.e., they have been recruited from different pathways. To evaluate the probable common origin of these enzymes, their amino acid sequences were analyzed using the domain identification system based on Superfamily database assignations [20]. In brief, Superfamily is a database devoted to identification of structural domains at the level of the Superfamily. This definition is based on the SCOP database classification [12]. Therefore, two proteins share a common ancestor if they contain domains belonging to the same Superfamily. In Fig. 6 an illustrative example is shown, with only the structural domains identified for at least two enzymes in the alignment of enzymatic steps previously described. From this analysis, three different structural domains were identified as repeated in at least two enzymes in the alignment of cluster 21: the actin-like ATPase domain, the aldolase domain, and AraD/HMP-PK domain-like. Therefore, the enzymatic steps common to the pathways included in cluster 21 may have been recruited posterior to an event of gene duplication. In addition, similar enzymatic steps identified in the glycolysis/gluconeognesis, fructose and mannose metabolism, and pentose phosphate and glucuronate interconversion metabolic maps (Glk, FucK, and XilB, respectively) and the different

evolutionary origins for some of their protein sequences suggest an example of *Patchwork Evolution* [16], in which the ancestor of these enzymes might be associated with diverse duplication events with posterior recruitment to different metabolic pathways. The catalytic activity associated with EC 2.7.1 was mainly recruited to perform a link between two isomerization reactions with the EC numbers 5.x.x.

Based on this analysis we identified that three proteins (FbaB, FbaA, and GatZ) share a common structural domain, identified by superfamily, the aldolase domain, and two else (FucA and AraD) the AraD/HMO-PK domain-like, which catalyze dissimilar reactions. These proteins might represent two cases of divergent evolution, in which two enzymes from a common ancestor diverged at different functions. Therefore, according to our data, it is clear that convergent evolution events are also frequently associated with enzymes devoted to metabolism; for instance, four nonhomologous enzymes (Pgi, FucI, AgaI, and XylA) catalyze the isomerization reaction 5.3.1.x, conserved in all the sequences of the cluster. This finding correlates with previous reports that showed that catalytic convergent evolution is a common phenomenon [13,24,25].

Finally, it is interesting that in the case of the sequences representing the fructose and mannose metabolism pathways and the pentose and glucoronate metabolism pathways, the FucK–FucA and XylB–AraD enzymatic steps are catalyzed by enzymes sharing a common structural domain, suggesting that the approach described here is able to identify probable patchwork events as well as duplication and convergent enzymes associated with the evolution of a microorganisms's metabolism.

4. Discussion and Conclusions

In general, two main approaches to compare metabolic pathways have been proposed. In one approach entails methods based on the alignment of complete or fragmented networks [3,10,18,22,29], that take into account the structure of the metabolic network and integrate more information about the system. However, an increase in complexity also increases the computational cost, making it more difficult the possibility to create multiple alignments. Via the other approach, entailing methods based on the alignment of linear pathways [7,27]; and that was utilized in this work, the complexity of the comparisons is reduced, and they can be executed and analyzed relatively easily.

In this regard, sequence alignment is a problem in bioinformatics where it is important to identify conserved regions in a set of sequences, and as a consequence, to find the differences between them based on a criterion to assess the quality of an alignment. Thus, a GA was proposed

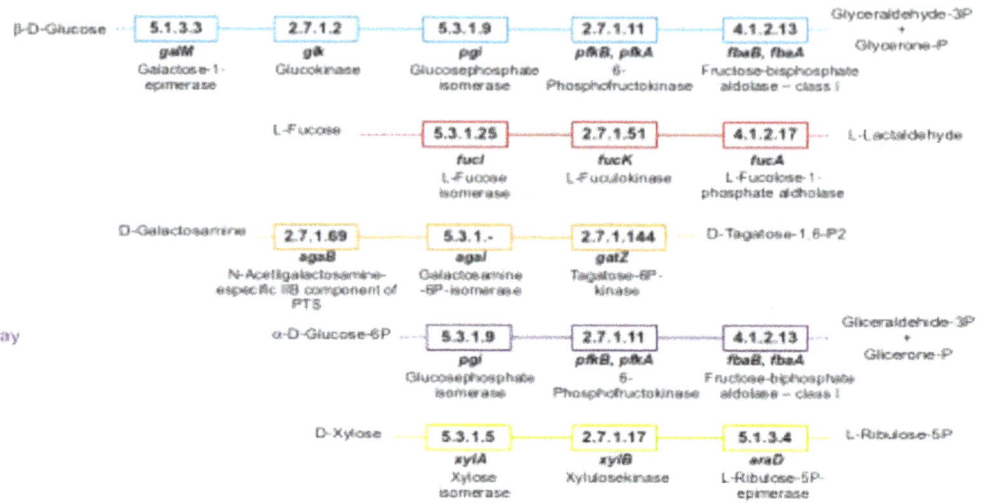

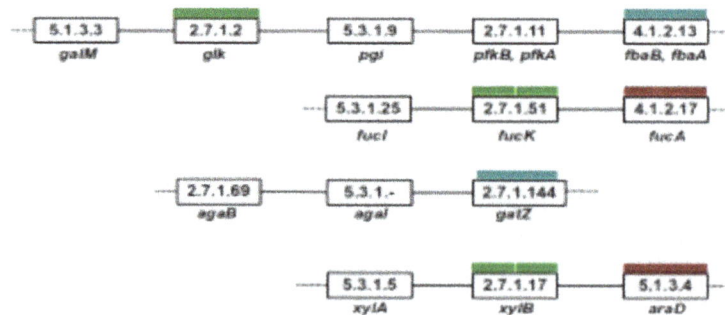

Fig. 6. Structural domain assignment according to the Superfamily database for proteins aligned in cluster 21. In panel A, all of the enzymes in the aignment were mapped in the corresponding metabolic map. Panel B are the results of the Superfamily domain assignations. Only the similar domains are indicated.

in order to optimize the quality criterion of potential alignments considering maximization of homogeneity and, at the same time, minimization of gaps. The proposed GA used a variable binary representation of individual enzymes (potential alignments) where 1 represents EC numbers and 0 indicates the gaps.

In the present work, the strategy to transform enzymatic reaction sequences from a metabolic pathway allowed us to implement a sequence alignment method in an efficient and easy fashion, where regions sharing a similar succession of EC steps were identified, suggesting common catalysis that is not easy to identify when using traditional computational tools. The progressive GA proposed method has shown efficient results, providing good alignments after less than 100 iterations. These results allow us to make comparative studies of metabolic pathways to elucidate the functions of newly discovered pathways, increase our understanding of evolutionary traits, and identify potential missing pathway elements. It is interesting that some of the enzymatic steps common to various sequences may, indeed, represent the same

reactions, such as the first 10 sequences in Fig. 6, which have a different representation of the same region of the map of the glycolysis/ gluconeogenesis pathway, although they represent the same reaction sequence steps. Moreover, these sequences are not identical, and they differ mainly in the beginnings or in the ends of the sequences. This implies that the proposed method introduced in this paper may have the capability to show alternative metabolic pathways not previously described.

In summary, we consider that the strategy described in this work allowed us to identify similar ESS from different metabolic maps, as shown in Figs. 5 and 6. Thus, maps associated with the metabolism of similar compounds also contain similar ESS, reinforcing the *Patchwork Model* for studying the evolution of metabolism in *E. coli* K-12, where similar consecutive enzymatic steps may have different origins, in agreement with biochemical restrictions in enzymatic recruitment [2,8], an observation that can be expanded to other organisms for which metabolism information is available.

Acknowledgments

We acknowledge fruitful discussions with Pedro Miramontes, Georgina Hernandez-Montes, and Mario Alberto Martinez-Nuñez. Authors gratefully acknowledge the support provided by DGAPA-UNAM grant numbers IN-204714 and IN-107214, and CONACYT grant number 155116.

References

[1] Armenta-Medina D, Perez-Rueda E, Segovia L. Identification of functional motions in the adenylate kinase (ADK) protein family by computational hybrid approaches. Proteins 2011;79:1662–71.

[2] Armenta-Medina D, Segovia L, Perez-Rueda E. Comparative genomics of nucleotide metabolism: a tour to the past of the three cellular domains of life. BMC Genomics 2014;15:800.

[3] Ay F, Kellis M, Kahveci T. SubMAP: aligning metabolic pathways with subnetwork mappings. J Comput Biol 2011;18:219–35.

[4] Caetano-Anolles G, Kim HS, Mittenthal JE. The origin of modern metabolic networks inferred from phylogenomic analysis of protein architecture. Proc Natl Acad Sci U S A 2007;104:9358–63.

[5] Caspi R, Altman T, Billington R. The MetaCyc database of metabolic pathways and enzymes and the BioCyc collection of Pathway/Genome Databases. Nucleic Acids Res 2014;42:D459–71.

[6] Chandrasekhar K, Wang J, Arjunan P. Insight to the interaction of the dihydrolipoamide acetyltransferase (E2) core with the peripheral components in the Escherichia coli pyruvate dehydrogenase complex via multifaceted structural approaches. J Biol Chem 2013;288:15402–17.

[7] Chen M, Hofestadt R. PathAligner: metabolic pathway retrieval and alignment. Appl Bioinformatics 2004;3:241–52.

[8] Diaz-Mejia JJ, Perez-Rueda E, Segovia L. A network perspective on the evolution of metabolism by gene duplication. Genome Biol 2007;8:R26.

[9] Durbin R, Eddy S, Krogh A, Mitchison G. Biological sequence analysis: probabilistic models of proteins and nucleic acids. Cambridge University Press; 1998 356.

[10] Flannick J, Novak A, Srinivasan BS, McAdams HH, Batzoglou S. Graemlin: general and robust alignment of multiple large interaction networks. Genome Res 2006;16:1169–81.

[11] Forst CV, Schulten K. Phylogenetic analysis of metabolic pathways. J Mol Evol 2001;52:471–89.

[12] Fox NK, Brenner SE, Chandonia JM. SCOPe: structural classification of proteins—extended, integrating SCOP and ASTRAL data and classification of new structures. Nucleic Acids Res 2013;42:D304–9.

[13] Gherardini PF, Wass MN, Helmer-Citterich M, Sternberg MJ. Convergent evolution of enzyme active sites is not a rare phenomenon. J Mol Biol 2007;372:817–45.

[14] Hernandez-Montes G, Diaz-Mejia JJ, Perez-Rueda E, Segovia L. The hidden universal distribution of amino acid biosynthetic networks: a genomic perspective on their origins and evolution. Genome Biol 2008;9.R95.

[15] Hla T. Genomic insights into mediator lipidomics. Prostaglandins Other Lipid Mediat 2005;77:197–209.

[16] Jensen RA. Enzyme recruitment in evolution of new function. Annu Rev Microbiol 1976;30:409–25.

[17] Klein CC, Cottret L, Kielbassa J, Charles H, Gautier C, Ribeiro de Vasconcelos AT, et al. Exploration of the core metabolism of symbiotic bacteria. BMC Genomics 2012;13:438.

[18] Kolar M, Meier J, Mustonen V, Lassig M, Berg J. GraphAlignment: Bayesian pairwise alignment of biological networks. BMC Syst Biol 2012;6:144.

[19] Light S, Kraulis P, Elofsson A. Preferential attachment in the evolution of metabolic networks. BMC Genomics 2005;6:159.

[20] Oates ME, Stahlhacke J, Vavoulis DV. The SUPERFAMILY 1.75 database in 2014: a doubling of data. Nucleic Acids Res 2014;43:D227–33.

[21] Okuda S, Yamada T, Hamajima M, Itoh M, Katayama T, Bork P, et al. KEGG Atlas mapping for global analysis of metabolic pathways. Nucleic Acids Res 2008;36:W423–6.

[22] Pinter RY, Rokhlenko O, Yeger-Lotem E, Ziv-Ukelson M. Alignment of metabolic pathways. Bioinformatics 2005;21:3401–8.

[23] Reed LJ, Hackert ML. Structure–function relationships in dihydrolipoamide acyltransferases. J Biol Chem 1990;265:8971–4.

[24] Ryan A, Wang CJ, Laurieri N, Westwood I, Sim E. Reaction mechanism of azoreductases suggests convergent evolution with quinone oxidoreductases. Protein Cell 2011;1:780–90.

[25] Schilling S, Wasternack C, Demuth HU. Glutaminyl cyclases from animals and plants: a case of functionally convergent protein evolution. Biol Chem 2008;389:983–91.

[26] Silvela J, Portillo J. Breadth-first search and its application to image processing problems. IEEE Trans Image Process 2001;10:1194–9.

[27] Tohsato Y, Matsuda H, Hashimoto A. A multiple alignment algorithm for metabolic pathway analysis using enzyme hierarchy. Proc Int Conf Intell Syst Mol Biol 2000;8:376–83.

[28] Wang J, Nemeria NS, Chandrasekhar K. Structure and function of the catalytic domain of the dihydrolipoyl acetyltransferase component in Escherichia coli pyruvate dehydrogenase complex. J Biol Chem 2014;289:15215–30.

[29] Wernicke S, Rasche F. Simple and fast alignment of metabolic pathways by exploiting local diversity. Bioinformatics 2007;23:1978–85.

[30] Yang Q, Sze SH. Path matching and graph matching in biological networks. J Comput Biol 2007;14:56–67.

19

Possible Biomarkers for the Early Detection of HIV-associated Heart Diseases: A Proteomics and Bioinformatics Prediction

Suraiya Rasheed *, Rahim Hashim, Jasper S. Yan

Laboratory of Viral Oncology and Proteomics Research, Keck School of Medicine, University of Southern California, Cancer Research Laboratory Building, 1303 North Mission Rd, Los Angeles, CA 90033, USA

ARTICLE INFO

ABSTRACT

Keywords:
HIV-cardiomyopathy
Ryanodine-receptors
ITPR
PI3K
Fetal-cardiac myosin
Myosin light-chain kinase

The frequency of cardiovascular disorders is increasing in HIV-infected individuals despite a significant reduction in the viral load by antiretroviral therapies (ART). Since the CD4$^+$ T-cells are responsible for the viral load as well as immunological responses, we hypothesized that chronic HIV-infection of T-cells produces novel proteins/enzymes that cause cardiac dysfunctions. To identify specific factors that might cause cardiac disorders without the influence of numerous cofactors produced by other pathogenic microorganisms that co-inhabit most HIV-infected individuals, we analyzed genome-wide proteomes of a CD4$^+$ T-cell line at different stages of HIV replication and cell growth over >6 months. Subtractive analyses of several hundred differentially regulated proteins from HIV-infected and uninfected counterpart cells and comparisons with proteins expressed from the same cells after treating with the antiviral drug Zidovudine/AZT and inhibiting virus replication, identified a well-coordinated network of 12 soluble/diffusible proteins in HIV-infected cells. Functional categorization, bioinformatics and statistical analyses of each protein predicted that the expression of cardiac-specific Ca^{2+} kinase together with multiple Ca^{2+} release channels causes a sustained overload of Ca^{2+} in the heart which induces fetal/cardiac myosin heavy chains (MYH6 and MYH7) and a myosin light-chain kinase. Each of these proteins has been shown to cause cardiac stress, arrhythmia, hypertrophic signaling, cardiomyopathy and heart failure ($p = 8 \times 10^{-11}$). Translational studies using the newly discovered proteins produced by HIV infection alone would provide additional biomarkers that could be added to the conventional markers for an early diagnosis and/or development of specific therapeutic interventions for heart diseases in HIV-infected individuals.

1. Introduction

The incidence of the acquired immunodeficiency syndrome (AIDS) has been successfully halted by a combination of antiviral therapies (ART), which suppress the human immunodeficiency virus (HIV)-load in the plasma of HIV-infected patients. However, despite the reduction of virus load, a significant number of HIV-infected individuals develop multiple immune-response and metabolic diseases including multiple cardiac disorders [1,3,21,26,31,32,38,45,49]. Global studies and Data Collection on Adverse Events of Anti-HIV Drugs (D: A:D) have indicated that although the traditional cardiovascular risk factors such as elevated lipid levels, diabetes and hypertension are common among both the HIV-infected and uninfected controls, a 26% increased risk of myocardial infarction and dyslipidemia could be associated to ART exposure [14,15,67]. This phenomenon was increased among women even when adjusted for other risk factors [17,62].

While many different cell types in the body are susceptible to HIV-replication, the bulk of the virus load in the blood comes from the peripheral mononuclear cells including CD4$^+$ T-cells, macrophage and dendritic cells. Earlier studies by our group had demonstrated that infants who died of "sudden death" syndrome had significant quantities of HIV-infected T-cells in the heart, and HIV could be cultured from pericardial effusions of these infants [27]. Our earlier studies had also indicated that HIV-infected cells produce novel proteins/enzymes, which protect the HIV-infected cells from apoptotic pathways, enhance fatty acid synthesis and induce a wide range of metabolic abnormalities which are involved in increased dyslipidemia in HIV-infected individuals [56].

Adult postmortem studies have also indicated that HIV-infected-activated CD4$^+$ T-cells are recruited in the left ventricular myocardium in hearts from HIV-infected subjects [36]. The HIV-infected, T-cells remain in an activated state due to constant antigenic stimulation in both treated and untreated individuals [2]. A subset of T-cells present in the heart could be stained by immunocytochemistry for cyclooxygenase 2 (COX-2) and inducible nitric oxide synthase, indicating that HIV-infected T-cells present in the heart played an important role in the progression of myocarditis [36].

* Corresponding author
 E-mail address: srasheed@usc.edu (S. Rasheed).

The T-cell activation has been associated with dysregulation of intracellular Ca^{2+} signaling and cell survival with suppression of apoptotic pathways [4,8,9,33]. While only a brief spike of Ca^{2+} is needed for most normal cellular functions, an increased intracellular calcium (Ca^{2+}) is essential for the contraction and relaxation of heart muscles [6,28,64]. We therefore tested a hypothesis that HIV-infected/activated CD4$^+$ T-cells express abnormal protein factors/enzymes that can infect heart tissue locally, and cause myocarditis, inflammatory responses and cardiac dysfunctions resulting in arrhythmia, cardiomyopathy and associated heart diseases.

In this report, we present experimental evidence that HIV-infection alone can enhance a well-coordinated expression of different types of Ca^{2+} release channels and kinases which have been shown to trigger overexpression of both fast and slow cardiac/fetal myosins together with a myosin light-chain kinase (MYH6-alpha, MYH7-beta and MYLK respectively) that are critical for the contraction and relaxation of heart muscles. Each of these proteins has been predicted by multiple bioinformatics tools including the binomial probability distribution to define the statistical significance of proteins associated with the development of arrhythmia, heart failures and cardiomyopathy in *HIV-uninfected* population groups ($p = 8 \times 10^{-11}$). Inclusion of these unconventional biomarkers with the conventional risk factors would help in translational studies leading to early detection of HIV-associated cardiac diseases and possibly novel therapeutic interventions for specific disorders.

2. Study Design, Materials and Methods

2.1. Cells and Virus

A major consideration in the design of the present study has been to choose an *in vitro* system in which molecular changes in the signaling proteins solely due to long-term HIV-infection (at least 6–7 months) could be evaluated sequentially at different time points. While numerous cell types including the freshly isolated peripheral blood mononuclear cells (PBMCs) or cardiomyocytes can be infected with HIV *in vitro*, these cells are not suitable for long-term infections as they die due to cytotoxicity or apoptosis within 7–10 days of infection. Further, the cytopathic indices of HIV-infections in primary cells vary from person to person due to genetic heterogeneity of human population groups and differences in the susceptibility of different human cell types to HIV replication (Rasheed, personal observations). We also could not conduct proteomics studies directly in HIV-infected individuals without first identifying protein profiles of cells infected by HIV alone, because most HIV-patients are co-infected with other pathogenic viruses and microorganisms that may produce factors similar to those induced by HIV. We therefore chose a genetically stable, single-cell-clone of a human T-cell line (RH9) [55] and infected it with a biologically cloned North American HIV-Clade B (HBX) *in vitro*. The differentially-regulated proteins expressed in HIV-infected and counterpart uninfected cells were evaluated at numerous time points from 1.5-, 3,6-, 12- and 24-h, followed by weekly, then 4–6 weeks samplings for 6–7 months post-infection. In addition, proteomes were also studied from a chronically-infected RH9 cell line that has been maintained in our lab for decades as it can sustain HIV-replication for prolonged periods *in vitro* without exhibiting major cytopathic effects or apoptosis [54, 56]. If these infected cultures showed excessive apoptotic cells due to overproduction of virus particles or viral proteins (gp120, Tat), in a crowded culture we harvested an aliquot of these cells for testing and added 1–2 $\times$ 10^6 per ml uninfected RH9 cells from the same cell-stock to maintain virus infected cultures. The uninfected cells were added to the infected cells approximately after 2–3 months of continuously maintaining these cells *in vitro*. An advantage of this regimen of analysis was that both newly induced cellular proteins and those produced by chronically infected cells were detected reproducibly multiple times. If

a protein was not detected at least twice, it was not included in the analyses.

2.2. Protein Isolation and Proteomics Analyses

We have used a rapid lysis method that was developed in our laboratory several years ago [54,56]. A sequential protein extraction technique has been extremely helpful in studying the dynamics expression of proteins associated with the plasma membrane, extracellular matrix, cytoplasm and nucleus. The membrane and extracellular matrix proteins were isolated by lysing cells for 15 s in solution containing 8 M Urea, 2% (w/v) CHAPS, 2% mercaptoethanol, 2.5% protease inhibitor cocktail, and 150 units/200 μl endonuclease. Each lysate was gently sonicated for 2 s then clarified first at 14,000 rpm followed by centrifugation at high speed (100,000 $\times$g) for 90 min. Soluble proteins in the *supernatants* were used to fractionate by 2-dimensional gradient (6–18%) gel electrophoresis (2DE). Cell-pellets were solubilized again with 9 M urea, 2 M thiourea and 4% CHAPS with other reagents being the same as above. This process was repeated one more time. More than 200 gels were evaluated and differentially expressed proteins from HIV-infected and uninfected controls cells were compared by using PDQuest software (BioRad). The presence or absence of each protein was confirmed by the matrix-assisted laser desorption ionization time of flight mass spectrometry (MALDI-TOF-MS). Only those gel-spots that reproducibly confirmed proteins at least two times by MALDI-TOF-MS were included in bioinformatics and statistical analyses. We also preferred to use MALDI-MS for this analysis than the liquid chromatography (LC) MS because proteins of interest could be reproducibly confirmed by re-sampling protein spots *manually* from multiple gels and both the high molecular weight proteins and small peptide fragments could be identified by MALDI-MS accurately. Thus, the specificity and reproducibility of membrane and extracellular matrix protein identifications were favored in this study over the total number of proteins (sensitivity) detected within a proteome.

2.3. Inhibition of Virus Replication

To validate protein expression profiles before and after infection, we inhibited chronic virus replication (in RH9 cell line that was infected for several months), by treating cells with a potent anti-reverse transcriptase agent, Zidovudine/AZT for 48 h *in vitro*. Protein profiles from HIV-infected, uninfected and AZT-treated and untreated counterpart cells were compared with each other by the use of a software (PDQuest-BioRad) and differentially regulated proteins in AZT-treated/-untreated cells, with and without HIV-infection were identified by mass spectrometry.

2.4. Bioinformatics and Statistical Analyses

Presence or absence of differentially expressed (upregulated, downregulated and *de novo* expressed) proteins, functional categorization, bioinformatics and statistical analyses were conducted by the use of multiple tools (Ingenuity Pathway Analysis (IPA) Systems, Expasy/UniProt), and various analytical programs available through the National Center of Biotechnology Institute (NCBI), and other public databases. The expected or E-values for all proteins associated with the calcium-release channels and cardiac myosins are from the Global Functional Analysis Program embedded in the IPA. The total number of proteins that have been shown to be statistically significant was also calculated by using the right-tailed Fisher Exact Test and IPA systems analyses.

3. Results

Replication of HIV in CD4$^+$ T-cell line (RH9) was highly efficient and proteomes could be analyzed sequentially in the absence of excessive cytotoxicity or cell death. These proteins represented changes due to numerous phases of virus replication and cell growth over >6 months. The protein expression profiles from HIV-infected cells were compared with counterpart uninfected cells (with and without treatment with antiretroviral drug Zidovudine/AZT) by subtractive proteomics analysis and differentially expressed proteins were categorized as follows: 1) *de novo* induced proteins were produced only by HIV-infected cells; 2) upregulated proteins are those that were expressed in uninfected control cells in small amounts but were increased in quantity (overexpressed) by HIV-replication and 3) proteins that are downregulated or their expression was completely turned off after HIV-infection. All differentially regulated proteins between HIV-infected and uninfected cells were identified in multiple gels and proteins that were not confirmed at least twice by mass spectrometry were discarded from the analyses.

Functional categorization of several hundred proteins identified 12 proteins that have been shown to be involved in regulating calcium (Ca^{2+}) homeostasis and affect heart muscle functions (MYH6 and MYH7) (Table 1). The Ca^{2+} regulating proteins ($n = 9$) were either upregulated or induced *de novo* post-HIV infection. Each of these proteins could be divided into distinct classes of Ca^{2+}-release channels, Ca^{2+} regulatory kinases, Ca^{2+}-binding and Ca^{2+} cycling proteins (Table 1). In addition, we have identified embryonic cardiac myosin heavy chains (myosin-6 and myosin-7) and light chain myosin kinase (MYLK) ($n = 3$) that were coexpressed in HIV-infected cells only but were not detected in any of the uninfected cells or in cells that were treated with AZT. These newly identified proteins have been localized to the extracellular matrix, plasma membrane or endoplasmic reticulum although some of these proteins have been shown to shuttle back and forth from the cytoplasm to other organelles including the nucleus. Full names, abbreviations and accession numbers of all proteins are according to the Swiss-PROT/Uni-PROT global databases (Table 1).

Bioinformatics and statistical analyses of each of the 12 proteins indicated that the calcium signaling proteins are critical for regulating Ca^{2+} influx and efflux; and cardiac myosins and myosin kinase are involved in cardiac morphogenesis in the embryo and are essential for controlling the contraction and relaxation of the heart muscle.

In the following sections, we have described the properties and functionalities of each of these 12 proteins and discussed their associations with T-cell activation, Ca^{2+} regulation, Ca^{2+} signaling and development

of heart diseases. Each of these proteins has been *independently* linked to arrhythmia, heart failure, cardiomyopathy or sudden cardiac death in *HIV-uninfected* individuals ($p = 6.1 \times 10^{-3} - 1.8 \times 10^{-7}$).

3.1. Dysregulation of Calcium-release Channels and Regulatory Kinase

Calcium is an essential element for life and its expression is controlled by extremely intricate mechanisms of Ca^{2+} release channels and pumps embedded in the plasma membranes and endoplasmic reticuli of all cell types in the body. While only small amounts of Ca^{2+} are needed for the sustenance of most cell types, the muscle and nerve cells utilize higher amounts of Ca^{2+} for contraction and relaxation. Among the main Ca^{2+} regulatory proteins, 9 were dysregulated during HIV-replication, and the significance of each of these proteins has been defined in relation to the cardiac functions by multiple bioinformatics and statistical tools.

3.1.1. The Ryanodine Receptors or the Type-1 Calcium-Release Channels

The type-1 calcium release channels include the ryanodine receptors RyR1, RyR2 and RyR3 that are responsible for the release of Ca^{2+} intracellularly. While all three isomers of RyRs were detected in the HIV-infected and uninfected $CD4^+$ T-cells, the expression of RyR1 was the same or very slightly downregulated (not significant) in HIV-infected cells compared to the uninfected control cells; and both RyR2 and RyR3 were upregulated post-HIV-infection (Fig. 1).

The RyR2 calcium release channel was detected more frequently (19 times) in HIV-infected T-cells compared to the uninfected counterpart cells in which it was detected only 5-times. RyR2 is the cardiac muscle-type receptor, which is essential for the regulation of the excitation–contraction coupling (ECC) or the contraction and relaxation rhythms of the heart muscle [5,11,12].The ECC depends directly on the opening and closing of RyR-Ca^{2+} release channels at specific rates while these proteins also regulate systolic and diastolic functions of the striated cardiac and skeletal muscles [50,65,69]. RyR2 has also been shown to mediate arrhythmia and sudden-death syndrome [5]. RyR3 protein mobilizes stored Ca^{+2} in both cardiac and skeletal muscle to initiate muscle contraction and dysregulation of both RyR2 and RyR3 that have been associated with cardiac diseases [11,12].

Table 1
HIV-modulated proteins associated with cardiovascular functions and dysfunctions.

Protein name	Abbrev	Accession #	Function/diseases	p-Value
Myosin heavy chain, cardiac muscle alpha isoform	MYH6	P13533	"Fast" ATPase used to hydrolyze ATP in the heart and causes high-velocity muscle 2	0.000135
Myosin heavy chain, cardiac muscle beta isoform	MYH7	P12883	"Slow" ATPase used to hydrolyze ATP in the heart and causes slow-velocity muscle contraction	0.000440
Myosin light chain kinase, smooth muscle and non-muscle isozymes	MYLK	Q15746	Calcium/calmodulin-dependent enzyme involved in smooth muscle contraction via phosphorylation of myosin light chains; essential in gap junction formation and permeability	0.000440
Ryanodine receptor 1	RYR1	P21817	Malignant hyperthermia, central core disease (increased heart rate, respiratory insufficiency), arrhythmia, heart failure	0.000722
Ryanodine receptor 2	RYR2	Q92736	Heart failure, atrial fibrillation, catecholaminergic polymorphic ventricular tachycardia	0.000879
Ryanodine receptor 3	RYR3	Q15413	Abnormal contraction of skeletal muscle	0.00615
Inositol 1,4,5-trisphosphate receptor type 1	ITPR1 (or IP3R1)	Q14643	Myocardial hypertrophy, leaky channels, altered calcium signaling, contractile dysfunction and cardiac arrhythmias	0.00615
Inositol 1,4,5-trisphosphate receptor type 2	ITPR2 (or IP3R2)	Q14571	Cardiac hypertrophy, ventricular arrhythmia, atrial fibrillation	0.00615
Sarcoplasmic/endoplasmic reticulum calcium ATPase 2	AT2A2 (SERCA2)	P16615	Abnormal contraction/relaxation cycles, dilated cardiomyopathy, heart failure, cardiac death	0.00615
Calumenin	CALU	O43852	Ca^{2+} binding; regulates RYRs and coagulates blood	0.0109
Calcium/calmodulin-dependent protein kinase type II alpha chain	CaMKII	Q96RR4	Regulates SERCA-related Ca^{2+} release	Detected in small quantities
Phosphatidylinositol-4-phosphate 3-kinase C2 beta	PI3K (P3C2B)	O00750	Binds to its receptors ITPR1 and ITPR2; regulates myocardial contractility	0.00288

Table 1: Full names, abbreviations, accession numbers (UniProt) and functional significance of HIV-modulated proteins identified to be associated with cardiovascular functions dysfunctions.
All 12 proteins are located in the plasma membrane/endoplasmic reticulum membrane. An overexpression of these proteins leads to cardiac muscle damage, tachycardia, heart failure, contractile dysfunction, cardiac hypertrophy, ventricular arrhythmia, heart failure and/or cardiac death.

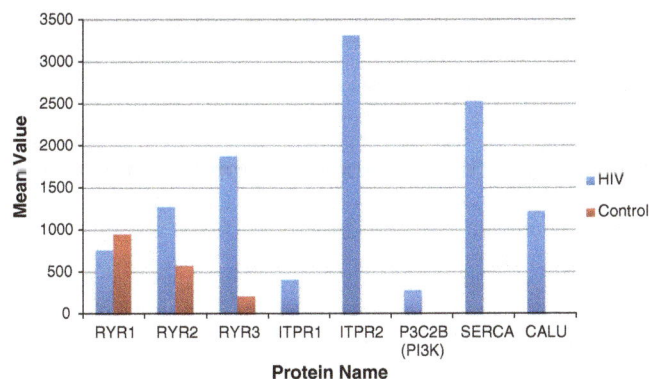

Fig. 1. Proteomics analyses of Ca^{2+} release channels and signaling proteins; average quantity of each protein detected in HIV-infected cells (blue bars) versus uninfected counterpart controls (red bars). From left to right Type 1 Calcium Release Channels or Ryanodine Receptors 1,2 &3.: RYR1, 2 & 3 = ITPR1 and ITPR2 followed by Type 2 Calcium Release Channels, PI3K = phosphatidylinositol-4-phosphate 3-kinase C2 beta; SERCA = sarcoplasmic/endoplasmic-reticulum calcium ATPase2; CALU = Calumenin, a calcium-binding protein. ITPR1, ITPR2, PI3K, SERCA and CALU were expressed exclusively in HIV-infected cells and were not detected in any of the numerous counterpart uninfected cells tested.

Whereas the expression of RyR2 and RyR3 has not been reported previously in relation to HIV-infection of T-cells, HIV infection of T-cells is activated due to increased intracellular Ca^{2+} levels [33]. An increased intracellular Ca^{2+} level in the muscles has also been shown to affect excitation–contraction coupling in cardiac muscle in experimental mice [16]. Our proteomics and bioinformatics analyses predict that the presence of HIV-infected-activated T-cells in the heart would enhance the release of Ca^{2+} levels *in vivo* and a sustained overload of calcium would dysregulate expression of RyR2 and RyR3 in the heart. This would conceivably be predicted to induce oxidative stress and enhanced susceptibility to myocarditis, abnormal ECC, arrhythmia, ventricular dysplasia, heart failure and/or hypertrophy and cell death in HIV-infected individuals [20]. These functionalities are statistically significant ($p = 5.0 \times 10^{-3}$).

3.1.2. The Type-2 Ca^{2+} Release Channels or the Inositol 1,4,5-Trisphosphate Receptors

The type-2 Ca^{2+} release channels belong to a distinct class of inositol 1,4,5-trisphosphate receptors (ITPR1 and ITPR2; together known as IP3Rs). While these channels have been reported to be expressed on different types of human and animal cells including human T-cells [60,65], we have identified both ITPR1 and ITPR2 to be expressed exclusively in HIV-infected cells, and these receptors were not detected among several thousands of protein spots tested by mass spectrometry from the uninfected counterpart cells at different stages of cell growth overtime (Fig. 1). The ITPR1 enhances intracellular Ca^{2+} stores and has been shown to interact with the HIV-encoded Nef protein in primary human peripheral mononuclear cells [37].

Inositol 1,4,5-trisphosphate (IP3) is a soluble factor that binds to IP3Rs present on the SERCA located in the endoplasmic reticulum (see Section 3.1.3 below). IP3 also triggers the release of Ca^{2+} into the cytosol and increases its quantity in the cell. Both the type 1 and type 2 classes of Ca^{2+} release channels (*i.e.* RyRs and ITPRs) are located in close proximity of each other and therefore they are phosphorylated during T-cell activation in the channel pore and work in harmony to maintain and regulate cardiac rhythm [19,20,24,65]. Another study suggested that primary macrophages exposed to HIV-encoded Tat protein also release intracellular calcium from IP3-regulated calcium stores which increases production of TNF-alpha [40]. However, these studies have not reported detection of ITPR2, RyRs, Ca^{2+} regulatory kinases, Ca^{2+} binding proteins identified in the present research (Fig. 1). Thus, our bioinformatic analyses of the co-expressed type 1 and type 2 classes

of Ca^{2+} release channels (*i.e.* RyR2 and ITPR2) in the T-cells and overexpression of PI3K and Ca^{2+}-kinase (CamK11) (see below) and other Ca^{2+} signaling or Ca^{2+} binding proteins would be predicted to dysregulate Ca^{2+} cycling in the heart of HIV-infected individuals ($p = 10^{-4}$).

3.1.3. Sarcoplasmic/Endoplasmic Reticulum Ca^{2+} ATPase2 (SERCA)

Sarcoplasmic/Endoplasmic Reticulum (SER) Ca^{2+} ATPase2 (SERCA2 or ATP2A2) was expressed exclusively in our experimentally HIV-infected T-cells and was not detected in any of the thousands of protein-spots tested from uninfected counterpart T-cells at various stages of cell growth (Fig. 1). SERCA are Ca^{2+} ATPase pumps involved in regulating Ca^{2+} release for cellular homeostasis in health and disease [13,22,39,63]. This highly specialized ATPase controls the amount of free calcium in cardiac muscle fibers via Ca^{2+} release-channels that are located in SERCA [28,33,41].

The stored Ca^{2+} is released from SERCA through two different classes of calcium-release channels described above (*i.e.* IP3R and RyRs). Under normal conditions, Ca^{2+} is pumped into SERCA and is released by IP3R or RyRs. The ratio between the expression levels of ITPRs to RyRs is therefore important for cardiac functions. Depolarization of sarcoplasmic/endoplasmic reticulum membrane triggers the contraction of muscles and when the Ca^{2+} pumps re-accumulate Ca^{2+} in the store the muscles relax [28,41,50]. Thus, dysregulation of SERCA causes defective interactions with Ca^{2+} release channels (RyRs or ITPRs), which results in a wide range of abnormalities in several organs including dysregulated contraction/relaxation cycles, dilated cardiomyopathy, heart failure and cardiac death [22] ($p = 0.00615$).

3.1.4. Calcium/Calmodulin-dependent Serine–Threonine Protein Kinase Type II Alpha Chain

The calcium/calmodulin-dependent serine–threonine protein kinase type II alpha chain (CaMKII) was detected only once in HIV-infected cells but in none of the counterpart uninfected cells. While we have not included CaMK11 in any statistical analysis, our bioinformatics analyses suggest that the expression of CaMKII may have been *suppressed* in HIV-infected cells by the overexpression of protein kinase C (PKC) a highly efficient serine/threonine kinase that we described earlier to be expressed exclusively in HIV-infected cells [54]. PKC could also compensate the functions of CaMKII especially in the presence of PI3K and multiple other kinases that were expressed concomitantly in HIV-infected cells [54,56]. However, it has been shown that even extremely small quantities of CaMKII can activate RyR2 by phosphorylation, which has been implicated in arterial fibrillation, cardiac arrhythmia, cardiac hypertrophy and heart failure [35].

3.1.5. Calumenin (CALU), a Calcium-Binding Protein

Calcium-binding proteins are critical for modulating Ca^{2+} release through RyRs or IP3Rs [25]. Calumenin (CALU) was upregulated exclusively in HIV-infected T-cells and it was not detected in the uninfected cells (Table 1; Fig. 1). An important characteristic of CALU is that it contains six Ca^{2+}-binding motifs in the amino-terminus and a tetrapeptide His-Asp-Glu-Phe, both of which are essential for maintaining protein-protein interactions in SERCA [58,68]. CALU is normally expressed in the endoplasmic reticulum of many cell types as it is involved in protein-folding and sorting in SERCA [25]. Persistent contact of cardiac cells with Ca^{2+}-regulatory proteins such as CALU is therefore critical for modifying contraction-relaxation signals of the heart muscle ($p = 0.0109$) [58].

3.1.6. Phosphatidylinositol-4-Phosphate 3-Kinase C2 Beta (P3C2B/PI3Kinase)

The phosphatidylinositol-4-phosphate 3-kinase C2 beta (P3C2B) was upregulated in HIV-infected T-cells (Fig. 1). This enzyme contains three critical protein–protein interaction domains (phosphoinositide-3 kinases (PI3Ks), C2 and phox-homology domain) that are important for regulating cell signaling, myocardial contractility and cell survival [47,48,52,70]. PI3K is an important enzyme for normal heart functions as it binds to its receptors ITPR1 and ITPR2, both of which were expressed only in HIV-infected but not in uninfected T-cells (Fig. 1). The overexpression of several different tyrosine kinases in HIV-infected cells raises intracellular calcium concentration by activating phospholipases which generate inositol 1,4,5-trisphosphate (ITP) [24], that binds to its receptors. Concomitant expression of PI3K and its interacting receptors (ITPR1 and ITPR2) in the HIV-infected T-cells are therefore predicted to alter myocardial functions through dysregulated Ca^{2+} leaks in the heart [66].

3.2. Dysregulation of Cardiac Muscle Myosin Heavy Chains and a Light Chain Kinase

3.2.1. Cardiac Muscle Myosin Heavy Chains Alpha MYH6 and Beta MYH7

Myosins are molecular motors that form complexes with actin and move along the actin filaments while activating myosin ATPases and hydrolyzing ATP. These interactions convert chemical energy into mechanical forces essential for the contraction and relaxation of muscles [34]. While multiple myosins were expressed exclusively in HIV-infected T-cells, two of the most prominent cardiac related myosin proteins produced post-HIV-infection were fetal cardiac muscle myosin heavy chains alpha MYH6 and beta MYH7 (also known as MHC-alpha and MHC-beta respectively) (Fig. 2). These myosins were not detected in numerous samples examined from uninfected T-cells or in any of the HIV-infected cells treated with AZT. MYH6 is a "Fast" ATPase which hydrolyzes ATP in the heart and causes high-velocity muscle contraction. MYH7 is a "Slow" ATPase that is associated with slow-velocity muscle contraction. MYH6 also uses greater amounts of energy for shorter muscle contractions while MYH7 uses smaller amounts of energy for long-term contractions [34]. Thus, the cardiac energy levels depend on the amount of MYH6 and MYH7 expressed and a controlled regulation of these myosins is essential for maintaining a balanced level of the excitation–contraction coupling.

While the exact mechanisms by which these fetal genes affect cardiac functions are not well-understood, comparative studies in the coexpression networks of cardiac genes in the fetal (developing) myocardium, failing myocardium and cardiac hypertrophy, identified MYH6, MYH7 and SERCA to be the most critical fetal genes to be expressed in fetal heart and diseased tissues [10]. The fetal genes MYH6 and MYH7, are essential for the growth and morphogenesis of cardiac tissue during embryonic development; and an overexpression of both MYH6 and MYH7 has been shown to be significantly associated with atrial septal-defects, arrhythmia, cardiomyopathy, myocarditis,

microvascular-dysfunctions, hypertrophy, and heart failure [7,18,46, 51]. Both cardiac hypertrophy and heart failure have been associated with the re-activation of the set of fetal cardiac genes, which are suppressed postnatally and are substituted by the adult cardiac genes [10]. The levels of MYH6 and MYH7 expression have also been considered as markers for the severity of myopathy [23,30,43,53].

These observations and our bioinformatics analyses of differentially regulated proteins not only validate our hypothesis that HIV induces unique (fetal) genes MYH6 and MYH7 in chronically infected T-cells, but also demonstrate that the upregulation of both MYH6 (alpha MHC) and MYH7 (beta MHC) in HIV-infected cells causes an overabundance of Ca^{2+} and ATPase in the heart muscles, accounting for the cardiac stress and heart failure, particularly in patients whose hearts may be infiltrated with HIV-infected T-cells. These proteins could be predicted to be used as unconventional biomarkers of arrhythmia and cardiomyopathy.

3.2.2. Myosin Light Chain Kinase

Myosin light chain kinase (MYLK) is a calcium/calmodulin-dependent enzyme which was co-expressed with MYH6 and MYH7 in association with multiple Ca^{2+} release channels described above. These proteins were expressed exclusively in HIV-infected cells but not in the numerous uninfected cells tested (Fig. 2). The calcium/calmodulin-dependent enzyme is involved in smooth muscle contraction via phosphorylation of myosin light chains and is essential in gap junction formation and permeability ($p = 0.000440$). MYLK also activates multiple cellular processes including muscle contraction, cell morphology, cell movement and it affects coronary, endothelial and microvascular dysfunctions.

Overexpression of MYLK has also been directly associated with numerous abnormalities in the vasculature as it compromises the integrity of endothelium in the coronary microvasculature [61]. Most importantly, the expression of MYLK is critical for transferring energy to MYH6 such that it can perform both smooth muscle and non-muscle functionalities and maintains basal levels of contractions via phosphorylation of cardiac myosin light chains.

3.3. Inhibition of HIV-Replication Prevents Expression of Myosins and Ca^{2+} Signaling Proteins

To define the specificity of HIV-modulated proteins and to validate the relationship between HIV-replication and induction of Ca^{2+} signaling and cardiac myosins, we treated both HIV-infected and uninfected RH9 T-cells with Zidovudine/AZT, a potent inhibitor of reverse transcriptase and HIV-replication [42,44]. Comparisons of proteomes from

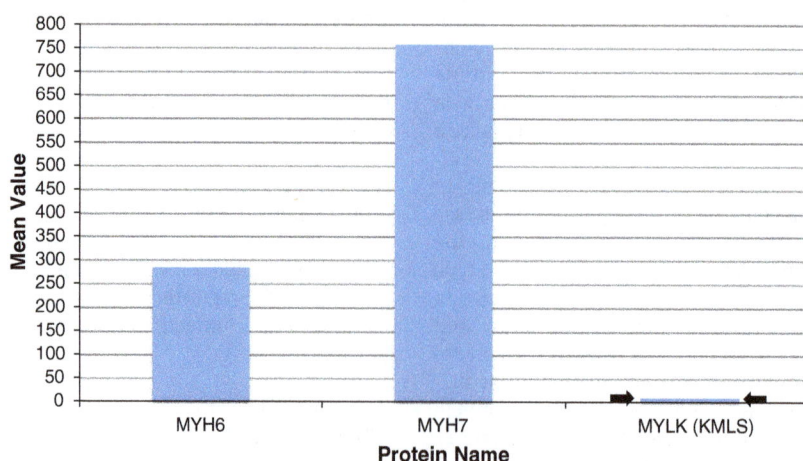

Fig. 2. Mean values of cardiac myosins. Cardiac/fetal myosin 6 (heavy chain—fast ATPase), myosin 7 (heavy chain—slow ATPase) and myosin light chain kinase (MYLK). These cardiac myosins and associated kinase MYLK were expressed exclusively in HIV-infected cells (i.e. not detected in numerous samples tested from uninfected or AZT-treated infected and uninfected cells).

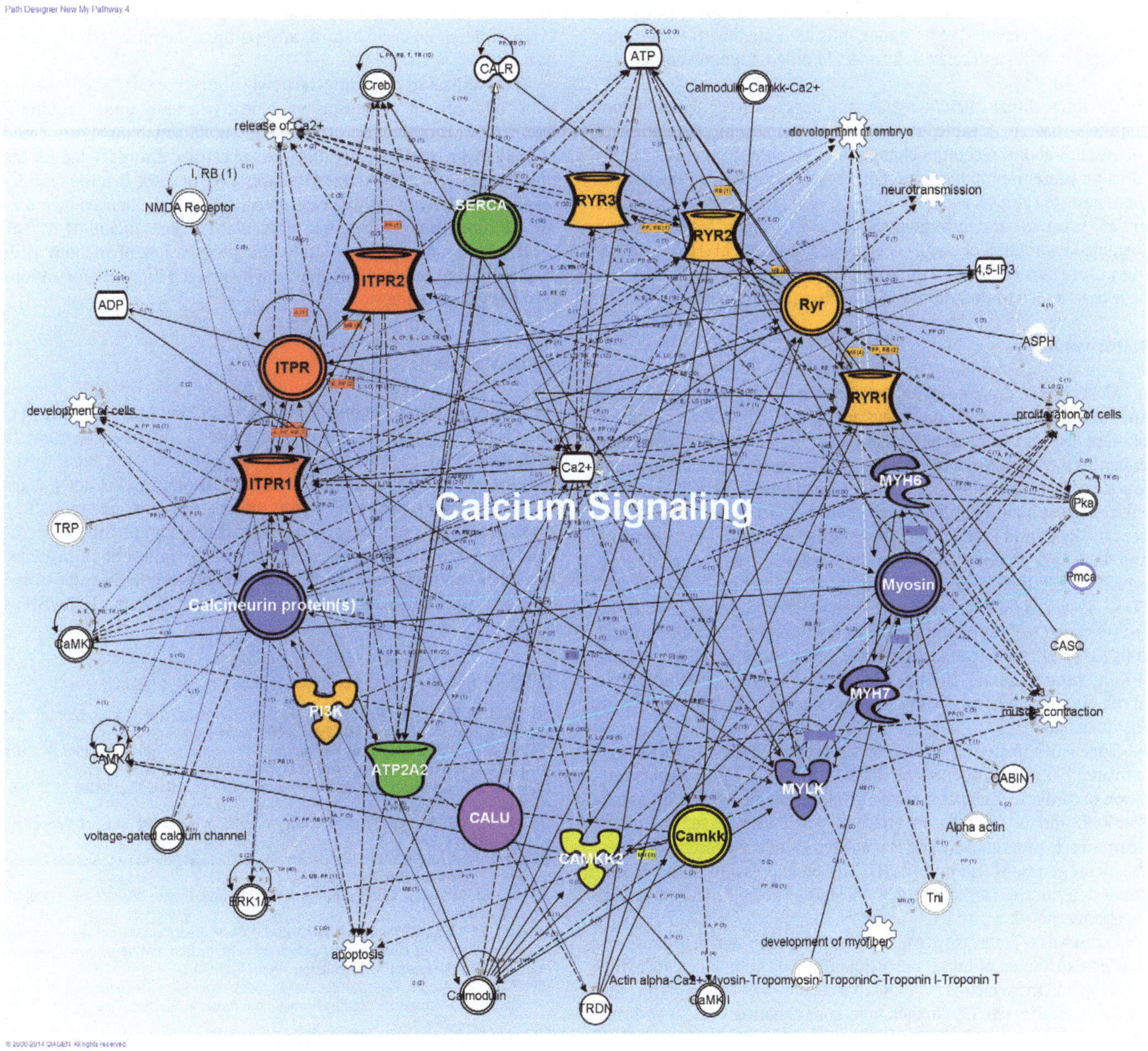

Fig. 3. Calcium signaling and canonical pathways were constructed using Ingenuity Pathway Analyses System. Protein–protein interaction pathways showing a well-coordinated network of interactions between proteins that were dysregulated post-HIV-infection (RYRs) or expressed exclusively in HIV-infected cells. All 12 proteins identified to be expressed or overexpressed in HIV-infected cell were found within the calcium signaling protein–protein interaction networks. All proteins are highlighted in different colors and the symbols represent the different families described in the following diagram.

Figure	▭	○	ℰ	✧	⊓	✂	⛢	○
Family	Chemical	Complex	Enzyme	Functions	Ion Channel	Kinase	Transporter	Other

AZT-treated, -untreated, HIV-infected and -uninfected RH9 T-cells, indicated that both HIV-infected and uninfected cells treated with AZT had distinct profiles not related to HIV-infected or uninfected control cells. With the exception of small quantities of PI3K, none of the HIV-modulated myosins, Ca^{2+} signaling cellular proteins or HIV-p24 antigen was detected in AZT-treated, HIV-infected cells. This is not surprising because PI3K is essential for cell-survival and it protects the AZT-treated cells (at least for short periods) from undergoing apoptosis [59]. These data indicate that HIV-replication was responsible for the modulation of multiple Ca^{2+} regulatory proteins and enzymes that

have been previously shown to be associated with cardiac muscle contraction.

3.4. Protein–Protein Interaction Pathways

To test if the newly discovered Ca^{2+} related proteins are structurally and functionally related, we constructed protein–protein-interaction maps using Ingenuity Pathway Analysis (IPA) bioinformatics program. As seen in Fig. 3 all differentially regulated Ca^{2+} receptors and Ca^{2+} signaling proteins identified herein interact with each other and with numerous other cellular proteins along the Ca^{2+} canonical pathways that have been reported to be significantly associated with arrhythmia, cardiomyopathies and related disorders in different population groups ($p = 2 \times 10^{-16}$).

The intracellular calcium-release and uptake are regulated by ion-channels that are located in the plasma membrane. Activation of membrane-bound receptors changes the membrane potential, which can alter protein–protein interaction pathways initiated at the plasma membrane to cytoplasm and the nucleus. We predict that abnormalities in Ca^{2+} channels and dysregulation of Ca^{2+}-cycling proteins induce a sustained expression of cardiac fetal gene MYH6 and MYH7 which can then cause cardiac dysfunctions, hypertrophy and heart failure that have been linked to alterations in ECC, arrhythmia and cardiomyopathy.

4. Discussion

Most chronic diseases including cardiac disorders are multifactorial and numerous molecular pathways are operative in the development of these diseases. Our earlier studies had indicated that HIV replication enhances production of free fatty acids and many key proteins including low density lipoproteins that disrupt normal lipid metabolism [56]. In this report we have identified a well-coordinated network of cardiac receptors, RyR2 and RyR3, IP3Rs Ca^{2+} release channels, Ca^{2+} signaling and regulatory proteins (Ca^{2+}-binding proteins, enzymes, and kinases) which induce fetal/cardiac MYH6 and MYH7 in chronically infected $CD4^+$ T-cells. While the ITPR1 has been shown to interact with the HIV-encoded Nef protein in primary human peripheral mononuclear cells [37] and HIV-encoded Tat protein has been reported to increase IP3 and production of TNF-alpha in primary macrophages [40], none of the other proteins has been shown previously to be associated with HIV-infection.

Since HIV-infected $CD4^+$ T-cells are present in the circulation and activated T-cells are routinely recruited to the heart, a perpetual stimulation of cardiac cells by abnormal Ca^{2+} leaks by the expression of cardiac RyR2 and Ca^{2+}-regulatory enzymes, would be expected to weaken heart muscle progressively and is predicted to induce expression of cardiac fetal genes, MYH6 and MYH7. The co-expression of these fetal genes deteriorates cardiac muscle, and causes arrhythmia and cardiomyopathy ($p = 2 \times 10^{-16}$).

Excessive Ca^{2+} release in the heart has also been shown to cause enhanced expression of numerous cytokines and chemokines post HIV-infection and these factors promote co-infections with other pathogenic viruses/microbes [29,57]. Further, numerous abnormal cellular and viral factors (HIV envelope-gp120 and Tat proteins) released in the hearts of HIV-infected individuals would cause oxidative stress locally and enhance susceptibility to myocarditis, arrhythmia, ventricular dysplasia and heart failure ($p = 5.0 \times 10^{-3}$). Thus, most chronic diseases are related to multiple etiological factors that are produced by different cell types in the body which alter our immune responses and affect the entire body's metabolon.

While *in vitro* findings cannot be directly implicated to the development of diseases *in vivo*, there is no precedence of a coordinated expression of Ca^{2+} release channels, Ca^{2+} regulatory enzymes and embryonic MYH6 and MYH7 genes/proteins in HIV-infected cells *in vitro or in vivo*. This could be because we studied the dynamics of protein expression over a long period of 6–7 months and identified proteins by a comprehensive and subtractive proteomics analyses in comparisons to the counterpart uninfected proteomes at different time points.

Our data have been supported by published reports in experimental animals in which MYH6 levels were significantly upregulated concurrently with the induced tachycardia and the left ventricular hypertrophy in mutant mice with familial hypertrophic cardiomyopathy [16]. While MYH6 and MYH7 have been associated with myocarditis, arrhythmias, heart failure, and related heart diseases in the non-HIV population groups ($p = 8 \times 10^{-11}$), our studies have provide strong experimental evidence that HIV infection *alone*, without any co-infection or treatment

can initiate an abnormal expression of novel myosin MYH6, MYH7 & MYLK proteins, involved in cardiomyopathy, arrhythmia and/or premature death ($p = 0.000440$).

Translational studies using the newly discovered Ca^{2+} signaling proteins and cardiac/fetal myosins would lead to a better understanding of how a new combination of clinically significant biomarkers could potentially be used for early diagnosis of cardiac disorders and for the future development of targeted therapies. We propose that the specificity and sensitivity of early disease detection can be enhanced by profiling additional unconventional molecular markers including MYH6, MYH7, RyRs and ITPRs to study the progression of in both HIV-infected and uninfected individuals who are at a high risk for various cardiovascular events.

Acknowledgment

We acknowledge technical assistance of Zisu Mao and Jane M.C. Chan with two-dimensional gel electrophoresis and mass spectrometry respectively. S.R. conceived, designed and performed the experiments with the help of technicians and wrote the manuscript. R.H. and J. Y. analyzed the original proteomics data, and made figures and graphs. R.H. constructed Ca^{2+} signaling canonical pathway and Jihyun (Rebecca) Shim modified the pathway and confirmed p-values of all proteins. The study was supported by the Rasheed Research Endowment Fund # 7256 at USC. R. H. and J. Y. were recipients of Summer Fellowships at USC.

References

[1] Aberg JA. Cardiovascular complications in HIV management: past, present, and future. J Acquir Immune Defic Syndr 2009;50:54–64.

[2] Bangs SC, McMichael AJ, Xu XN. Bystander T cell activation—implications for HIV infection and other diseases. Trends Immunol 2006;27:518–24.

[3] Barbaro G. Cardiovascular manifestations of HIV infection. Circulation 2002;106: 1420–5.

[4] Berridge MJ. The biology and medicine of calcium signalling. Mol Cell Endocrinol 1994;98:119–24.

[5] Blayney LM, Lai FA. Ryanodine receptor-mediated arrhythmias and sudden cardiac death. Pharmacol Ther 2009;123:151–77.

[6] Brini M, Carafoli E. Calcium pumps in health and disease. Physiol Rev 2009;89: 1341–78.

[7] Carniel E, Taylor MR, Sinagra G, Di LA Ku L, Fain PR, Boucek MM, et al. Alpha-myosin heavy chain: a sarcomeric gene associated with dilated and hypertrophic phenotypes of cardiomyopathy. Circulation 2005;112:54–9.

[8] Chami M, Oules B, Paterlini-Brechot P. Cytobiological consequences of calcium-signaling alterations induced by human viral proteins. Biochim Biophys Acta 2006; 1763:1344–62.

[9] Dadsetan S, Zakharova L, Molinski TF, Fomina AF. Store-operated Ca^{2+} influx causes Ca^{2+} release from the intracellular Ca^{2+} channels that is required for T cell activation. J Biol Chem 2008;283:12512–9.

[10] Dewey FE, Perez MV, Wheeler MT, Watt C, Spin J, Langfelder P, et al. Gene coexpression network topology of cardiac development, hypertrophy, and failure. Circ Cardiovasc Genet 2011;4:26–35.

[11] Durham WJ, Wehrens XH, Sood S, Hamilton SL. Diseases associated with altered ryanodine receptor activity. Subcell Biochem 2007;45:273–321.

[12] Eisner DA, Kashimura T, O'Neill SC, Venetucci LA, Trafford AW. What role does modulation of the ryanodine receptor play in cardiac inotropy and arrhythmogenesis? J Mol Cell Cardiol 2009;46:474–81.

[13] Erkasap N. SERCA in genesis of arrhythmias: what we already know and what is new? Anadolu Kardiyol Derg 2007;7(Suppl. 1):43–6.

[14] Friis-Moller N, Reiss P, Sabin CA, Weber R, Monforte A, El-Sadr W, et al. Class of antiretroviral drugs and the risk of myocardial infarction. N Engl J Med 2007;356: 1723–35.

[15] Friis-Moller N, Weber R, Reiss P, Thiebaut R, Kirk O, d'Arminio MA, et al. Cardiovascular disease risk factors in HIV patients—association with antiretroviral therapy. Results from the DAD study. AIDS 2003;17:1179–93.

[16] Gao WD, Perez NG, Seidman CE, Seidman JG, Marban E. Altered cardiac excitation-contraction coupling in mutant mice with familial hypertrophic cardiomyopathy. J Clin Invest 1999;103:661–6.

[17] Grinspoon SK. Metabolic syndrome and cardiovascular disease in patients with human immunodeficiency virus. Am J Med 2005;118(Suppl. 2):23S–8S.

[18] Guo Q, Xu Y, Wang X, Guo Y, Xu R, Sun K, et al. Exome sequencing identifies a novel MYH7 p.G407C mutation responsible for familial hypertrophic cardiomyopathy. DNA Cell Biol 2014;10:699–704.

[19] Harnick DJ, Jayaraman T, Ma Y, Mulieri P, Go LO, Marks AR. The human type 1 inositol 1,4,5-trisphosphate receptor from T lymphocytes. Structure, localization, and tyrosine phosphorylation. J Biol Chem 1995;270:2833–40.

[20] Harzheim D, Movassagh M, Foo RS, Ritter O, Tashfeen A, Conway SJ, et al. Increased InsP3Rs in the junctional sarcoplasmic reticulum augment Ca^{2+} transients and arrhythmias associated with cardiac hypertrophy. Proc Natl Acad Sci U S A 2009; 106:11406–11.

[21] Herskowitz A, Wu TC, Willoughby SB, Vlahov D, Ansari AA, Beschorner WE, et al. Myocarditis and cardiotropic viral infection associated with severe left ventricular dysfunction in late-stage infection with human immunodeficiency virus. J Am Coll Cardiol 1994;24:1025–32.

[22] Hovnanian A. SERCA pumps and human diseases. Subcell Biochem 2007;45:337–63.

[23] James J, Hor K, Moga MA, Martin LA, Robbins J. Effects of myosin heavy chain manipulation in experimental heart failure. J Mol Cell Cardiol 2010;48:999–1006.

[24] Jayaraman T, Ondrias K, Ondriasova E, Marks AR. Regulation of the inositol 1,4,5-trisphosphate receptor by tyrosine phosphorylation. Science 1996;272:1492–4.

[25] Jung DH, Mo SH, Kim DH. Calumenin, a multiple EF-hands Ca^{2+}-binding protein, interacts with ryanodine receptor-1 in rabbit skeletal sarcoplasmic reticulum. Biochem Biophys Res Commun 2006;343:34–42.

[26] Khunnawat C, Mukerji S, Havlichek Jr D, Touma R, Abela GS. Cardiovascular manifestations in human immunodeficiency virus-infected patients. Am J Cardiol 2008;102: 635–42.

[27] Kovacs A, Hinton DR, Wright D, Xu J, Li XL, Rasheed S, et al. Human immunodeficiency virus type 1 infection of the heart in three infants with acquired immunodeficiency syndrome and sudden death. Pediatr Infect Dis J 1996;15:819–24.

[28] Kranias EG, Bers DM. Calcium and cardiomyopathies. Subcell Biochem 2007;45:523–37.

[29] Krentz HB, Kliewer G, Gill MJ. Changing mortality rates and causes of death for HIV-infected individuals living in Southern Alberta, Canada from 1984 to 2003. HIV Med 2005;6:99–106.

[30] Krenz M, Robbins J. Impact of beta-myosin heavy chain expression on cardiac function during stress. J Am Coll Cardiol 2004;44:2390–7.

[31] Langston C, Cooper ER, Goldfarb J, Easley KA, Husak S, Sunkle S, et al. Human immunodeficiency virus-related mortality in infants and children: data from the pediatric pulmonary and cardiovascular complications of vertically transmitted HIV (P(2)C (2)) Study. Pediatrics 2001;107:328–38.

[32] Leonard EG, McComsey GA. Antiretroviral therapy in HIV-infected children: the metabolic cost of improved survival. Infect Dis Clin North Am 2005;19:713–29.

[33] Lewis RS. Calcium signaling mechanisms in T lymphocytes. Annu Rev Immunol 2001;19:497–521.

[34] Li M, Zheng W. All-atom molecular dynamics simulations of actin–myosin interactions: a comparative study of cardiac alpha myosin, beta myosin, and fast skeletal muscle myosin. Biochemistry 2013;52:8393–405.

[35] Ling H, Zhang T, Pereira L, Means CK, Cheng H, Gu Y, et al. Requirement for Ca^{2+}/calmodulin-dependent kinase II in the transition from pressure overload-induced cardiac hypertrophy to heart failure in mice. J Clin Invest 2009;119: 1230–40.

[36] Liu QN, Reddy S, Sayre JW, Pop V, Graves MC, Fiala M. Essential role of HIV type 1-infected and cyclooxygenase 2-activated macrophages and T cells in HIV type 1 myocarditis. AIDS Res Hum Retroviruses 2001;17:1423–33.

[37] Manninen A, Saksela K. HIV-1 Nef interacts with inositol trisphosphate receptor to activate calcium signaling in T cells. J Exp Med 2002;195:1023–32.

[38] Marchetti G, Casana M, Tincati C, Bellistri GM, Monforte A. Abacavir and cardiovascular risk in HIV-infected patients: does T lymphocyte hyperactivation exert a pathogenic role? Clin Infect Dis 2008;47:1495–6.

[39] Massaeli H, Austria JA, Pierce GN. Overexpression of SERCA2 Atpase in vascular smooth muscle cells treated with oxidized low density lipoprotein. Mol Cell Biochem 2000;207:137–41.

[40] Mayne M, Holden CP, Nath A, Geiger JD. Release of calcium from inositol 1,4,5-trisphosphate receptor-regulated stores by HIV-1 Tat regulates TNF-alpha production in human macrophages. J Immunol 2000;164:6538–42.

[41] Michalak M, Opas M. Endoplasmic and sarcoplasmic reticulum in the heart. Trends Cell Biol 2009;19:253–9.

[42] Mitsuya H, Weinhold KJ, Furman PA, St Clair MH, Lehrman SN, Gallo RC, et al. 3′-Azido-3′-deoxythymidine (BW A509U): an antiviral agent that inhibits the infectivity and cytopathic effect of human T-lymphotropic virus type III/lymphadenopathy-associated virus in vitro. Proc Natl Acad Sci U S A 1985;82:7096–100.

[43] Nakao K, Minobe W, Roden R, Bristow MR, Leinwand LA. Myosin heavy chain gene expression in human heart failure. J Clin Invest 1997;100:2362–70.

[44] Nakashima H, Matsui T, Harada S, Kobayashi N, Matsuda A, Ueda T, et al. Inhibition of replication and cytopathic effect of human T cell lymphotropic virus type III/lymphadenopathy-associated virus by 3′-azido-3′-deoxythymidine in vitro. Antimicrob Agents Chemother 1986;30:933 7.

[45] Nanavati KA, Fisher SD, Miller TL, Lipshultz SE. HIV-related cardiovascular disease and drug interactions. Am J Cardiovasc Drugs 2004;4:315–24.

[46] Nunez L, Gimeno-Blanes JR, Rodriguez-Garcia MI, Monserrat L, Zorio E, Coats C, et al. Somatic MYH7, MYBPC3, TPM1, TNNT2 and TNNI3 mutations in sporadic hypertrophic cardiomyopathy. Circ J 2013;77:2358–65.

[47] Oudit GY, Kassiri Z. Role of PI3 kinase gamma in excitation–contraction coupling and heart disease. Cardiovasc Hematol Disord Drug Targets 2007;7:295–304.

[48] Oudit GY, Sun H, Kerfant BG, Crackower MA, Penninger JM, Backx PH. The role of phosphoinositide-3 kinase and PTEN in cardiovascular physiology and disease. J Mol Cell Cardiol 2004;37:449–71.

[49] Pao V, Lee GA, Grunfeld C. HIV therapy, metabolic syndrome, and cardiovascular risk. Curr Atheroscler Rep 2008;10:61–70.

[50] Phrommintikul A, Chattipakorn N. Roles of cardiac ryanodine receptor in heart failure and sudden cardiac death. Int J Cardiol 2006;112:142–52.

[51] Posch MG, Waldmuller S, Muller M, Scheffold T, Fournier D, Andrade-Navarro MA, et al. Cardiac alpha-myosin (MYH6) is the predominant sarcomeric disease gene for familial atrial septal defects. PLoS ONE 2011;6:e28872.

[52] Pretorius L, Owen KL, McMullen JR. Role of phosphoinositide 3-kinases in regulating cardiac function. Front Biosci 2009;14:2221–9.

[53] Pummerer CL, Grassl G, Sailer M, Bachmaier KW, Penninger JM, Neu N. Cardiac myosin-induced myocarditis: target recognition by autoreactive T cells requires prior activation of cardiac interstitial cells. Lab Invest 1996;74:845–52.

[54] Rasheed S, Yan J, Hussain A, Lai B. Proteomic characterization of HIV-modulated membrane receptors, kinases and signaling proteins involved in novel angiogenic pathways. J Transl Med 2009;7:75.

[55] Rasheed S, Gottlieb AA, Garry RF. Cell killing by ultraviolet-inactivated human immunodeficiency virus. Virology 1986;154:395–400.

[56] Rasheed S, Yan JS, Lau A, Chan AS. HIV replication enhances production of free fatty acids, low density lipoproteins and many key proteins involved in lipid metabolism: a proteomics study. PLoS ONE 2008;3:e3003.

[57] Reingold J, Wanke C, Kotler D, Lewis C, Tracy R, Heymsfield S, et al. Association of HIV infection and HIV/HCV coinfection with C-reactive protein levels: the fat redistribution and metabolic change in HIV infection (FRAM) study. J Acquir Immune Defic Syndr 2008;48:142–8.

[58] Sahoo SK, Kim dH. Calumenin interacts with SERCA2 in rat cardiac sarcoplasmic reticulum. Mol Cells 2008;26:265–9.

[59] Sasaki T, Irie-Sasaki J, Jones RG, Oliveira-dos-Santos AJ, Stanford WL, Bolon B, et al. Function of PI3Kgamma in thymocyte development, T cell activation, and neutrophil migration. Science 2000;287:1040–6.

[60] Taylor CW, Rahman T, Tovey SC, Dedos SG, Taylor EJ, Velamakanni S. IP3 receptors: some lessons from DT40 cells. Immunol Rev 2009;231:23–44.

[61] Tinsley JH, Hunter FA, Childs EW. PKC and MLCK-dependent, cytokine-induced rat coronary endothelial dysfunction. J Surg Res 2009;152:76–83.

[62] Triant VA, Lee H, Hadigan C, Grinspoon SK. Increased acute myocardial infarction rates and cardiovascular risk factors among patients with human immunodeficiency virus disease. J Clin Endocrinol Metab 2007;92:2506–12.

[63] Venetucci LA, Trafford AW, O'Neill SC, Eisner DA. The sarcoplasmic reticulum and arrhythmogenic calcium release. Cardiovasc Res 2008;77:285–92.

[64] Verkhratsky A. Calcium and cell death. Subcell Biochem 2007;45:465–80.

[65] Wehrens XH, Lehnart SE, Marks AR. Intracellular calcium release and cardiac disease. Annu Rev Physiol 2005;67:69–98.

[66] Wheeler M, Domin J. Recruitment of the class II phosphoinositide 3-kinase C2beta to the epidermal growth factor receptor: role of Grb2. Mol Cell Biol 2001;21:6660–7.

[67] Worm SW, Sabin CA, Reiss P, El-Sadr W, Monforte A, Pradier C, et al. Presence of the metabolic syndrome is not a better predictor of cardiovascular disease than the sum of its components in HIV-infected individuals: data collection on adverse events of anti-HIV drugs (D:A:D) study. Diabetes Care 2009;32:474–80.

[68] Yabe D, Nakamura T, Kanazawa N, Tashiro K, Honjo T. Calumenin, a Ca^{2+}-binding protein retained in the endoplasmic reticulum with a novel carboxyl-terminal sequence, HDEF. J Biol Chem 1997;272:18232–9.

[69] Zalk R, Lehnart SE, Marks AR. Modulation of the ryanodine receptor and intracellular calcium. Annu Rev Biochem 2007;76:367–85.

[70] Zima AV, Blatter LA. Inositol-1,4,5-trisphosphate-dependent Ca(2+) signalling in cat atrial excitation–contraction coupling and arrhythmias. J Physiol 2004;555: 607–15.

20

A computational design approach for virtual screening of peptide interactions across K$^+$ channel families☆

Craig A. Doupnik [a,*], Katherine C. Parra [b], Wayne C. Guida [b,c]

[a] Department of Molecular Pharmacology & Physiology, University of South Florida College of Medicine, 12901 Bruce B. Downs Boulevard, Tampa, FL 33612, United States
[b] Department of Chemistry, University of South Florida, 4202 E. Fowler Avenue, Tampa, FL 33620, United States
[c] Drug Discovery Department, H. Lee Moffitt Cancer Center and Research Institute, 12902 Magnolia Drive, Tampa, FL 33612, United States

ARTICLE INFO

Keywords:
Homology modeling
Computational docking
Virtual screening
Ion channels
Venom peptides
Protein–protein interactions

ABSTRACT

Ion channels represent a large family of membrane proteins with many being well established targets in pharmacotherapy. The 'druggability' of heteromeric channels comprised of different subunits remains obscure, due largely to a lack of channel-specific probes necessary to delineate their therapeutic potential *in vivo*. Our initial studies reported here, investigated the family of inwardly rectifying potassium (Kir) channels given the availability of high resolution crystal structures for the eukaryotic constitutively active Kir2.2 channel. We describe a 'limited' homology modeling approach that can yield chimeric Kir channels having an outer vestibule structure representing nearly any known vertebrate or invertebrate channel. These computationally-derived channel structures were tested *in silico* for 'docking' to NMR structures of tertiapin (TPN), a 21 amino acid peptide found in bee venom. TPN is a highly selective and potent blocker for the epithelial rat Kir1.1 channel, but does not block human or zebrafish Kir1.1 channel isoforms. Our Kir1.1 channel-TPN docking experiments recapitulated published *in vitro* findings for TPN-sensitive and TPN-insensitive channels. Additionally, *in silico* site-directed mutagenesis identified 'hot spots' within the channel outer vestibule that mediate energetically favorable docking scores and correlate with sites previously identified with *in vitro* thermodynamic mutant-cycle analysis. These 'proof-of-principle' results establish a framework for virtual screening of re-engineered peptide toxins for interactions with computationally derived Kir channels that currently lack channel-specific blockers. When coupled with electrophysiological validation, this virtual screening approach may accelerate the drug discovery process, and can be readily applied to other ion channels families where high resolution structures are available.

1. Introduction

Structural determination of the *Streptomyces lividans* K$^+$ channel (KcsA) by X-ray crystallography ushered in a new era in ion channel biology, where channel function such as ion selectivity can be understood mechanistically at atomic level resolution [1,2]. Subsequent structures of other prokaryotic and eukaryotic channels belonging to the two transmembrane (2-TM) K$^+$ channel family indicate high structural conservation in key channel domains across phyla [3–5]. Pharmacological control of ion channel activity, either increased or decreased, has tremendous therapeutic potential given the ubiquitous role of these membrane proteins in human physiology and disease [6]. With the inward rectifier Kir2.2 channel from chicken (cKir2.2) having ~90% amino acid sequence identity with the human isoform, we began to assess the feasibility of using this high resolution structure (3.1 Å resolution), together with emerging structure-based computational tools, for virtual screening of known and novel peptides that can modify channel activity in a channel-selective manner.

The cKir2.2 channel served as a template structure with homology modeling restricted to the outer vestibule region where peptide venom toxins are known to bind and block channel conductance [7]. Several 'chimeric' Kir channels were constructed having outer vestibules known to be either 'sensitive' or 'insensitive' to block by tertiapin (TPN), a 21 amino acid peptide (ALCNCNRIIIPHMCWKKCGKK) produced by the venom gland of the European honey bee *Apis mellifera* [8–10]. The rest of the chimeric Kir channel remained cKir2.2 structure, including transmembrane and intracellular domains. The different homology modeled Kir channels were then tested *in silico* for their energetic docking characteristics to NMR-derived solution structures of TPN [11]. The *in silico* results recapitulate previously published *in vitro* observations for TPN sensitivity and block of different Kir channels. Moreover, the interface of the docked TPN-Kir1.1 channel complex revealed a novel molecular mechanism for TPN channel block at the

☆ The authors declare no conflict of interest.
 * Corresponding author
   E-mail address: cdoupnik@health.usf.edu (C.A. Doupnik).

'GYG' K$^+$ selectivity filter, which was stabilized by multiple salt bridges and hydrogen bonding along the walls of the channel outer vestibule.

Peptide toxins from venomous snails, snakes, scorpions, and spiders, have a long tradition of providing valuable tools for assessing physiological roles of ion channels, and in some instances have provided new therapeutic agents [7]. Virtual screening of interactions between homology modeled ion channels and computationally re-engineered venom peptides, may accelerate the drug discovery process where *in silico* 'hits' can then be validated (or invalidated) using standard *in vitro* electrophysiology or other cell-based assays. We believe that the novel approach described here for Kir channels may also extend more broadly for other ion channels (voltage and ligand-gated) where high resolution structures are increasingly becoming available.

2. Materials and methods

2.1. Homology modeling

The cKir2.2 crystal structure (3JYC.pdb) served as the template for homology modeling the outer vestibule of all Kir channels constructed in this study [4]. The approximately 50 amino acid 'outer vestibule' sequence connecting the 1st and 2nd transmembrane domains of each Kir channel was substituted for the corresponding sequence in the cKir2.2 channel (His108-Pro156, 49 a.a.). The resulting chimeric sequence (Kir2.2/Kirx.y channel) was then used to generate homology Kir channel subunit structures using the Swiss-Model homology-modeling server [12–14]. Each homology-modeled Kir channel subunit was then assembled as a tetramer based on the macromolecular I4 space group determined for the assembled cKir2.2 tetramer [4]. Both homo- and hetero-tetramers could be assembled using the PDBe PISA program (Protein Interfaces, Surfaces, and Assemblies: http://pdbe.org/pisa/), though our results reported here are only for homo-tetrameric constructs. All structural rendering was performed using either PyMol v1.6 (Schrödinger) or the Swiss PdbViewer.

For single-residue and multiple-residue changes (i.e. site-specific mutagenesis), amino acids were changed in the 'outer vestibule' linker sequence, homology modeled against the cKir2.2 template structure, and assembled as a homo-tetrameric channel as described above.

2.2. Computational Kir channel-TPN docking simulations

The NMR solution structures of TPN (1TER.pdb) [11] were used for *in silico* docking to the homology-modeled Kir channels using ZDOCK 3.0.2 [15,16]. Rigid-body searches of docking orientations between TPN and the Kir channel outer vestibule returned 2000 complexes for each Kir channel examined, ranked by an initial-stage scoring function that computes optimized pairwise shape complementarity, electrostatic energies, and a pairwise statistical energy potential for interface atomic contacts energies [17]. The calculated TPN docking score profiles were then quantitatively compared among each Kir channel tested, and then referenced to known *in vitro* TPN binding affinities or IC$_{50}$ values reported in the literature [9,10,18,19].

2.3. Interface analysis

To evaluate the interface between homology modeled Kir channels and the docked TPN peptide, we first used the Cluspro2.0 program that performs pairwise RMSD analysis with 'greedy clustering' to derive a refined and energetically-favored complex for subsequent interface analysis [20–22]. The PDBePISA program was then used to evaluate the predicted interface features and key atoms between the two docked structures, including putative hydrogen bonds and salt bridges contributing to the favored energetics [23,24].

2.4. Kir channel expression in Xenopus oocytes

To test *in vitro* TPN sensitivity of the cKir2.2 channel, we heterologously expressed cKir2.2 in *Xenopus* oocytes for electrophysiological recordings. All procedures for the use and handling of *Xenopus laevis* (Xenopus Express, Plant City, FL) were approved by the University of South Florida Institutional Animal Care and Use Committee and have been described in detail elsewhere [25]. Isolated stage V–VI oocytes were maintained for 1–7 days at 17–19 °C in the following solution (in mM); 82.5 NaCl, 2.5 KCl, 1.0 CaCl$_2$, 1.0 MgCl$_2$, 1.0 NaHPO$_4$, 5.0 HEPES, 2.5 Na pyruvate, pH 7.5 (NaOH), with 5% heat-inactivated horse serum.

Oocytes were injected with cRNA transcribed *in vitro* by T7 RNA polymerase (mMessage mMachine, Ambion, Austin, TX, USA) from a linearized cDNA-containing vector, and maintained for 3–5 days at 17–19 °C prior to electrophysiological recording. The chicken Kir2.2 cDNA construct used for structural determination (XP_004945226.1, fragment encoding residues 38-369) was tested and compared to the TPN-sensitive rat Kir1.1 cDNA (GenBank: X72341.1) as a positive control [26].

2.5. Oocyte electrophysiology

Kir channel currents were recored ""by two-electrode voltage clamp techniques [25]. Oocytes were initially superfused with ND98 solution (in mM); 98 NaCl, 1 MgCl$_2$ and 5 HEPES at pH 7.5 (NaOH). Glass electrodes having tip resistances of 0.8–1.0 MΩ were used to clamp oocytes at a holding membrane potential of −80 mV. Voltage ramps from −80 to +20 mV (200 ms in duration) were evoked periodically to assess the inward rectification characteristics of Kir channel currents during changes in the recording solutions.

After establishing a baseline holding current, the perfusion solution was changed to a 'high K$^+$ solution' that was comprised of an equal molar substitution of NaCl for KCl. The K$^+$ concentration varied depending on channel expression levels and the ability to voltage clamp inward K$^+$ currents, where the 'high K$^+$ solution' ranged from 20 to 98 mM KCl. Application and washout of 100 nM TPN$_Q$ (Tocris Bioscience, Bristol, UK) or 1 mM BaCl$_2$ were performed by perfusion barrels located adjacent to the oocyte [25]. All recordings were performed at room temperature (21–23 °C). The Kir currents were digitized, stored, and analyzed using an A/D acquisition board and PC computer (pCLAMP software, Digidata 1200 acquisition system, Axon Instruments). Experiments were replicated in 3–4 oocytes.

3. Results

3.1. Homology modeling the outer vestibule of different Kir channels

The cKir2.2 channel 'core' domains share ~90% sequence identity to human Kir channel orthologs suggesting high structural conservation among the Kir channel family isoforms [4]. We reasoned that restricting the homology modeling to the outer vestibule region, where TPN block is known to occur, would minimize any computationally introduced structural rearrangements during the modeling procedures (e.g. global energy minimization changes).

This method, illustrated in Fig. 1, effectively yields Kir channel chimeras that consist largely of the cKir2.2 structure, but with a channel-specific outer vestibule structure (~50 a.a.) for screening and scoring TPN docking interactions *in silico*. The homology modeled Kir channel subunits can be assembled as either homo- or hetero-tetrameric channels, enabling the replication of Kir channel diversity that exists *in vivo*. This computational approach assumes that isoform-specific sequence differences within the transmembrane domains and intercellular N- and C-termini, do not significantly influence the outer vestibule structure to an extent that would impact TPN binding and channel block. Ramachandran plot analysis of the homology modeled

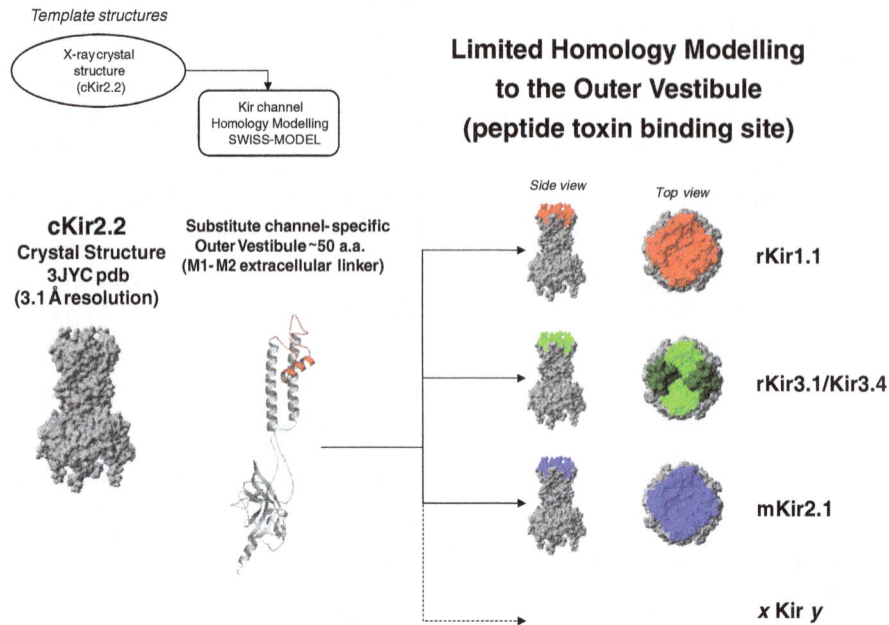

Fig. 1. Computational design approach using 'Limited' homology Kir channel structural modeling. The chicken Kir2.2 channel crystal structure (3JYC.pdb) served as the template for homology modeling. Shown are the assembled cKir2.2 homotetrameric channel (gray), and a single Kir2.2 subunit where the outer vestibule structure has been replaced with the rat Kir1.1 sequence (red) and homology modeled as described in the Methods. The modeled chimeric Kir channel subunit (rKir1.1/cKir2.2) was then reassembled as a homo-tetrameric channel (side and top views, red indicating rKir1.1 structure, gray indicating cKir2.2 structure). The same 'limited' homology modeling approach could also be applied for different heteromeric (Kir3.1/3.4, green) or homomeric Kir channels (Kir2.1, blue).

Kir channels confirmed that ~90% of the amino acids were located within the favored alpha helix (A), beta strand (B) and left alpha helix (L) allowable regions (data not shown).

3.2. Energetics of TPN-Kir channel docking recapitulates in vitro TPN channel block

To test our outer vestibule homology modeling approach, we compared the rigid-body TPN docking energetics for a Kir channel isoform known to be blocked by TPN *in vitro* at nanomolar affinity (rat Kir1.1), with a Kir channel isoform expected to be insensitive to TPN block (chicken Kir2.2). We confirmed that cKir2.2 channels expressed in *Xenopus* oocytes are insensitive to 100 nM TPN$_Q$, in contrast to rKir1.1 channels that are completely blocked by 100 nM TPN$_Q$ (Fig. 2). The TPN insensitivity of cKir2.2 channels is consistent with other Kir2.0 channels reported in the literature, and in contrast to the nanomolar affinity of rat Kir1.1 channels [9,10]. These two Kir channels therefore represent good examples of TPN-insensitive and TPN-sensitive

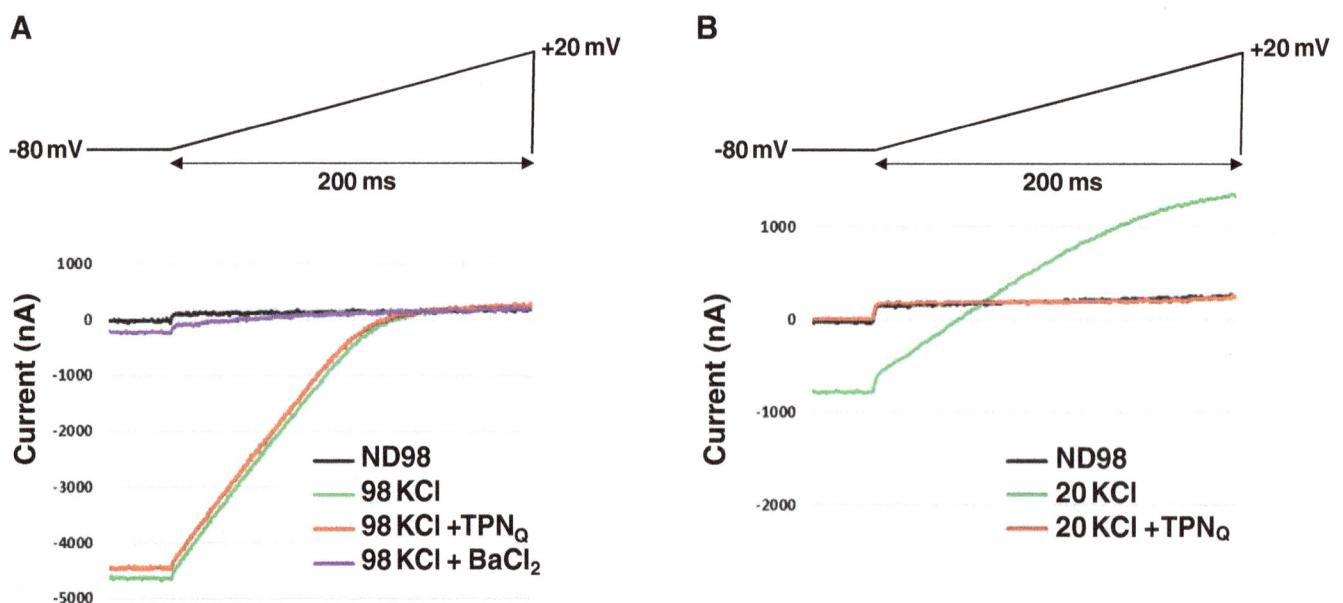

Fig. 2. Differential sensitivity of cKir2.2 and rKir1.1 channels to TPN$_Q$ block. A. Electrophysiological recordings of cKir2.2 channel currents evoked by voltage ramps from -80 to $+20$ mV in *Xenopus* oocytes. Strong inwardly rectifying K$^+$ currents in 98 mM external [K$^+$] (green trace) were insensitive to 100 nM TPN$_Q$ (red trace), but completely blocked by 1 mM Ba^{2+} (purple trace). B. In contrast, the inward rectifier rKir1.1 channel current (green trace) was completely blocked by 100 nM TPN$_Q$ (red trace). The 20 mM external [K$^+$] was used to lower current amplitudes that were typically greater with rKir1.1 expression.

channels for assessing the predictive capacity of our virtual screening approach.

Each docked complex was scored with an energy function that incorporates pairwise shape complementarity, desolvation contact energy, and electrostatic interactions [17]. Unbiased computational docking of the TPN 'ligand' (rotational/translational sampling ~10^9 positions) to the stationary 'receptor' (Kir channel outer vestibule) generated thousands of docked complexes, where the relative difference in the 'top scored' complexes was compared. The top 2000 scored complexes for each Kir channel were ranked to produce the TPN docking profiles shown in Fig. 3.

As shown in the Fig. 3 plot, the rKir1.1–TPN1 docked complexes exhibited significantly greater docking scores compared to the cKir2.2–TPN1 complexes, indicating a more energetically favorable interaction between TPN1 and rKir1.1. These simulation results are therefore in good agreement with the relative in vitro TPN sensitivity differences for these two Kir channels.

3.3. Docking profile of different TPN conformers

The derived NMR solution structure of TPN includes a 'bundle' of 21 peptide conformers, where 2 Cys–Cys disulfide bridges structurally constrain the number of peptide conformations [11]. An overlay of the 21 TPN conformers is shown in Fig. 4A and illustrates significant differences in the coordinates of both the α-carbon backbone and the amino acid side chains that would be expected to impact rigid-body docking results. Shown in Fig. 4B, the root mean squared deviation (RMSD) of the TPN α-carbon backbone indicates 2 peptide regions have significant mobility; 1) the amino terminal alanine residue (Ala1), and 2) the highly basic carboxyl terminal region (KKCGKK).

Our initial docking experiments shown in Fig. 3 utilized the first 'conformer' designated TPN1. However since there is no a priori knowledge of which of the 21 possible TPN conformers dock, bind, and block

the rKir1.1 channel, we evaluated the rigid-body docking characteristics of each conformer (TPN1–TPN21) to both rKir1.1 and cKir2.2. Shown in Fig. 4C, all 21 TPN conformers had greater docking scores to the TPN-sensitive rKir1.1 channel, when compared to those derived from docking to the TPN-insensitive cKir2.2 channel. The TPN12 conformer docked to rKir1.1 with the greatest scores, while TPN20 docked with the lowest (see Fig. 4D). Given that TPN12 yielded the most energetically favorable complex with the rKir1.1 channel, we evaluated TPN12 with TPN1 (as a secondary basis of comparison) in subsequent simulation experiments.

3.4. Refinement of the docked rKir1.1–TPN12 complex

To further analyze the docked complex, we used Cluspro2.0 to derive the most likely and energetically-favored orientation of TPN12 docked to the rKir1.1 outer vestibule. This program performs pairwise RMSD analysis with 'greedy clustering' using a 9 Å Cα RMSD radius, where radii <2.5 Å yield highly predictive near-native results based on analysis of known native complexes [27]. Given the highly basic nature of TPN (+4.9 net charge at pH 7), we utilized an electrostatically-favored energy function to score the approximately 10^9 rotational positions generated by the fast Fourier transform (FFT) algorithm [28].

Greedy clustering of the top 1000 scored TPN12 positions returned 9 clusters with the following number of members; 374, 294, 136, 110, 40, 14, 13, 13, and 6; where the cluster having the highest number of members represents the most likely and favored 'pose'. Initial evaluation of the 'center' complex from the top 2 clusters (374, 294) indicated that they were nearly identical in overall orientation of TPN docked to the Kir1.1 outer vestibule. This is expected with the 'dimer-of-a-dimer' symmetry of the assembled homo-tetrameric channel structure [4]. We therefore chose the docked complex shown in Fig. 5 for further analysis of the channel–peptide interface, which represented the 'center' of the top cluster of docked complexes.

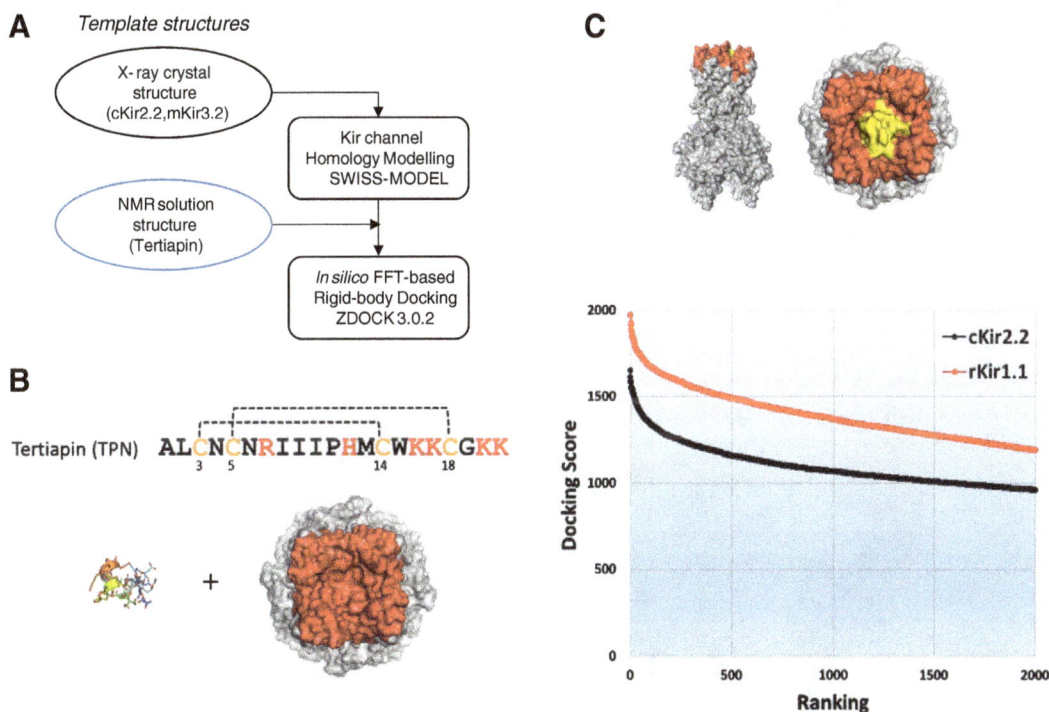

Fig. 3. Computational docking of TPN to homology-modeled Kir channels. A. Workflow diagram showing initial-stage computational docking of an NMR solution structure of tertiapin (1TER.pdb) to the homology modeled Kir channel using ZDOCK3.0.2. B. The 21 amino acid primary sequence of TPN is shown, including the 2 disulfide bonds (C3–C14, C5–C18) that constrain the structural conformations of the peptide. The six basic residues that contribute to electrostatic interactions are also shown in red. C. Side and top views of TPN (yellow) docked to the outer vestibule of the homology modeled rat Kir1.1 channel (rat structures shown in red). The TPN Docking Scores for rKir1.1 (red line) and cKir2.2 (black line) are shown in the lower panel plot, where the top 2000 docking scores for each are ranked highest-to-lowest to produce the docking profile curves.

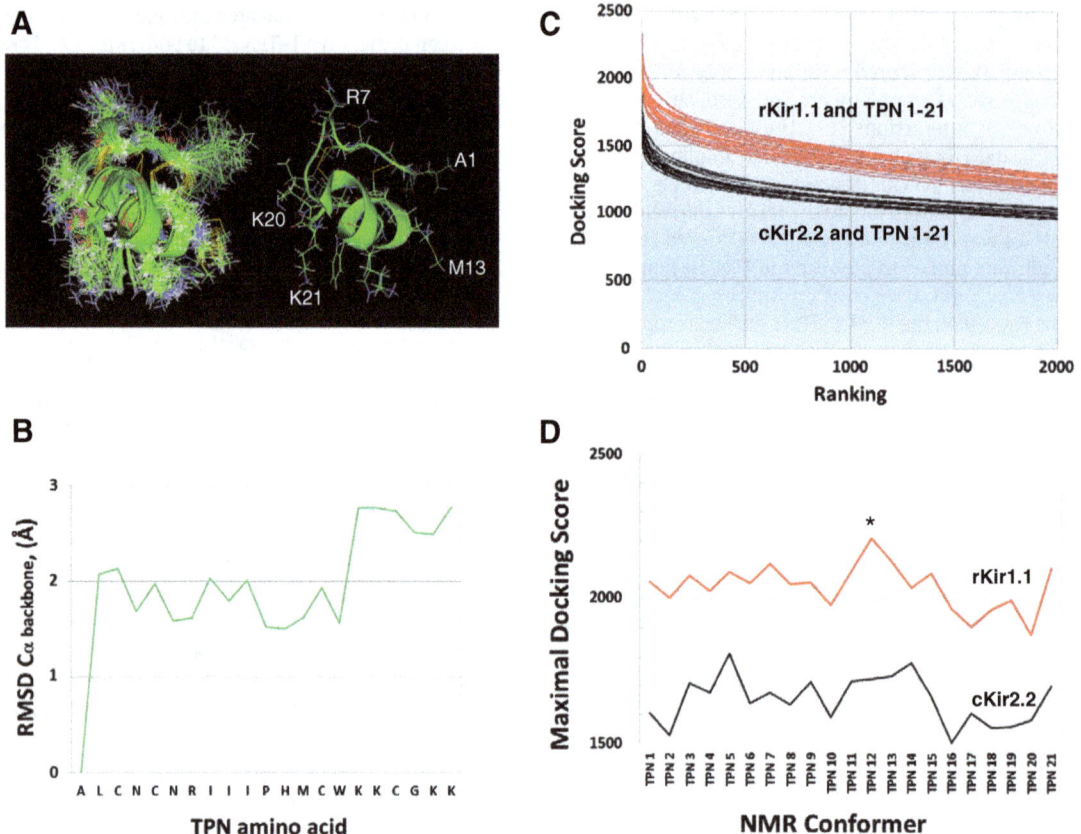

Fig. 4. Different TPN conformations reveal mobile peptide domains effecting rigid-body docking to Kir channels. A. Overlay of the 21 conformers of TPN resolved from the NMR solution structure bundle (left image). For comparison, a single TPN conformation is shown (right image) with selected residues labeled. B. Alpha-carbon RMSD analysis for the 21 TPN conformers. The 21 TPN structures were analyzed where the coordinates for the N-terminal alanine residue served as the initial spatial reference point. C. Docking Score profile for each TPN conformer (TPN1-21) docked independently to either the cKir2.2 channel outer vestibule (black lines), or the homology modeled rat Kir1.1 channel outer vestibule (red lines). D. Plot of the Maximal Docking Score profile for each TPN conformer (TPN1-21) docked to the homology modeled rat Kir1.1 channel outer vestibule (red plot) and the chicken Kir2.2 channel outer vestibule (black plot). The average docking score for the top 5 complexes for each TPN conformer is shown. The most energetically favored rKir1.1–TPN complex (i.e. highest score) was produced by the TPN12 conformer (indicated by the asterisk).

3.5. Characterization of the docked rKir1.1–TPN12 interface

Interface analysis of the docked complex was performed using the PDBe PISA server. The 'footprint' of TPN12 on the rKir1.1 outer vestibule is shown in Table 1, where 3 'hotspots' were readily apparent; 1) the 'signature' GYG pore region, 2) a glutamic acid 'Glu ring', and 3) the channel turrets. Hydrogen bonding and salt bridges were predicted in each of these interface 'hotspots' and are listed in Table 2. Overall, a

total of 13 putative H-bonds and 9 salt bridges stabilize the docked TPN complex asymmetrically around the solvent-accessible vestibule formed by the four rKir1.1 channel subunits.

The H-bond network is arranged essentially as three 'contact rings' at different depths within the outer vestibule (see Fig. 6). The 3 contact rings consist of, 1) a negatively charged narrow 'pore ring' involving the Tyr111 carbonyl groups that form part of the K^+ selectivity filter, 2) a negatively charged 'mid-level ring' formed by Glu90 side chains

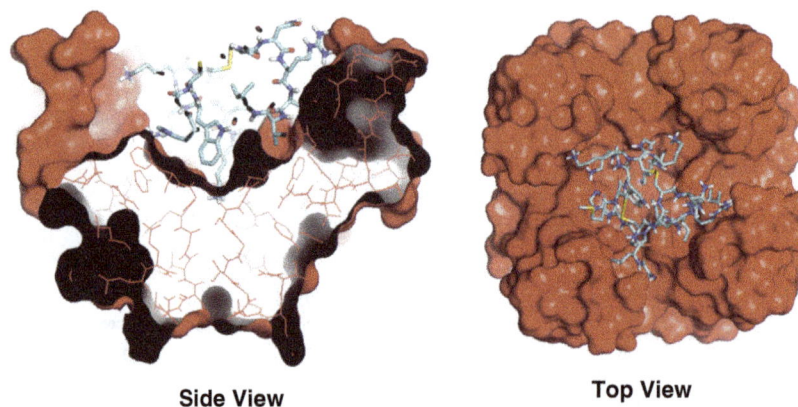

Fig. 5. Asymmetric docking of TPN12 to the rat Kir1.1 channel outer vestibule. Left panel. Surface-rendering, cut-away side view of the rat Kir1.1 channel outer vestibule (red) containing the docked TPN12 peptide (blue). TPN12 was docked to the homology modeled rat Kir1.1 channel using the Cluspro2.0 program. Right panel. Top view of the rat Kir1.1 channel outer vestibule (red) containing the docking TPN12 peptide (blue), illustrating asymmetric interactions with the channel turrets.

Table 1
Docked TPN 'Footprint' within the rat Kir1.1 outer vestibule.

Pos #	rKir1.1 subunit A	rKir1.1 subunit B		rKir1.1 subunit C		rKir1.1 subunit D		
1	A:HIS 73	B:HIS 73		C:HIS 73		D:HIS 73		
2	A:LYS 74	B:LYS 74		C:LYS 74		D:LYS 74		
3	A:ASP 75	B:ASP 75		C:ASP 75		D:ASP 75		
4	A:LEU 76	B:LEU 76		C:LEU 76		D:LEU 76		
5	A:PRO 77	B:PRO 77		C:PRO 77		D:PRO 77		
6	A:GLU 78	B:GLU 78		C:GLU 78		D:GLU 78		
7	A:PHE 79	B:PHE 79		C:PHE 79		D:PHE 79		
8	A:TYR 80	B:TYR 80	H	C:TYR 80		D:TYR 80		
9	A:PRO 81	B:PRO 81	H	C:PRO 81		D:PRO 81		
10	A:PRO 82	B:PRO 82		C:PRO 82		D:PRO 82		
11	A:ASP 83	B:ASP 83		C:ASP 83	HS	D:ASP 83	HS	Turret
12	A:ASN 84	B:ASN 84		C:ASN 84		D:ASN 84		
13	A:ARG 85	B:ARG 85		C:ARG 85		D:ARG 85		
14	A:THR 86	B:THR 86		C:THR 86		D:THR 86		
16	A:PRO 87	B:PRO 87		C:PRO 87		D:PRO 87		
17	A:CYS 88	B:CYS 88		C:CYS 88		D:CYS 88		
18	A:VAL 89	B:VAL 89		C:VAL 89		D:VAL 89		
19	A:GLU 90	S	B:GLU 90	HS	C:GLU 90	H	D:GLU 90	H
20	A:ASN 91	B:ASN 91		C:ASN 91		D:ASN 91		Glu Ring
21	A:ILE 92	B:ILE 92		C:ILE 92		D:ILE 92		
22	A:ASN 93	B:ASN 93		C:ASN 93		D:ASN 93		
23	A:GLY 94	B:GLY 94		C:GLY 94		D:GLY 94		
24	A:MET 95	B:MET 95		C:MET 95		D:MET 95		
25	A:THR 96	B:THR 96		C:THR 96		D:THR 96		
26	A:SER 97	B:SER 97		C:SER 97		D:SER 97		
27	A:ALA 98	B:ALA 98		C:ALA 98		D:ALA 98		
28	A:PHE 99	B:PHE 99		C:PHE 99		D:PHE 99		
29	A:LEU 100	B:LEU 100		C:LEU 100		D:LEU 100		
30	A:PHE 101	B:PHE 101		C:PHE 101		D:PHE 101		
31	A:SER 102	B:SER 102		C:SER 102		D:SER 102		
32	A:LEU 103	B:LEU 103		C:LEU 103		D:LEU 103		
33	A:GLU 104	B:GLU 104		C:GLU 104		D:GLU 104		
34	A:THR 105	B:THR 105		C:THR 105		D:THR 105		
35	A:GLN 106	B:GLN 106		C:GLN 106		D:GLN 106		
36	A:VAL 107	B:VAL 107		C:VAL 107		D:VAL 107		
37	A:THR 108	B:THR 108		C:THR 108		D:THR 108		
38	A:ILE 109	B:ILE 109		C:ILE 109		D:ILE 109		
39	A:GLY 110	B:GLY 110		C:GLY 110		D:GLY 110		
40	A:TYR 111	H	B:TYR 111	H	C:TYR 111	H	D:TYR 111	H
41	A:GLY 112	B:GLY 112		C:GLY 112		D:GLY 112		Pore
42	A:PHE 113	B:PHE 113		C:PHE 113		D:PHE 113		
43	A:ARG 114	B:ARG 114		C:ARG 114		D:ARG 114		
44	A:PHE 115	B:PHE 115		C:PHE 115		D:PHE 115		
45	A:VAL 116	B:VAL 116		C:VAL 116		D:VAL 116		
46	A:THR 117	B:THR 117		C:THR 117		D:THR 117		
47	A:GLU 118	B:GLU 118		C:GLU 118		D:GLU 118		
48	A:GLN 119	B:GLN 119		C:GLN 119		D:GLN 119		
49	A:CYS 120	B:CYS 120		C:CYS 120		D:CYS 120		

- Inaccessible residues.
- Solvent-accessible residues.
- Interfacing residues.
- HB Residues making a hydrogen bond or salt bridge link.

Table 2
Putative hydrogen bonds and salt bridges that stabilize TPN to the rat Kir1.1 channel outer vestibule and block ionic conduction.

Hydrogen bonds

rKir1.1 subunit A	Dist. [Å]	TPN
TYR 111[O]	2.04	LYS 21[HZ2]

rKir1.1 subunit B	Dist. [Å]	TPN
GLU 90[OE1]	1.84	LYS 16[HZ2]
GLU 90[OE2]	1.74	LYS 16[HZ1]
PRO 81[O]	1.75	LYS 17[HZ1]
TYR 80[O]	1.71	LYS 17[HZ2]
TYR 111[O]	2.03	LYS 21[HZ2]

rKir1.1 subunit C	Dist. [Å]	TPN
ASP 83[OD1]	1.89	ARG 7[HH12]
GLU 90[OE2]	2.02	ILE 9[H]
TYR 111[O]	1.72	LYS 21[HZ3]

rKir1.1 subunit D	Dist. [Å]	TPN
ASP 83[OD1]	2.15	LEU 2[H]
GLU 90[OE1]	1.97	HIS 12[H]
GLU 90[OE2]	2.33	MET 13[H]
TYR 111[O]	1.69	LYS 21[HZ1]

Salt bridges

rKir1.1 subunit A	Dist. [Å]	TPN
GLU 90[OE1]	2.55	LYS 20[NZ]
GLU 90[OE2]	2.66	LYS 20[NZ]

rKir1.1 subunit B	Dist. [Å]	TPN
GLU 90[OE1]	2.48	LYS 16[NZ]
GLU 90[OE2]	2.72	LYS 16[NZ]

rKir1.1 subunit C	Dist. [Å]	TPN
ASP 83[OD1]	2.80	ARG 7[NH1]
ASP 83[OD2]	3.28	ARG 7[NH1]
ASP 83[OD1]	2.78	ARG 7[NH2]
ASP 83[OD2]	2.72	ARG 7[NH2]

rKir1.1 subunit D	Dist. [Å]	TPN
ASP 83[OD1]	2.95	ALA 1[N]

contributed by all four channel subunits, and 3) a negatively charged 'upper ring' formed by Asp83 side chains in 2 adjacent channel 'turrets', and negatively charged carbonyl groups from 2 neighboring residues (Tyr80, Pro81) in a third turret. The fourth channel turret did not significantly contribute to the H-bond network; however Tyr80 did contribute to the TPN-turret interface ring that included each Tyr80 of the 4 turrets (see Table 1).

In addition to the electrostatic interaction network between basic residues of TPN and the negatively charged atoms of the channel vestibule, a hydrophobic phenylalanine ring at the pore entryway (formed by Phe113 and Phe115) also contributed to the docked interface. Solvent exposed hydrophobic residues in TPN (Ile8, Ile9, Trp15) were in close proximity to this hydrophobic 'Phe ring' which may help orient TPN within the vestibule as part of a 'functional dyad' (see Discussion).

Notably, the C-terminal lysine of TPN (Lys21) was found to descend deepest into the channel vestibule where it formed a putative H-bond with the carbonyl groups of the tyrosine residues (Tyr111) located in the 'GYG' signature sequence (i.e. the 'pore' ring). In doing so, TPN Lys21 would effectively disrupt or block K$^+$ occupancy at the selectivity filter and ostensibly K$^+$ conductance through the channel pore.

From this interface analysis of the docked TPN-rKir1.1 complex, we hypothesized that the electrostatic contact network formed by the middle 'Glu ring' and upper 'Turret ring', provide the primary contact energy that stabilizes TPN within the outer vestibule and is largely responsible for the high binding affinity. When TPN is stably bound at nanomolar affinity to the outer vestibule, the TPN C-terminal Lys21 residue then prevents channel K$^+$ conductance through a direct interaction with

the lower pore ring Tyr111 carbonyl groups that constitute the exposed part of the K$^+$ selectivity filter.

3.6. Site-directed mutagenesis in silico recapitulates in vitro rKir1.1 channel block kinetics

To test this hypothesis, we performed site-directed mutagenesis of Kir1.1 residues in silico, and evaluated the impact of single vestibule residues on 1) the TPN12 docking scores and 2) the position of TPN Lys21within the outer vestibule. We initially took advantage of the reported in vitro differences in Kir1.1 channel sensitivity to TPN among different species, where both human and zebrafish Kir1.1 channel isoforms are relatively insensitive to TPN block in contrast to the nanomolar affinity exhibited by the rat Kir1.1 channel [18,19]. Moreover, amino acid determinants within the rat Kir1.1 outer vestibule affecting high affinity binding of TPN and subsequent block of rKir1.1 channels have been previously mapped by alanine scanning mutagenesis and thermodynamic mutant-cycle analysis [9,10].

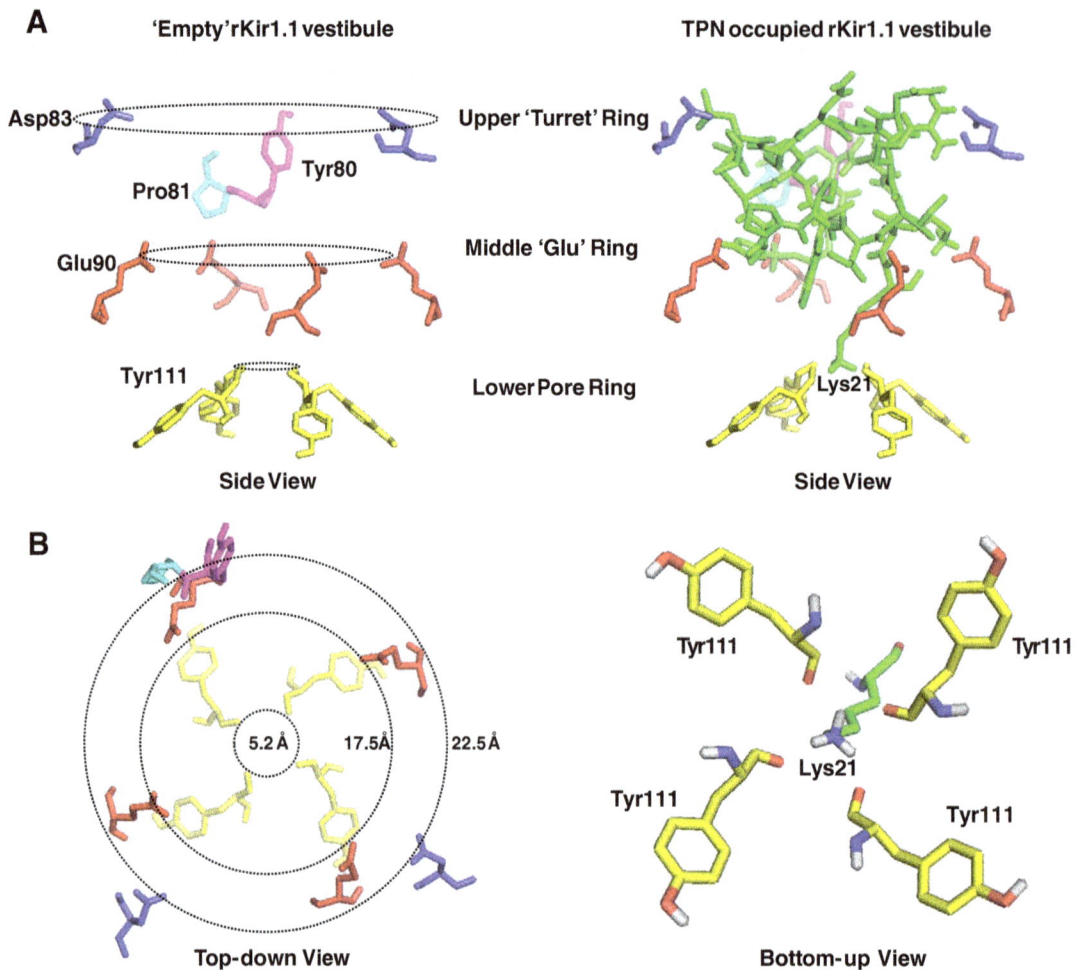

Fig. 6. TPN12 contact rings in the rat Kir1.1 outer vestibule. A. Side views of the rat Kir1.1 outer vestibule amino acids comprising the upper 'Turret' ring, middle Glu ring, and lower pore ring, that together form multiple contact sites for interaction with the TPN peptide. The left panel shows the channel contact rings 'empty' (no TPN), and the right panel included the docked TPN12 peptide shown in green. The TPN C-terminal Lys21 is indicated and is positioned to form H-bonds with the Tyr111 carbonyl groups of the Kir1.1 channel GYG selectivity filter. B. Top-down view of the empty rat Kir1.1 outer vestibule residues forming the TPN contact rings, with ring diameter distances indicated (right panel). The left panel illustrates a bottom-up 'zoomed in' view of the four rKir1.1 tyrosine residues (Tyr111), with their carbonyl groups facing the pore, juxtaposed to the TPN Lys21 side chain. The putative H-bonds and contact distances between TPN Lys21 atoms and the rKir1.1 Tyr111 carbonyl group atoms are provided in Table 2.

Shown in Fig. 7A is the multiple sequence alignment of the outer vestibule sequence of rat, human, and zebrafish Kir1.1 channels, with chicken Kir2.2 included for comparison. Computational docking of TPN to each homology modeled Kir1.1 channel, indicated only the rat isoform produced docking scores significantly greater than cKir2.2. This finding is consistent with the *in vitro* reports for species-dependent Kir1.1 channel sensitivity and block by TPN.

Comparison of the closely related outer vestibule sequences for the TPN-sensitive rat, versus TPN-insensitive human Kir1.1 channel (Fig. 8A), indicates only 6 residue differences exist and are necessarily responsible for the observed TPN docking and binding differences between rat and human Kir1.1 channels. Four of these residue differences reside in the turret structure. To further test our hypothesis, we examined individually and in combination, the six residue differences using *in silico* site-directed mutagenesis on TPN12 docking scores (indicated by vestibule position #: H8Y, S10P, A11D, H13R, L24M, C44F).

When each of the six residues in the human Kir1.1 outer vestibule was individually mutated to the corresponding rat amino acid residue, TPN docking energetics remained significantly less than that observed for the rKir1.1 channel (Fig. 8B). Therefore multiple residues (not just one) necessarily contribute to the species-dependent TPN docking differences. Individually, the C44F site produced the greatest increase in the TPN docking scores followed by A11D and S10P (cf. Fig. 8B).

We next evaluated double mutations, introducing two rat-specific turret residues into the hKir1.1 outer vestibule. Shown in Fig. 8C, there was a synergistic effect of H8Y with either S10P or A11D on the TPN docking scores. In contrast, S10P with A11D did not significantly enhance the TPN docking scores above those observed with the single point mutations (cf. Fig. 8B). These findings are consistent with the upper turret ring contacts (Y8, P10, and D11) that were evident in the TPN–rKir1.1 interface analysis. The TPN docking energetics for the H8Y + S10P and H8Y + A11D double mutations were still significantly less than that observed for the rKir1.1 channel, and the triple mutation (H8Y + S10P + A11D) did not significantly improve the TPN docking energetics (Fig. 8D).

When the C44F mutation was included with either the H8Y + S10P or H8Y + A11D double mutations, TPN docking closely mirrored the rKir1.1 docking energetics. The H8Y + A11D + C44F triple mutation was essentially indistinguishable from rKir1.1 (Fig. 8D). These computational experiments revealed that 3 of the residues in rat Kir1.1, when introduced into human Kir1.1, are both necessary and together sufficient to recapitulate the TPN docking scores for rKir1.1. Two of these sites (turret A11D, and pore C44F) were the same residues identified previously by Felix et al. using site-directed mutagenesis of Kir1.1 and *in vitro* binding of a TPN derivative [18], and implicated by Jin et al. using mutant-cycle analysis [10].

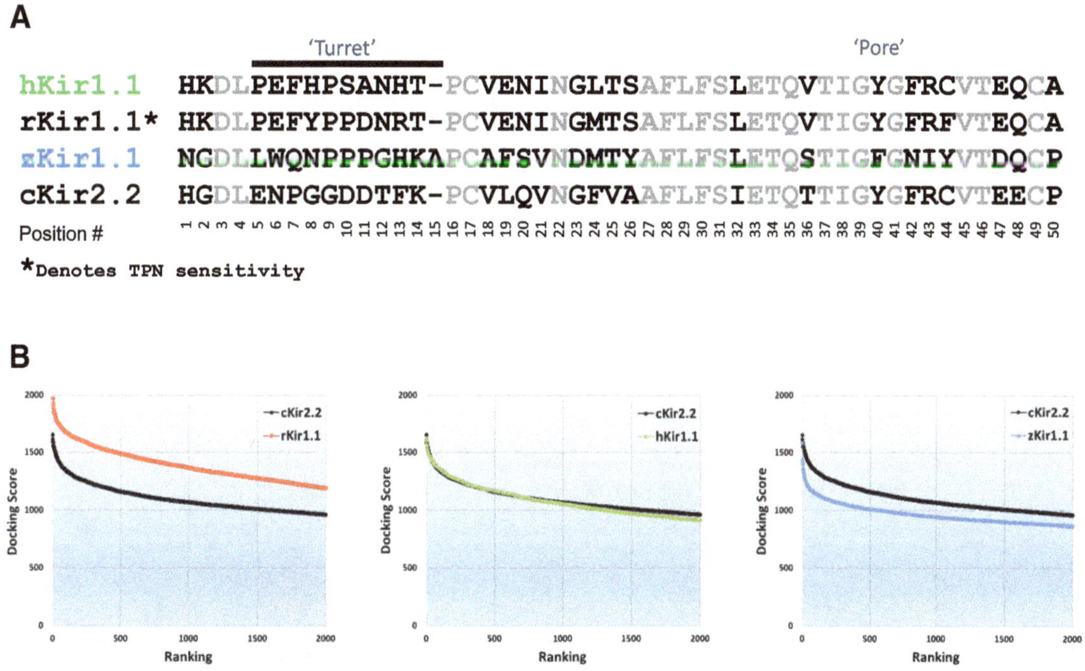

A

*Denotes TPN sensitivity

B

Fig. 7. Species-dependent differences in TPN docking to the Kir1.1 channel outer vestibule. A. Multiple sequence alignment of amino acid residues forming the human, rat, and zebrafish Kir1.1 channel outer vestibules. Identical residues between species are shown in gray, divergent residues in black. The turret and pore regions are indicated. Only the rat isoform is functionally sensitive to TPN block, and denoted by the asterisk. B. TPN docking score profile plots are shown for rat Kir1.1 (red, left plot), human Kir1.1 (green, center plot), and zebrafish Kir1.1 (blue, right plot). The docking score profile to the TPN-insensitive chicken Kir2.2 channel is shown in each panel for comparison (black). The docking profiles for the TPN-insensitive hKir1.1 and zKir1.1 channels are not significantly greater than the TPN-insensitive cKir2.2 channel, whereas the TPN-sensitive rKir1.1 displays significantly greater TPN docking energetics.

Fig. 8E shows the structural locations for Y8, P10, D11, and F44 within the outer vestibule, where Y8, P10 and D11 reside in the 'Turret contact ring' that faces the pore, and F44 resides at the base of the vestibule as the 'Phe contact ring' near the pore entrance. These residues correspond to Y80, P82, D83, and F115 in the homology modeled rKir1.1 structure (cf. Fig. 6), and Y113, P115, D116, and F148 in the native full-length rKir1.1 primary sequence.

Taken altogether, these computational results validate our *in silico* homology modeling and docking approach, by consistently reproducing the reported *in vitro* effects of TPN on different Kir1.1 channels.

4. Discussion and conclusions

The growing structural library of ion channels and other membrane proteins, that now includes G protein coupled receptors (GPCR's) [29], creates new computational opportunities for drug development and discovery [6]. When coupled with *in vitro* validation assays, computational approaches that can accurately predict target interactions, relative binding affinity, and discriminate channel specificity, have significant cost-effective potential for identifying new therapeutics and experimentally useful biologic probes (e.g. peptide inhibitors and activators).

We have demonstrated here a novel 'limited' homology modeling approach that introduces only the extracellularly exposed outer vestibule of various Kir channels, within the structural constraints of an already resolved Kir channel by X-ray crystallography (i.e. cKir2.2). Our findings indicate this computational design approach can accurately reproduce the reported *in vitro* sensitivities for TPN block of the modeled Kir1.1 channel isoforms, and therefore suggests a good predictive capability. Since turret structures between Kir channels are expected to vary with sequence divergence, the reliability of our limited homology modeling approach beyond Kir1.1 remains to be determined. However in preliminary simulations (data not shown), homology-modeled Kir3 channels dock TPN consistent with the nanomolar affinity (IC_{50} ~ 8 nM) of native Kir3 channels [9], suggesting reliability beyond Kir1.1. Moreover, the feasibility of this approach is further supported

by the successful transfer of high affinity *in vitro* kaliotoxin (KTN) binding (a scorpion toxin peptide) to a KcsA channel chimera containing the outer vestibule of Kv1.3 (a KTN-sensitive Kv channel) [30].

Recent studies published during our investigation similarly explored the interactions of TPN with Kir1.1 channels *in silico* [31,32]. Notably however, the binding interface of TPN with the Kir1.1 outer vestibule was reportedly different then we describe here, where the toxin histidine residue (His12) was found to be juxtaposed to the K^+ selectivity filter versus the C-terminal Lys21 in our study. Significant methodological differences are likely to explain the different results. First, the homology modeling approach used by Hilder and Chung (2013) involved the entire sequence (372 a.a., NP_722449.2) of the human Kir1.1b channel that is TPN-insensitive, whereas our approach was limited to the outer vestibule region of the TPN-sensitive rat isoform (NP_058719.1). Second, we found significant rigid-body docking differences among the 21 TPN conformers, with TPN12 being the most energetically favored. The TPN conformer used in the Hilder and Chung study was not specified, and incorporated the M13Q mutation (TPN_Q) [33].

Although Hu et al. used the TPN-sensitive rat Kir1.1 sequence in their studies, they too homology modeled the entire rKir1.1 channel structure (55% sequence identity between full-length rKir1.1 and cKir2.2) that included a refinement step that remodeled the turret structures using an additional 'segment-assembly' homology modeling method [32]. This may yield coordinate differences in the rKir1.1 outer vestibule when compared to our limited homology approach. They also performed rigid body TPN_Q docking using an unspecified TPN conformer, whereas we assessed all 21 TPN conformers, evaluating the favored TPN12 with model refinement using Cluspro2.0 [27].

Our results thus highlight a significant impact of peptide conformation on rigid-body docking and 'top score' selection. TPN12 and TPN20 yielded the highest and lowest docking scores, respectively, among the 21 conformers resolved by NMR spectroscopy [11]. The docking score differences were attributable to the highly mobile C-terminal region of TPN (KKCGKK), where the basic lysine residues form most of the H-bond/salt bridge network seen in the TPN12-docked complex.

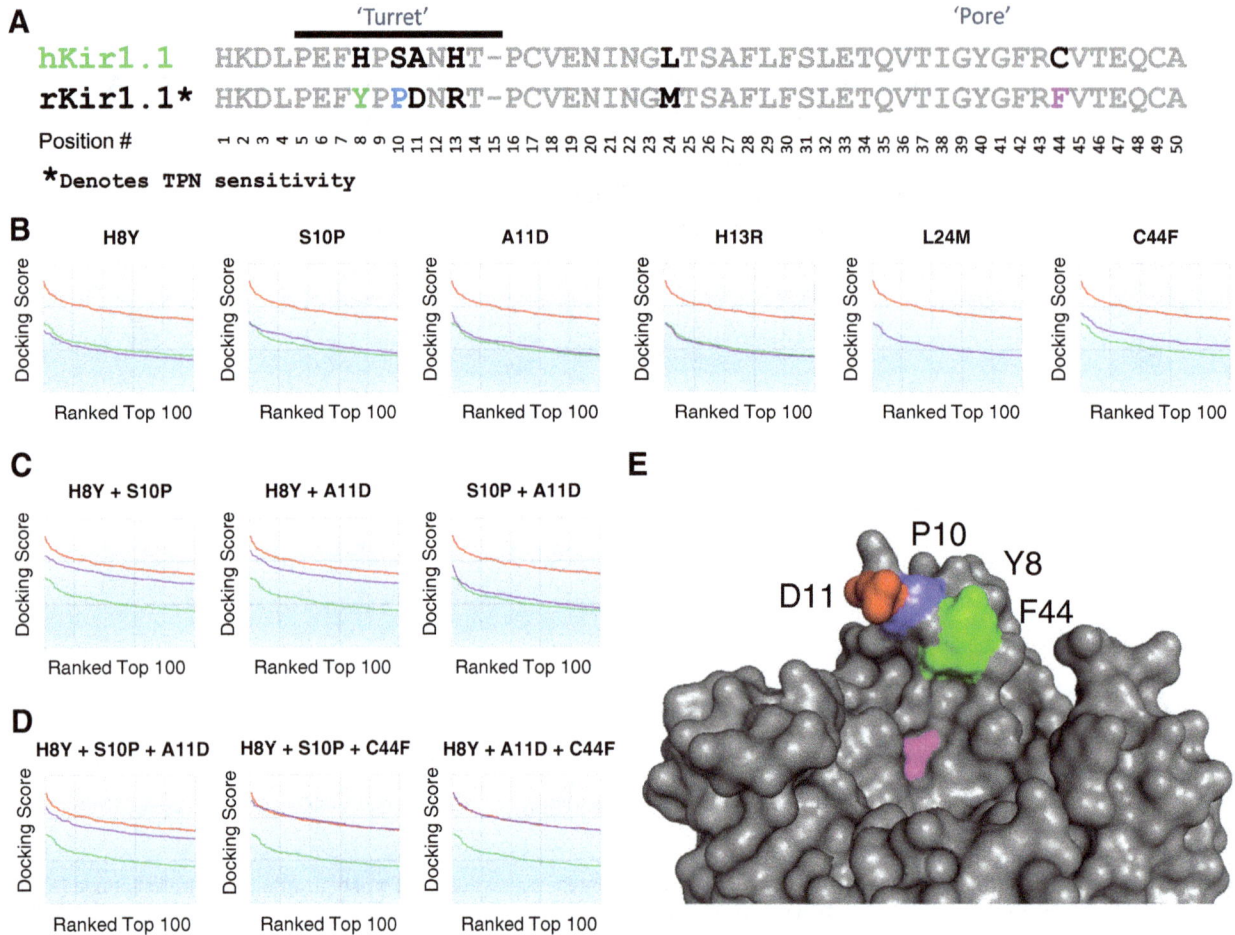

Fig. 8. Mapping 'hotspots' that confer species-dependent differences in TPN docking to the Kir1.1 channel outer vestibule. A. Multiple sequence alignment of the TPN-sensitive rat Kir1.1, and TPN-insensitive human Kir1.1 channel outer vestibules. Highlighted are the six residue differences, with the conserved identical residues shown in gray. B. Single amino acid *in silico* mutagenesis of hKir1.1. Using the homology modeled hKir1.1 channel as a beginning template, each of the six residue differences shown in panel A, were individually mutated to the corresponding residue in the rKir1.1 channel. The TPN docking profile plots for the mutant hKir1.1 channels possessing the single point mutations are shown in purple. For reference, each graph also displays the TPN docking score profile for rat Kir1.1 (red plot) and human Kir1.1 (green plot). The C44F point mutation produced the most significant increase in TPN docking energetics. C. Double amino acid *in silico* mutagenesis of hKir1.1. Two residues in the hKir1.1 turret structure were mutated to corresponding rKir1.1 amino acids. Note the synergistic action of H8Y with either S10P or A11D, on the TPN docking energetics (cf. panel B). D. Triple amino acid *in silico* mutagenesis of hKir1.1. Three residues in the hKir1.1 outer vestibule were mutated to the corresponding rKir1.1 amino acids. Three residue changes were both necessary and sufficient to reproduce the TPN sensitivity differences between rKir1.1 versus hKir1.1, two within the turret (H8Y with either S10P or A11D) and one in the pore region (C44F). E. Surface rendering of the rat Kir1.1 outer vestibule with the locations of the identified 'hotspot' residues mediating differences in TPN docking sensitivity to the human Kir1.1 channel.

When comparing TPN20 to TPN12 coordinates, the TPN20 lysine residues and side chains are in significantly different spatial locations, which apparently are less favorable for electrostatic interactions with the rat Kir1.1 contact rings and thus lower the calculated docking score.

Ultimate validation of docked peptide–channel interactions will require resolution of the bound complex by X-ray crystallography, similar to that recently reported for a charybdotoxin (CTX)-bound Kv channel construct [34]. Interestingly, the CTX-Kv channel complex indicates a similar molecular mechanism of block as reported here, where a lysine residue (Lys27) in the docked CTX peptide interacts with the Kv channel 'GYG' selectivity filter (tyrosine carbonyl groups) effecting K^+ ion occupancy. This molecular mechanism for toxin block had been proposed from earlier CTX and Kv channel mutagenesis studies [35,36]. Analogous toxin blocking mechanisms have been reproduced in several other toxin-Kv channel mutagenesis studies and docking simulations, implicating a conserved 'functional dyad' mechanism where a selectivity filter plugging lysine is assisted by an aromatic residue, and may additionally involve turret interactions [37–42].

Extending our computational approach to a broader 'palette' of ion channels coupled with peptide re-engineering, offers promise for

rationale design of new selective peptides that can block different ion channels with high affinity and specificity [42]. For Kir channels, the hyper-variable turret structures are clearly major determinants for channel specificity in TPN block [10,39,40,43,44]. The Kir channel turrets present unique opportunities for molecular engineering, as was recognized originally by MacKinnon's group with the resolved structure of the cKir2.2 channel [4]. Interrogation of these structures *in silico* with re-engineered peptides offers the potential to yield novel 'virtual probes' that can then be readily synthesized using modern solid state chemistry and tested *in vitro* for validation. The findings reported here will help set the stage for advancing that effort.

Acknowledgments

We thank Rod MacKinnon (Rockefeller University) for providing the chicken Kir2.2 cDNA clone for testing *in vitro* TPN sensitivity, and Jahanshah Amin (University of South Florida) for help with oocyte preparation. This work was supported by internal research funds from the University of South Florida College of Medicine (CD).

References

[1] Doyle DA, Morais Cabral J, Pfuetzner RA, Kuo A, Gulbis JM, Cohen SL, et al. The structure of the potassium channel: molecular basis of K+ conduction and selectivity. Science 1998;280:69–77.

[2] MacKinnon R. Potassium channels and the atomic basis of selective ion conduction (Nobel Lecture). Angew Chem Int Ed Engl 2004;43:4265–77.

[3] Kuo A, Gulbis JM, Antcliff JF, Rahman T, Lowe ED, Zimmer J, et al. Crystal structure of the potassium channel KirBac1.1 in the closed state. Science 2003;300:1922–6.

[4] Tao X, Avalos JL, Chen J, MacKinnon R. Crystal structure of the eukaryotic strong inward-rectifier K+ channel Kir2.2 at 3.1 A resolution. Science 2009;326:1668–74.

[5] Whorton MR, MacKinnon R. Crystal structure of the mammalian GIRK2 K+ channel and gating regulation by G proteins, PIP2, and sodium. Cell 2011;147:199–208.

[6] Kaczorowski GJ, McManus OB, Priest BT, Garcia ML. Ion channels as drug targets: the next GPCRs. J Gen Physiol 2008;131:399–405.

[7] Dutertre S, Lewis RJ. Use of venom peptides to probe ion channel structure and function. J Biol Chem 2010;285:13315–20.

[8] Gauldie J, Hanson JM, Rumjanek FD, Shipolini RA, Vernon CA. The peptide components of bee venom. Eur J Biochem 1976;61:369–76.

[9] Jin W, Lu Z. A novel high-affinity inhibitor for inward-rectifier K+ channels. Biochemistry 1998;37:13291–9.

[10] Jin W, Klem AM, Lewis JH, Lu Z. Mechanisms of inward-rectifier K+ channel inhibition by tertiapin-Q. Biochemistry 1999;38:14294–301.

[11] Xu X, Nelson JW. Solution structure of tertiapin determined using nuclear magnetic resonance and distance geometry. Proteins 1993;17:124–37.

[12] Guex N, Peitsch MC. SWISS-MODEL and the Swiss-PdbViewer: an environment for comparative protein modeling. Electrophoresis 1997;18:2714–23.

[13] Schwede T, Kopp J, Guex N, Peitsch MC. SWISS-MODEL: an automated protein homology-modeling server. Nucleic Acids Res 2003;31:3381–5.

[14] Arnold K, Bordoli L, Kopp J, Schwede T. The SWISS-MODEL workspace: a web-based environment for protein structure homology modelling. Bioinformatics 2006;22:195–201.

[15] Chen R, Li L, Weng Z. ZDOCK: an initial-stage protein-docking algorithm. Proteins 2003;52:80–7.

[16] Pierce BG, Wiehe K, Hwang H, Kim BH, Vreven T, Weng Z. ZDOCK server: interactive docking prediction of protein–protein complexes and symmetric multimers. Bioinformatics 2014;30:1771–3.

[17] Mintseris J, Pierce B, Wiehe K, Anderson R, Chen R, Weng Z, et al. Integrating statistical pair potentials into protein complex prediction. Proteins 2007;69:511–20.

[18] Felix JP, Liu J, Schmalhofer WA, Bailey T, Bednarek MA, Kinkel S, et al. Characterization of Kir1.1 channels with the use of a radiolabeled derivative of tertiapin. Biochemistry 2006;45:10129–39.

[19] Abbas L, Hajihashemi S, Stead LF, Cooper GJ, Ware TL, Munsey TS, et al. Functional and developmental expression of a zebrafish Kir1.1 (ROMK) potassium channel homologue Kcnj1. J Physiol 2011;589:1489–503.

[20] Comeau SR, Gatchell DW, Vajda S, Camacho CJ. ClusPro: a fully automated algorithm for protein-protein docking. Nucleic Acids Res 2004;32:W96–9.

[21] Kozakov D, Hall DR, Beglov D, Brenke R, Comeau SR, Shen Y, et al. Achieving reliability and high accuracy in automated protein docking: ClusPro, PIPER, SDU, and stability analysis in CAPRI rounds 13-19. Proteins 2010;78:3124–30.

[22] Kozakov D, Beglov D, Bohnuud T, Mottarella SE, Xia B, Hall DR, et al. How good is automated protein docking? Proteins 2013;81:2159–66.

[23] Krissinel E, Henrick K. Inference of macromolecular assemblies from crystalline state. J Mol Biol 2007;372:774–97.

[24] Krissinel E. Crystal contacts as nature's docking solutions. J Comput Chem 2010;31:133–43.

[25] Doupnik CA, Jaen C, Zhang Q. Measuring the modulatory effects of RGS proteins on GIRK channels. Methods Enzymol 2004;389:131–54.

[26] Ho K, Nichols CG, Lederer WJ, Lytton J, Vassilev PM, Kanazirska MV, et al. Cloning and expression of an inwardly rectifying ATP-regulated potassium channel. Nature 1993;362:31–8.

[27] Kozakov D, Clodfelter KH, Vajda S, Camacho CJ. Optimal clustering for detecting near-native conformations in protein docking. Biophys J 2005;89:867–75.

[28] Kozakov D, Brenke R, Comeau SR, Vajda S. PIPER: an FFT-based protein docking program with pairwise potentials. Proteins 2006;65:392–406.

[29] Shoichet BK, Kobilka BK. Structure-based drug screening for G-protein-coupled receptors. Trends Pharmacol Sci 2012;33:268–72.

[30] Legros C, Pollmann V, Knaus HG, Farrell AM, Darbon H, Bougis PE, et al. Generating a high affinity scorpion toxin receptor in KcsA-Kv1.3 chimeric potassium channels. J Biol Chem 2000;275:16918–24.

[31] Hilder TA, Chung SH. Conduction and block of inward rectifier K+ channels: predicted structure of a potent blocker of Kir2.1. Biochemistry 2013;52:967–74.

[32] Hu J, Qiu S, Yang F, Cao Z, Li W, Wu Y, et al. Unique mechanism of the interaction between honey bee toxin TPNQ and rKir1.1 potassium channel explored by computational simulations: insights into the relative insensitivity of channel towards animal toxins. PLoS ONE 2013;8:e67213.

[33] Jin W, Lu Z. Synthesis of a stable form of tertiapin: a high-affinity inhibitor for inward-rectifier K+ channels. Biochemistry 1999;38:14286–93.

[34] Banerjee A, Lee A, Campbell E, Mackinnon R. Structure of a pore-blocking toxin in complex with a eukaryotic voltage-dependent K(+) channel. Elife (Cambridge) 2013;2:e00594.

[35] Park CS, Miller C. Mapping function to structure in a channel-blocking peptide: electrostatic mutants of charybdotoxin. Biochemistry 1992;31:7749–55.

[36] Park CS, Miller C. Interaction of charybdotoxin with permeant ions inside the pore of a K+ channel. Neuron 1992;9:307–13.

[37] Dauplais M, Lecoq A, Song J, Cotton J, Jamin N, Gilquin B, et al. On the convergent evolution of animal toxins. Conservation of a diad of functional residues in potassium channel-blocking toxins with unrelated structures. J Biol Chem 1997;272:4302–9.

[38] Gao YD, Garcia ML. Interaction of agitoxin2, charybdotoxin, and iberiotoxin with potassium channels: selectivity between voltage-gated and Maxi-K channels. Proteins 2003;52:146–54.

[39] Jouirou B, Mouhat S, Andreotti N, De Waard M, Sabatier JM. Toxin determinants required for interaction with voltage-gated K+ channels. Toxicon 2004;43:909–14.

[40] Mouhat S, Mosbah A, Visan V, Wulff H, Delepierre M, Darbon H, et al. The 'functional' dyad of scorpion toxin Pi1 is not itself a prerequisite for toxin binding to the voltage-gated Kv1.2 potassium channels. Biochem J 2004;377:25–36.

[41] Chen R, Chung SH. Molecular dynamics simulations of scorpion toxin recognition by the Ca(2+)-activated potassium channel KCa3.1. Biophys J 2013;105:1829–37.

[42] Gordon D, Chen R, Chung SH. Computational methods of studying the binding of toxins from venomous animals to biological ion channels: theory and applications. Physiol Rev 2013;93:767–802.

[43] Ramu Y, Klem AM, Lu Z. Short variable sequence acquired in evolution enables selective inhibition of various inward-rectifier K+ channels. Biochemistry 2004;43:10701–9.

[44] Ramu Y, Xu Y, Lu Z. Engineered specific and high-affinity inhibitor for a subtype of inward-rectifier K+ channels. Proc Natl Acad Sci U S A 2008;105:10774–8.

21

The role of water in protein's behavior: The two dynamical crossovers studied by NMR and FTIR techniques

Francesco Mallamace [a,c,*], Carmelo Corsaro [a], Domenico Mallamace [b], Sebastiano Vasi [a], Cirino Vasi [c], Giacomo Dugo [b]

[a] Dipartimento di Fisica e Scienze della Terra, Università di Messina, Viale F. Stagno D'Alcontres 31, 98166 Messina, Italy
[b] Dipartimento di Scienze dell'Ambiente, della Sicurezza, del Territorio, degli Alimenti edella Salute, Università di Messina, Viale F. Stagno d'Alcontres 31, 98166 Messina, Italy
[c] CNR-IPCF, Istituto per i Processi Chimico-Fisici, Viale F. Stagno D'Alcontres 37, 98158 Messina, Italy

ARTICLE INFO

Keywords:
Protein dynamic transition
Amide bending mode
Lysozyme unfolding
Hydration water
HR-MAS

ABSTRACT

The role the solvent plays in determining the biological activity of proteins is of primary importance. Water is the solvent of life and proteins need at least a water monolayer covering their surface in order to become biologically active. We study how the properties of water and the effect of its coupling with the hydrophilic moieties of proteins govern the regime of protein activity. In particular we follow, by means of Fourier Transform Infrared spectroscopy, the thermal evolution of the amide vibrational modes of hydrated lysozyme in the temperature interval 180 K < T < 350 K. In such a way we are able to observe the thermal limit of biological activity characterizing hydrated lysozyme. Finally we focus on the region of lysozyme thermal denaturation by following the evolution of the proton Nuclear Magnetic Resonance (NMR) spectra for 298 K < T < 366 K with the High-Resolution Magic Angle Spinning probe. Our data suggest that the hydrogen bond coupling between hydration water and protein hydrophilic groups is crucial in triggering the main mechanisms that define the enzymatic activity of proteins.

1. Introduction

Protein activity is connected with their hydration water [1]. In fact, at least, a monolayer of water molecules called the first hydration shell or directly hydration water, extended over the protein surface is needed for the execution of the enzymatic activity [2,3]. The key factor of protein hydration is the H-bonding between protein surface polar groups and hydration water. Furthermore, the coupling between the hydration water and the hydrophilic moieties of the protein surface triggers the search for the correct native state (protein folding): the complex heteropolymeric amino acid sequences spontaneously fold up into organized three-dimensional structures. The spontaneity of the folding process depends on the occurrence of non-functional, or promiscuous, interactions between any suitable pair of residues that can provoke the formation of transient intermediate structures, through "non-native" interactions or frustration [4–6], that can be interpreted as roughness on the folding energy landscape [7]. The native state of a protein corresponds to a global free energy minimum that the protein reaches in a time range from microseconds to seconds [8]. If the protein should find its native state just by random

searching (Levinthal paradox [9]) among the huge number of possible conformations, this search could take longer than the age of the universe.

It is noteworthy that already in 1936, Mirsky and Pauling discriminated between the fundamentally different character of the native and denatured states of proteins [10]. They argued that the folding process is no more (nor less) miraculous than is the formation of a crystal from a supersaturated solution [10]. The reason lies in the cooperative nature of the denaturation and in the large magnitude of the corresponding enthalpy change. Indeed the native state must be nearly unique in structure, like a crystal, whereas the denatured state has a much higher entropy, reflecting the numerous disordered conformations that a chain molecule could take on [10]. However one has to consider that the protein is a finite system and the extension of concepts such as those of nucleation and growth mechanism cannot be easily applied. From a physical point of view, proteins are hard matter at low temperature, whereas they are soft matter at high temperature depending on the competition between the enthalpic end entropic contributions [8]. The highly directional and polar character of the hydrogen bond seems to be the key to understand the microscopic mechanisms occurring during protein folding [10,11]. In fact, the physical and chemical properties of proteins depend on the characteristic of the hydrogen bonds formed within the protein residuals and with its hydration water.

* Corresponding author.
  E-mail address: francesco.mallamace@unime.it (F. Mallamace).

For lysozyme in particular it has been shown how the hydrogen bond network that hydration water develops on the protein surface is stable at atmospheric pressure within a temperature interval going from $\approx$ 225 K to $\approx$ 320 K [12–18]. The temperature of 225 K has been identified as the temperature of the protein glass-transition but its nature is up to now the subject of many controversies [17–22]. In fact, it was pointed out that the glass transition temperature of hydrated lysozyme depends on the hydration level, and the time scale of measurement [23,24]. Below 225 K, the water hydrogen bond network is highly rigid being fully developed and protein side-chains motion is hindered. On the contrary, above 320 K the lifetime of the hydrogen bond is too short (less than picoseconds) and does not allow to keep together the protein residuals giving rise to the unfolding process [16,25–27]. The native state of lysozyme does not evolve directly into the completely unfolded (or denatured) state but passes through an intermediate state (within which the unfolding process is reversible) where rapid conformational changes occur and can provoke alteration of the folding. The alteration of the folding (or misfolding) of proteins is the source of neurodegenerative illnesses such as Alzheimer's and Parkinson's diseases [7,28].

In this paper we use two different but complementary techniques such as Fourier Transform Infrared (FTIR) and Nuclear Magnetic Resonance (NMR) spectroscopies to investigate how the coupling between the hydration water and the protein residuals evolves as a function of the temperature and determines the limits of biological activity of hydrated lysozyme. In particular, with FTIR we were able to probe the interval from 180K to 360K with a 10K step, whereas with NMR we focused on the unfolding process from 298K to 366K with a 2K step.

2. Materials and methods

Lysozyme is a small protein of 14.4 kDa; it is constituted by 129 amino acid residuals and in the native state has a globular shape. Lysozyme is easily found in animal tissues and displays anti-inflammatory and antibacterial properties. Protein samples were obtained from Fluka (L7651 three times crystallized, dialyzed, and lyophilized) and used without further purification. Samples were dried, hydrated isopiestically, and controlled by means of a precise procedure [12]. Our aim is to study the first monolayer of water molecules surrounding the protein surface and this corresponds to a hydration level, h (grams of water per gram of dry protein) equals to 0.3. Fourier Transform Infrared (FTIR) absorption measurements were performed by means of a Bomem DA8 Fourier transform spectrometer, operating with a Globar source, in combination with a KBr beamsplitter and a DTGS/KBr detector. We operated in the attenuated total reflection (ATR) geometry to avoid saturation effects. Spectra were recorded with a resolution of 4 cm^{-1}, automatically adding 200 repetitive scans in order to obtain a good signal-to-noise ratio and highly reproducible spectra; then they are normalized by taking into account the effective number of absorbers [13]. Measurements were performed at ambient pressure in the spectral region from 1300 cm^{-1} to 1750 cm^{-1}, in the temperature range from 180 K to 350 K. Proton NMR experiments were performed at atmospheric pressure in the temperature range 298 K < T < 366 K by using a 700 MHz Bruker Avance spectrometer equipped with the Magic Angle Spinning (MAS) probehead. Hydrated protein samples were placed in a 50 µl rotor and spun at 4000 Hz at the magic angle to increase the spectral resolution. By tilting samples of a precise angle with respect to the applied magnetic field, the hamiltonian term corresponding to the dipolar interactions vanishes and NMR peaks become narrower. Furthermore, by spinning the rotor at the magic angle by few thousands of Hertz, line broadening effects due to susceptibility differences within the sample are removed resulting in high resolution quality spectra. The sample temperature was controlled by a cold N_2 flow and a heating element, calibrated by using the frequency shift between ethylene glycol peaks. The duration

of the hard pulse was 8 µs with a relative attenuation of 3 dB; the spectral width was 10 kHz, the acquisition time 2.9 s, the points in the time domain 64 k, the number of transient 128 and the relaxation time 2 s for a total time of about 10 min per experiment. All spectra were processed (line broadening, Fourier transform, phase correction and baseline adjustment), by using the standard routines of the Bruker software Xwinnmr version 3.5.

3. Results and discussions

InfraRed and NMR spectroscopies are probably the most used experimental techniques able to study protein structure and properties [29]. In particular, FTIR spectroscopy permits a detailed analysis of the structure and stability of proteins, using peptide backbone and side-chain marker bands as conformation-sensitive monitors. Specific information on the secondary structure of proteins is obtained from the analysis of the various amide bands which are indeed sensitive to the protein conformation.

In details, IR spectra of hydrated proteins provide useful structural information especially in the region of Amide I (the most intense band centered at $1600 - 1700$ cm^{-1}) which is mostly a carbonyl (C=O) stretching [30,31]. In particular, the amide I band is sensitive to hydrogen bond pattern, dipole–dipole interaction and the geometry of the polypeptide backbone. It consists of several overlapping bands of different structural elements that could be studied separately by means of a peak deconvolution [30,32].

Other intense and important Amide bands are Amide II and Amide III extending respectively from 1480 to 1580 cm^{-1} and from 1300 to 1450 cm^{-1}. The Amide II mode is essentially the combination of the N–H in plane bending and of the C–N stretching, while Amide III consists of more complex vibrational modes [31,33]. The different Amide contributions to the IR bending region are reported in Fig. 1 with different colors. In the figure, the Amide I and II vibrational modes are also represented on a peptide fragment using the same color of the corresponding IR frequency regions.

The hydrogen bond coupling is a complex phenomenon that can be studied by analyzing the trend that IR spectra show as a function of the temperature. In particular, the behavior of the Amides peak intensity, on increasing the temperature in all the studied range, is not monotonic. In Fig. 2 we use three panels to separate the three important thermal regions within which the spectral behavior is monotonic with temperature. In each panel, for clarity we report only three significant temperatures able to describe the thermal behavior; the intermediate

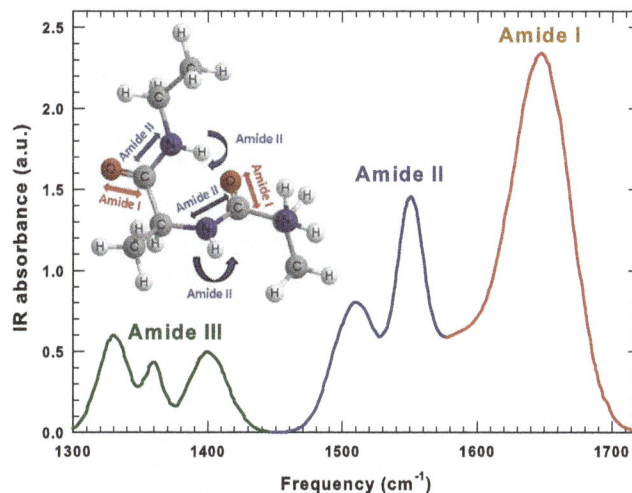

Fig. 1. The different Amide contributions to the IR bending region are reported with different colors in the interval $1300 - 1720$ cm^{-1}. The Amide I and II vibrational modes are also represented on a peptide fragment using the same color of the corresponding IR frequency regions.

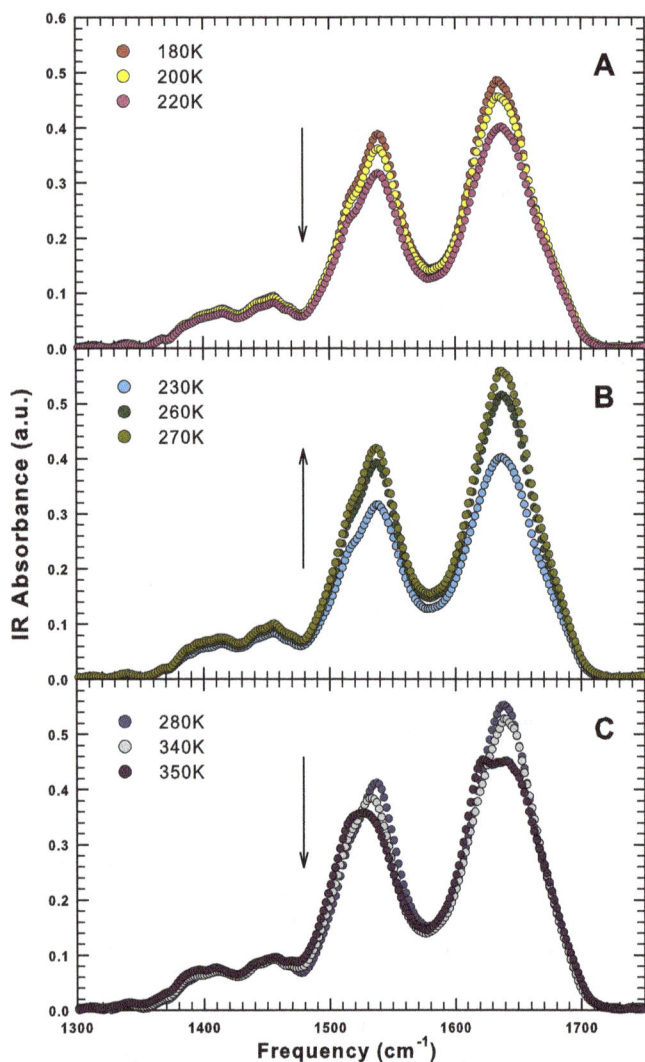

Fig. 2. The Infrared spectra of hydrated lysozyme (h = 0.3) in the interval 1300 − 1750 cm^{-1} for 180 K < T < 220 K (panel A), for 230 K < T < 270 K (panel B) and for 280 K < T < 350 K (panel C). The arrows indicate the evolution of the signal intensity with temperature.

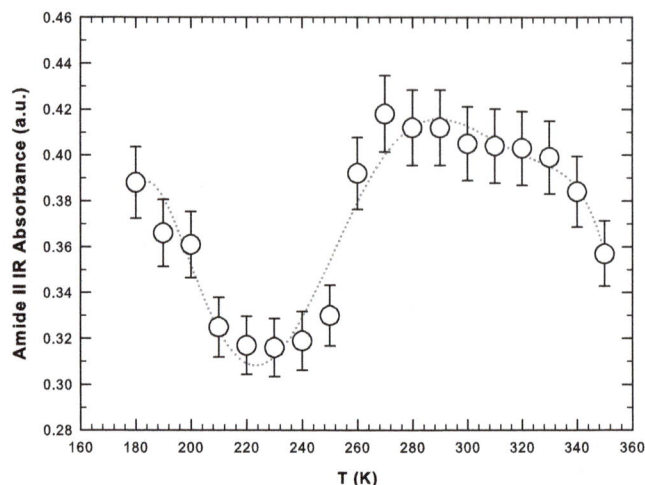

Fig. 3. The intensity of the Amide II infrared band as a function of the temperature for hydrated lysozyme (h = 0.3). The dotted line is a polynomial best fit as a guide for the eye.

temperatures follow the same trend. In particular, panel A of Fig. 2 shows the IR spectra in the low temperature region from 180 K to 220 K. The signal intensity decreases on increasing the temperature as indicated by the arrow. Panel B of Fig. 2 shows the IR spectra in the intermediate temperature region, from 230 K to 270 K. Note that, in the intermediate temperature region, the signal intensity increases with the temperature. Panel C of Fig. 2 shows the IR spectra in the high temperature region, from 280 K to 350 K. Again, in this high temperature region, the signal intensity decreases on increasing the temperature. The shoulder at about 1500 cm^{-1}, associated with the amide II N–H residual, is well evident at low temperature and disappears upon denaturation [34].

One suitable approach for the characterization of temperature-induced conformational changes in protein is to construct intensity/ temperature profiles for selected IR bands. In such a way one can determine standard thermodynamic properties of the system such as transition or crossover temperatures [35]. Fig. 3 shows the intensity of the Amide II band as a function of the temperature in the considered thermal range. The dotted line is a polynomial best fit as a guide for the eye. The Amide II band, representing the N–H bending contribution, reflects directly the coupling between hydration water and protein residuals. Note that, on increasing the temperature, the intensity of the Amide II band shows a minimum at ≈ 225 K. This temperature

coincides with that of the dynamical crossover observed by means of neutron scattering [12,36] and NMR spectroscopy [13,27,37]. It is noteworthy that even though these experimental techniques encompass different time scales, all lead to the same temperature range for the occurrence of the dynamical transition. Above 225 K, when the motional amplitude (i.e., the flexibility) of hydrated lysozyme sharply increases, the intensity of the Amide II band increases with temperature up to ≈ 270 K. Then, the intensity of the Amide II band starts to slowly decrease on increasing the temperature with a smooth inflection point at ≈ 320 K. This is the "magic" temperature above which water behaves as a normal liquid since the HB lifetime becomes too small (less than picoseconds) and all water molecules are essentially free [38]. Furthermore at this temperature the refolding rate constant assumes the maximum value [8]. Above 320 K the intensity of the Amide II band sharply decreases on increasing the temperature (Fig. 3); water is no more a "good solvent" and lysozyme loses its globular structure evolving toward a linear chain of amino acids. Thus, two temperatures appear to be relevant for the onset of different dynamical regimes (and so to the functioning) of hydrated lysozyme. The values of these temperatures agree with those already found for the same system and described in the introduction, which are 225 K and 320 K. All the described changes and their thermal borders are well described by the intensity of the Amide II band reported in Fig. 3 confirming the ability of the FTIR technique in detecting the structural conformational behavior of proteins and the corresponding role of hydration water. The spectrum at 350 K, which is the highest measured temperature, shows (panel C) a clear peak bifurcation relatively to the Amide I band at 1650 cm^{-1}. This is due to the onset of aggregation processes in which α -helices transform into β-sheets that tend to self-aggregate. The unfolding process is reversible in character up to ≈ 346 K and then becomes irreversible [26].

We performed a detailed NMR experiment in order to get a precise insight into the thermal denaturation of hydrated lysozyme. In fact, as shown in Fig. 3, aside from the smooth inflection point at about 320 K, the thermal trend of the intensity of the Amide II band for 280 K < T < 330 K is quite flat. NMR spectra of hydrated proteins allow studying the different chemical groups of protein and water separately. In Fig. 4 we present the stacked plot of NMR spectra for hydrated lysozyme (h = 0.3) in the high temperature region measured by means of the HR-MAS set-up. We started the measurement at 298 K and reached 366 K with steps of 2 K.

The most intense signal, which is cut off in the figure, belongs to hydration water protons and is more than four orders of magnitude larger than the other contributions. Note that, except for the peak at 1 ppm (assigned to the methyl functional group), all the peaks belonging to

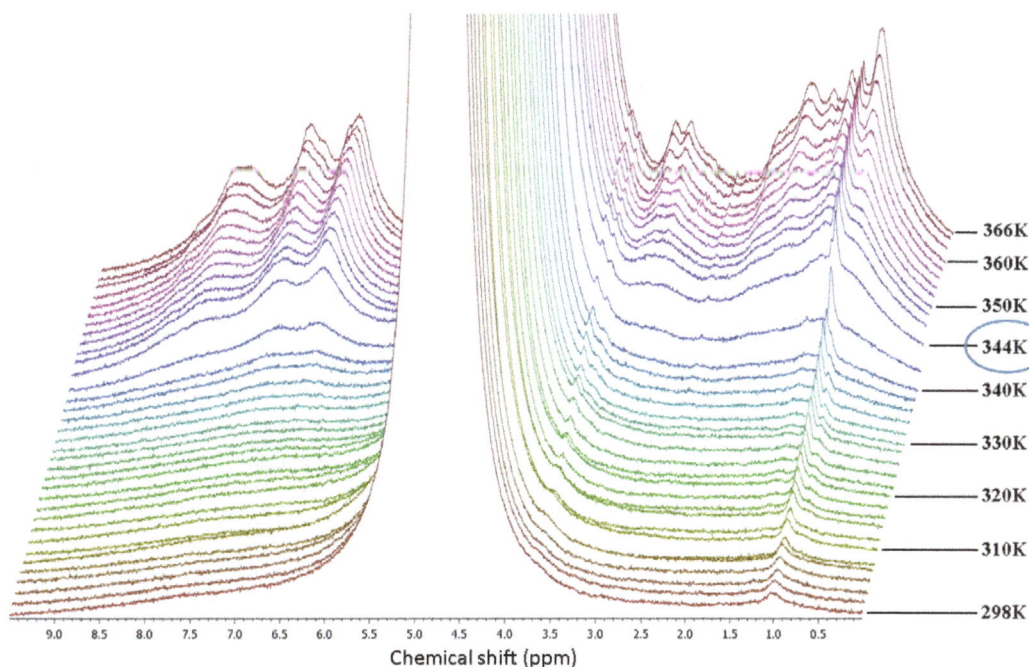

Fig. 4. The stacked plot of NMR spectra for hydrated lysozyme (h = 0.3) in the high temperature region measured by means of the HR-MAS set-up. The temperature of 344 K is highlighted because it marks the irreversibility of the unfolding process.

the protons of lysozyme, begin to appear at $\approx$ 320 K. This is a clear sign that the protons of lysozyme are immobile (protein side-chains are not flexible) on the NMR timescale at this hydration level up to $\approx$ 320 K. This temperature is on the border between the native state and the intermediate region where the unfolding process starts [26,27].

The protein side-chains become more mobile on increasing the temperature, and their contribution to the free induction decay of the magnetization is indeed detectable. The temperature that marks the irreversibility of the unfolding process is $\approx$ 344 K where the magnetization signal suddenly increases and all protein contributions are clearly visible (Fig. 4). The decreasing of their peak width reflects the enhanced protein mobility due to the almost complete hydrogen bonding breakage.

4. Conclusions

In this paper we have studied the coupling of water with the hydrophilic moieties of hydrated lysozyme (h = 0.3) by means of Fourier Transform Infrared and NMR spectroscopy. In particular, by looking at selected vibrations we study how the hydrogen bond interaction governs the regime of protein activity. The intensity of the amide vibrational modes of hydrated lysozyme in the temperature interval 180 K < T < 350 K is able to reflect the thermal limit of biological functioning characterizing the studied system. Two temperatures, 225 K and 320 K, appear to be relevant for the onset of different dynamical regimes (and so to the functioning) of hydrated lysozyme. Their values agree with those already found for the same system and described in the introduction. Above 225 K the flexibility of hydrated lysozyme sharply increases due to the softening of the hydrogen bond network that the hydration water develops on its surface. Above 320 K water behaves as a normal liquid since the HB lifetime becomes too small (less than picoseconds). All water molecules become essentially free and water is no more a "good solvent". The refolding rate constant assumes the maximum value [8] and lysozyme tends to lose its globular structure evolving toward a linear chain of amino acids.

Finally, in order to get precise insight into the thermal denaturation of hydrated lysozyme we followed the evolution of the proton NMR spectra for 298 K < T < 366 K with the High-Resolution Magic Angle

Spinning probe. In the obtained spectra, all the peaks belonging to the protons of lysozyme, begin to appear at $\approx$ 320 K, except that at 1 ppm (assigned to the methyl functional group). This means that at this hydration level up to $\approx$ 320 K, the protein side-chains are not so flexible to be revealed by NMR. At higher temperatures, the protein side-chains become more mobile and their contribution is indeed well detectable. The temperature ($\approx$344 K) at which the magnetization signal suddenly increases and all protein contributions are clearly visible signs the irreversibility of the unfolding process. The corresponding decrease of the peaks width reflects the increasing protein mobility provoked by the almost complete hydrogen bonding breakage.

In conclusion the experimental data we have presented in this work, suggest that the main mechanisms that define the thermal limits of the enzymatic activity of proteins are given by the hydrogen bond coupling between hydration water and protein hydrophilic groups.

Acknowledgments

CC thanks the Centro Siciliano di Fisica Nucleare e Struttura della Materia (Grant no. 01/14) (Catania, Italy) for its support.

References

[1] Gregory RB. Protein solvent interaction. New York: Marcel Dekker; 1995.
[2] Rupley JA, Careri G. Protein hydration and function. Adv Protein Chem 1991;41: 37–172.
[3] Teeter MM. Water–protein interactions: theory and experiment. Annu Rev Biophys Biophys Chem 1991;20:577–600.
[4] Bryngelson JD, Wolynes PG. Spin glasses and the statistical mechanics of protein folding. Proc Natl Acad Sci U S A 1987;84:7524–8.
[5] Bowman GR, Pande VS. Protein folded states are kinetic hubs. Proc Natl Acad Sci 2010;107:10890–5.
[6] Ferreiro DU, Hegler JA, Komives EA, Wolynes PG. Localizing frustration in native proteins and protein assemblies. Proc Natl Acad Sci U S A 2007;104:19819–24.
[7] Chiti F, Dobson CM. Amyloid formation by globular proteins under native conditions. Nat Chem Biol 2009;5:15–22.
[8] Karplus M. Behind the folding funnel diagram. Nat Chem Biol 2011;7:401–4.
[9] Levinthal C. How to fold graciously. In: Debrunner P, Tsibris JCM, Münck E, editors. Mossbauer spectroscopy in biological systems, proceedings of a meeting held at Allerton House, Monticello, Illinois. Urbana: University of Illinois Press; 1969.
[10] Mirsky AE, Pauling L. On the structure of nature, denatured and coagulated proteins. Proc Natl Acad Sci U S A 1936;22:439–47.

[11] Ben-Naim A. The role of hydrogen bonds in protein folding and protein association. J Phys Chem 1991;95:1437–44.
[12] Chen S-H, Liu L, Fratini E, Baglioni P, Faraone A, Mamontov E. Observation of fragile-to-strong dynamic crossover in protein hydration water. Proc Natl Acad Sci U S A 2006;103:9016.
[13] Mallamace F, Chen SH, Broccio M, Corsaro C, Crupi V, Majolino D, et al. Role of the solvent in the dynamical transitions of proteins: the case of the lysozyme-water system. J Chem Phys 2007;127:045104.
[14] Mallamace F, Corsaro C, Broccio M, Branca C, González-Segredo N, Spooren J, et al. NMR evidence of a sharp change in a measure of local order in deeply supercooled confined water. Proc Natl Acad Sci U S A 2008;105:12725.
[15] Mallamace F, Branca C, Corsaro C, Leone N, Spooren J, Stanley HE, et al. Dynamical crossover and breakdown of the Stokes-Einstein relation in confined water and in methanol-diluted bulk water. J Phys Chem B 2010;114(5):1870–8.
[16] Zhang Y, Lagi M, Liu D, Mallamace F, Fratini E, Baglioni P, et al. Observation of high-temperature dynamic crossover in protein hydration water and its relation to reversible denaturation of lysozyme. J Chem Phys 2009;130:135101.
[17] Lagi M, Chu X, Kim C, Mallamace F, Baglioni P, Chen SH. The low-temperature dynamic crossover phenomenon in protein hydration water: simulations vs experiments. J Phys Chem B 2008;112(6):1571–5.
[18] Kumar P, Yan Z, Xu L, Mazza MG, Buldyrev SV, Chen SH, et al. Glass transition in biomolecules and the liquid-liquid critical point of water. Phys Rev Lett 2006;97:177802.
[19] Angell CA. Formation of glasses from liquids and biopolymers. Science 1995;267:1924.
[20] Iben IET, Braunstein D, Doster W, Frauenfelder H, Hong MK, Johnson JB, et al. Glassy behaviour of a protein. Phys Rev Lett 1989;62:1916.
[21] Ngai KL, Capaccioli S, Shinyashiki N. The protein "glass" transition and the role of the solvent. J Phys Chem B 2008;112(12):3826–32.
[22] Ngai KL, Capaccioli S, Paciaroni A. Nature of the water specific relaxation in hydrated proteins and aqueous mixtures. Chem Phys 2013;424:37–44.
[23] Capaccioli S, Ngai KL, Ancherbak S, Paciaroni A. Evidence of coexistence of change of caged dynamics at T_g and the dynamic transition at T_d in solvated proteins. J Phys Chem B 2012;116:1745–57.
[24] Ngai KL, Capaccioli S, Paciaroni A. Change of caged dynamics at T_g in hydrated proteins: trend of mean squared displacements after correcting for the methyl-group rotation contribution. J Chem Phys 2013;138:235102.
[25] Ball P. Water as an active constituent in cell biology. Chem Rev 2008;108:74–108.
[26] Salvetti G, Tombari E, Mikheeva L, Johari GP. The endothermic effects during denaturation of lysozyme by temperature modulated calorimetry and an intermediate reaction equilibrium. J Phys Chem B 2002;106:6081–7.
[27] Mallamace F, Corsaro C, Mallamace D, Baglioni P, Stanley HE, Chen S-H. A possible role of water in the protein folding process. J Phys Chem B 2011;115:14280–94.
[28] Selkoe DJ. Folding proteins in fatal ways. Nature 2003;426:900–4.
[29] Spectroscopic methods for analysis of protein secondary structure. Anal Biochem 2000;277:167–76.
[30] Barth A, Zscherp C. What vibrations tell us about proteins. Q Rev Biophys 2002;35:369–430.
[31] Kong J, Yu S. Fourier transform infrared spectroscopic analysis of protein secondary structures. Acta Biochim Biophys Sin 2007;39:549–59.
[32] Mallamace D, Corsaro C, Vasi C, Vasi S, Dugo G, Mallamace F. The protein irreversible denaturation studied by means of the bending vibrational mode. Phys A Stat Mech Appl 2014;412:39–44.
[33] Banker J. Amide modes and protein conformation. Biochim Biophys Acta 1992;1120:123–43.
[34] Mallamace F, Baglioni P, Corsaro C, Chen S-H, Mallamace D, Vasi C, et al. The influence of water on protein properties. J Chem Phys 2014;141:165104.
[35] Fabian H, Mantele W. Infrared spectroscopy of proteins. In: Chalmer Griffiths, editor. Handbook of vibrational spectroscopy, 5. Wiley; 2002. p. 3426–52.
[36] Schiro G, Natali F, Cupane A. Physical origin of anharmonic dynamics in proteins: new insights from resolution-dependent neutron scattering on homomeric polypeptides. Phys Rev Lett 2012;109:128102.
[37] Corsaro C, Mallamace D. A nuclear magnetic resonance study of the reversible denaturation of hydrated lysozyme. Phys A Stat Mech Appl 2011;390:2904–8.
[38] Mallamace F, Corsaro C, Stanley HE. A singular thermodynamically consistent temperature at the origin of the anomalous behavior of liquid water. Sci Rep 2012;2:993.

Permissions

The contributors of this book come from diverse backgrounds, making this book a truly international effort. This book will bring forth new frontiers with its revolutionizing research information and detailed analysis of the nascent developments around the world.

We would like to thank all the contributing authors for lending their expertise to make the book truly unique. They have played a crucial role in the development of this book. Without their invaluable contributions this book wouldn't have been possible. They have made vital efforts to compile up to date information on the varied aspects of this subject to make this book a valuable addition to the collection of many professionals and students.

This book was conceptualized with the vision of imparting up-to-date information and advanced data in this field. To ensure the same, a matchless editorial board was set up. Every individual on the board went through rigorous rounds of assessment to prove their worth. After which they invested a large part of their time researching and compiling the most relevant data for our readers.

The editorial board has been involved in producing this book since its inception. They have spent rigorous hours researching and exploring the diverse topics which have resulted in the successful publishing of this book. They have passed on their knowledge of decades through this book. To expedite this challenging task, the publisher supported the team at every step. A small team of assistant editors was also appointed to further simplify the editing procedure and attain best results for the readers.

Apart from the editorial board, the designing team has also invested a significant amount of their time in understanding the subject and creating the most relevant covers. They scrutinized every image to scout for the most suitable representation of the subject and create an appropriate cover for the book.

The publishing team has been an ardent support to the editorial, designing and production team. Their endless efforts to recruit the best for this project, has resulted in the accomplishment of this book. They are a veteran in the field of academics and their pool of knowledge is as vast as their experience in printing. Their expertise and guidance has proved useful at every step. Their uncompromising quality standards have made this book an exceptional effort. Their encouragement from time to time has been an inspiration for everyone.

The publisher and the editorial board hope that this book will prove to be a valuable piece of knowledge for researchers, students, practitioners and scholars across the globe.

List of Contributors

Christofer S. Tautermann, Daniel Seeliger and Jan M. Kriegl
Boehringer Ingelheim Pharma GmbH & Co. KG, Lead Identification and Optimization Support, Birkendorfer Str. 65, D-88397 Biberach a.d. Riss, Germany

Yusuke Kanematsu and Masanori Tachikawa
Graduate School of Information Science, Hiroshima City University, 3-4-1 Ozuka-Higashi, Asa-Minami-Ku, Hiroshima 731-3194, Japan
Quantum Chemistry Division, Yokohama City University, Seto 22-2, Kanazawa-ku, Yokohama 236-0027, Japan

Hironari Kamikubo, Mikio Kataoka and Masanori Tachikawa
Nara Institute of Science and Technology, 8916-5 Takayama, Ikoma, Nara 630-0192, Japan

Krithika Bhuvaneshwar and Robinder Gauba
Innovation Center for Biomedical Informatics (ICBI), Georgetown University, Washington, DC 20007, USA

Dinanath Sulakhe, Alex Rodriguez and Ravi Madduri
Computation Institute, University of Chicago, Argonne National Laboratory, 60637, USA
Globus Genomics, USA

Geneviève Hélie, Marie Parat, Thomas P. Loisel and Allan Matte
Protein Purification, Human Health Therapeutics, National Research Council Canada, 6100 Royalmount Ave. Montreal, QC H4P 2R2, Canada

Frédéric Massé
Primary Assays, Human Health Therapeutics, National Research Council Canada, 6100 Royalmount Ave. Montreal, QC H4P 2R2, Canada

Cory J. Gerdts
Protein BioSolutions Inc., Suite 280, 401 Professional Drive, Gaithersburg, MD, 20879, USA

Hervé Seligmann
Unité de Recherche sur les Maladies Infectieuses et Tropicales Émergentes, Faculté de Médecine, URMITE CNRS-IRD 198 UMER 6236, Université de la Méditerranée, Marseille, France

Zied Gaieb and Dimitrios Morikis
Department of Bioengineering, University of California, Riverside 92521, USA

Carter T. Butts and Xuhong Zhang
Department of Sociology, UC Irvine, USA
Department of Statistics, UC Irvine, USA
Department of Electrical Engineering and Computer Science, UC Irvine, USA
Calit2, UC Irvine, CA, 92697, UCI.

JohnE. Kelly, KyleW. Roskamp, Megha H. Unhelkar, J. Alfredo Freites, Seemal Tahir and Rachel W. Martin
Department of Chemistry, UC Irvine, USA
Department of Molecular Biology & Biochemistry, UC Irvine, Irvine, CA, 92697 USA

Rong Zhang, Tong Zhang, Ali Muhsen Ali, Mohammed Al Washih, Benjamin Pickard and David G. Watson
Strathclyde Institute of Pharmacy and Biomedical Sciences, 161, Cathedral Street, Glasgow G4 0RE, Scotland, UK
Institute of Clinical Pharmacology, Guangzhou University of Chinese Medicine, No. 12 Jichang Road, Guangzhou 510405, China
Department of Clinical Biochemistry/Diabetes and Endocrinology Centre, Thi-Qar Health Office, Thi-Qar, Nassiriya, Iraq
General Directorate of Medical Services, Ministry of Interior, Riyadh 13321, KSA

Anthony C. Dona, Michael Kyriakides and Kirill Veselkov
Department of Surgery and Cancer, Faculty of Medicine, Imperial College London, SW7 2AZ, United Kingdom
Institute of Structural and Molecular Biology, University College London, London WC1E 6BT, United Kingdom
Medway Metabonomics Research Group, University of Greenwich, Chatham Maritime, Kent ME4 4TB, United Kingdom

Sevasti Filippidou, Thomas Junier, TinaWunderlin and Pilar Junier
Laboratory of Microbiology, Institute of Biology, University of Neuchatel, CH-2000, Neuchâtel, Switzerland
Vital-IT group, Swiss Institute of Bioinformatics, CH-1015 Lausanne, Switzerland

Chien-Chi Lo, Po-E Li and Patrick S. Chain
Bioscience Division, Los Alamos National Laboratory, Los Alamos, NM 87545, USA

Duncan Ayers
Centre for Molecular Medicine and Biobanking, University of Malta, Msida, Malta

Pieter Mestdagh, TomVan Maerken and Jo Vandesompele
Center for Medical Genetics Ghent, Ghent University Hospital, Ghent, Belgium

Koushik Das, Punam Chowdhury and Sandipan Ganguly
Division of Parasitology, National Institute of Cholera and Enteric Diseases, P-33, CIT Road, Scheme XM, Beliaghata, Kolkata 700010, India

Difei Wang, Lei Song, Varun Singh, Shruti Rao, Lin An and Subha Madhavan
Department of Oncology, Lombardi Comprehensive Cancer Center, Georgetown University Medical Center,Washington, DC 20007, USA
Innovation Center for Biomedical Informatics, Georgetown University Medical Center, Washington, DC 20007, USA
Department of Biochemistry and Molecular & Cellular Biology, Georgetown University, Washington, DC 20007, USA

Thomas Scior and Bertin Paiz-Candia
Facultad de Ciencias Químicas, Universidad Autónoma de Puebla, Puebla, Mexico

Ángel A. Islas, Alfredo Sánchez-Solano, Claudia Mancilla-Simbro and Eduardo M. Salinas-Stefanon
Laboratorio de Biofísica, Instituto de Fisiología, Universidad Autónoma de Puebla, Puebla, Mexico

Lourdes Millan-Perez Peña
Centro de Química, Instituto de Ciencias, Universidad Autónoma de Puebla, Puebla, Mexico

Eleftherios Pilalis, Theodoros Koutsandreas, Ioannis Valavanis and Aristotelis Chatziioannou
Metabolic Engineering and Bioinformatics Programme, Institute of Medicinal Chemistry and Biotechnology, National Hellenic Research Foundation, Athens, Greece
Emmanouil Athanasiadis and George Spyrou

Søs Torpenholt, Mats H.M. Olsson, Jan H. Jensen and Leila Lo Leggio
Department of Chemistry, University of Copenhagen, Universitetsparken 5, 2100 Copenhagen, Denmark

Leonardo DeMaria, Lars H. Christensen and Michael Skjøt
Novozymes A/S, Smørmosevej 25, 2880 Bagsværd, Denmark

Peter Westh
NSM, Research Unit for Functional Biomaterials, University of Roskilde, Universitetsvej 1, 4000 Roskilde, Denmark

Patricia Ortegon, Augusto C. Poot-Hernández and Katya Rodriguez-Vazquez
Departamento de Ingeniería de Sistemas Computacionales y Automatización, IIMAS, Universidad Nacional Autónoma de México, Mexico
Departamento de Ingeniería Celular y Biocatálisis, Instituto de Biotecnología, Universidad Nacional Autónoma de México, Cuernavaca, Morelos, Mexico

Ernesto Perez-Rueda
Departamento de Ingeniería Celular y Biocatálisis, Instituto de Biotecnología, Universidad Nacional Autónoma de México, Cuernavaca, Morelos, Mexico
Unidad Multidisciplinaria de Docencia e Investigación, Sisal Facultad de Ciencias, Sisal, Yucatán, UNAM, Mexico

Suraiya Rasheed, Rahim Hashim and Jasper S. Yan
Laboratory of Viral Oncology and Proteomics Research, Keck School of Medicine, University of Southern California, Cancer Research Laboratory Building, 1303 North Mission Rd, Los Angeles, CA 90033, USA

Craig A. Doupnik
Department of Molecular Pharmacology & Physiology, University of South Florida College of Medicine, 12901 Bruce B. Downs Boulevard, Tampa, FL 33612, United States

Katherine C. Parra and Wayne C. Guida
Department of Chemistry, University of South Florida, 4202 E. Fowler Avenue, Tampa, FL 33620, United States
Drug Discovery Department, H. Lee Moffitt Cancer Center and Research Institute, 12902 Magnolia Drive, Tampa, FL 33612, United States

Francesco Mallamace, Carmelo Corsaro and Sebastiano Vasi
Dipartimento di Fisica e Scienze della Terra, Università di Messina, Viale F. Stagno D'Alcontres 31, 98166 Messina, Italy
CNR-IPCF, Istituto per i Processi Chimico-Fisici, Viale F. Stagno D'Alcontres 37, 98158 Messina, Italy

Domenico Mallamace
Dipartimento di Scienze dell'Ambiente, della Sicurezza, del Territorio, degli Alimenti edella Salute, Università di Messina, Viale F. Stagno d'Alcontres 31, 98166 Messina, Italy

Index